AIR FORCES OF THE WORLD
OF THE
WORLD

**An illustrated directory of
all the world's military air powers**

AIR FORCES OF THE WORLD

An illustrated directory of all the world's military air powers

Mark Hewish · Bill Sweetman · Barry C. Wheeler · Bill Gunston

PEERAGE BOOKS

First published in Great Britain in 1979 by
Salamander Books Ltd

This edition published in 1984 by
Peerage Books
59 Grosvenor Street
London W1

© 1979 Salamander Books

ISBN 0 907408 93 1

Printed in Hong Kong

EDITOR'S ACKNOWLEDGEMENTS

A great many people have given advice and assistance during
the preparation of this book, and I would like to thank
them all. In particular I am grateful to the various
ministries of defence, air attaches and other embassy staff
of several of the nations whose air forces are described
here, to the manufacturers of many of the aircraft and other
weapons systems illustrated and also to the eagle-eyed
aircraft "spotters" who have scoured their collections of
photographs to illustrate this unique volume.

Ray Bonds

CREDITS

Editor: Ray Bonds

Designer: Lloyd Martin

Maps: Base maps © George Philip & Son Ltd.
Special air force information © Salamander Books Ltd.

Three-view drawings: © Pilot Press Ltd; three-views of
Agusta A 109, Sikorsky UH-60A, MBB BO 106, Hughes YAH-64,
Kaman SH-2F, Mil Mi-24 Hind D, DHC Twin Otter and Shenyang
F-6bis were prepared with reference to drawings by Pilot
Press Ltd., with their kind permission.

THE AUTHORS

Chief author and technical adviser:
Mark Hewish, BA
A former member of the defence staff of *Flight International*, he is now editor of *Defence Materiel* and a contributor to many aerospace and defence journals in Europe and the United States. He is the author of several books on aerospace topics.

Bill Sweetman
Bill Sweetman has established an international reputation for his detailed, objective analyses of the air forces of the Communist powers, and is a former member of the editorial staff of *Flight International*. He is co-author (with Bill Gunston) of the previous Salamander title, *Soviet Air Power*, and a contributor to their *The Chinese War Machine*. He has contributed technical articles on aviation affairs to several authoritative publications, including *Aeroplane Monthly* and *Middle East Economic Digest*, and is a frequent radio broadcaster.

Barry C. Wheeler
Barry Wheeler is an author and contributor to technical defence journals, and is also an adviser to a manufacturer of accurate scale models of aircraft and other weapons. He was formerly on the staff of *Flight International*, and compiles the journal's annual "Air Forces Review".

Bill Gunston
Bill Gunston is an internationally respected author on aviation, scientific and defence oriented subjects. Among his numerous books are the Salamander titles *The Illustrated Encyclopedia of the World's Modern Military Aircraft*, *The Encyclopedia of the World's Combat Aircraft* and *The Illustrated Encyclopedia of the World's Rockets and Missiles*. He is also co-author (with Bill Sweetman) of Salamander's three titles, *The Soviet War Machine*, *The US War Machine* and *The Chinese War Machine*. He is former Technical Editor of *Flight International*, and is an assistant compiler of *Jane's All the World's Aircraft*.

THE CONSULTANT

John W. R. Taylor, FRHistS, MRAeS, FSLAET
John Taylor has been the Editor of *Janes's All the World's Aircraft* since 1959, and is the author of over 200 books. He also edited *Air BP* magazine from 1956 to 1973, and was Technical Consultant to Salamander's *Soviet Air Power*. In 1941 he elected to give up a place at Cambridge, having been invited to join Sir Sydney Camm's design staff at Hawker Aircraft Ltd. At the age of 20 he produced all the drawings to convert the Typhoon into a prototype night fighter. He was also technical publicity and press officer for the Fairey Aviation Group.

FOREWORD by John W. R. Taylor, FRHistS, MRAeS, FSLAET

AIR FORCES have lost none of their significance in a missile age. Statesmen of East and West may spend time arguing about balanced forces of ballistic rockets, carrying varied permutations of nuclear warheads and poised ready for instant despatch from underground silo-launchers or submarines; but these weapons are so destructive that their use is unthinkable. By comparison, there has been hardly a day for three-quarters of a century when air forces have not been in action somewhere in our world.

The easy way to compile a book of this kind is simply to catalogue the equipment, personnel and deployment of each individual air force, as far as governments and inspired guesswork will permit, and then leave the reader to assess relative capability. Such a concept has little practical value. Numbers, by themselves, never guarantee success in air warfare. Time after time, victory has resulted from a sudden technological breakthrough, inspired tactics, better training, even the sheer physical endurance of pilots and civilians who refuse to be beaten.

In 1915–16 the German Army Air Service almost shot the Allied air forces from the sky over the Western Front in France, using Fokker monoplanes, of which a mere 425 were built during a war in which Germany, Britain and France manufactured a total of 170,712 military aeroplanes. Just two B-29 bombers ended World War II a generation later, by dropping atomic bombs on Japan. Neither weapon killed as many people as did a raid by B-29s which showered incendiary bombs on Tokyo five months earlier, but nuclear weapons introduced such terrifying possibilities that the world has never since been subjected to a global war.

For that reason, the people of NATO's member nations need not despair on learning that aircraft of the Warsaw Pact air forces outnumber their own by nearly two to one along the key central front in Europe. More dangerous is unquestioning acceptance of the assumption that the first onslaught in any confrontation would come from the East.

An air force which strikes first gains an immense advantage, as the Japanese proved at Pearl Harbor, and as the Israelis demonstrated again in 1967. Six years later it was Israel's turn to be taken by surprise at the start of the Yom Kippur War; yet its air force achieved remarkable results in subsequent air-to-air combat, losing only five of its own air-superiority fighters against more than 250 Arab aircraft, due mainly to availability of a small guided missile named Shafrir. Not since

the time when the Fokker monoplanes flew into action with the first synchronised forward-firing machine-guns had a new weapon been responsible for such disproportionate losses in air fighting.

The more one studies every aspect of modern air power, the more fascinating it becomes, and there is no better basis for study than a book of this kind, compiled by persons adept at both research and analysis. The importance of such knowledge cannot be overemphasised. If our generation appreciates fully the capability of modern weapons, and ensures that East/West military might remains in balance, we should continue to live in relative peace. Despite all the problems, not least the expenditure of vast sums of money for which more humane uses could be found, this is preferable to the kind of war that would eliminate whole areas of population as well as the problems.

Contents

9

WORLD MAP OF AIR FORCE BASES

These maps (covering the northern hemisphere on this side and the southern hemisphere overleaf) show the locations of major bases operated by the world's air arms, where such (reliable) information is available. Larger-scale maps of particular countries or areas are provided when many bases are located in a small area, such as in Western Europe.

Each base is represented by a circle containing a number which can be matched with the accompanying list to find the name of that airfield. To find which aircraft are known to be located at a particular base, turn to the table in the appropriate air force entry.

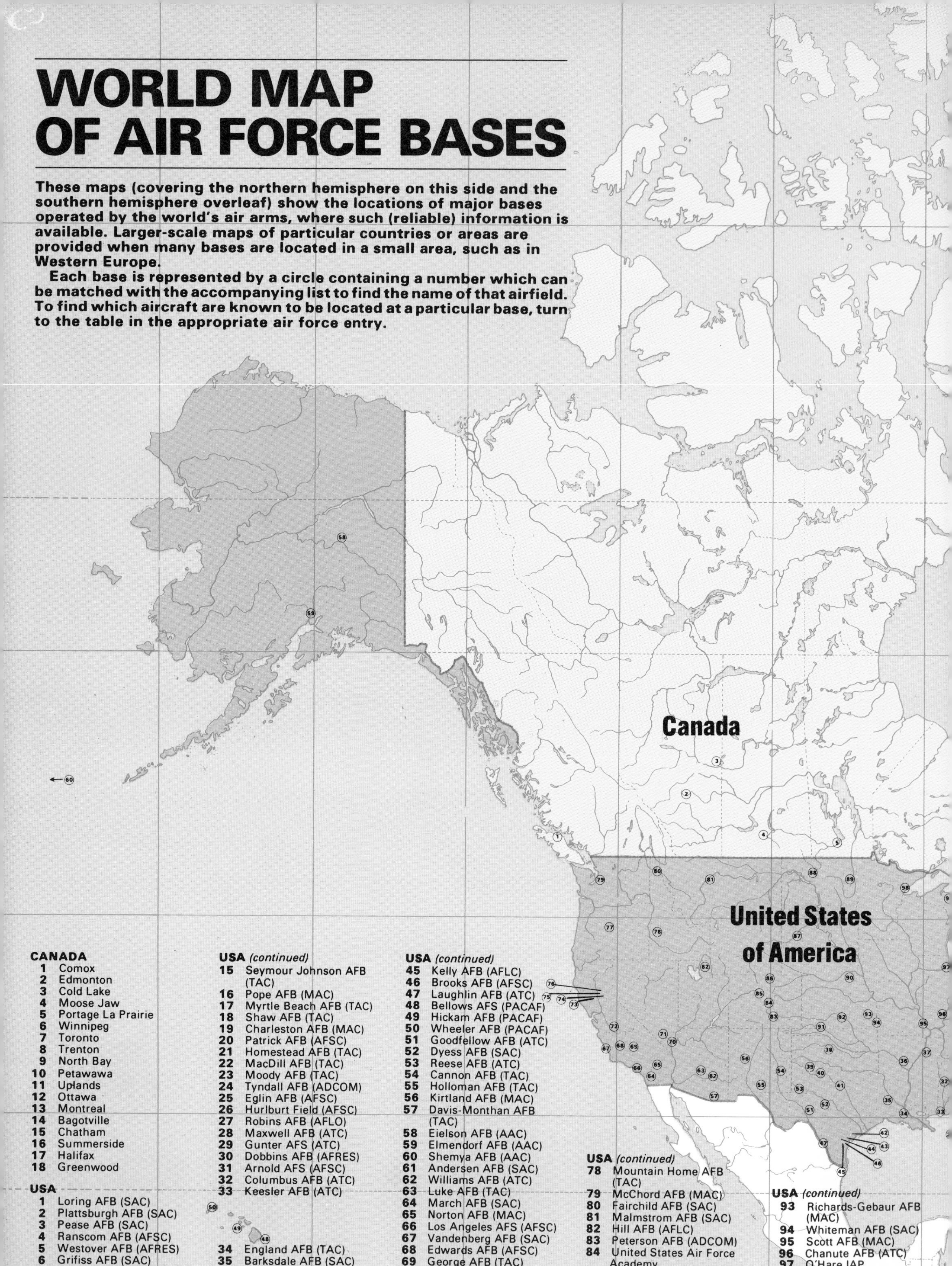

Canada

United States of America

CANADA
1 Comox
2 Edmonton
3 Cold Lake
4 Moose Jaw
5 Portage La Prairie
6 Winnipeg
7 Toronto
8 Trenton
9 North Bay
10 Petawawa
11 Uplands
12 Ottawa
13 Montreal
14 Bagotville
15 Chatham
16 Summerside
17 Halifax
18 Greenwood

USA
1 Loring AFB (SAC)
2 Plattsburgh AFB (SAC)
3 Pease AFB (SAC)
4 Ranscom AFB (AFSC)
5 Westover AFB (AFRES)
6 Grifiss AFB (SAC)
7 Hancock Field (ADCOM)
8 Niagara Falls IAP
9 McGuire AFB (MAC)
10 Dover AFB (MAC)
11 Washington DC (Hq USAF)
12 Bolling AFB (MAC)
13 Andrews AFB (MAC)
14 Langley AFB (TAC)

USA (continued)
15 Seymour Johnson AFB (TAC)
16 Pope AFB (MAC)
17 Myrtle Beach AFB (TAC)
18 Shaw AFB (TAC)
19 Charleston AFB (MAC)
20 Patrick AFB (AFSC)
21 Homestead AFB (TAC)
22 MacDill AFB (TAC)
23 Moody AFB (TAC)
24 Tyndall AFB (ADCOM)
25 Eglin AFB (AFSC)
26 Hurlburt Field (AFSC)
27 Robins AFB (AFLO)
28 Maxwell AFB (ATC)
29 Gunter AFS (ATC)
30 Dobbins AFB (AFRES)
31 Arnold AFS (AFSC)
32 Columbus AFB (ATC)
33 Keesler AFB (ATC)
34 England AFB (TAC)
35 Barksdale AFB (SAC)
36 Little Rock AFB (MAC)
37 Blytheville AFB (SAC)
38 Tinker AFB (AFLC)
39 Altus AFB (MAC)
40 Sheppard AFB (ATC)
41 Carswell AFB (SAC)
42 Bergastrom AFB (TAC)
43 Randolph AFB (ATC)
44 Lackland AFB (ATC)

USA (continued)
45 Kelly AFB (AFLC)
46 Brooks AFB (AFSC)
47 Laughlin AFB (ATC)
48 Bellows AFS (PACAF)
49 Hickam AFB (PACAF)
50 Wheeler AFB (PACAF)
51 Goodfellow AFB (ATC)
52 Dyess AFB (SAC)
53 Reese AFB (ATC)
54 Cannon AFB (TAC)
55 Holloman AFB (TAC)
56 Kirtland AFB (MAC)
57 Davis-Monthan AFB (TAC)
58 Eielson AFB (AAC)
59 Elmendorf AFB (AAC)
60 Shemya AFB (AAC)
61 Andersen AFB (SAC)
62 Williams AFB (ATC)
63 Luke AFB (TAC)
64 March AFB (SAC)
65 Norton AFB (SAC)
66 Los Angeles AFS (AFSC)
67 Vandenberg AFB (SAC)
68 Edwards AFB (AFSC)
69 George AFB (TAC)
70 Nellis AFB (TAC)
71 Indian Springs AF Auxiliary Field (TAC)
72 Castle AFB (SAC)73
73 Travis AFB (MAC)
74 Mather AFB (ATC)
75 McClellan AFB (AFLC)
76 Beale AFB (SAC)
77 Kingsley Field (ADCOM)

USA (continued)
78 Mountain Home AFB (TAC)
79 McChord AFB (MAC)
80 Fairchild AFB (SAC)
81 Malmstrom AFB (SAC)
82 Hill AFB (AFLC)
83 Peterson AFB (ADCOM)
84 United States Air Force Academy
85 Lowry AFB (ATC)
86 Francis E Warren AFB (SAC)
87 Ellsworth AFB (SAC)
88 Minot AFB (SAC)
89 Grand Forks AFB (SAC)
90 Offutt AFB (SAC)
91 Vance AFB (ATC)
92 McConnell AFB (SAC)

USA (continued)
93 Richards-Gebaur AFB (MAC)
94 Whiteman AFB (SAC)
95 Scott AFB (MAC)
96 Chanute AFB (ATC)
97 O'Hare IAP
98 Duluth IAP
99 K I Sawyer AFB (SAC)
100 Wurtsmith AFB (SAC)
101 Grisson AFB (SAC)
102 Wright-Patterson AFB (AFLC)
103 Rickenbacker AFB (SAC)
104 Youngstown Municipal AP

Guam

10

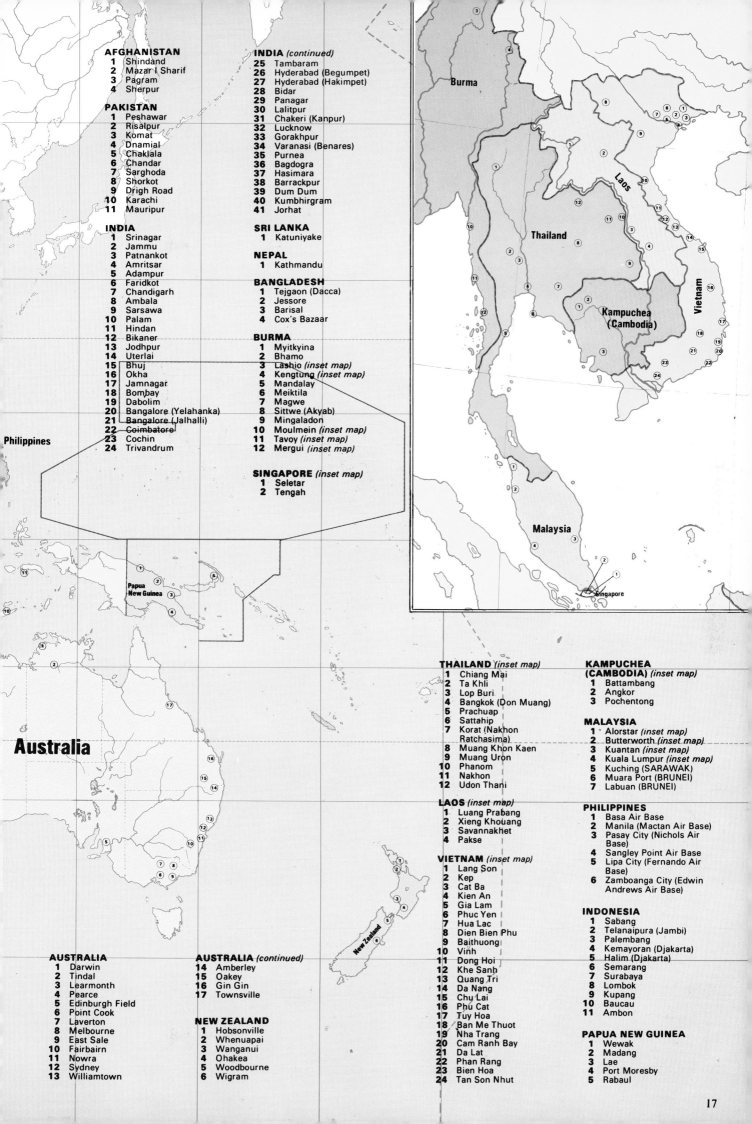

AFGHANISTAN
1 Shindand
2 Mazar I Sharif
3 Pagram
4 Sherpur

PAKISTAN
1 Peshawar
2 Risalpur
3 Komat
4 Dnamial
5 Chaklala
6 Chandar
7 Sarghoda
8 Shorkot
9 Drigh Road
10 Karachi
11 Mauripur

INDIA
1 Srinagar
2 Jammu
3 Patnankot
4 Amritsar
5 Adampur
6 Faridkot
7 Chandigarh
8 Ambala
9 Sarsawa
10 Palam
11 Hindan
12 Bikaner
13 Jodhpur
14 Uterlai
15 Bhuj
16 Okha
17 Jamnagar
18 Bombay
19 Dabolim
20 Bangalore (Yelahanka)
21 Bangalore (Jalhalli)
22 Coimbatore
23 Cochin
24 Trivandrum

INDIA *(continued)*
25 Tambaram
26 Hyderabad (Begumpet)
27 Hyderabad (Hakimpet)
28 Bidar
29 Panagar
30 Lalitpur
31 Chakeri (Kanpur)
32 Lucknow
33 Gorakhpur
34 Varanasi (Benares)
35 Purnea
36 Bagdogra
37 Hasimara
38 Barrackpur
39 Dum Dum
40 Kumbhirgram
41 Jorhat

SRI LANKA
1 Katuniyake

NEPAL
1 Kathmandu

BANGLADESH
1 Tejgaon (Dacca)
2 Jessore
3 Barisal
4 Cox's Bazaar

BURMA
1 Myitkyina
2 Bhamo
3 Lashio *(inset map)*
4 Kengtung *(inset map)*
5 Mandalay
6 Meiktila
7 Magwe
8 Sittwe (Akyab)
9 Mingaladon
10 Moulmein *(inset map)*
11 Tavoy *(inset map)*
12 Mergui *(inset map)*

SINGAPORE *(inset map)*
1 Seletar
2 Tengah

AUSTRALIA
1 Darwin
2 Tindal
3 Learmonth
4 Pearce
5 Edinburgh Field
6 Point Cook
7 Laverton
8 Melbourne
9 East Sale
10 Fairbairn
11 Nowra
12 Sydney
13 Williamtown

AUSTRALIA *(continued)*
14 Amberley
15 Oakey
16 Gin Gin
17 Townsville

NEW ZEALAND
1 Hobsonville
2 Whenuapai
3 Wanganui
4 Ohakea
5 Woodbourne
6 Wigram

THAILAND *(inset map)*
1 Chiang Mai
2 Ta Khli
3 Lop Buri
4 Bangkok (Don Muang)
5 Prachuap
6 Sattahip
7 Korat (Nakhon Ratchasima)
8 Muang Khon Kaen
9 Muang Uron
10 Phanom
11 Nakhon
12 Udon Thani

LAOS *(inset map)*
1 Luang Prabang
2 Xieng Khouang
3 Savannakhet
4 Pakse

VIETNAM *(inset map)*
1 Lang Son
2 Kep
3 Cat Ba
4 Kien An
5 Gia Lam
6 Phuc Yen
7 Hua Lac
8 Dien Bien Phu
9 Baithuong
10 Vinh
11 Dong Hoi
12 Khe Sanh
13 Quang Tri
14 Da Nang
15 Chu Lai
16 Phu Cat
17 Tuy Hoa
18 Ban Me Thuot
19 Nha Trang
20 Cam Ranh Bay
21 Da Lat
22 Phan Rang
23 Bien Hoa
24 Tan Son Nhut

KAMPUCHEA (CAMBODIA) *(inset map)*
1 Battambang
2 Angkor
3 Pochentong

MALAYSIA
1 Alorstar *(inset map)*
2 Butterworth *(inset map)*
3 Kuantan *(inset map)*
4 Kuala Lumpur *(inset map)*
5 Kuching (SARAWAK)
6 Muara Port (BRUNEI)
7 Labuan (BRUNEI)

PHILIPPINES
1 Basa Air Base
2 Manila (Mactan Air Base)
3 Pasay City (Nichols Air Base)
4 Sangley Point Air Base
5 Lipa City (Fernando Air Base)
6 Zamboanga City (Edwin Andrews Air Base)

INDONESIA
1 Sabang
2 Telanaipura (Jambi)
3 Palembang
4 Kemayoran (Djakarta)
5 Halim (Djakarta)
6 Semarang
7 Surabaya
8 Lombok
9 Kupang
10 Baucau
11 Ambon

PAPUA NEW GUINEA
1 Wewak
2 Madang
3 Lae
4 Port Moresby
5 Rabaul

17

EUROPEAN AIR BASES excluding the United Kingdom

PORTUGAL
1 Ovar (BA-8)
2 Sao Jacinto/Aveiro (BA-7)
3 Monte Real (BA-5)
4 Tancos (BA-3)
5 OTA (BA-2)
6 Alverca
7 Sintra (BA-1)
8 Lisbon
9 Montijo (BA-6)
10 Beja (BA-11)
11 Lajes (BA-4)

SPAIN
1 Burgos
2 Villanubla
3 Matacan
4 Talavera/La Real
5 Seville
6 Tablada
7 Rota
8 Jerez
9 Moron
10 Malaga
11 Murcir/San Javier
12 Albacete (Los Llanos)
13 Manises
14 Cuatro Vientos
15 Getafe
16 Torrejon
17 Zaragoza
18 Agoncillo

FRANCE
1 Landivisiau
2 Lanveoc-Poulmic
3 Lann-Bihoue
4 Cambrai
5 Reims
6 Creil
7 Evreux
8 Le Bourget
9 Villacoublay
10 Poissy
11 St Dizier
12 Toul
13 Metz
14 Nancy
15 Strasbourg
16 Colmar
17 Luxeuil
18 Dijon
19 Orléans
20 Avord
21 St Yan
22 Clermont-Ferrand
23 Chambéry
24 Hyères
25 Aix-en-Provence
26 APT
27 Nîmes-Garons
28 Marseille
29 Salon
30 Orange
31 Istres
32 Toulouse
33 Mont-de-Marsan
34 Cazaux
35 Bordeaux
36 Cognac

BELGIUM
1 Coxyde
2 Brasschaat
3 Melsbroek
4 Chievres
5 Florennes
6 Beauvechain
7 Goetsenhoven
8 Brustem
9 Kleine Brogel
10 Aachen

NETHERLANDS
1 De Kooy
2 Valkenburg
3 Ypenburg
4 Gilze Rijen
5 Eindhoven
6 De Peel
7 Volkel
8 Soesterberg
9 Deelen
10 Twenthe
11 Leeuwarden

GERMANY
1 Sylt
2 Leck
3 Eggebeck
4 Husum
5 Schleswig
6 Hohn
7 Kiel-Holtenau
8 Helgoland
9 Itzehoe-Hungriger Wolf
10 Utersen

GERMANY (continued)
11 Nordholz
12 Jever
13 Wittmundhafen
14 Borkum
15 Oldenburg
16 Ahlhorn
17 Fassberg
18 Celle
19 Wunstorf
20 Hildesheim
21 Buckeburg
22 Rheine-Hopsten
23 Rheine-Bentlage
24 Eritzlar
25 Rotenburg
26 Geilenkirchen
27 Norvenich
28 Köln-Wahn
29 Buchel
30 Pferdsfeld
31 Niederstetten
32 Roth
33 Neuburg
34 Manching
35 Oberschleissheim
36 Furstenfeldbruck
37 Lechfeld
38 Landsberg
39 Kaufbeuren
40 Memmingen
41 Laupheim
42 Bremgarten

DENMARK
1 Aalborg
2 Karup
3 Vandel
4 Skrydstrup
5 Avnø
6 Vaerløse
7 Copenhagen
8 Kastrup

SWITZERLAND
1 Grenchen
2 Bern/Belp
3 Payerne
4 Interlaken
5 Meiringen
6 Alpnach
7 Stans
8 Emmen
9 Dubendorf
10 Samedan/St Moritz
11 Locarno
12 Sion

AUSTRIA
1 Schwaz
2 Aigen
3 Klagenfurt-Annibiehl
4 Graz
5 Horsching
6 Langenlebarn

ITALY
1 Ronchi dei Legionari
2 Rivolto (Udine)
3 Aviano
4 Treviso-San Angelo
5 Venice/San Nicolo
6 Istrana (Treviso)
7 Padova
8 Vicenza
9 Bolzano
10 Bergamo/Orio Al Serio
11 Milan (Gallarate)
12 Cameri
13 Ghedi (Brescia)
14 Verona-Villafranca
15 Piacenza/San Domiano
16 Bologna/Borgo Panigale
17 Forli
18 Cervia-San Giorgio
19 Florence/Peretol
20 Pisa-San Giusto
21 Rimini-Miramare
22 Ancona/Falconara
23 Grosseto
24 Viterbo
25 Pescara (Liberi)
26 Guidonia
27 Ciampino
28 Pratica Di Mare
29 Latina
30 Frosinone
31 Grazzanise (Caserta)
32 Naples/Capodichino
33 Foggia/Gino Lisa
34 Foggia-Amendola   ← [11] Portugal
35 Bari-Palese
36 Gioia Del Colle
37 Grottaglie
38 Brindisi
39 Lecce-Galatine
40 Crotone (Sanna)
41 Reggio Calabria
42 Catania-Fontanarossa

ITALY (continued)
43 Catania/Sigonella
44 Comiso
45 Palermo/Boccadifalco
46 Trapani/Birgi
47 Cagliari Elmas
48 Decomimannu
49 Alghero
50 Pantelleria
51 Lampedusa

GREECE
1 Komotini
2 Kavala
3 Thessaloniki
4 Kastoria
5 Kozani
6 Ioannina
7 Larisa
8 Nea Ankhialos
9 Preveza
10 Agrinion
11 Araxos
12 Andravidha
13 Tanagra
14 Elefsis
15 Megara
16 Hellenikon
17 Tripolis
18 Kalamata
19 Souda
20 Iraklion
21 Kastellion
22 Timbakion

EAST GERMANY
1 Putznitz
2 Sanitz
3 Peenemunde
4 Neubrandenburg
5 Parchim
6 Rechlin
7 Wittstock
8 Neuruppin
9 Finow
10 Oranienburg
11 Werneuchen
12 Stendal
13 Brandenburg-Briesen
14 Drewitz
15 Furstennaldes
16 Wunsdorf
17 Zerbst
18 Juterbog
19 Kothen
20 Grossenhain
21 Merseburg
22 Altenburg
23 Dresden-Klotzsche
24 Spremburg
25 Cottbus

POLAND
1 Slupsk
2 Stettin
3 Poznan
4 Sroda
5 Legnica
6 Jelenia Gora
7 Breslau
8 Cracow
9 Krosno

POLAND (continued)
10 Rzeszow
11 Mielec
12 Radom
13 Deblin
14 Lublin
15 Brest
16 Siedlce
17 Warsaw
18 Kutno
19 Inowrocław
20 Torun
21 Bydgoszcz
22 Grudziadz
23 Gdansk
24 Elblag
25 Olsztyn
26 Rastenburg

CZECHOSLOVAKIA
1 Zatec
2 Cheb
3 Pilsen
4 Dobrany
5 Ruzyne
6 Klecany
7 Muchovo
8 Gakovice
9 Letnamy
10 Kbely
11 Mimon
12 Milovice
13 Caslav
14 Pardubice
15 Chocen
16 Bechyne
17 Ceske Budejovice
18 Brno
19 Prerov
20 Otrokovice
21 Kunovice
22 Piestany
23 Bratislava

CZECHOSLOVAKIA (continued)
24 Lucenec
25 Ivanka
26 Sliac
27 Barca
28 Presov
29 Spisska Nova Ves
30 Poprad
31 Mosnov

HUNGARY
1 Gyor
2 Papa
3 Szombathely
4 Veszprem
5 Tapolca
6 Szekesfehervar
7 Pecs
8 Szeged
9 Kecskemet
10 Tokol
11 Ferihegy
12 Budaors
13 Matyasfold
14 Miskolc
15 Debrecen

Belg[ium]

France

Portugal

Spain

Base maps: © George Philip & Son Ltd.
Special air force information: © Salamander Books Ltd.

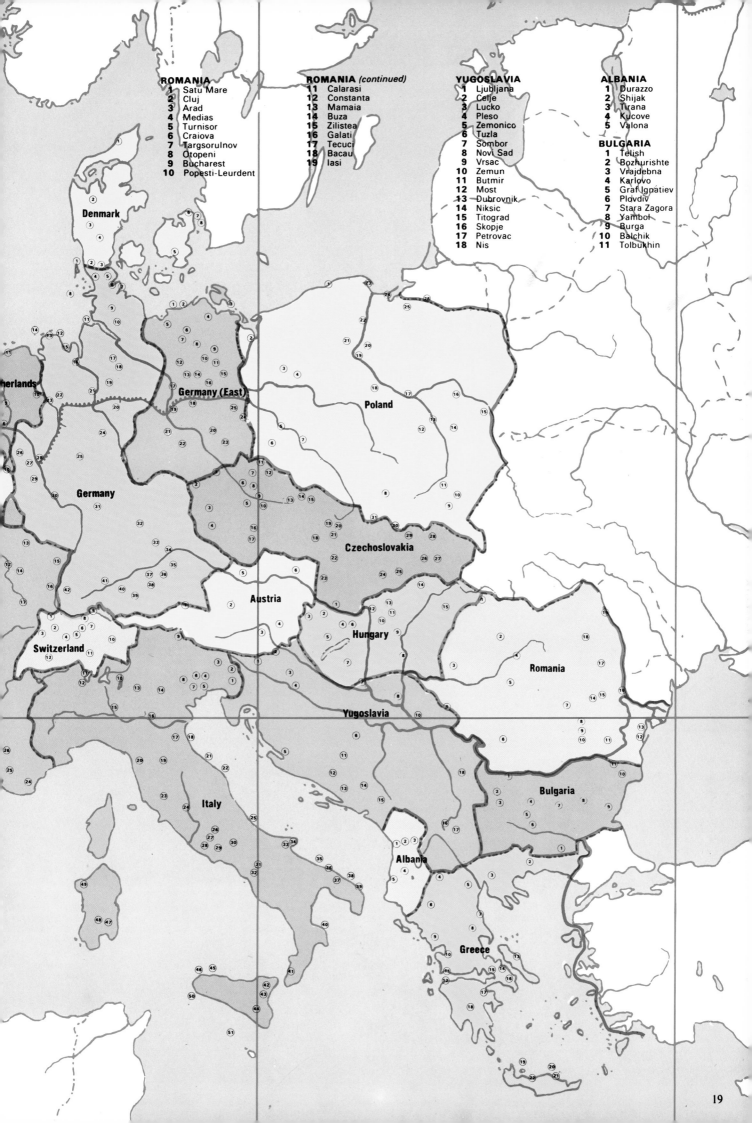

ROMANIA
1 Satu Mare
2 Cluj
3 Arad
4 Medias
5 Turnisor
6 Craiova
7 Targsorulnov
8 Otopeni
9 Bucharest
10 Popesti-Leurdent

ROMANIA *(continued)*
11 Calarasi
12 Constanta
13 Mamaia
14 Buza
15 Zilistea
16 Galati
17 Tecuci
18 Bacau
19 Iasi

YUGOSLAVIA
1 Ljubljana
2 Celje
3 Lucko
4 Pleso
5 Zemonico
6 Tuzla
7 Sombor
8 Novi Sad
9 Vrsac
10 Zemun
11 Butmir
12 Most
13 Dubrovnik
14 Niksic
15 Titograd
16 Skopje
17 Petrovac
18 Nis

ALBANIA
1 Durazzo
2 Shijak
3 Tirana
4 Kucove
5 Valona

BULGARIA
1 Telish
2 Bozhurishte
3 Vrajdebna
4 Karlovo
5 Graf Ignatiev
6 Plovdiv
7 Stara Zagora
8 Yambol
9 Burga
10 Balchik
11 Tolbukhin

Denmark

Netherlands

Germany (East)

Poland

Germany

Czechoslovakia

Austria

Switzerland

Hungary

Romania

Yugoslavia

Italy

Bulgaria

Albania

Greece

US AIR FORCE BASES IN EUROPE

1	Upper Heyford (RAF)	**12**	Wiesbaden AB
2	Woodbridge (RAF)	**13**	Rhein Main AB
3	Alconbury (RAF)	**14**	Sembach AB
4	Mildenhall (RAF)	**15**	Zweibrucken AB
5	Lakenheath (RAF)	**16**	Aviano AB
6	Bentwaters (RAF)	**17**	Zaragoza AB
7	Camp New Amsterdam AB	**18**	Torrejon AB
8	Hahn AB	**19**	Hellenikon AB
9	Bitburg AB	**20**	Izmir AB
10	Spangdahlem AB	**21**	Ankara Asn
11	Ramstein AB	**22**	Incirlik AB

US AIR FORCE BASES IN THE PACIFIC

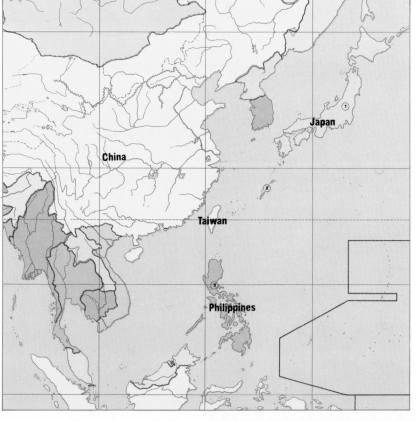

1 Yokota AB
2 Kadena AB
3 Clark AB

In addition to the bases shown, there are USAF bases at Keflavik Airport, Iceland (ADCOM); Thule AB, Greenland (ADCOM); Kunsan AB (ADCOM), Osan AB (PACAF) and Taegu AB (PACAF) in South Korea; Lajes Field (MAC) the Azores; and Lindsey AS (Support, USAFE) and Tempelhof Airport (Support, USAFE) in Germany.

20

INTRODUCTION

For the purposes of this book the world has been divided into 13 regions made up of countries which have some common tie — geographical, political or ethnic. This means that the members of an alliance such as NATO may be listed in more than one section, but to have arranged the sections in groups of countries with military affiliations would have resulted in unnecessary fragmentation and difficulties in relating adjacent states.

Within the 13 regions, each with its own introduction setting the local scene, every country with a military air arm of any appreciable size has its own entry listing the state's area, population, gross national production and defence expenditure. This is accompanied by a description of the organisation, structure, inventory, method of training, combat experience, future plans and other aspects of the country's air forces including air arms which may be operated by the army, navy, marine corps, coast guard or other military-style organisation.

Air Forces of the World is unique, however, in that it also contains tables showing the unit to which a country's aircraft belong, what their type and role are, where they are based, how many are operational, and other information vital to the serious student. Obviously there are omissions in a work of this complexity, but the text and tables — when cross-referred to the comprehensive maps of major bases on pages 10 to 20 — provide the most detailed information published in any generally available reference work.

The roles of various aircraft types listed in the tables have been abbreviated as follows: Att, attack; Bomb, bomber; Casevac, casualty evacuation; Coin, counter-insurgency; FAC, forward air control; FGA, fighter/ground-attack; Fight, fighter; Int, interceptor; Mar pat, maritime patrol; Obs, observation; Recce, reconnaissance; SAR, search and rescue; Train, trainer; Trans, transport; Util, utility.

At the back of the book is an inventory of the major types of aircraft operated by the world's air arms, including line drawings and details of the dimensions, speed, range and users, the last-named allowing aircraft to be cross-referred with their operators in each air force entry, so that readers can determine which particular type of that aircraft is used by that air force.

Mark Hewish

North America

The North American continent comprises only two states, the United States and Canada, although Greenland – which is Danish territory – lies in the same geographical area. Both the US and Canada are members of NATO, with forces deployed in Europe as well as on home soil, and the United States – as one of the two 'superpowers' – has extensive interests to protect throughout the world. Canada has not been substantially involved in any conflict since World War II but its southern neighbour – which has until comparatively recently adopted a role of 'world policeman' – has become involved in action at all levels from diplomatic persuasion to virtually full-scale war in Southeast Asia.

Canada and the US collaborate closely on air defence of the continent, this co-operation extending to operation of similar equipment as well as integrating alerting, command and control facilities. The Canadian Armed Forces seem certain to order a US-designed interceptor to succeed their Voodoos which are tasked with home defence, yet paradoxically the US Aerospace Defense Command is less sure of receiving a modern fighter in the foreseeable future. Canada has deployed no long-range surface-to-air missiles on its own territory since the withdrawal of Bomarc, and the US has similarly run its home-based Nike Hercules and Hawk batteries down to a minimal level, so both countries would rely heavily on manned aircraft for defence against a long-range bomber threat from the Soviet Union.

The US Army has also deactivated its Safeguard anti-ballistic missile system, so the United States now has no defence against intercontinental ballistic missiles. Relationships with China have become increasingly warm over the past few years, however, which means that the US does not have to worry about a second potential major aggressor – at least in the short term.

Both Canada and the United States have extensive forces deployed in Europe, with the consequent heavy reliance on long supply routes for reinforcement and replenishment. Regular exercises, and the more pressing Nickel Grass operation to supply Israel during the 1973 October war, have shown that the USAF's massive force of heavy transport aircraft – backed up by CRAF (Civil Reserve Air Fleet) airliners – could run a trans-atlantic shuttle in time of war.

The US maintains an air-power presence in South Korea, the Philippines, Japan and other countries, but is withdrawing from Taiwan and will increasingly turn its eyes towards Europe, where there exists a massive market for arms as well as the need to match the might of the Warsaw Pact.

CANADA

Area: 3·85 million sq. miles (9·97 million sq. km.).
Population: 23·3 million.
Total armed forces: 80,000.
Estimated GNP: $197·9 billion (£99 billion).
Defence expenditure: $3·64 billion (£1·82 billion).

Comprising the northern half of the North American continent, Canada continues to be a member of the British Commonwealth and plays an important role in world stability through its geographical position between the United States and the Soviet Union. The cornerstones of Canadian defence policy continue to be the country's participation in Nato, its valued involvement in European and North American defence, and peacekeeping operations for the United Nations. The defence forces are a single body called Canadian Armed Forces, a unification first planned in 1964 and firmly established in September 1975, under which there are three distinct commands for sea, land and air operations. Commander-in-Chief of the CAF is the Governor General of Canada and below him is the relevant Minister of Defence. The Canadian Forces are grouped into the following major functional organisations: National Defence HQ, Ottawa; Maritime Command, Halifax; Mobile Command, Montreal; Air Command, Winnipeg; Canadian Forces Europe, Lahr in West Germany; and Canadian Forces Training System, Trenton. In addition the CAF is also organised into six geographical Regions: Atlantic with HQ at Halifax; Eastern, Montreal; Central, Trenton; Prairie, Winnipeg; Pacific, Esquimalt; and Northern, Yellowknife.

Air Command consists of four operational groups (Maritime Air Group, Air Defence Group, 10 Tactical Air Group and Air Transport Group) and exercises command and control over the Air Training Schools and the Air Reserve. It also has training and reinforcement responsibilities for one Canadian Air Group in Europe. Sixteen bases and 24 radar stations stretching from coast to coast support flying operations of the Canadian-based groups.

Although there was Canadian military aviation involvement during the First World War, albeit limited, it was only in 1919 that an Air Board was set up to control part-time flying instruction; this expanded the following year into the Canadian Air Force. With RAF assistance, the new force received numbers of war-surplus aircraft and its status within the Commonwealth became assured on April 1, 1924, when the prefix 'Royal' was applied to the name. With the enormous distances involved, the Royal Canadian Air Force soon found itself assisting in many operations of a civil nature across the country. Aircraft types flown by the RCAF during the 1920s included Armstrong-Whitworth Siskin fighters and Atlas liaison machines, Ford Trimotor transports, and de Havilland Gipsy Moth and Avro Tutor trainers. The economic situation forced defence cuts in Canada in 1932, but this was reversed three years later when the worsening international situation prompted an

Below: A rocket-firing ground-attack run, in this instance by a Canadair/Northrop CF-5A. These Freedom Fighters were built in a collaborative programme with Northrop and the Netherlands.

Left: Practice ground-attack run by a Canadair CF-104 using spin-stabilized rockets; aircraft belongs to No 1 Group from Canadian Forces, Europe (4 ATAF).

Inventory

One of the most important future equipment programmes is Canada's New Fighter Aircraft requirement under which the existing McDonnell CF-101 Voodoos and Lockheed CF-104 Starfighters will be replaced by a new type in the early 1980s; the CF-5 will also be replaced and relegated to training duties. Aircraft under consideration by the end of 1978 included the Northrop F-18 and the General Dynamics F-16. The order is worth Can$2300m and the type chosen is likely to be built either wholly or partially in Canada. The

expansion programme which by 1939 resulted in the RCAF operating modest numbers of Bristol Blenheim bombers, Hawker Hurricane fighters and Blackburn Shark torpedo-bombers, to add to some 200 more obsolescent aircraft. A British Air Mission visited Canada in 1939 and laid the foundation of the country's participation in the Empire Air Training Scheme, in which Canada trained aircrew for the RAF.

The advent of the Second World War found Canada gearing up to produce aircraft and equipment for Britain; by 1941 no fewer than 16 RCAF squadrons were operating from the UK against Germany. Many of the most famous names in aircraft were flown by the Service including Spitfires, Typhoons, Mosquitoes, Beaufighters, Wellingtons, Lancasters and Halifaxes. In the unsteady peace that followed the war, Canada provided forces of occupation in Germany and at the same time gradually reduced her strength, which as far as the RCAF was concerned meant a cut from a wartime peak of 90 squadrons to eight regular and 15 reserve units. Post-war equipment included Mitchell light bombers, Mustang fighters and Canadian-built Douglas C-54s, known as Canadair North Stars, for transport work.

In 1949 Canada became one of the founder members of Nato and had by that time received the first jets in RCAF service, de Havilland Vampires. The important maritime-patrol task was undertaken by Avro Lancasters in those early years, these later being replaced by Lockheed Neptunes which in turn gave way to Canadair Arguses, developed from the Bristol Britannia airliner. The country's aircraft industry has provided designs to meet RCAF requirements since the late 1940s, including the Avro Canada CF-100 all-weather fighter, the Arrow interceptor (subsequently cancelled) and the range of de Havilland transports and trainers, as well as building a number of types under licence from US manufacturers.

As well as being a full member of Nato, Canada has a defence agreement with the United States known as NORAD (North American Air Defence). This agreement provides for the protection of Canadian and US urban centres and military installations although it does not specify any level of forces, equipment or facilities.

Above: Boeing CC-137, a multirole version of the civil 707-320C, with wing-tip inflight-refuelling pods supplied by Beech Aircraft; a CF-5 is being refuelled.

Below: A CP-121 Tracker, originally called a DH Canada CS2F-1, built under Grumman licence.

Voodoo force of interceptors was supplied by the USA in 1969 and provides air defence under the NORAD agreement, while the bulk of the Starfighters operate with the 1st Canadian Air Group in Germany. 10 Tactical Air Group has two squadrons of Canadair-built Northrop CF-5As for ground support, although most of the units in the group have helicopters for Army support and liaison totalling more than 130 machines in some six units.

Maritime Air Group has Argus and Tracker fixed-wing patrol aircraft and Sea King helicopters charged with the surveillance and control of Canadian territorial waters. A replacement for the long-serving Argus has been ordered in the form of 18 Lockheed P-3C Orions, designated CP-140 Aurora in CAF service, and delivery of these is planned for 1980–81. The Aurora will be a highly sophisticated aircraft with a Univac digital computer, APS-116 radar (as used in the S-3A Viking), infra-red sensors and cameras, enabling the machine to undertake many more missions than just anti-submarine patrol. The Sea Kings operate in the form of detachments from destroyers and replenishment ships of Maritime Command.

Canada's air commitment to Nato is exemplified in the three squadrons of CF-104 Starfighters based in Germany and forming the 1st Canadian Air Group, which is part of Canadian Forces Europe based at Lahr and Baden-Soellingen in the southern part of the country. An integral part of Nato's 4th Allied Tactical Air Force, the 1st CAG is tasked with offensive air-support operations in the event of a conflict in Europe. Pilots attend advanced air-to-ground and air-to-air training courses at Cold Lake in Canada as well as taking part in Red Flag exercises with the US Air Force at Nellis AFB, Nevada. Supplying the bases in Europe are the squadrons of the Air Transport Group flying Lockheed Hercules and Boeing CC-137s, some of the latter fitted with air-to-air refuelling equipment for the tanker role and used to support the CF-5s. Search-and-rescue duties are performed by a number of squadrons flying helicopters and some fixed-wing types including Buffalos and Twin Otters. To comply with government directions to improve the country's SAR capability, some Labrador and Voyageur helicopters are being upgraded for day-and-night, all-weather operation by 1980.

Above: A Bell-Vertol CH-113A Voyageur tandem-rotor rescue helicopter (CAF No 11308); in the background are a CC-130 and a CC-115, strongly suggesting that the photograph was taken at CAF Station Trenton, near Ottawa.

Below: A CSR-123 Otter.

Training

Training of aircrews for the Canadian Forces is the responsibility of Air Command. Because all training bases are situated in the Prairie provinces, reasonably close to the Command Headquarters at Winnipeg, the schools are administered directly through the HQ rather than through an air training group. No 3 Flying Training School at Portage La Prairie, Man, handles pilot selection and grading with graduates moving to 2 FTS at Moose Jaw, Sask, for training to wings standard on Canadair Tutors. The Air Navigation School, Winnipeg, trains navigators for service with the Maritime and Air Transport Groups. Four Hercules fitted out as trainers are used for this purpose, each aircraft having a removable module that fits into the forward half of the cargo compartment.

The CAF has an Air Reserve operating four wings and seven squadrons on light transport and SAR duties.

CANADA

Unit	Type	Role	Base	No	Notes
409 Sqn, 410 Sqn, 416 Sqn, 425 Sqn	CF-101B	Int	Comox, Bagotville, Chatham	59	
419 TFS, 433, 434 Sqn	CF-5A	GA	Bagotville, Cold Lake	24	Further 25 in storage
417, 421, 439, 441 Sqn	CF-104G	Strike	Baden Soellingen	85	Used by 1 CAG; 417 Sqn acts as OTU in Canada
	CF-104D	Train	Baden Soellingen	22	
407 Sqn	CP-140 Aurora	Mar pat	Comox	18	First due in May 1980, completion in March 1981
404, 405, 407, 415 Sqn	CP-107 Argus	Mar pat	Greenwood, Comox, Summerside	26	Also used by Maritime Evaluation Unit
VU-33, 406 OTU, 880 Sqn	CP-121 Tracker	Mar pat	Comox, Halifax, Shearwater	20	Also used by 420 Reserve Sqn
VT-406, 423, 443 Sqn	CH-124A Sea King	Mar pat	Halifax	32	Updated in 1978
414 Sqn	CF-100 Mk 5 Canuck	ECM	North Bay	—	Few remain; replaced by Falcons
435, 436 Sqn	CC-130E/H	Trans	Edmonton, Trenton	19/5	
437 Sqn	CC-137	Trans	Trenton	5	Two converted to tankers for CF-5 support
412 Sqn	CC-109 Cosmopolitan	Trans	Upland	7	Two aircraft with 1 CAG, Germany
	CC-117 Fan Jet Falcon	VIP	Upland	7	Three used by 414 Sqn, North Bay, as ECM trainers
413, 424, 442 Sqn	CC-115 Buffalo	SAR	Comox, Trenton, Summerside	14	
440 Sqn	CC-138 Twin Otter	SAR	Edmonton	8	
400, 401, 411, 438 Sqn	CSR-123 Otter	Trans	Montreal, Toronto	30	Reserve units
	CH-118 Iroquois	SAR	Cold Lake, Moose Jaw, Bagotville, Chatham	10	SAR flights at four bases
430, 450 Sqn	CH-135 Twin Huey	SAR	Valcartier, Ottawa	46	
413, 424, 442 Sqn	CH-113/CH-113A Labrador/Voyageur	SAR	Summerside, Comox, Trenton	14	
402, 429, 440 Sqn	CC-129 Dakota	Trans	Winnipeg	8	
1 CAG	CC-132 Dash Seven	Trans	Baden-Soellingen	2	
408, 427, 430, 444 Sqn, 3 FTS	CH-136 Kiowa	Liaison	Edmonton, Petawawa, Portage La Prairie, Valcartier	72	
450 Sqn	CH-147 Chinook	Trans	Ottawa	7	
3 FTS	CT-134 Musketeer	Train	Portage La Prairie	25	
2 FTS	CT-114 Tutor	Train	Moose Jaw	113	
414, 429 Sqn	CT-133 Silver Star	Train	Winnipeg, North Bay	22	

UNITED STATES

Area: 3·54 million sq. miles (9·16 million sq. km.).
Population: 207 million.
Total armed forces: 2,068,800.
Estimated GNP: $1,890 billion (£945 billion).
Defence expenditure: $115·2 billion (£57·6 billion).

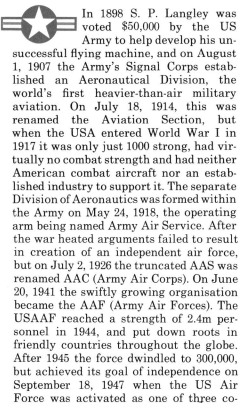

In 1898 S. P. Langley was voted $50,000 by the US Army to help develop his unsuccessful flying machine, and on August 1, 1907 the Army's Signal Corps established an Aeronautical Division, the world's first heavier-than-air military aviation. On July 18, 1914, this was renamed the Aviation Section, but when the USA entered World War I in 1917 it was only just 1000 strong, had virtually no combat strength and had neither American combat aircraft nor an established industry to support it. The separate Division of Aeronautics was formed within the Army on May 24, 1918, the operating arm being named Army Air Service. After the war heated arguments failed to result in creation of an independent air force, but on July 2, 1926 the truncated AAS was renamed AAC (Army Air Corps). On June 20, 1941 the swiftly growing organisation became the AAF (Army Air Forces). The USAAF reached a strength of 2.4m personnel in 1944, and put down roots in friendly countries throughout the globe. After 1945 the force dwindled to 300,000, but achieved its goal of independence on September 18, 1947 when the US Air Force was activated as one of three co-equal major military arms (the others, the Navy and Army, also deploy airpower, as described later).

Prior to its formation the USAF had decided its basic peacetime structure, and despite ceaseless worldwide activity and dramatic changes in size and missions this structure has changed little since. Its major operating commands are: SAC (Strategic Air Command), TAC (Tactical Air Command), ADCOM (Aerospace Defense Command), MAC (Military Airlift Command), USAFE (USAF Europe), PACAF (Pacific Air Forces) and AAC (Alaskan Air Command). Supporting these are ATC (Training Command), AFSC (Systems Command), AFLC (Logistics Command), AFCS (Communications Service), USAFSS (Security Service) and AU (Air University), as well as numerous separate operating agencies. Like all US armed forces, the USAF is an all-volunteer force; there is no conscripted training whatever. Total personnel strength at the end of the 1970s is about 570,000, plus some 230,000 civilians of whom half are in AFLC and AFSC. In addition there is an important combat-ready reserve force in the ANG (Air National Guard), numbering some 93,000, as well as over 100,000 in the Air Force Reserve of whom about half are paid.

Above: Dramatic photograph of a Boeing B-52G leaving light-diffusing contrails; under the inner-wing pylons are four triplets of Lockheed AGM-69A SRAM missiles.

Below: Two of the latest American strategic missiles, the Boeing Minuteman III (left) and Lockheed Trident (right). Minuteman is expected to be replaced by M-X.

Gradually, and remarkably painlessly, the US armed forces have been transformed since 1945, and have maintained world-wide peacekeeping commitments despite progressive budgetary pressures and escalation in costs. 7% of the USAF active strength is female, and the ratio is rising; and included in this total are numerous pilots and other aircrew, though women pilots are not assigned (yet) to combat units. About 15% of the total force is black, and there are now no barriers, real or imagined, to black personnel reaching the highest rank.

Essentially, the USAF is the prime instrument of US airpower, and the only service to deploy strategic striking forces other than Navy submarines with counter-value missiles. NORAD (North American Air Defense Command) once included USAF missiles, but today all land-based SAMs (surface-to-air missiles) belong to the Army. Likewise the Navy fighter squadrons are assigned to fleet duties, and in fact air defence of North America has been one area pared to the bone. Again, the former MATS (Military Air Transport Service) was a joint global logistic arm shared with the Navy, but today all responsibility for global movement of US material and troops is vested in the Air Force's MAC. Almost all air operations of the Navy (and subsidiary Marine Corps) and Army are in support of sea and land warfare, whereas the Air Force has a free-ranging duty to apply airpower anywhere. From its inception the Air Force has enjoyed political power at least as great as that of any other service, and this has resulted in its winning political battles such as denial to the Army of major fixed-wing airpower or any kind of strategic missile. In fact even the deployment by the Army of specially configured attack helicopters was resisted, the Air Force arguing that it alone had the right to engage in air combat.

Largest by far of the operating commands, SAC has a uniformed strength of some 105,000. Formed in 1946, its most important task is to deploy a visibly credible nuclear deterrent so that external aggression and war will not take place. In addition it is responsible for global strategic reconnaissance of many kinds, for long-range air-refuelling tanker support to US and Allied aircraft, and for all degrees of conventional and nuclear support to theatre or naval commanders.

During the Cold War in the 1950s SAC's piston and jet bombers flew throughout the world, often carrying live nuclear weapons, and were routinely rotated on a wing or squadron basis to prolonged deployment overseas, usually in England or North Africa. By 1960 vulnerability of the high-flying bomber had resulted in urgent training in low-level missions, together with prolonged and costly modifications to the structure of the B-47 and B-52 to fit them to this arduous duty. In 1961–63, a time of particular crisis, SAC even maintained a round-the-clock airborne alert to ensure that at any given time a substantial proportion of the bombers would be airborne with bombs on board. These policies had to be terminated to prolong structure life and reduce cost, and other means have been sought to disperse bombers and avoid destruction of the force in surprise attack on SAC airfields.

Quite contrary to previous planning, there has been no new bomber to replace the B-52, a noble but old bomber first flown in April 1952. The backbone of SAC's manned bomber strength comprises 24 squadrons with a total of 432 aircraft. Of these, 15 squadrons, organised into an approximately equal number of wings, have nominal strength of 241 B-52G and H. Each type has been so often strengthened as to be largely rebuilt, and a long programme of updating and augmenting the defensive and offensive electronics is not only continuing but being augmented, and B-52 modification funds have risen from $68.7m in Fiscal Year 1977 to $129.3m in 1978, $202.5m in 1979 and $437.2m proposed for 1980. At present the only weapons available are the AGM-69 SRAM (Short-Range Attack Missile) and free-fall bombs, and though each aircraft could carry 20 SRAMs there are only 1250 available in all. In 1979 competitive trials were to be held between the AGM-86B ALCM and AGM-108 Tomahawk to select a cruise missile for carriage by the B-52, with range of at least 1727 miles. Unlike the defunct AGM-86A neither of these cruise missiles is interchangeable with SRAM, and it is not publicly known how the B-52 will be modified to carry them, nor in what

Below: One of the fantastic Lockheed SR-71A "Blackbird" strategic reconnaissance platforms operated by the USAF 9th SR Wing from Beale AFB, California.

number. The intention is to modify about 120 aircraft, which at any one time might by the late 1980s be able to launch about 1400 of whichever missile is chosen.

SAC also has 75 of the older, shorter-range B-52D bombers, organised in five squadrons. These even more extensively reworked aircraft were rebuilt by Boeing-Wichita to carry up to 70,000lb of conventional bombs, and they could also carry such precision-guided weapons as the laser-homing winged GBU-15 when this becomes available. The only other operational bombers are 66 FB-111A in two two-squadron wings, each able to carry four SRAMs but requiring overseas deployment for conceivable two-way operational missions. The SAC air-breathing force might also later be dramatically augmented by purchase of a small number of transports – anything from a 747 to a YC-14 – modified to carry 'several dozen' cruise missiles; but this possibility remains a series of paper studies.

SAC uses about 20 B-52Fs for crew training, and a B-52D wing is based on Guam. A further 125 bombers, nearly all B-52D or F models, are inactive.

By far the largest single element in SAC's manned force is 30 Air Refueling Squadrons equipped with 487 KC-135s.

Above: Four photographs of the General Dynamics F-111 swing-wing attack bomber, one of the very few aircraft at present in service that can make a blind first-pass attack on a point target; the aircraft lower left is a TAC or USAFE F-111F (now based in UK), while the others are SAC FB-111As.

Left: Formation picture of a test AGM-109 Tomahawk air-launched cruise missile, heading inland across the California coast. There are land- and sea-launched versions.

These support all kinds of aircraft of all US services and Allied nations that are equipped with Flying Boom receptacles. Future tanker/airlift capability will be improved by acquisition of the enormously capable KC-10A, based on the DC-10-30CF, but these will be bought in very small numbers and the eventual total is unlikely to exceed 20.

A substantial part of SAC's aircraft force comprises special reconnaissance and electronic platforms, and command and control aircraft. One Strategic Reconnaissance Wing operates ten of the unique SR-71A with multiple sensors and the ability to cruise at over Mach 3; these are supported with JP-7 fuel from KC-135Q tankers. About ten U-2s of various sub-types fly high-altitude reconnaissance

missions with special sensors, and there are in the order of 55 EC-135 and RC-135 in about a dozen sub-types used for long-range surveillance, sampling, electronic missions and airborne command posts. Most C-135 versions are having their lower wing skins replaced to add a further 27,000hr flight time.

Since late 1975 SAC has also been responsible for operating the E-4 advanced airborne national command post (AABNCP) force of a probable total of six aircraft all to be updated to E-4B standard with F103 engines. The most costly aircraft in the world, these 747-type vehicles are based at SAC HQ at Offutt AFB and in time of crisis would be the seat of government and the command structure of the United States.

SAC is also manager of the American land-based intercontinental ballistic missile (ICBM) force. Largest missiles are 54 LGM-25C Titan IIs, emplaced in hardened silos (designed to resist nearby nuclear explosions) but over 15 years old and nothing like as effective as newer missiles despite electronic improvements to raise reliability and reduce costs. The main force comprises the much smaller solid-fuel Minuteman; SAC has 450 LGM-30F Minuteman II missiles and 550 LGM-30G Minuteman III. These are dispersed throughout the Midwest in silos much simpler and harder than those of Titan II, but recent Soviet ICBM developments have made them increasingly vulnerable. Capability of any missiles that survive a first-strike is being increased by providing

Above: The Fairchild (Republic) A-10A Thunderbolt II has been viewed by some as nonsensical, designed to fly slowly over the battlefield near thousands of hostile guns and missiles. TAC and USAFE still regard it as a most valuable platform.

Left: Two Thunderbolt IIs seen against the sky at dusk. New two-seat and other versions are under development to back up the spartan daylight A-10A now in use.

the LGM-30F with Mk 11C re-entry vehicles with upgraded penetration aids and a number of the LGM-30G force with triple 370-kiloton Mk 12A MIRV warheads with the NS-20 main vehicle guidance giving excellent accuracy. But SAC's future deterrent credibility inevitably rests on missiles whose position at any time cannot be known by an enemy, and despite several years of study of the all-new MX missile and a possibly less-costly ICBM based partly on the Navy C4 or D5 Trident there has been no decision to build any new missile by mid-1979. This means that SAC cannot restore any credible missile force until after the mid-1980s.

Second-largest element in the USAF is TAC, with uniformed personnel numbering about 84,000. This command has a multitude of functions, embracing command of the air over friendly ground forces or ships, close air support, deep interdiction/strike, multi-sensor reconnaissance, defence suppression and electronic warfare, with substantial transport and AWACS (airborne warning and control system) support. Of the 26 active TFWs (Tactical Fighter Wings), 15 belong to TAC, the other 11 being outside Conus (Continental United States). In fact the term 'fighter' is often a misnomer, because though most TAC aircraft are small jets able to give at least some showing in air combat, nearly half are attack aircraft whose equipment and weapons are oriented entirely towards air/ground missions, for example, the A-10, A-7 and F-111.

Unique in its single-minded concentration upon the close air support mission, with especial capability against armour and other point targets, the A-10A is a relatively large and slow weapon-platform with tremendous lethality and unusual qualities of survivability; it also has STOL rough-field performance for basing near forward troops (though it is hard to hide like a Harrier). Production of a planned 733 is proceeding at 14 per month, and two TFWs were operational by mid-1979 (one in USAFE). The highly cost/effective Vought A-7D has many weapons and precision delivery in almost all kinds of weather but is tapering off as the A-10 force builds up; there were still eight 24-aircraft squadrons in early 1979 but this will be halved by 1980. The costly swing-wing F-111 continues as the only aircraft in service (except possibly the Soviet Su-19) with complete blind first-pass attack capability. TAC has a wing of the original F-111A model and another of the F-111D, which has greater electronic capability but is proving a maintenance headache. Various F-111E and F aircraft serve at the busy Nellis AFB, mainly in the 57th Tac Training Wing.

Still the most numerous aircraft in TAC, the Phantom force numbers 600, down from 1,000 in 1976, in F-4C, D and E versions. This famous aircraft is an all-rounder. So is the F-15 Eagle, like the Phantom planned as a fighter and modified for air/ground delivery. By mid-1979 the number of 24-aircraft squadrons had risen

to 14, compared with nine a year earlier. Total buy is to be 729, the last squadron forming around August 1981, but very high costs of procurement and modification may reduce the total size of the force. It was chiefly because of high unit price that the USAF launched the lightweight-fighter programme that resulted in the F-16, powered by an engine identical to that used in the twin-engined F-15. The first F-16 TFW is now working up, and production of a planned total of 1388 aircraft is proceeding at a rate rising to 180 per year, to equip ten TFWs in TAC and 2½ reserve wings by summer 1986. (This is the USAF programme; other F-16s are being built in collaboration with European industry for other customers.)

Standard TAC multi-sensor reconnaissance aircraft is the RF-4C, but TAC will almost certainly be the operator of the planned force of 20 Lockheed TR-1 high-altitude multi-sensor 'standoff surveillance' aircraft to meet a joint USAF/Army need. A total of 25 TR-1s were budgeted in 1978 at $551.5m, much of the high cost being due to the advanced synthetic-aperture radar and all-weather SLAR (side-looking airborne radar). TAC may eventually receive another version, likewise based on the large U-2R airframe, to carry the PLSS (precision location strike system) originally to have been carried by the Compass Cope RPV (remotely piloted vehicle) cancelled in 1977; the PLSS carriers could number 34.

TAC has made great efforts to match the prowess of Soviet Frontal Aviation in EW (electronic warfare). In 1979 the main platforms for the Wild Weasel defence-suppression role were still two squadrons of F-105G Thunderchiefs and two of F-4G Phantoms, the latter having suffered development snags and not becoming fully effective with 48 aircraft until mid-1979. For the 1980s TAC hopes to rebuild 40

F-111s (at present used for combat crew training and needing modification to clear them for operational duty) to the EF-111A configuration now on extensive test, with the Navy-derived ALQ-99 tactical jammer and other systems installed internally, and retaining an operating crew of only two. The EF-111A would support both TAC's attack aircraft and the F-4G Wild Weasels, but in mid-1979 no decision on the EF-111A had been taken.

TAC is also progressively trying to improve the weapons delivery capability of all its combat aircraft. 'Smart' weapons of the Paveway and Hobos types have been widely supplemented by AGM-65A and later types of Maverick, and among new sensor and designator equipments are the Pave Spike laser pod retrofitted on almost all TAC F-4D and E aircraft and the sophisticated FLIR (forward-looking infra-red) and laser pod known as Pave Tack retrofitted to the F-111F and selected F-4Es.

Almost the whole operational strength of TAC is a quick-reaction force able to

pack its bags and go anywhere at a few hours' notice. All the combat aircraft have air refuelling probes or boom receptacles, and a little progress has been made in improving interoperability with NATO and other friendly air forces. Most of TAC's missions are oriented towards the European theatre, and a near-duplicate of TAC, from time to time reinforced by transfer of TFWs from TAC, is USAFE, which in terms of uniformed strength rates next in the combat hierarchy with 49,000. Three Air Forces based in Britain, West Germany and Spain deploy 28 squadrons, most of them on a 24-aircraft basis and equipped with the latest equipment. This is shown in the organizational chart, notable presences being two TFWs of F-111s, one (rising to three) TFW of A-10s and $1\frac{1}{2}$ (rising to a probable $3\frac{1}{2}$) TFWs of F-15s. The F-16 will also be an important type in USAFE, which with four European air forces may use a large computerised F-16 maintenance depot planned since 1975 but not yet built. USAFE still uses

large numbers of F-4 and RF-4 Phantoms, a few F-5Es for 'Aggressor' air combat training, and a heterogeneous collection of transports, helicopters and other support types. USAFE also expects to deploy the GLCM (Ground-Launched Cruise Missile), a version of the strategic type of BGM-109 Tomahawk. This will be used from mobile land launchers to cover important fixed-location European targets, releasing manned tactical aircraft for other duties.

Unlike the tactical forces, Adcom normally operates permanently from fixed Conus bases. Owing to a supposed dwindling of the Soviet bomber threat this command has shrunk more than any other, from 40 squadrons in 1964 to 28 in 1968 and six today. For nine years these have all been equipped with the F-106 Delta Dart, though the F-4 and F-101 are still flown by the ANG, and as the last F-106 was delivered in July 1960 there remains a suspicion that perhaps a new all-weather interceptor is needed. Following a decade playing with the YF-12 and other IMIs (improved manned interceptor) the decision has at last been taken to assign some of TAC's F-15s, the equivalent of one TFW, to have an additional interceptor function to allow the F-106 force to be phased out. Paradoxically, this withdrawal of defence of Conus against bombers was justified on the grounds that it could not be defended against missiles, with termination of the

Army ABM (anti-ballistic missile). This is an interim solution which robs Peter to pay Paul; should a bomber or cruise-missile threat develop Adcom would have to buy a new interceptor very quickly, the F-15 being one possibility (Tornado ADV is another).

All combat-aircraft operations in Conus will be controlled by a joint network of Adcom/FAA radars and by the AWACS force managed by TAC, initially by the 552nd Wing at Tinker AFB. A force of 34 of the extremely costly E-3A Sentry aircraft has been authorised, to support an operating force of 29, of which 16 had been delivered by mid-1979 (plus three test aircraft). These are gradually being phased in together with the JSS (Joint Surveillance System), which allows both the EC-121 Warning Star and the costly SAGE radar/computer ground network to be withdrawn. An EC-121 detachment in Iceland is likely to be replaced by the AWACS, but the equally old EB-57 Canberra remains in service (one Adcom and two ANG squadrons) as a sophisticated electronics platform to simulate hostile aircraft threats.

In fact, Adcom is increasingly concerned with missiles and space. It manages the global Spacetrack network of radar and optical sensors, the BMEWS (Ballistic-Missile Early-Warning System), several much newer and extremely large phased-array radars, and is building an

all-weather day/night Deep Space Surveillance System for instant global surveillance of all hardware out to distances beyond that for geostationary orbit at 22,350 miles. Adcom also has responsibility for all space warfare, including the development of ASAT (anti-satellite) systems; but, unlike the Soviet Union, it has no funds to conduct actual testing. Since the mid-1970s the murmurs from the Air Force about its total lack of space-warfare capability – to parallel that growing in the Soviet Union – have become increasingly audible.

Adcom's uniformed personnel number only 23,000, compared with over 70,000 for MAC. Possibly 'the world's largest airline', MAC has over 1000 aircraft, all gas-turbine powered and including 77 C-5A Galaxies, 271 C-141 StarLifters (currently being rebuilt with longer fuselage, air refuelling receptacle and other changes) and about 300 C-130s. MAC has total responsibility for all US government airlift, which not only means the armed forces but also government itself (mainly

by the VIP aircraft of the 89th MAG), emergency services (such as flying in snow clearance gear to a blizzard, or helping a country hit by earthquake, or handling air transport necessary after a major air disaster) and the widespread functions of the Aerospace Rescue and Recovery Service which has equipment ranging from helicopters to C-135s. Routine passenger airlift is assigned chiefly to the commercial carriers.

PACAF, with uniformed strength similar to Adcom at 23,000, has HQ in Hawaii and covers more than half the globe. The F-4 is the chief combat type, though a few F-5Es are used for Aggressor training and the OV-10 is used for FAC (forward Air

Below: USAF No 11407 was one of the first Boeing E-3A Sentry (AWACS) aircraft to enter service with the 552nd wing at Tinker AFB, Oklahoma.

Foot of the page: No 50748 was an early F-16A air-superiority fighter, delivered from General Dynamics, Fort Worth, and assigned to Hill AFB, Utah.

Left: Langley AFB, Virginia, is the home of these McDonnell Douglas F-15A Eagle fighters, carrying AIM-9J Sidewinders and with provision for various marks of AIM-7 Sparrow missile. Unit: 1st TFW, TAC.

Control) as in Europe. AAC (Alaskan Air Command) numbers about 7000 military and comprises 13 ACW (Aircraft Control and Warning) squadrons and a mixed bag of tankers, trainers, FAC and helicopters, as well as two operational F-4 squadrons.

Backing up the operational commands are a diverse array of other commands and services. The largest are the gigantic AFLC (9000 uniformed and 82,000 civilian) which keeps the Air Force running, AFSC (25,000 military and 26,000 civilian) which handles all management of advanced technology, research, design and procurement of hardware (spending over $10bn a year, about 30% of the Air Force budget,

in the process), ATC (64,500 plus 14,000) which now incorporates the AU (7000 plus 2000) and carries out all training not only for the USAF but for many friendly air forces, and the Air Force Intelligence Center and Air Force Communications Service, which are the largest in a long list of supporting commands and agencies.

US Navy

Like the US Army, the Navy was ahead of all others in getting into aviation; Eugene Ely made the first takeoff from a ship (the cruiser *Birmingham*) on November 14, 1910, following with an arrested landing, with transverse ropes braked by sandbags, on the *Pennsylvania* on January 18, 1911. By America's entry to World War I the Navy had nearly 140 aircraft, most of them small seaplanes, on order but few in use. Eighteen months later, at the Armistice, it had thousands, including 1172 large ocean-patrol flying boats and a strong force of airships. Just after the war four of the biggest patrol flying boats then

Above left: A late-model F-4E Phantom, with slatted wing and improved electronics, and carrying a multi-sensor pod for ground attack.
Centre: Four Northrop F-5E Tiger II fighters serving with an Aggressor unit at RAF Alconbury, England (simulating hostile aircraft, such as the MiG-21).
Below left: The C-130 is numerically the chief transport in MAC.

Below: Sole Aerospace Defense Command interceptor is the F-106A Delta Dart.

built, the Curtiss NC type, were assigned to make the first heavier-than-air flight across the Atlantic, and one of them, NC-4, flew in stages from Newfoundland to the Azores, Lisbon and Plymouth in May 1919. Subsequently the Navy commissioned a series of carriers, beginning with *Langley* (1922), and pioneered dive bombing and a policy of making even single-seat fighters carry bombs. From World War I the US Marine Corps has also had airpower, as described later, with the Navy Bureau of Aeronautics (now Naval Air Systems Command) handling development and procurement on behalf of the smaller service.

After World War I, as in Britain, there was severe rivalry and friction between the aviation aspirations of the Navy and Army (and even between the 'air' and 'sea' branches of the Navy). After World War II this rivalry was intensified by the formation of the US Air Force, which not only gained a monopoly in the new global nuclear deterrent but sought such funds in a time of financial stringency that the Navy's seagoing force suffered. A particular battle grew up between the B-36 global bomber and the proposal to deploy global airpower aboard a new fleet of super-carriers far larger than any previously constructed. Eventually one monster, CVA-58 *United States*, was authorised as the result of efforts by the Secretary of Defence (and former Navy Secretary) James V. Forrestal. In 1949 the ship was laid down, but Forrestal died and the giant carrier was cancelled. Need for such ships, however, was recognised following the outbreak of war in Korea in 1950, and with it the need of the Navy for strategic nuclear attack aircraft as exemplified by the Douglas A3D (later A-3) Skywarrior. Almost a seagoing B-47, such an aircraft demanded a super-carrier, and in the 1952 budget the first was again authorised. CVA-59 was named *Forrestal* and commissioned in October 1955. She has since been followed by 11 further giant carriers, four of them nuclear-powered, which represent by far the greatest investment

in seagoing (and truly global) airpower the world has seen. There are also two of the smaller *Midway* class still active, and a growing force of LHA and LPH V/STOL carriers for amphibious warfare.

Indisputably the US Navy has for many years been world leader in the design and operation of fixed-wing airpower at sea, despite the fact that nearly all the significant improvements in carriers came from Britain. The engineering design of US carrier-based aircraft has reached a remarkable pitch, not even approached elsewhere, and in the unique case of the F-4 Phantom resulted in a machine superior to land-based rivals. Likewise the F-14 Tomcat, despite prolonged problems with engines, structure and systems, and severe cost-overruns, demonstrates long-range interception capability possessed by no other aircraft. This is due to its AWG-9 radar fire control, infra-red and electro-optics for long-range target identification, and long-range Phoenix missiles, which can be fired by the AWG-9 against up to six different targets almost simultaneously. These two types make up the seagoing fighter force, comprising 14 VF (fighter squadrons) with the F-14A (168 aircraft) and 12 with 144 F-4J and other Phantom versions. Despite continuing problems with the F-14 it is hoped to continue procurement to a total of 18 VFs by 1983 and also buy enough for later attrition to allow this potentially outstanding machine to replace the RF-8G and RA-5C in the reconnaissance role.

Largest part of the combat strength comprises 36 VA (attack squadrons) comprising 11 with 110 A-6E Intruders and 25 with 300 A-7E Corsairs, plus a handful of

Below: Vought A-7E Corsair II attack aircraft from the Carrier Air Group embarked aboard USS *America*. By 1983 these will be replaced by the F-18 Hornet.

Right: CVA-65 *Enterprise* was the world's first large nuclear-powered warship and is still the only platform of its type. Her Carrier Air Wing includes F-14s.

earlier A-7A or B versions. The existing front-line force of the F-4, F-14, A-6 and A-7 appears excellent but has worried the Navy which has for many years accepted that the soaring price of new acquisition was leading to smaller procurement and an increasingly ageing force. To try to update the Navy and Marine Corps (as presently described) in a cost/effective way the decision was taken in May 1975 to buy a large number, an initial 800, of a supposed less-costly but modern aircraft derived from the Northrop YF-17 (the

Above: This Grumman F-14A Tomcat is armed with the full spectrum of missiles — AIM-9 Sidewinder, AIM-7 Sparrow and the extremely long-range AIM-54 Phoenix — plus the internal 20mm gun.

loser to the F-16 in the USAF lightweight fighter competition). This aircraft became the McDonnell Douglas/Northrop Hornet, being bought with a nearly common airframe in A-18, F-18 and TF-18 versions. The F-18 multi-role fighter will replace the F-4 in the roles of fighter escort and interdiction and is designed to bring a dramatic increase in operational readiness and reduced maintenance burden. The first F-18 rolled out in September 1978, and after very substantial funding at $870/920m/year the first Navy F-18 squadron is to become operational in 1984. The A-18, with different sensors and displays configured for attack, will replace the A-7 (and, in the Marines, other types). The tandem-seat TF-18 has dual controls.

Standard fixed-wing ASW (anti-submarine warfare) aircraft for carrier operation is the Lockheed S-3A Viking, perhaps the greatest-ever exercise in packaging into a compact airframe. This twin-turbofan four-seater carries essentially the same kit of radar, sonics and MAD sensors, with advanced digital processor, as larger

shore-based types; in fact the Viking's ASW systems were adopted complete for the Canadian CP-140 Aurora in preference to standard P-3 equipment in the same airframe. The S-3 equips 13 VS squadrons, each with ten aircraft. Standard shipboard ASW helicopter is the familiar SH-3 Sea King series, which in various subtypes equips 12 HS squadrons of eight aircraft each. Eventually a later ASW helicopter, the SH-60B, will supplement and then replace the SH-3 and smaller SH-2 Seasprite, chiefly operating aboard frigates, destroyers and cruisers in the ASW and anti-cruise-missile roles.

Electronic warfare at sea is handled by the four-seat EA-6B Prowler, a costly but proven aircraft whose two rear-seat crew

are managers of the power supplies and various EW systems, chief among which is the ALQ-99 system whose main jamming transmitters are carried in external pods, tuned to different wavelengths. This is a more powerful system than any in USAF service (it is installed in the EF-111A now on test). Airborne early warning is handled by the extremely useful E-2 Hawkeye, which equips 13 AEW squadrons of four aircraft; all at sea have the updated E-2C with ARPS (Advanced Radar Processing System), the E-2B being relegated to the Reserve squadrons.

These varied aircraft are arranged into 12 multi-purpose air wings (MPAW) embarked aboard the 12 large-deck carriers. Each MPAW has the following composition: 2 VF fighter squadrons; two VA light attack squadrons (A-7); and one each of VA medium attack (A-6); VS anti-sub-

Below: Scheduled to play a large role in tomorrow's Navy and Marine squadrons, the F-18 Hornet is under development by McDonnell Douglas and Northrop Aircraft.

Above: One of the earlier E-2A series of Grumman Hawkeye airborne early-warning platforms airborne from CVA-43 *Coral Sea* in 1967 (no longer in commission).

marine fixed-wing; HS rotary-wing; VAQ electronic warfare; VAW early warning; and RVAH all-weather reconnaissance. The attack squadrons are supported by four KA-6D air-refuelling tankers on each carrier. The Navy pioneered the digital transmission of multi-sensor reconnaissance data, on a secure basis (ie, not interfered with by an enemy), from aircraft at any altitude direct to surface vessels or shore bases. The usual aerial platform in this system has been the capable RA-5C Vigilante, used operationally in this vital role in Southeast Asia. The proposed interim reconnaissance model of the Tomcat, possibly to be designated RF-14A, will not fully meet requirements, and neither could an adaptation of the F-18.

By far the most important element of shore-based Navy airpower is the powerful force of 24 VP (patrol) squadrons,

organised into five wings, equipped with 280 Lockheed P-3 Orion long-range turbo-prop aircraft. Though many P-3B Orions remain in service, the main active model is the P-3C, with Univac digital computer linking A-NEW detection and weapon-control systems and many other new equipment items. These efficient aircraft have been improved by Update and Update II modification programmes, and in the early 1980s are to be given largely new ASW electronics and an advanced signal-processing system under the Update III programme. The recurrence of crises in distant locations, and gradual whittling down of available bases (for example, Bandar Abbas in Iran became unavailable to the US early in 1979), have belatedly prompted the Navy to follow the RAF and fit its long-range patrol aircraft with flight-refuelling capability. This will increase the radius at which a 4hr patrol can be maintained from 1400 to 2400 nautical miles (4450km). Orions are still being delivered at the rate of 12 per year.

The Navy maintains a large and capable support and training force, including C-130 Hercules in many variants, C-9B (DC-9 with long-range tanks and cargo doors) Skytrain II transports, CT-39 Sabre-liner multi-role jets, F-5E Tiger IIs for Aggressor training at NAS Miramar, 19 other training squadrons and a major rotary-wing force including the very powerful RH-53D mine-countermeasures helicopter for neutralising mechanical, magnetic and acoustic mines.

There are powerful and immediately operational Reserve Forces, including two fighter wings, one (squadrons VF-201 and 202) at Dallas and the other (VF-301 and 302) at Miramar, both flying the F-4N Phantom. There are no fewer than 13 VP squadrons with the P-3 Orion (mostly earlier models), six with the A-7, three ECM units with the EKA-3B Skywarrior and two-seat EA-6A, two composite squad-rons with the A-4/TA-4J Skyhawk, two E-2B AEW units, two reconnaissance squadrons and seven helicopter squad-rons.

US Marine Corps

Combining the requirements of both sea and land warfare, the USMC deploys a brand of airpower unique in the world. It is strongly oriented towards amphi-bious beach assault, and the need to trans-port attack forces by V/STOL vehicles while maintaining air superiority over-head. Unlike all other branches of the US armed forces the Marines have shown great interest in the British-developed Harrier as an available V/STOL combat aircraft, and not only put a number into combat service but sponsored the develop-ment of a chiefly American improvement offering double the range or weapon load. This aircraft, the McDonnell Douglas AV-8B, could be a vital factor in giving the Marines greatly enhanced multi-role air-power without airfields, for it has a STOL lift increased by 6,700lb and an internal fuel capacity increased from 630 to 1130 Imp gal. The Marines have a requirement for 342, and the Navy has announced interest in purchasing a similar number for multiple V/STOL roles aboard various types of surface ship. Unfortunately, the fact that the basic concept is not American has adversely influenced Congress and the Pentagon, and the AV-8B has had to struggle for funds and suffered a pro-tracted development (the Navy view, and it is the Navy that buys aircraft for the Marine Corps, is that V/STOL is merely 'a potentially attractive weapons system concept for the 1990s and beyond'). The AV-8B could have been in service in 1980, but at the time of writing is – surely tem-porarily – not even funded, through the lobbying of its detractors.

At present, apart from the AV-8A force, the chief Marine Corps combat types are the F-4N and F-4S Phantom and two attack types, the A-4M and A-4Y Skyhawk II and the A-6E (plus a few A-6A) Intruder. These will all in due course probably be replaced by the A-18 and F-18 versions of the Hornet, which McDonnell Douglas has naturally developed faster than the AV-8B because the conventional Hornet has been funded at the rate of $341–920m/

Top: The Navy's only shore-based ASW platform is the Lockheed P-3 Orion; this one is with squadron VP-91.

Above: EA-6B Prowler and (rear) EA-6A from squadron VMAQ-2.

year (1977–80) while the V/STOL aircraft has received only $33–85m over the same period, smallest by far of all major Navy and Marine aircraft programmes.

Main Corps airpower is organised into three MAW (Marine Air Wings), with headquarters at El Toro, California; Cherry Point, North Carolina; and Oki-nawa. Each wing is associated with a force-reinforced Marine division to oper-ate as a so-called division-wing team. It will be noted that one is based on the US East Coast while both the others are oriented towards the Pacific. A fourth division-wing team is based at Glenview, Illinois, as a reserve force, able to become active at short notice and move wherever it may be needed. All these MAWs use the types listed above, there being 12 VMFA squadrons with 144 F-4s (two more in the

Reserve), 13 with 80 AV-8As, five with 60 A-4s (five more in Reserve), five with 60 A-6s, and a supporting ECM squadron (10 EA-6B) and reconnaissance squadron (the unique RF-4B, similar in sensors to the USAF RF-4C). There is also a single detachment operating the two-seat EA-6A aboard *Midway*, planned for retirement in about 1985. While the AV-8B languishes, the evergreen A-4 continues in limited production, the latest variant being the A-4Y with improved Angle-Rate Bombing System, updated HUD (head-up display) and new landing gear. Exactly how the existing fixed-wing force will be replaced by the F-18 fighter and A-18 attack versions of the Hornet is not clear, nor how the future fixed-wing force will be able to operate from short unprepared airstrips. It has long been Marine Corps policy to go ashore carrying a transportable airfield (said to be quickly assembled, but in fact needing a substantial ship to carry it and a large labour force one to three weeks to make it ready for sustained operations).

Backing up the three active and one reserve MAWs are a rather larger number of squadrons equipped with helicopters, transports and other types. The standard fixed-wing airlifter is the KC-130 Hercules in two variants, both equipped as air re-fuelling tankers, equipping three 12-aircraft squadrons (another in the Reserve). All can have heated ski landing gear and assisted-takeoff rockets. The unusual STOL OV-10A Bronco, which has also been used by the Navy and Air Force, remains popular in the Marine Corps as an observation aircraft, equipping two 18-aircraft squadrons (another in the Reserve). This small twin-turboprop proved its worth in Vietnam as a FAC (Forward Air Control) platform, and also doubles as a light close-support aircraft and casevac (casualty-evacuation) transport.

The Marine helicopter force is divided into heavy, medium and light categories, called Assault Support Squadrons, plus three Attack Helicopter squadrons. The Heavy Helicopter is the Sikorsky S-65 Sea Stallion, operated in the CH-53A, more powerful CH-53D and considerably further uprated three-engined CH-53E forms. The CH-53E can carry 93% of a Marine division's combat equipment and could retrieve 98% of Marine combat aircraft without dismantling, but the number being delivered has been cut from 70 to 49. The E can carry 55 troops, compared with 37 for the earlier Sea Stallions. Standard Medium Helicopter is the Boeing Vertol CH-46 Sea Knight, of which 162 equip nine squadrons. The Light Helicopter is the UH-1 in various versions, of which 96 equip four squadrons. The Attack Helicopter is the specially developed SeaCobra, most of which were originally built to AH-1J configuration differing from the AH-1G HueyCobra chiefly in having a T400 twin-engined power unit. Recent deliveries have been of the Improved SeaCobra AH-1T, with TOW missile system and sight, and most Marine SeaCobras are being rebuilt to this standard. The USMC Reserve deploys an AH-1G squadron, two with the UH-1E, three with the CH-53A and three with the CH-46.

Top: This Lockheed S-3A Viking is from squadron VS-22 (the Fighting Red Tails) embarked aboard USS *Saratoga*. It can carry a mixture of mines, bombs, missiles and drop tanks. The ship is the destroyer *Charles F. Adams*. The Viking has a very good record in sea service.

Above: British Aerospace AV-8A Harriers operating 'in the rough' with the Marine Corps. These relatively simple but effective V/STOL jets have spurred the Marines and Navy to press for the AV-8B, with improved payload/range. The AV-8B is not certain to go ahead, however.

35

US Army

Throughout the United States' participation in World War II there were aviation units attached to all Army divisions and especially to major theatre headquarters, the most numerous aircraft being the Piper L-4 Grasshopper which could operate from any 100-yard level surface with reasonable approaches. On September 18, 1947 the US Army (abb. USA) retained only the nucleus of a future aviation force for battlefield reconnaissance, liaison, and light utility transport, with the accent strongly on V/STOL and with fixed-wing equipment limited to 12,500lb gross weight. This restriction was later removed, but the Army has never been able to show a need for large aircraft and has to rely totally upon the USAF's MAC for airlift of all heavy equipment and for all strategic movements. Even the fitting of weapons in Army aircraft resulted in political battles with the Air Force, and though the latter has at times tried to maintain that all tactical airpower – interpreted as aircraft carrying weapons – is its own exclusive preserve, the Army has managed to build up a strong force of attack helicopters.

Though the Army has an Aviation Centre at Fort Rucker, Alabama, it has hardly any separate 'Army Aviation' force or command structure; its aviation is, to use a favourite US Army word, 'organic' to the Army itself. Thus an infantry battalion may be Light, Ranger, Airborne (in which case they are airlifted, usually by fixed-wing aircraft) or Air Assault (in which case they drop on the enemy by helicopter). Again, cavalry are the mobile forces for various battlefield duties including reconnaissance and pricking the enemy either

to delay him or force him to reveal his locations. There are just two species, Armored Cavalry and Air Cavalry, and the latter uses not AFVs (armoured fighting vehicles) but helicopters. Air Cavalry squadrons and troops are organised into three groups. Aerorecon provides a limited ground reconnaissance capability. Aeroscouts fly the aerial reconnaissance missions to provide the whole troop, and higher headquarters, with fullest information. Aeroweapons provide firepower, with attack helicopters.

The US Army pioneered sophisticated attack (gunship) helicopters from the early 1960s, reaching a very high pitch of training and technique in Southeast Asia in which various types of gunship, scout and transport helicopter flew integrated missions, often even collaborating with fixed-wing aircraft. Since then the standard attack helicopter has been the Bell AH-1 HueyCobra, usually deployed in each Air Cavalry troop in the ratio of five regular AH-1Gs and four AH-1S TOW-Cobras for anti-armour use. The Army still has a four-figure total of Cobras, and many of the G-models are being converted to TOW armament with the designation AH-1Q. The intention is to deploy 987 AH-1S TOWCobras, with certain sensors and other items that are not yet available, by 1984.

While the big Cobra force fills the 'low' side of the attack helicopter requirement, the 'high' side (and some armies, notably the British, have not recognised an attack helicopter need at all) will be met by the more powerful Hughes AH-64A, selected as the future AAH (Advanced Attack Helicopter) in December 1976. The AH-64A will carry up to 16 of the precision-homing Hellfire missiles for destroying

AFVs and other small targets, as well as up to 76 rockets and a 30mm XM230 Chain Gun, using the same ammunition as the British Aden and French DEFA. Among more extensive all-weather sensing and weapon-aiming systems than ever before put into service on a helicopter, chief equipment will be the TADS (Target Acquisition and Designation System) and the PNVS (Pilot's Night Vision System), as well as the IHADSS (Integrated Helmet And Display Sighting System). For three years the integration of these systems has gone ahead, and procurement of the first 18 AH-64A for the inventory is due in FY81 (fiscal year 1981, ending on June 30, 1981). Total procurement is planned to be 536, including three prototypes, and development funding has averaged $170m annually.

Only just coming into the big-money phase of development, the ASH (Advanced Scout Helicopter) is an uncertain programme that may not come to fruition. The Army stated a requirement for an ASH as the smallest possible platform able to carry a laser designator and communications (and possibly some night/all-weather sensors). Its purpose would be to find targets for the attack helicopters and designate them for homing weapons.

Basic transport helicopters comprise the ubiquitous UH-1 'Huey' and heavy-lift CH-47 Chinook and CH-54 Tarhe. Probably the most numerous single type in any air arm in the world, the USA's fleet of UH-1 versions numbers not far short of 4,000, though not all are currently active. The UH-1H is still being delivered, but the whole force is scheduled for eventual replacement with the much more powerful Sikorsky UH-60A Black Hawk, developed under the UTTAS (Utility Tactical Trans-

UNITED STATES OF AMERICA Air Force

Division/Wing	Type	Role	Base	No	Notes
Strategic Air Command, HQ Offutt AFB, Nebraska					
8th Air Force, HQ Barksdale AFB, Louisiana:					
19th Air Division, HQ Carswell AFB, Texas:					
2nd Bomb Wing	B-52/KC-135	Bomb/tank	Barksdale AFB, La	18/38	
7th Bomb Wing	B-52/KC-135	Bomb/tank	Carswell AFB, Tex	18/38	
340th Air Refueling Group	KC-135	Tank	Altus AFB, Okla	19	Tenant unit, MAC base
381st Strategic Missile Wing	Titan II	ICBM	McConnell AFB, Kan	18	
384th Air Refueling Wing	KC-135	Tank	McConnell AFB, Kan	38	
40th Air Division, HQ Wurtsmith AFB, Michigan:					
305th Air Refueling Wing	KC-135	Tank	Grissom AFB, Ind	38	
351st Strategic Missile Wing	Minuteman II	ICBM	Whiteman AFB, Mo	150	
379th Bomb Wing	B-52/KC-135	Bomb/tank	Wurtsmith AFB, Mich	18/38	
410th Bomb Wing	B-52/KC-135	Bomb/tank	K.I.Sawyer AFB, Mich	18/38	
42nd Air Division, HQ Blytheville AFB, Arkansas:					
19th Bomb Wing	B-52/KC-135	Bomb/tank	Robins AFB, Ga	18/38	Tenant unit, AFLC base
68th Bomb Wing	B-52/KC-135	Bomb/tank	Seymour Johnson AFB, NC	18/38	Tenant unit, TAC base
97th Bomb Wing	B-52/KC-135	Bomb/tank	Blytheville AFB, Ark	18/38	
301st Air Refueling Wing	KC-135	Tank	Rickenbacker AFB, Ohio	38	
308th Strategic Missile Wing	Titan II	ICBM	Little Rock AFB, Ark	18	
45th Air Division, HQ Pease AFB, New Hampshire:					
42nd Bomb Wing	B-52/KC-135	Bomb/tank	Loring AFB, Me	18/38	
380th Bomb Wing	FB-111/KC-135	Bomb/tank	Plattsburgh AFB, NY	30/38	
416th Bomb Wing	B-52/KC-135	Bomb/tank	Griffiss AFB, NY	18/38	
509th Bomb Wing	FB-111/KC-135	Bomb/tank	Pease AFB, NH	30/38	
15th Air Force, HQ March AFB, California:					
4th Air Division, HQ Francis E. Warren AFB, Wyoming:					
28th Bomb Wing	B-52/KC-135	Bomb/tank	Ellsworth AFB, SD	18/38	
44th Strategic Missile Wing	Minuteman II	ICBM	Ellsworth AFB, SD	150	
55th Strategic Recon Wing	EC/RC-135	Recce	Offutt AFB, Neb	c14	
90th Strategic Missile Wing	Minuteman III	ICBM	F.E.Warren AFB, Wyo	200	
12th Air Division, HQ Dyess AFB, Texas:					
22nd Bomb Wing	B-52/KC-135	Bomb/tank	March AFB, Calif	18/38	
96th Bomb Wing	B-52/KC-135	Bomb/tank	Dyess AFB, Tex	18/38	
390th Strategic Missile Wing	Titan II	ICBM	Davis-Monthan AFB, Ariz	18	Tenant unit, TAC base
14th Air Division, HQ Beale AFB, California:					
9th Strategic Recon Wing	SR-71/U-2	Recce	Beale AFB, Calif	c.18/c.12	
93rd Bomb Wing	B-52/KC-135	Bomb/tank	Castle AFB, Calif	18/38	
100th Air Refueling Wing	KC-135Q	Tank	Beale AFB, Calif	c.9	Carry JP-7 fuel for SR-71
307th Air Refueling Group	KC-135	Tank	Travis AFB, Calif	19	
320th Bomb Wing	B-52/KC-135	Bomb/tank	Mather AFB, Calif	18/38	
47th Air Division, HQ Fairchild AFB, Washington:					
6th Strategic Wing	RC-135	Recce	Eielson AFB, Alaska	c.14	Tenant unit, AAC base
92nd Bomb Wing	B-52/KC-135	Bomb/tank	Fairchild AFB, Wash	18/38	
341st Strategic Missile Wing	Minuteman II/III	ICBM	Malmstrom AFB, Mont	150/50	

port Aviation System) programme as the future standard vehicle for assault-helicopter, air cavalry and aeromedical evacuation units. Though its 11 passenger seats are actually fewer than in some UH-1 versions, the UH-60A is vastly more capable and better equipped, though the unit price is around $3m. Numbers bought in 1977–80 inclusive were 15, 56, 129 and 168, towards an eventual total of 1107.

Above: The Sikorsky UH-60A Black Hawk is now in volume production as the Army's UTTAS (utility tactical transport aviation system). Though modest in size, it has fair airlift capability.

UNITED STATES OF AMERICA   Air Force—continued

Division/Wing	Type	Role	Base	No	Notes
57th Air Division, HQ Minot AFB, North Dakota:					
5th Bomb Wing	B-52/KC-135	Bomb/tank	Minot AFB, ND	18/38	
91st Strategic Missile Wing	Minuteman III	ICBM	Minot AFB, ND	150	
319th Bomb Wing	B-52/KC-135	Bomb/tank	Grand Forks AFB, ND	18/38	
321st Strategic Missile Wing	Minuteman III	ICBM	Grand Forks AFB, ND	150	
1st Strategic Aerospace Division, HQ Vandenberg AFB, California				—	Space missions
3rd Air Division, HQ Andersen AFB, Guam:					
43rd Strategic Wing	B-52D/KC-135	Bomb+train/tank	Andersen AFB, Guam	45/38	
376th Strategic Wing	KC-135	Tank	Kadena AB, Okinawa	c.38	Tenant unit, PACAF base
Direct reporting:					
306th Strategic Wing	various	various	Ramstein AB, Germany	—	Tenant unit, USAFE base
544th Aerospace Recon Tech Wing	various	Research	Offutt AFB, Ne	—	
1st Combat Eval Group	various	Research	Barksdale AFB, La	—	
3902nd Air Base Wing	various	Support	Offutt AFB, Neb	—	
Tactical Air Command, HQ Langley AFB, Virginia:					
9th Air Force, HQ Shaw AFB, South Carolina:					
1st Tactical Fighter Wing	F-15/EC-135/UH-1	FGA	Langley AFB, Va	72++	
4th Tactical Fighter Wing	F-4E/F-15	FGA	Seymour Johnson AFB	72	
23rd Tactical Fighter Wing	A-7D/A-10A	Att	England AFB, La	72	
31st Tactical Fighter Wing	F-4E	FGA	Homestead AFB, Fla	72	
33rd Tactical Fighter Wing	F-4E/F-15	FGA	Eglin AFB, Fla	72	
56th Tactical Fighter Wing	F-4D, 4E/UH-1	FGA	MacDill AFB, Fla	72+	
347th Tac. Fighter Wing	F-4E	FGA	Moody AFB, Ga	72	
354th Tac. Fighter Wing	A-10A	FGA	Myrtle Beach, SC	72	
355th Tac. Fighter Wing	A-7D/A-10A	FGA	Davis-Monthan AFB, Ariz	72	Also HQ Tactical Trg
363rd Tac. Recon Wing	RF-4C	Recce	Shaw AFB, SC	54	
1st Special Ops Wing	AC-130/helos	various	Hulburt (Eglin Aux No 9)	—	Also USAF Spec Ops Sch
12th Air Force, HQ Bergstrom AFB, Texas:					
27th Tactical Fighter Wing	F-111D	Att	Cannon AFB, NM	72	
35th Tactical Fighter Wing	F-4/F-105G/UH-1	FGA	George AFB, Calif	—	Also HQ Tac Trg George
49th Tactical Fighter Wing	F-15	FGA	Holloman AFB, NM	72	
58th Tac. Training Wing	F-4/F-15/TF-104	Train	Luke AFB, Ariz	—	
67th Tac. Recon/Wing	RF-4C	Recce	Bergstrom AFB, Tex	54	
366th Tac. Fighter Wing	F-111A	Att	Mountain Home AFB, Idaho	72	
388th Tac. Fighter Wing	F-16/F-4D	FGA	Hill AFB, Utah	72	
432nd Tactical Drone Grp	DC-130/RPVs	RPV	Davis-Monthan AFB, Ariz	—	
474th Tac. Fighter Wing	F-4D	FGA	Nellis AFB, Nev	72	Also see Nellis below
479th Tac. Training Wing	T-38		Holloman AFB, NM	—	
602nd Tac. Air Control Wing	OV-10/O-2/CH-3		Bergstrom AFB, Tex	—	
Direct reporting:					
552nd AWAC Wing	E-3A		Tinker AFB, Okla	—	Also EC-135, EC-130, etc
USAF Southern Air Division	Inter-American Air Forces Academy, Albrook AFS, CZ				
	24th Composite Wing, Howard AFB, CZ				

Standard LOH (Light Observation Helicopter) was originally planned in 1961 to be the Hughes OH-6A Cayuse, of which almost 1000 remain of 1434 produced. Later the Bell OH-58A Kiowa won an even larger order for 2200. By far the most important heavy-lift force is the 600-odd Boeing Vertol CH-47 Chinooks, built in CH-47A, B and C standards of increasing capability and all now being rebuilt to an even better common CH-47D standard to maintain the Army's medium-lift logistical support force. The heavy lifter remains the Sikorsky CH-54 Tarhe, of 8000hp, the much larger XCH-62 having been cancelled on cost grounds. There are several hundred light training and liaison helicopters, mainly the Hughes TH-55 Osage and Bell H-13 Sioux.

A unique fixed-wing aircraft, and the only "combat" fixed-wing type deployed by the Army, the Grumman OV-1 Mohawk is a twin-turboprop STOL two-seater of which about 200 remain in use as OV (battlefield surveillance), JOV (armed reconnaissance), EV (electronic warfare) and RV (electronic reconnaissance) versions. The Army has about 200 twin-turboprop Beech U-21 Utes (versions of the civil King Air), RU-21 (Cefly Lancer) electronic platforms, C-12A Hurons and piston-engined U-8 Seminoles. Standard fixed-wing trainers are the Cessna T-41 Mescalero and twin-engined Beech T-42 Cochise.

Below: This UH-1H was one of the last of the original families of 'Huey' transport helicopter to be delivered, with a large cabin but less power than T400 or T35 versions. The photograph was taken during the Vietnam War; hundreds were lost here.

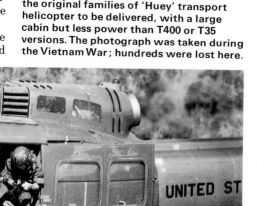

Recent combat experience

It is sometimes overlooked that the forces of the United States, in terms of numbers of personnel and equipment items, have more recent combat experience than any others in the world. This is almost entirely because of their long 'involvement' in Southeast Asia (the SEA theatre), which, short of nuclear war and fullscale actions at sea, encompassed every kind of military activity, often on a very large scale.

American involvement in SEA was almost continuous from 1941 until 1973. In 1950 the young USAF sent military advisors, combat crews, maintenance personnel and supply experts to help the French in Indo-China, and when the situation at Dien Bien Phu became grave in the spring of 1954 the US planned active intervention with both Air Force and Navy aircraft (including B-29s). Congress said that it would agree only if the British also took part, and Prime Minister Churchill refused. This was the end of French rule in the area, but the US military presence continued and the Military Assistance Advisory Group (MAAG) gave arms and training to the new states of South Vietnam, Cambodia and Laos. By the end of the 1950s US forces were falling victim to Viet Cong attacks, and the official US armed involvement dated from October 1959. The first active air supplies were 25 AD-6 Skyraiders and 11 H-34 Choctaw helicopters, in September 1960.

In 1961, under President Kennedy, there was a major across-the-board expansion in American involvement in SEA, and in particular in devising better methods of waging defensive war against guerrilla forces in primitive environments. Almost all the first-line aircraft in American use

UNITED STATES OF AMERICA Air Force—continued

Division/Wing	Type	Role	Base	No	Notes
USAF Tactical Fighter Weapons Center, Nellis AFB, Nevada:					
57th Tac Trg Wing, F-15/F-111A,E,F,/F-4C,D,E/F-5E/A-10A/UH-1					
USAF Air Demo Sqn, T-38					
4440th Tac Ftr Trg Group (Red Flag)				—	
Tac Ftr Weaps Center Range Group					
USAF Tactical Air Warfare Center, Eglin AFB, Florida:				—	
USAF Air-Ground Ops School, Hurlburt Field					
4441st Tactical Trg Group (Blue Flag)					Also see Spec Ops in 9th AF
USAF Europe, HQ Ramstein AB, Germany:					
3rd Air Force, HQ RAF Mildenhall, England:					
10th Tactical Recon Wing	RF-4C/F-5E	Recce	RAF Alconbury	54/24	F-5 Aggressors 527th Sqn
20th Tactical Fighter Wing	F-111E	Att	RAF Upper Heyford	72	
48th Tactical Fighter Wing	F-111F	Att	RAF Lakenheath	72	
81st Tactical Fighter Wing	A-10A	Att	RAF Bentwaters/Woodbridge	108	Forward op locations: Noervenich, Sembach, Ahlhorn, Leipheim
16th Air Force, HQ Torrejon AB, Spain:					
401st Tactical Fighter Wing	F-4D/E	FGA	Torrejon AB, Spain	72	
406th Tactical Ftr Trg Wing	various	Train/Supp	Zaragoza AB, Spain	—	Plus SAC KC-135, MAC UH-1
40th Tactical Group	various	Cmd/Control	Aviano AB, Italy	—	
7206th Air Base Group	various	Supp/Comms	Hellenikon AB, Greece	—	
HQ TUSLOG	various	Cmd/Control	Ankara AB, Turkey	—	
Detachment 10, TUSLOG	various	Cmd/Control	Incirlik AB, Turkey	—	Also Izmir AB
17th Air Force, HQ Sembach AB, Germany:					
26th Tactical Recon Wing	RF-4C	Recce	Zweibrücken AB, Germany	54	
32nd Tactical Fighter Sqn	F-15	Fight	Camp New Amsterdam, Neth	24	
36th Tactical Fighter Wing	F-15	FGA	Bitburg AB, Germany	72	
50th Tactical Fighter Wing	F-4E	FGA	Hahn AB, Germany	72	
52nd Tactical Fighter Wing	F-4E	FGA	Spangdahlem AB, Germany	72	
86th Tactical Fighter Wing	F-4/UH-1/T-39, etc	Multi	Ramstein AB, Germany	—	
435th Tactical Airlift Wing	C-5/C-141/C-130/C-9	Trans	Rhein-Main AB, Germany	—	MAC unit
600th Tactical Control Group	various	Cmd/Control	Hessisch-Oldendorf AS	—	
601st Tactical Control Wing	OV-10A/CH-53	FAC, etc	Sembach AB, Germany	—	
7100th Air Base Group	—	CCC	Lindsey AS, Germany	—	USAFE Support
7350th Air Base Group	—	Supp/Comms	Berlin, Germany	—	
Aerospace Defense Command, HQ Peterson AFB, Colorado:					
Combat Ops Center		Cmd	Cheyenne Mountain Complex	—	US/Canadian HQ
20th Air Division	F-106	Int	Ft Lee AFS, Va	18	
21st Air Division	F-106	Int	Hancock Field, NY	18	
23rd Air Division	F-106	Int	Duluth Airport, Minn	18	
24th Air Division	F-106	Int	Malmstrom AFB, Mont	18	One squadron each
25th Air Division	F-106	Int	McChord AFB, Wash	18	
26th Air Division	F-106	Int	Luke AFB, Ariz	18	
Air Forces Iceland	F-4/EC-121	Int	IDF Keflavik	24/9	
Air Defense Weapons Center	various	Train	Tyndall AFB, Fla		

Above: 750lb bombs cascading from a B-52D during the Vietnam War. Remaining B-52s, of later versions, are not equipped for large conventional bomb loads.

RB-26, T-28 and SC-47. At that time all US combat aircraft in SEA carried VNAF (South Vietnam Air Force) markings. Meanwhile, design went ahead rapidly on special types or modifications of aircraft for war in that theatre, among them STOL FAC (Forward Air Control) machines, multi-sensor OV-1 Mohawk twin-turbo-props, the multi-role OV-10 Bronco LARA (Light Armed Reconnaissance Aircraft), C-123 transports equipped for defoliation spraying, B-52 strategic bombers with the 'Big Belly' modification for carrying a load of conventional bombs increased from 27,000lb to 80,000lb, a great variety of

Below: A Lockheed C-130B Hercules amidst war cargo in Southeast Asia. Warfare in that theatre was a stern test of the USAF Military Airlift Command, which is now being augmented by rebuilding its C-141 force.

were wholly unsuited to this kind of operation, being intended instead for major nuclear and conventional war between advanced powers. USAF Chief of Staff Gen LeMay set up the 'Jungle Jim' 4400th Combat Crew Training Squadron at Eglin AFB. Subsequently this unit was a fountain-head of Co-In (counter-insurgency) training and techniques, and also of combat formations which were assigned to active operations in SEA. The first such detachment arrived at Bien Hoa on November 4, 1961, equipped with the

UNITED STATES OF AMERICA Air Force—continued

Division/Wing	Type	Role	Base	No	Notes
10th Aerospace Def Sqn	—	Space	Vandenberg AFB, Calif	—	
14th Missile Warning Sqn	—	EW	MacDill AFB, Fla	—	
46th Aerospace Def Wing	—	Cmd/Cont	Peterson AFB, Colo	—	
425th Munitions Support Sqn	—	Supp	Peterson AFB, Colo	—	
4754th Radar Eval Sqn	—	Res	Hill AFB, Utah	—	
Military Airlift Command, HQ Scott AFB. Illinois:					
21st Air Force, HQ McGuire AFB, New Jersey:					
76th Military Airlift Wing	various	Trans	Andrews AFB, Md	—	
89th Military Airlift Grp	—	Trans	Andrews AFB, Md	—	
317th Tactical Airlift Wing	C-130, etc	Trans	Pope AFB, NJ	30	
435th Tactical Airlift Wing	C-130, etc	Trans	Rhein-Main AB, Germany	30	
436th Military Airlift Wing	C-5/C-141, etc	Trans	Dover AFB, Del	17/18+	
437th Military Airlift Wing	C-5/C-141, etc	Trans	Charleston AFB, SC	17/18+	
438th Military Airlift Wing	C-141, etc	Trans	McGuire AFB, NJ	18+	
1605th Air Base Wing	various	Supp	Lajes Field, szores	—	
1100th Air Base Group	various	Supp	Bolling AFB, DC	—	
22nd Air Force, HQ Travis AFB, California:					
60th Military Airlift Wing	C-5/C-141, etc	Trans	Travis AFB, Calif	17/18+	
61st Military Airlift Suppt Wing	various	Supp	Hickam AFB, Hawaii	—	
62nd Military Airlift Wing	C-141, etc	Trans	McChord AFB, Wash	18+	
63rd Military Airlift Wing	C-141, etc	Trans	Norton AFB, Calif	18+	
314th Tactical Airlift Wing	C-130	Trans	Little Rock AFB, Ark	30+	
374th Tactical Airlift Wing	C-130, etc	Trans	Clark AB,, Philippines	—	
443rd Military Airlift Wing	C-5/C-141, etc	Trans/Train	Altus AFB, Okla	17/18+	
463rd Tactical Airlift Wing	C-130, etc	Trans	Dyess AFB, Tex	30	
1606th Air Base Wing	various	Supp	Kirtland AFB, NM	—	

Direct reporting:
Aerospace Rescue & Recovery Service, Scott AFB, Illinois, HC-130/HH-1/HH-3/HH-53, etc, many locations
Air Weather Service, Scott AFB, III, WC-135, WC-130, etc, several locations
Aerospace Audio-Visual Service, Norton AFB, Calif, photo/video production, libraries, etc
375th Aeromedical Airlift Wing, Scott AFB, III, C-9/C-141/C-130

Pacific Air Forces, HQ Hickam AFB, Hawaii:
5th Air Force, HQ Yokota AB, Japan:

8th Tactical Fighter Wing	F-4D/E	Int	Kunsan AB, S Korea	72	
18th Tactical Fighter Wing	F-4/RF-4/T-39/etc	Multi	Kadena AB, Okinawa	c.80	
51st Compo Wing (Tactical)	F-4D/E/OV-10	FGA	Osan AB, S Korea	60+	
475th Air Base Wing	T-39/UH-1/H-3	Supp	Yokota AB, Japan	—	

Note: The above are administered by two Air Divisions, 313th at Kadena, 314th at Osan
13th Air Force, HQ Clark AB, Philippines:

3rd Tactical Fighter Wing	F-4/F-5/T-33/T-38/T-39	Multi	Clark AB, Philippines	c.80	
Detachment 1	varies	Multi	Taipei AS, Taiwan	—	Being withdrawn

Various units attached to Pacaf from other commands
Direct reporting:

15th Air Base Wing	EC-135/RC-135/T-33	Multi	Hickam AFB, Hawaii	c.8/8/20	
326th Air Division	O-2, etc	Cmd/cont	Wheeler AFB, Hawaii	—	

helicopters for various tasks (most in connection with a land battle) and armed 'gunship' versions of the C-47, C-119 and C-130 transports.

One of the first land engagements that rated as a full-scale battle was at Ap Bac, in January 1963. Despite superior forces the ARVN (South Vietnamese Army) with strong US support suffered a defeat, entirely because of grossly inept planning, inappropriate field tactics and bad morale, plus hopelessly inadequate communications. Despite prolonged efforts, this ill-starred action was to characterise many subsequent engagements, and on top of this the US Congress laid down unprecedented 'rules' for the conduct of all US operations in SEA. These removed almost all the crushing effect of what was to become the world's greatest concentration of air power. A particular example of these rules was to prohibit any engagement by US aircraft of suspected hostile aircraft until the latter had been closed upon and positively identified visually. This made it impossible for the F-4 Phantom – which was to become by far the most important US air-combat aircraft in SEA – to use its medium-range Sparrow missiles, and often precluded use even of the close-range Sidewinder. It forced closure to the point where the nimble MiG-17 and MiG-21 fighters of North Vietnam, armed with heavy cannon, had all the advantages. At this time the F-4 did not even have a gun at all, unless it was hung as an external item of ordnance on a weapon pylon.

Despite gross inefficiency in many oper-

ations, in part caused by the strict rules – according to one pilot 'The White House wants to know the enemy pilot's name before I can join combat' – the US forces gradually learned how to fight this extremely tough and different kind of warfare. For much of the time a policy of 'brute force and ignorance' was adopted in absence of any better method. More than 25,000 air strikes were flown against unseen jungle targets, many carefully and closely directed by a FAC who eventually could collaborate with such skilled tactical crews that cannon fire, free-fall bombs and rockets could be made to hit particular trees. This was at a time when angle/rate bombing systems were in their infancy, and laser-guided 'smart' bombs non-existent. Immense skill and expertise were gained by large numbers of USAF, Navy and Marines aircrew engaged in such attacks, the only mitigating factor being absence in the 1960s of intense anti-aircraft defences.

Long-range missions by large aircraft grew month by month. The remoteness of SEA from the United States put a very heavy burden on Military Airlift Command, in particular on the C-141 StarLifter squadrons which were the workhorses on the '10,000-mile pipeline' across the Pacific. The quantities of material shipped were awesome. In 1965 it ran at around 100,000 tons, but by 1969 had climbed to almost 2m tons. This was combined with finding better ways to make airstrips, maintain aircraft, and fetch back and repair those shot down. RASS (Rapid Area Supply Support) teams were formed to bring order

out of chaos in a region where – predictably – all kinds of supplies tended simply to disappear. RAM (Rapid Area Maintenance) teams repaired or salvaged for spares billions of dollars-worth of downed aircraft, more than 380 of which were bodily uplifted and flown back by Army CH-54 Tarhe heavy-lift helicopters. As for building airstrips, this reached a fine art in BEEF (Base Engineering Emergency

Below: These F-4D Phantoms are carrying laser-guided 'smart bombs', precision ground-attack weapons of the Paveway family.

Force) teams, more than 40 of which were in operation by 1968. Helicopter landing grounds were constructed by the quick process of dropping 10,000lb high-blast bombs, which levelled the area.

In 1963 the B-52 was studied as a CBC (Conventional Bomb Carrier), and modification of the first 30 B-52Fs to carry 108 bombs each began in October 1964. These aircraft deployed to Guam in February 1965, and it was typical of the Vietnam war that they should have flown 12-hour missions, requiring at least one KC-135 air refuelling, in order to rain down bombs on targets that were often unseen and

Above, left: A KC-135A photographed during air-refuelling of a group of F-105D Thunderchief single-seaters, probably during the SE Asia campaign.

Above: A Boeing Vertol CH-47A Chinook heavy transport helicopter of the US Army photographed in SE Asia in 1967 dropping a water trailer to the 4th Inf Div.

sometimes not there at all. Sometimes, however, they were, and one of many actions decided by the sheer firepower of the CBC was the siege of Khe Sanh in early 1968. US C-in-C General Westmoreland reported: 'Without question the amount of firepower put on that piece of real estate exceeded anything that had ever been seen before in history by any foe. . . .' By this time B-52 operations were increasingly being transferred to bases in Thailand, only two to five hours from targets, but in the closing stages of the war in the winter 1972–73 the North Vietnamese defences, including more SAMs (surface-to-air missiles) than had ever before been emplaced in one area, caused severe casualties to B-52s flying relatively low with heavy loads of conventional bombs. In December 1972 SAMs claimed 15 B-52s, but the SAM sites were then all knocked out by low-level operations with F-111 swing-wing bombers.

Only small numbers of F-111s were deployed to SEA, the first six in March 1968 with discouraging results. Later this advanced all-weather attacker amply justified itself with outstanding results against the extremely heavily defended area round Hanoi and Haiphong, attacking specific point targets (often surrounded by 'off-limits' residential zones) that dared not be assigned to any other aircraft. Equipped with terrain-following radar/ autopilot, the F-111 could skate along the surface of the terrain at about Mach 0.9 even in the worst weather, and put its bombs precisely on target. This is how all advanced tactical air forces will go to war in future, and the USAF is the only armed force in the world actually to have done it – and on a substantial scale over a long period.

There were countless other facets of the

US involvement in SEA that led to new techniques and weapons. One of the most difficult problems was how to find the enemy. At first a FAC in a light aircraft (usually an L-19 or O-2) would search visually just over the treetops at 100 knots, calling up attack aircraft with detailed radio instructions. This became hazardous, and the 'fast mover' FAC then became important in the F-100 and F-4, while 'brute force and ignorance' level bombing continued with such aircraft as the F-105 and F-4 (sometimes guided by special lead ships such as EB-66 versions with complex precision navaids and ECM systems). Greater precision was gradually achieved by locally built ground navaids, such as versions of Loran, a radio aid of the hyperbolic area-coverage type, which was fitted to certain F-4s and other tactical bombers which used Pathfinder techniques. To detect trucks a major system called Igloo White involved radio emissions from various kinds of packaged sensors – acoustic, seismic and combined – dropped along the suspected roads and trails. Hard to detect on the ground, they sent back signals showing passage of vehicles or even foot soldiers, so that night attacks could be mounted on the truck parks hidden in deep jungle. Various light aircraft and even EC-121 Warning Star (Super Constellation) platforms were used to relay information to ground processing centres. Much other information was brought back by RF-101 and RF-4C multi-sensor reconnaissance aircraft, backed up by the RA-5C and RF-8A of the Navy, and a wide range of special platforms for intelligence gathering included the U-2 and RPV drones.

The scale of the US air operations in the period 1962–73 is shown by the loss of more than 3600 aircraft in the SEA theatre, the majority of them in action. The expenditure on ordnance was prodigious. The USAF's consumption of 6.2 million tons of bombs, rockets, shells and napalm is approximately three times the weight of munitions dropped by the USAAF in World War II. From 1968 an increasing proportion comprised new cluster bombs, laser-guided smart bombs and other advanced weapons.

Central America and the Caribbean

The number of independent states in the area has steadily increased since World War II as former colonies have gained home rule, but few of these have established air forces of any size. Cuba has been a dominant military force in the region since Fidel Castro came to power in the 1950s and began receiving substantial aid from the Soviet Union, a process which led to the 'Cuba crisis' of October 1962 when the United States demand that Russian-built intermediate-range ballistic missiles which could deliver nuclear warheads on targets in the southern US should be removed. The Soviet Union acceded to these demands, but has nevertheless since supplied Cuba with a variety of modern tactical weapons.

Long-standing disagreements between Guatemala and Belize have resulted in British forces — including Harrier V/STOL fighter-bombers and Rapier low-level surface-to-air missiles — being deployed in the latter country, and traditional enmity between Honduras and El Salvador has likewise led to skirmishes in the recent past.

Many of the smaller air arms are gradually acquiring more modern equipment such as Cessna A-37 light attack aircraft to replace their vintage types, and this upgrading of capability could possibly lead to more aggression. The United States has also announced its intention to withdraw from the Canal Zone, and this is a further potential source of conflict.

MEXICO

Area: 761,600 sq. miles (1·97 million sq. km.).
Population: 61 million.
Total armed forces: 340,000.
Estimated GNP: $83·8 billion (£41·9 billion).
Defence expenditure: $557 million (£278·5 million).

Although the *Fuerza Aérea Mexicana* is one of the largest air forces in Latin America it has only a modest combat potential, operating mainly on civil aid, training and transport tasks. Headquarters of the FAM is at Lomas de Sotelo and control comes from the Secretary of National Defence in Mexico City. Mexico has an armed force totalling 340,000 men, of which some 6000 are FAM personnel.

As well as the air force, the Navy has a small force of some 27 aircraft used for maritime reconnaissance, search and rescue, and communications. The basic operational organisation of the FAM is based on Air Groups and squadrons located at a number of airfields around the country, the squadrons each flying a dozen or more aircraft.

Military aviation in Mexico has its origins around the early 1900s when various revolutionary and counter-revolutionary groups employed mercenary pilots to conduct isolated bombing and reconnaissance sorties. In 1915 President Carrenza ordered the creation of an "Air Arm of the Constitutional Forces" and at the same time founded a flying school and an aircraft industry to overcome the shortage of aircraft caused by the First World War.

By 1920 the Mexican air force had nearly 50 indigenous types in use. When the *Fuerza Aérea Mexicana* was officially established in 1924 a few US-built D.H.4 bombers and Douglas O-2 observation aircraft were acquired, and in 1932 Mexico undertook the licence-production of 32 Vought Corsairs. During the years immediately before the Second World War the FAM remained a small compact force of two Air Regiments, both based at the Balbuena Air Base near the capital. In 1942 Mexico joined the Allies, putting some of her bases at their disposal and receiving in return a number of aircraft from the United States, including Dauntless dive-bombers plus some AOP and training machines. After the war more US

Right: The Mexican naval air arm operates a single Gates Learjet 24D as a VIP transport, in which role it complements an FH-227.

types arrived in the form of Thunderbolts and Mitchells for the combat units, C-45s and C-47s for the transport squadrons and Fairchild Cornells and Vultee Valiants for the training schools. Jets were acquired in the late 1950s when some Vampires and T-33s updated the combat squadrons, and the latter type continues in service.

Mexico is a signatory of the Charter of the Organisation of American States, which was drawn up in 1948, while in 1967 the country became one of 22 Latin American states to sign the Tlatelolco Treaty prohibiting nuclear weapons in the area. The United States has a military assistance agreement with Mexico under which aid and equipment are supplied.

Inventory

The *Fuerza Aérea Mexicana* has no pure air-defence force as such, the main combat element comprising one fighter-bomber squadron flying more than a dozen Lockheed T-33As. Internal policing and counter-insurgency are the tasks of six squadrons equipped with about 20 T-6s and 50 T-28 piston-engined aircraft fitted with underwing armament. If finances allow, some small light attack jets could replace much of this obsolescent equipment in the next few years. Recent procurement has been from Israel, where ten Aravas were bought, but the United States remains Mexico's main arms supplier. Transport duties are flown by two Groups: a heavy element with C-54s and DC-6s, and a light transport group with C-47s, a dozen Islanders, the Aravas and a Short Skyvan. There is a support squadron which uses both fixed-wing aircraft and helicopters, the former made up of 18 LASA-60 high-wing utility machines built by the Lockheed-Azcarate factory in Mexico and now mainly concerned with SAR duties, and the latter comprising some Alouette IIIs and a Hiller UH-12E.

Support

In recent years the Mexican air force has modernised its training fleet with the introduction of 40 Beech Musketeers and Bonanzas for the primary role, although the basic training task is still undertaken by T-6s and T-28s. A batch of Swiss Turbo-Trainers are being delivered in 1979 to replace the older types in use at the basic stage, while some B55 Barons conduct twin training. Advanced work and jet conversion is flown on the two-seat T-33s. The *Colegio del Aire* is at Zapopan Air Base, Jalisco, and includes the flying elements, ground trades and meteorological schools. Also controlled by the FAM are the airborne troops and the aeronautical engineering division.

Navy

Aviacion de la Armada de Mexico is the aviation arm of the Mexican Navy and has as its main role the protection of the country's coastline, search and rescue, and liaison. Furnished initially with a small number of Consolidated Catalinas by the United States towards the end of the Second World War for anti-U-boat

patrols, the Service has replaced them with eight Grumman Albatrosses. The remaining aircraft types are generally used for liaison and communications duties. Recent procurement has included a Learjet for VIP flights and a couple of Bonanzas for continuation training.

MEXICO   Navy

Type	Role	No
HU-16A	ASW/SAR	8
C-47	Trans	2+
FH-227	VIP	1
Learjet 24D	VIP	1
Alouette II	Liaison	4
Bell 47G	Liaison	5
Cessna 180	Liaison	2+
Bonanza	Train	2+
Cessna 150	Train	2+

MEXICO   Air Force

Unit	Type	Role	Base	No	Notes
202 *Esc Aérea*	T-33A	FGA	Satillo	15	
201, 205, 206, 207 *Esc Aérea*	T-28A	FGA/train	Ciudad, Puebla, Merida	50	Also used by *Colegio del Aire*
203, 204 *Esc Aérea*	T-6	FGA/train	El Cipres	20	
209 *Esc Aérea*	LASA-60	Util/SAR	Cozumel	18	
	Alouette III	Util/SAR	Cozumel	9	
	UH-12E	Util/SAR	Cozumel	1	
	DC-7	Trans		1	Operated by *Grupo Transporto Aéreo Pesante* (Heavy Air Transport Group)
	DC-6	Trans		2	
	C-54	Trans		5	
	C-47	Trans		7	Operated by *Grupo Leggero* (Light Air Transport Group). One C-47 used for VIP duties
	Skyvan	Trans		1	
	Islander	Trans		12	
	Aero Commander 500	Trans		20	
	Arava	Trans		10	
	One-Eleven	VIP	Mexico City	1	
	HS.125-400	VIP	Mexico City	1	
	JetStar	VIP	Mexico City	1	
	Boeing 727-100QC	VIP	Mexico City	2	
	Bell 47G	Train		14	
	Bell 205	Util		10	
	Bell 212	Util		1	
	JetRanger	Liaison		5	
	Aztec	Liaison		—	
	Bonanza F33C	Train	Zapopan	20	
	Musketeer Sports	Train	Zapopan	20	
	Baron	Train	Zapopan	—	Some used for twin training
	Vampire T.11	Train	Zapopan	3	
	PC-7 Turbo-Trainer	Train		12	Delivery 1979; for dual training-strike role

GUATEMALA

Area: 42,000 sq. miles (108,800 sq. km.).
Population: 5·4 million.
Total armed forces: 14,000.
Estimated GNP: $4·6 billion (£2·3 billion).
Defence expenditure: $58 million (£29 million).

Administered by the Army, the *Fuerza Aérea Guatemalteca* is almost entirely dependent on United States military aid in the form of aircraft and equipment, although Israel has recently supplied aircraft. Organised along American lines, the FAG has a total manpower of some 400 and has its headquarters at La Aurora air base, Guatemala City.

An air unit was formed by the Guatemalan Army after the First World War, using Avro 504 trainers and assisted by a French aviation mission. Nearly ten years elapsed before the official formation of a *Cuerpo de Aéronautica Militar* equipped with a few second-line French aircraft in 1929. A small number of American aircraft were received in 1937 and, following the

agreement allowing Allied Powers the use of Guatemalan bases during the Second World War, more US aircraft and equipment were supplied on a lease-lend basis. A squadron of F-51D Mustangs constituted the air arm's first major combat unit when it formed after the war and it was to be 1972 before the last Mustangs left the FAG front-line inventory. Other types used during the 1950s and 1960s included some Douglas C-47s, PT-17 trainers, B-26 Invaders and a few T-33As, the air force's first jet aircraft.

Guatemala is a signatory of the Charter of the Organisation of American States drawn up in 1948, and of the Tlatelolco Treaty prohibiting the use of nuclear weapons in Latin America. Apart from the US aid agreement, Guatemala also has an agreement with El Salvador, Honduras and Nicaragua to resist Communist aggression with combined force.

Inventory

A token combat force of one ground-attack squadron flying Cessna A-37Bs operates alongside what has been for some

years a mainly transport and training arm. The *Fuerza Aérea Guatemalteca* has recently introduced nearly a dozen Israeli Arava light transports which now link a number of distant bases in the country. With neighbouring Honduras and El Salvador operating Super Mystères and Ouragans respectively, it remains to be seen whether more modern jet aircraft are acquired by the FAG in the near future to equalise the imbalance in the area. Two of the types under consideration are the F-5E and Israeli Kfir C2.

Primary and basic training has been flown on the North American T-6s until superseded by the T23s. Jet conversion is carried out on the Cessna T-37Cs and T-33s, the latter operated mainly for advanced flying prior to pilot allocation to the Coin unit.

Below: The FAG's sole DC-6 is the largest aircraft in its four-type transport squadron, the others comprising the ever-present C-47 and C-54 operating alongside IAI Aravas. The Israeli aircraft has achieved considerable success in penetrating the Latin American market.

GUATEMALA

Type	Role	Base	No	Notes
A-37B	GA		13	
T-33A	Train	Los Cipresales	5	
T-37C	Train	Los Cipresales	3	
DC-6	Trans	La Aurora	1	One squadron (*Esc de Transporte*); some C-47s camouflaged for tactical role
C-54	Trans	La Aurora	1	
C-47	Trans	La Aurora	9+	
Arava	Trans	La Aurora	10	
T-6	Train	Los Cipresales	7	
T-23 Uirapuru	Train	Los Cipresales	10	
UH-1D	Trans	La Aurora	6	
Cessna 170	Liaison		6	
Cessna 180	Liaison		3	
Cessna U206C	Liaison		2	
UH-19	Util		3	
OH-23G	Util		1	
T-41D	Train		2+	

EL SALVADOR

Area: 8,200 sq. miles (21,200 sq. km.).
Population: 4·54 million.
Total armed forces: 7,150.
Estimated GNP: $2·55 billion (£1·275 billion).
Defence expenditure: $30 million (£15 million).

This republic on the Pacific coast of Central America supports a modest air force, the *Fuerza Aérea Salvadorena*, which in recent years has undergone an equipment modernisation programme. With headquarters in San Salvador, the capital, the FAS is manned by some 1000 personnel from a total armed forces figure of 7150. Main role of the air force is transport and communications, but there is a combat element which in the mid-1970s converted from Second World War-vintage aircraft to jets.

A Military Aviation Service was established in El Salvador in 1923, with five Aviatik trainers being followed by second-line Waco and Curtiss-Wright aircraft from the United States. Apart from four Caproni attack aircraft obtained in 1939 little expansion of the force occurred until

after the end of the Second World War, when small numbers of Beech AT-11s, Fairchild PT-19s and Vultee trainers were acquired from US stocks. A quantity of Beech T-34s arrived in the mid-1950s under the American Military Aid Programme for training duties, while a combat flight was formed with six Chance Vought/Goodyear Corsair fighter-bombers.

El Salvador is a signatory of the Charter of the Organisation of American States and also of the Tlatelolco Treaty prohibiting nuclear weapons in Latin America. She has a military assistance agreement with the US and in 1965 El Salvador, Guatemala, Honduras and Nicaragua agreed to form a bloc against communist aggression in the area.

After a minor war with neighbouring Honduras in 1969, Salvador signed a peace treaty in May 1970, and like Honduras reached agreement with Israel over the supply of military equipment to update the armed forces. The *Fuerza Aérea Salvadorena* received a squadron of 18 refurbished ex-Israeli Air Force Dassault Ouragan fighter-bombers plus half-a-dozen IAI-built Fouga Magister trainers. Five Arava light transports have also been procured

to add to the transport flight equipped with C-47s. The remaining aircraft in FAS service have been supplied by the US and any future orders are likely to be placed with either the United States or Israel, depending on budget allocations.

Training is undertaken at Ilopango initially on T-41s and then on the T-34s and T-6s. The Magisters provide jet training prior to assignment to the Ouragan squadron. Military missions from both the US and Israel have assisted the Salvadorean crews in recent years.

EL SALVADOR

Type	Role	Base	No
Ouragan	FGA	San Miguel (?)	18*
Magister	Strike/train	Ilopango	9**
C-47	Trans	San Salvador	17
DC-6	Trans	San Salvador	2
Arava	Trans	San Salvador	5
FH-1100	Liaison		1
Lama	SAR		3
Alouette III	Liaison		2
T-34	Train	Ilopango	3
T-41C	Train	Ilopango	3+
Cessna 180	Liaison	Ilopango	3
T-6	Trans	Ilopango	5

*Ex-Israeli AF; replaced Corsairs and Mustangs in single combat squadron
**Six are ex-Israeli AF

PANAMA

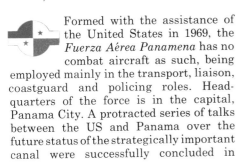

Area: 31,900 sq. miles (82,600 sq. km.).
Population: 1·72 million.
Total armed forces: 11,000.
Estimated GNP: $2·13 billion (£1·65 billion).
Defence expenditure: $22 million (£11 million).

Formed with the assistance of the United States in 1969, the *Fuerza Aérea Panamena* has no combat aircraft as such, being employed mainly in the transport, liaison, coastguard and policing roles. Headquarters of the force is in the capital, Panama City. A protracted series of talks between the US and Panama over the future status of the strategically important canal were successfully concluded in mid-1977, with the USA agreeing to relinquish sovereign rights over the Canal Zone by the end of the 1990s. The present level of the armed forces totals some 11,000 officers and men but a gradual increase of that figure is expected as the American withdrawal approaches.

Inventory

The Panamanian armed forces benefit from a military assistance agreement with the United States under which aircraft and equipment have been supplied since the *Fuerza Aérea Panamena* was formed. The country is also a member of the Organisation of American States. As well as acquiring a small number of fixed-wing aircraft and some helicopters of US origin, the FAP has also purchased a couple of Britten-Norman Islanders.

Below: The FAP's fleet of utility helicopters comprises four marks of UH-1 Hueys.

Bottom: The long-serving Otter continues to give good service in Panama.

PANAMA

Type	Role	Base	No
L-188 Electra	Trans	Panama City	1*
C-47	Trans	Panama City	4
Twin Otter	Trans	Panama City	2
Otter	Trans	Panama City	2
Islander	Liaison	Panama City	2
U-17B	Liaison	Panama City	2
UH-1B/D/H/N	Util	Panama City	12/2/2/1
Cessna 172	Train	Panama City	1

Ex-US airline aircraft bought in 1975

HONDURAS

Area: 43,000 sq. miles (112,000 sq. km.).
Population: 3·1 million.
Total armed forces: 14,200.
Estimated GNP: $1·3 billion (£650 million).
Defence expenditure: $31 million (£15·5 million).

 The *Fuerza Aérea Hondurena* comes under the Armed Forces Command, the headquarters of which is in the country's capital, Tegucigalpa. Manned by some 1200 personnel, the FAH operates mainly in the ground-attack role although the transport elements provide vital communications in this, the largest of the Central American states.

Two Bristol Fighters were the first equipment, in the early 1920s, but it was not until 1933 that the *Aviacion Militar* was officially established with the help of American personnel. A training school was formed with some old Stinsons at Tocontin air base. A variety of obsolete US aircraft were acquired during the war years, and it was 1948 before more modern types were obtained via military aid programmes. Lockheed P-38Ls, Bell P-63s and North American F-51Ds formed the combat force for some years, support being provided by C-47s and Beech C-45s.

Honduras is a signatory of the Rio Defence Treaty of 1947 and is also a member of the Charter of the Organisation of American States. The country has a military assistance agreement with the US and in 1965 Honduras, El Salvador, Guatemala and Nicaragua agreed to form a bloc against communist aggression.

Front-line strength was boosted in 1976 with the acquisition of a squadron of ex-Israeli Super Mystère B.2 single-seat supersonic fighter-bombers. Re-engined with Pratt & Whitney J52s and armed with two 30mm cannon plus underwing stores, these aircraft have significantly increased Honduran air strength and give the FAH marked ascendancy in air power over neighbouring countries. Another modern type in use is the Cessna A-37B

Right: This immaculate Cessna A-37B is one of six which form the FAH's main punch in the ground-attack role. With a weapon load of 5680lb, the A-37 gave the Honduran air arm a substantial increase in capability when it replaced Corsair fighter-bombers in the mid-1970s.

CUBA

Area: 44,000 sq. miles (114,000 sq. km.).
Population: 9·5 million.
Total armed forces: 189,000.
Estimated GNP: $4·5 billion (£2·25 billion).
Defence expenditure: $784 million (£392 million).

CUBA

Type	Role	Base	No	Notes
MiG-21F/MF	Int/FGA	Camaguey, San Antonio, Santa Clara	50/30	Five sqns; Atoll missiles carried
MiG-19	Fight		40	Two sqns
MiG-17	FGA		75	Four sqns
MiG-15	Fight/train		15	One second-line training unit
MiG-23/27	Int/FGA	San Julian/Guines	18	One sqn
An-2	Trans		20	
An-24	Trans			
An-26	Trans			
IL-14	Trans		20	
Mi-1	Liaison		30	
Mi-4	Trans		24	
Zlin 326	Train		30	
MiG-15UTI	Train		10	
SA-2	SAM		24 sites	

The Cuban Revolutionary Air Force or *Fuerza Aérea Revolucionaria* is the only totally Soviet-equipped air arm in the Americas and has its headquarters in Havana. Apart from the Navy, which has a small fleet of helicopters for coastal surveillance, the FAR is the sole user of military aircraft in the country's armed forces. Of the 189,000 personnel under arms, some 20,000 serve in the air force and include a substantial number of air-defence personnel manning the missile and radar sites around the island. In return for the military and economic aid supplied by Russia, Cuba provides bases for the use of Soviet Air Force and Navy long-range reconnaissance aircraft and port facilities for Soviet ships.

The first aircraft to arrive in Cuba for military use were six Curtiss trainers in 1917, followed by some DH.4 bombers and Vought reconnaissance aircraft six years later. The destruction of most of these aircraft in a hurricane in 1926 virtually ended the country's brief flirtation with air power until the mid-1930s, when the Aviation Corps was more firmly established into Army Aviation and Naval Aviation. US aircraft were acquired, including Waco general-purpose biplanes and Stearman trainers. In return for putting her bases at the disposal of the Allies in World War II, Cuba received a number of second-line types to swell her modest inventory. In 1948 she received her first front-line combat aircraft, North American Mustang fighters and Mitchell bombers. Jet equipment in the shape of US-supplied Lockheed T-33s was delivered in 1955 and in that year the force became a branch of the Army and was renamed *Fuerza Aérea Ejercito de Cuba.*

Guerrillas operating in the mountains of eastern Cuba since 1956 under their leader Fidel Castro became a major force by 1958, and after fighting a number of pitched battles with government forces they overthrew the regime of President Fulgencio Batista. Ties were broken with the United States and the new government under Castro announced that Cuba was to be a socialist state. The armed forces were entirely re-equipped, modernised and expanded with the help of Russia and Eastern Bloc countries. In 1962 a direct head-on confrontation between the United States and Russia over the basing of Soviet missiles in Cuba brought the world to the brink of nuclear war. Sense prevailed and the missiles were withdrawn.

Cuba receives military and economic aid from Russia and legally remains a member of the Organisation of American States, although she has been excluded since 1962. Cuba also supports left-wing Marxist regimes in Africa such as Angola and Mozambique.

The *Fuerza Aérea Revolucionaria* has a front-line combat force of some 210 air-craft, of which nearly half are MiG-21s. The last jet bombers left the island following the Cuban missile crisis and the only remaining attack aircraft are some MiG-17s, -21s and -23/27s. Soviet aid to Cuba has tailed off since the mid-1960s as the country's economy has stabilised, and the air force now has only a few Soviet advisers attached to it. The only surface-to-air missiles on the island are about 500 SA-2 Guidelines situated at some 24 sites around the country. A limited assistance programme is currently under way in Peru, with 12 ex-Cuban MiG-21s having been supplied to the Peruvian Air Force for familiàrisation training before that country's acquisition of 36 Sukhoi Su-22s ordered from Russia. Future involvement could see Cuban support for guerrilla operations against Rhodesia. The United States maintains the naval base at Guantanamo Bay on the south-eastern corner of the island under a treaty confirmed in 1934.

Basic training is undertaken on the fleet of Zlin 326s supplied to Cuba by Czechoslovakia, the pilots then taking an advanced course in the Soviet Union. In Cuba itself there is an operational conversion unit equipped with single- and two-seat MiG-15s, while the front-line squadrons have a number of two-seat MiG-21s and MiG-23s for familiarisation.

counter-insurgency jet bought from the US in 1975 and capable of carrying warloads of some 5680lb on eight hardpoints. Before this influx of new aircraft the FAH flew ten Chance-Vought Corsair fighter-bombers, bought in 1958, as front-line aircraft. Three Israeli Arava transports have been bought to modernise the force, the first being delivered in 1975.

Fuerza Aérea Hondurena combat experience includes fighter-bomber sorties by Corsairs against neighbouring El Salvador in 1969. In 1976 the air force was again alerted following a border clash between the two countries, but no combat operations ensued.

Support

A Military Aviation School is situated at Tocontin, where students are given basic training on Cessna T-41As and progress to T-28s and T-6Gs. Beech AT-11s are used for navigational and twin training. Maintenance of the FAH's aircraft is also undertaken at the base.

HONDURAS

Type	Role	Base	No	Notes
Super Mystère B.2	Fight		12	Ex-Israeli, delivered in two batches; re-engined with P&W J52s
A-37B	Coin	Tocontin ?	6	Delivered June-Nov 1975
F4U-5 Corsair	Int/GA	Tocontin	10	One wing; delivered 1958-59
B-26C Invader	Bomb	Tocontin/San Pedro	6	One squadron; acquired in 1969
RT-33A	Recce	San Pedro Sula	3	One squadron; also used for training
C-45/AT-11	Trans/train	Tocontin		
C-47	Trans		6	
C-54	Trans		2	
Arava	Trans		3	Delivery began in 1975
T-41A	Train	Tocontin	5	Delivered in 1973
T-28	Train	Tocontin	12	Eight ex-Moroccan aircraft acquired in 1978
Cessna 180/185	Liaison	Tocontin	4	
H-19	Liaison	Tocontin	3	
T-6	Train		6	

JAMAICA

Area: 4,400 sq. miles (11,400 sq. km.).
Population: 2·1 million.
Total armed forces: 1,000.
Estimated GNP: $2·97 billion (£1,485 billion).
Defence expenditure: $29·5 million (£14·75 million).

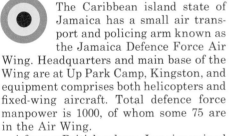

The Caribbean island state of Jamaica has a small air transport and policing arm known as the Jamaica Defence Force Air Wing. Headquarters and main base of the Wing are at Up Park Camp, Kingston, and equipment comprises both helicopters and fixed-wing aircraft. Total defence force manpower is 1000, of whom some 75 are in the Air Wing.

A former British colony, Jamaica gained independence in 1962 and formed a small armed force, the Jamaica Defence Force. Although the nucleus of an air arm was already in existence with a few privately owned light aircraft, the JDF required a more permanent air element and in July 1963 the Air Wing was officially established when the United States, as part of a military agreement, handed over four Cessna 185B Skywagons. Three months later a Bell 47G was delivered, and was followed by a second machine in March 1964. Assistance in forming the Wing was provided by British and Canadian personnel, with Jamaican nationals being trained in Canada and at the Army Air

Corps Centre in Britain. The first fixed-wing twin was acquired in the late 1960s when a Twin Otter was delivered for medium-range transport duties.

Jamaica is a member of the Organisation of American States and has received military assistance from Britain, Canada and the US.

The Air Wing undertakes roles with its fleet of 14 aircraft. Regular search-

Below: One of two JDF Air Wing Islanders use for light transport duties. Bottom: the wing's four Bell 206Bs fly medical-aid and other paramilitary missions.

and-rescue flights are flown, as are casualty evacuation (casevac) missions around the island. The ground forces also benefit from Air Wing support, as do the police and civil communities. The Twin Otter is used for VIP flights to neighbouring states in the area, along with the King Air and Duke.

Support

Pilot training is undertaken in Canada up to Wings standard, while ground technical training is carried out at the UK Army Air Corps Centre at Middle Wallop for technicians involved in helicopter operations.

JAMAICA

Type	Role	No
Twin Otter 300	Trans	1
King Air 100	Trans	1
Duke	Trans	1
Islander	Trans	2
Cessna 185B Skywagon	Liaison	2
Bell 206B	Liaison	4
Bell 212	Liaison	3

Note: All based at Up Park Camp

DOMINICAN REPUBLIC

Area: 19,300 sq. miles (50,000 sq. km.).
Population: 4 million.
Total armed forces: 18,000.
Estimated GNP: $4·3 billion (£2·15 billion).
Defence expenditure: $49·5 million (£24·8 million)

The Dominican Republic shares with some other Central and South American countries economic problems which prevent the support of anything more than a token armed force. Air power in Dominica comes under the command of the *Fuerza Aérea Dominicana*, with headquarters in the capital, Santo Domingo. Despite attempts in the past to modernise its front-line equipment, the FAD continues to operate a small number of out-of-date aircraft in the combat role. Total personnel numbers 18,000, of which 3500 are in the air force; defence expenditure for 1978 amounted to $49·5 million.

An air corps branch of the Dominican Army was formed in the early 1940s, a few Aeronca L-3 AOP aircraft being purchased together with some Boeing, Vultee and North American trainers. In 1948 the air arm, renamed *Aviacion Militar Dominicana*, embarked on a limited expansion programme which saw the introduction of a number of ex-RAF and USAF aircraft into the force including Bristol Beaufighters and Republic Thunderbolts. Some surplus Mustangs were purchased from Sweden in 1952 and these, together with some ex-Swedish Vampires bought a few years later, continue in use as the FAD's front-line combat force. Various negotia-

Below: One of an apparently unknown number of Aero Commander liaison/VIP transports used by the FAD. Beyond it can be seen a SOCATA Rallye and a T-28D Coin/trainer.

Bottom: The faithful C-47 forms the backbone of many air forces' transport fleets, the FAD being no exception.

tions for Hunters, Sabres and Mystères to replace the ageing Vampires met with little success in the late 1950s and early 1960s.

Dominica is a member of the Organisation of American States and is a recipient of limited military aid from the United States. She is also a signatory of the Tlatelolco Treaty for the prohibition of nuclear weapons in Latin America.

Inventory

About 30 aircraft form the main combat element in the Dominican air force – Vampires, Mustangs and Invaders – and as some of these machines have been in use for nearly 25 years the near future could see the FAD acquiring a small number of light attack jets from its main arms supplier, the United States. Most recent purchase has been a few Hughes liaison helicopters and some Cessna trainers. Only three active squadrons operate within the FAD, under a central air command.

REPUBLIC OF DOMINICA

Type	Role	No	Notes
Vampire F.1/FB.50	Fight	6	Bought from Sweden in 1955–56; ten originally purchased
F-51D	FGA	20	Total of 42 purchased from Sweden in 1950s; about 30 overhauled in USA in mid-1960s, remainder still in front-line use in single combat unit
B-26K	Bomb	7	Ex-USAF, not all are operational
Catalina	SAR/MR	2	
C-47	Trans	6	
C-46	Trans	6	
Beaver	Trans	3	
Alouette II/III	Liaison	3	
Bell 47G	Train	2	
UH-12E	Liaison	2	
OH-6A	Liaison	7	
UH-19	Liaison	2	
Cessna 172	Train	4	
T-6	Train	3+	
T-28D	Train/coin	6	
T-11	Train	2+	Used for twin conversion

NICARAGUA

Area: 57,100 sq. miles (147,900 sq. km.).
Population: 2·4 million.
Total armed forces: 7,100.
Estimated GNP: $2 billion (£1 billion).
Defence expenditure: $29·2 million (£14·6 million).

The *Fuerza Aérea de Nicaragua* is committed to civilian aid duties but retains a token combat element for internal policing and counter-insurgency work. Approximate strength of the FAN totals some 55 aircraft with a manpower of 1500. Control of the air force is vested in the Ministry of Defence, with headquarters in the capital, Managua. The air force is the sole operator of military aircraft in the country, much of the equipment being derived from US sources.

To protect the Panama Canal area, the Nicaraguan Air National Guard was formed in 1922 with US-supplied D.H.4s and Curtiss Jennies, but the force was largely ineffective until 1938 when it was re-organised with American help. Supplied at the time were a few Waco D armed biplanes and some Grumman G-23s; these continued in service until the late 1940s, when one squadron each of Mustangs and Thunderbolts were received under the US military aid programme. Second-line types acquired under the same source included C-47s, T-6s and T-11 Kansans.

Nicaragua signed the Charter of the Organisation of American States in 1948 and in 1967 signed the Tlatelolco Treaty. Recently, Nicaragua's dictatorship has been fighting a civil war, described as a war of national destruction by the US.

NICARAGUA

Type	Role	No	Notes
Cessna O-2A	FGA	6	Equips one sqn
T-33A	FGA	6	Equips one sqn
T-28D	Coin	6	Equips one sqn
C-47	Trans	3	Being replaced by C.212
C.212	Trans	5	
Arava	Trans	2	
HS.125	VIP	1	
Cessna 180	Liaison/util	7	
Cessna U-17	Liaison/util	2	
OH-6A	Liaison	4	
CH-34	Trans	3	
Otter	Util	5	
T-6	Train	4	
Super Cub	Train	3	

Note: All based at Managua

Inventory

Spasmodic military aircraft procurement has occurred over the last 20 years but few machines have been acquired. Most recent purchase has been from Spain in the form of an $8m order for five CASA C.212 Aviocars, following the purchase of a single Arava light transport and an HS.125 for VIP use. The combat element has recently received six Cessna O-2As for light strike duties and these fly alongside six armed T-33As and T-28Ds in three squadrons.

Below: Nicaragua's air force operates a mixed bag of light transports, the most numerous of which is the CASA C.212 Aviocar. Five have been ordered to replace long-serving C-47s, flying alongside a pair of IAI Aravas.

HAITI

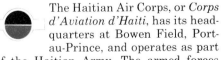

Area: 10,700 sq. miles (27,700 sq. km.).
Population: 4·8 million.
Total armed forces: 6,550.
Estimated GNP: $1·23 billion (£615 million).
Defence expenditure: $11·9 million (£5·95 million).

The Haitian Air Corps, or *Corps d'Aviation d'Haiti*, has its headquarters at Bowen Field, Port-au-Prince, and operates as part of the Haitian Army. The armed forces have a total manpower of some 6550, while defence spending is around $12m. For many years US aid, traditionally the source of equipment, was cut off but was resumed in 1973 after President Papa Doc Duvalier was succeeded by his son. Despite the resumption of this aid, the HAC remains poorly equipped and operates on a limited basis on communications, internal transport and policing/Coin duties.

With the assistance of the United States, Haiti formed an Air Corps in 1943 as a branch of the Army for carrying mail between the main towns. The red, white and blue insignia was appropriately chosen for the Service as the country is the only French-speaking republic in South America, while the shape of the marking shows the influence of the United States Air Force. With the acquisition of some US-supplied P-51D Mustangs, the HAC added combat patrols to its list of duties. By the mid-1950s, the Air Corps had established its method of organisation, the com-

HAITI

Type	Role	Base	No
Cessna 337	Coin/liaison	Bowen Field	8
Cessna 402	Liaison		1
Beaver	Trans		3
C-47	Trans		3
C-45	Trans/train		2
H-34	Trans	Bowen Field	4
S-55	Trans		5
Cessna 150	Train		3
Cessna 172	Train		1

manding officer having three assistant commandants – one each for operations, supply and maintenance, and administration and personnel. Aircraft supplied to the Corps in the 1950s included Fairchild PT-19s, Vultee BT-13s, Beech T-11s and a Boeing 307 airliner for the personal use of the President.

Haiti signed the Charter of the Organisation of American States in 1948, and is currently the recipient of limited American aid.

Inventory

A nominal personnel strength of some 250 makes up the Haitian Air Corps, which has about 32 aircraft flying para-military and civil aid duties. The fighter force of six Mustangs has recently been retired, replaced by Super Skymasters fitted for the COIN role with underwing hardpoints for rockets and machine-gun pods. Some of the older training aircraft have given way to a small number of more up-to-date Cessna-type machines.

South America

The South American continent is one of the few to have escaped major warfare over the past few decades, although relations between neighbours are in many cases strained at the best of times. Peru has recently introduced Russian combat equipment to the area in the form of Su-22 attack aircraft and SA-3 surface-to-air missiles, although Chile would probably have been the Soviet Union's first real foothold in the military market if the Marxist government of President Allende had not been overthrown.

Europe and the United States are also major arms suppliers to South American countries, while both Argentina and Brazil build military aircraft for their own uses and to earn foreign currency. Brazil, especially, has a booming engineering industry and is likely to establish itself as a considerable force in the world arms marketplace.

South American countries have traditionally had strong navies, and this holds good even more today than in the past. The aircraft carriers operated by Brazil and Argentina are obsolescent and are unlikely to be replaced as such, but several states in the area are acquiring or have expressed interest in smaller aircraft-carrying vessels, both new and second-hand. Offshore resources in the form of marine life and minerals are important to many South American countries, and new maritime-patrol aircraft are being introduced to safeguard areas of interest out to 200 miles from the coast.

The vast area of Brazil and to a lesser extent some of her neighbours means that there is little or no pressure to invade other countries in search of new land, despite the almost universal catholic religion and a consequent exploding birthrate. Other topographical extremes in the Andes require that an air force should also double as a national internal airline, so several states have modern transport fleets even when their combat element may be obsolescent and under-strength.

Despite the low standard of living suffered by much of the population, many South American countries have rich natural resources which can be exploited to pay for new arms – especially when a military government is in power. South America could be a major area of Russian influence in the future, and any move by the Soviet Union to foment revolution in the area – possibly with the support of the Cubans – might well precipitate a massive arms-buying spree, and would obviously be of great concern to North American near-neighbours.

COLOMBIA

Area: 440,000 sq. miles (1·14 million sq.km.).
Population: 25·8 million.
Total armed forces: 54,300.
Estimated GNP: $12·9 billion (£6·45 billion).
Defence expenditure: $110 million (£55 million).

Possessing only a handful of modern combat aircraft, the *Fuerza Aérea Colombiana* has as its primary task the operation of internal communications throughout the country which, with an area of some 440,000 square miles, has only tenuous road links in many areas. The FAC is the sole military operator of aircraft, the headquarters being in the capital, Bogota. Of the total armed forces of 54,300, some 6300 are in the air force; annual defence spending has been around $110m for some years. As with a number of South American countries, Colombia operates a military airline serving many points in the country's hinterland.

The present *Fuerza Aérea Colombiana* has its origins in the establishment of an army air arm at Flandes in April 1922, initial equipment being a Caudron G.III for training. At about the same time the Navy formed a flying-boat flight and both arms, administered by the Ministry of War, acquired a modest number of aircraft – mainly purchased from American sources. In 1943 the FAC was established as a separate entity on the recommendation of an American Aviation Mission and in 1948

Colombia became a member of the Organisation of American States, becoming eligible to receive American military aid. This included a squadron of Thunderbolt fighters, a squadron each of Fortress and Mitchell bombers and a number of transports and training aircraft. In March 1954 the air force received six Lockheed T-33s

Above: Dassault Mirage 5COA fighter-bombers have replaced Sabres as the FAC's main combat aircraft.

from the United States Air Force – the first transaction in which a South American nation purchased a US jet under the Reimbursable Assistance Programme. Two years later the single fighter squadron

GUYANA

Area: 83,000 sq. miles (215,000 sq. km.).
Population: 714,000.
Total armed forces: 2,200.
Estimated GNP: $500 million (£250 million).
Defence expenditure: $13·4 million (£6·7 million).

The Guyana Defence Force Air Command has liaison and light transport as its main tasks as a branch of the country's armed forces, with policing, anti-smuggling patrols and civil community support important operational roles. Headquarters of the GDFAC is at Camp Ayanganna, situated in the capital, Georgetown. Total manpower in the Defence Forces is some 2200 officers and men.

Formerly British Guiana, the small state of Guyana became independent in 1966 but remains within the British Commonwealth. Initial equipment purchased by the Defence Force Air Wing comprised two Helio Courier light STOL aircraft delivered in 1968 and subsequently flown on liaison and Army support duties. With the steady expansion of the air component, the Air Wing changed its name to Air Command in the early 1970s and in 1971 bought its first twin-engined type, the Britten-Norman Islander. Following delivery of the Islander, the Couriers were withdrawn from use.

Above: The Guyana Defence Force operates eight Islanders. No combat aircraft are on strength, the GDF's main roles being paramilitary in nature.

GUYANA

Type	Role	Base	No	Notes
Islander	Trans	Georgetown	8	First two delivered in 1971, others in 1975
King Air 200	Trans/VIP	Georgetown	1	Delivered 1975
U206F	Trans	Georgetown	1	Delivered 1976
Alouette III	Trans	Georgetown	2	
JetRanger	Trans	Georgetown	2	Delivered December 1976
Bell 212	Trans	Georgetown	3	Delivered 1976

Inventory

The GDF Air Command currently operates a mixed force of helicopters and fixed-wing types flown on a variety of duties, both para-military and civilian. The helicopter section has French and American types, some operated on government work in connection with hydro-electric survey and road-building projects in the more remote regions of the country. There has been no major effort to develop a serious military capability, and future procurement is likely to be limited to liaison and training aircraft.

---

received the first of a batch of Canadair and North American Sabres, which replaced the ageing Thunderbolts, and these were destined to remain the country's front-line combat element until the arrival of Mirages in 1970. To aid development of the Amazon and Orinoco regions of Colombia, the air force formed Satena (*Servicio de Aéronavegacion a Territorios Nacionales*) in September 1962, using a variety of transport aircraft including C-47s, Beech C-45s and Beavers.

Colombia is a member of the OAS, is one of the signatories of the Tlatelolco Treaty which prohibits nuclear weapons in Latin America, and in return has a bilateral military assistance agreement with the United States.

Inventory

The FAC, as with the neighbouring air arms of Brazil, Peru and Venezuela, operates Mirage fighter-bombers. The 18 aircraft were delivered between 1970 and 1972 to replace US and Canadian-supplied Sabres. There is also a single bomber squadron with B-26 Invaders supplied by the United States under the military aid programme, and in the attack role these aircraft can be supplemented by a few Lockheed AT-33s fitted with underwing weapon racks. Internal security is an important role for the air force, and for some years operations have been flown against guerrillas operating in the mountains.

To cater for maritime operations along the coast, which borders both the Pacific and Atlantic, the FAC has four DHC Twin

Otters for both land and water operation; four Convair Catalinas were withdrawn from use in 1976. Air transport forms the vast majority of FAC duties, there being two elements for this role: Satena and Transport Command. The former, with a para-military airline network throughout the country, operates British and American aircraft and the latter has a number of different types for logistic and communications work, including Lockheed Hercules and de Havilland Beavers. Heli-

copters are used for a variety of roles such as search and rescue, liaison and communications, training and transport. The FAC's main operational bases include Cali, Germon Olana, Melgar, Techo, Monteria, Eldorado and Berastegul.

Fixed-wing training and the air force academy are at the Marco Fidel Suarez air base at Cali, as is the ground training element which was established in 1960. Helicopter training and the main rotary-wing base in Colombia are at Melgar.

COLOMBIA

Unit	Type	Role	Base	No	Notes
Grupo de Combate	Mirage 5COA	FGA	Germon Olana	14 ⌉	Delivered 1970–72 to
	Mirage 5COR	Recce	Germon Olana	2 ⎬	single combat group,
	Mirage 5COD	Train	Germon Olana	2 ⌋	replacing Sabres
	AT-33A/T-33A	Att/train	Germon Olana	10/12	Mainly used in advanced trainer role, but with secondary Coin task
Grupo de Bombardeo	B-26K/RB-26C	Bomb	Luis F. Gomez Nino air base	8	Ex-MAP supplied. Due for replacement
	C-130B	Trans	Techo	1	Ex-Canadian, three delivered in 1971, two lost in accidents
	C-54	Trans	Techo	5	Two operated by Satena
	C-47	Trans	Techo	23	Seven operated by Satena
	F.28 Mk 100	VIP	Bogota	3	Delivered 1972; used by Satena
	HS.748	Trans	Bogota	1	For Presidential use
	C-45	Trans	Techo	3	
	Twin Otter	Mar pat/trans	Techo	4	
	Beaver	Lt trans	Techo	10	
	Porter/Turbo-Porter	Util		6	
	Lama	SAR		27	
	Bell 47G	Train	Melgar	16	
	UH-1B	Trans	Melgar	6	Also used for Coin
	Bell 212	VIP	Bogota	1	For Presidential use
	H-23	Util	Melgar	4	
	OH-6A	AOP	Melgar	12	
	Hughes 500C	AOP	Melgar	10	Delivered 1977
	TH-55A	Train	Melgar	6	
	HH-43B	SAR	Melgar	6	
Escuela Militar de Aviacion	T-41D	Train	Cali	30	Delivered 1968
	T-37C	Train	Cali	10	Delivered 1969 for advanced training
	T-34	Train	Cali	30	

ECUADOR

Area: 226,000 sq. miles (585,000 sq. km.).
Population: 6·5 million.
Total armed forces: 25,300.
Estimated GNP: $5·9 billion (£2·95 billion).
Defence expenditure: $114 million (£57 million).

The *Fuerza Aérea Ecuatoriana* has benefited in recent years from higher defence spending, and the 2600-man FAE now ranks as one of South America's most modern air forces. Headquarters are in the country's second-largest city, Guayaquil. Until recently Ecuador has relied on United States military aid, but orders placed in the past few years have been biased towards European and Middle East arms suppliers. Both the Army and Navy operate small air components, chiefly in the support and liaison roles.

Ecuador formed an aviation arm in 1920 with Italian aid. It was known as the *Cuerpo de Aviadores Militares*, initial equipment comprising aircraft of Italian and German origin. Little expansion occurred until the late 1930s, when the name changed to its present title and a number of American aircraft were procured. The Italian advisers were replaced by an American Air Mission in 1941 and a training scheme was initiated using North American NA-16s and Ryan PT-20s. Seversky P-35s gave the FAE its first combat aircraft and a number of ex-German Ju 52/3ms markedly improved the air force's transport potential. After the Second World War Ecuador became a member of the Rio Defence Pact and continued to receive US aid, including P-47D Thunderbolts, Catalina amphibians, Douglas C-47s and some T-6 trainers. In 1954 12 Gloster Meteor FR.9 fighters and six English Electric Canberra B.6 bombers arrived to give the FAE its first major jet equipment. Three Bell 47Gs

Right: The FAE has 11 remaining Cessna A-37B attack aircraft for use in the counter-insurgency role, complementing the Strikemasters which are additionally employed for training.

were acquired from the US in 1960 and American aid continued through the rest of the decade.

Ecuador is a signatory of the Charter of the Organisation of American States, and of the Tlatelolco Treaty prohibiting the use of nuclear weapons in Latin America.

Inventory

The *Fuerza Aérea Ecuatoriana* was, along with Oman, the first overseas customer for the Anglo-French Jaguar strike aircraft when an order for 12 aircraft was placed in 1974. Delivery began in January 1977 to replace nine F-80C Shooting Stars in an interceptor squadron. The Canberras remain in use with the *Escuadrilla de Bombardeo* and have been the main element in the FAE's offensive force for 25

years. The third BAC type in use is the Strikemaster, of which 16 aircraft have been ordered since 1972 for dual jet training and light-strike roles. To supplement the Strikemasters, 12 Cessna A-37B counter-insurgency jets were delivered in 1976. Ecuador initiated negotiations with Israel for 24 IAI Kfir fighter-bombers worth $150m in 1976, but the United States refused to sanction the deal on the grounds that the aircraft is too sophisticated for Latin American countries. Ecuador remained in the market for another strike aircraft, and an order for Mirage F.1s was announced in 1978.

Transportes Aéreos Nacionales Ecuatorianos or TAME is a branch of the Air Force formed in 1962 to operate scheduled passenger and cargo flights throughout the country and to the Galapagos Islands, 600 miles out in the Pacific; four Electras and a pair of HS.748s fly the TAME network. The FAE also has a Military Air Transport Group which operates a variety of types including DH Buffaloes, HS.748s and half-a-dozen Israeli Arava light transports. French machines predominate in the helicopter element, but American aircraft continue to provide the FAE with its liaison and training elements. Bases include Quito, Salinas, Riobamba and Guayaquil.

Aviacion Naval Ecuatoriana

The Ecuadorian Navy has a small liaison and support air section with nine aircraft including two Alouette III helicopters for ship-to-shore duties.

Aviacion del Ejercito Ecuatoriana

Like the Navy, the Ecuadorian Army has an air element for supply and support duties. Types include a single Short Skyvan and some Pilatus Turbo-Porters.

FAE training is run on established principles and is generally conducted from primary through to advanced jet flying at the Military Flying School and a Training Group. To modernise the training fleet, 20 Beech T-34C Turbo-Mentors were ordered in 1976. Type conversion is usually conducted in the country of origin involving both air and ground crews.

ECUADOR

Type	Role	No	Notes
Jaguar International	FGA*	12	Ten single-seat and two two-seat delivered beginning in January 1977; replaced nine F-80G
Canberra B.6	Bomb	3	Originally six
Strikemaster Mk 89	GA/train	10	Total of 16 purchased between 1972 and 1974; equip dual-role sqn
A-37B	GA	11	Delivered 1976
Mirage F.IJE/JB	FGA/train	16/2	
Military Air Transport Group			
DC-6	Trans	4	
HS.748	Trans	3	Two are Series 2A with large freight door delivered in 1976
DHC-5D Buffalo	Trans	2	Delivered 1976
C-130H	Trans	1	Two delivered, one lost in 1978
Twin Otter	Trans	1	Second aircraft operated by TAME (see below)
Miscellaneous			
SA.330 Puma	Trans	2	
Lama	SAR	4	
Alouette III	Liaison	6	
Bell 212		—	
King Air 90	Calibration	1	
Gates Learjet 25B	VIP	2	
SF.260ME	Train	12	Delivered in 1977
T-34C	Train	20	Ordered in 1976

Type	Role	No	Notes
T-33A	Train	12	
T-41	Train	18	
Cessna 150 Aerobat	Train	24	
Cessna 180	Liaison		
Cessna 320	Liaison		
Transportes Aereos Nacionales Ecuatorianos (TAME)			
Electra	Trans*	4	
HS.748	Trans*	2	
Twin Otter	Trans*	1	
Aviacion Naval Ecuatoriana			
Alouette III	Liaison	2	
Arava	Trans	3	
Cessna 320E		1	
Cessna 177		1	
T-41		2	
Aviacion del Ejercito Ecuatoriana			
Skyvan 3M	Trans	1	
Turbo-Porter	Trans	3	
Arava	Trans	6	
Instituto Geografico Militar			
Gates Learjet 25D	Survey	1	
Queen Air 80	Survey	1	
Lama	Survey	1	

Based at Guayaquil

VENEZUELA

Area: 354,000 sq. miles (916,800 sq. km.).
Population: 12 million.
Total armed forces: 44,000.
Estimated GNP: $36 billion (£18 billion).
Defence expenditure: $615 million (£307·5 million).

The Fuerza Aérea Venezolana operates modern combat aircraft obtained from a number of sources. These include the United States, which has played a large part in modernising the air force. Venezuela's armed forces total some 44,000 personnel, and both the army and navy have small aviation elements. The 8000-man FAV is administered by the *Comandancia General de la Aviacion* with headquarters in the capital, Caracas, and has three main commands: the *Comando Aéreo de Combate*, which controls all combat and transport units; the *Comando Aéreo de Instruccion*, controlling all training activities; and the *Comando Aéreo de Logistico*, dealing with supply.

Military aviation in Venezuela dates from April 1920, when a Presidential decree established a Military Air Service at Maracay. The first element was a flying school set up by a French Aviation Mission in 1921 and equipped with a few Caudron and Farman training aircraft. Little official interest was taken in the air arm and the force was all but disbanded until the late 1930s, when the arm was re-named the Military Aviation Regiment and re-established with Italian help, acquiring three Fiat CR.32 fighters and a single Fiat BR.20 bomber.

Venezuela was on the side of the Allies during the Second World War, and in return for the use of some of her bases the air arm received small numbers of American training and liaison aircraft. Venezuela's signature of the Rio Pact in 1947 sealed US assistance to the country, and P-47 Thunderbolts, B-25 Mitchells and C-47s were supplied to establish the air arm as a viable military force. The force was officially named *Fuerza Aérea Vene-*

zolana in 1949 and began an ambitious modernisation programme which saw orders placed or 24 Vampires in 1950, six Canberras in 1953 and 22 F-86F Sabres in 1955.

In 1956, 22 Venoms arrived and the transport force acquired 18 C-123 Providers for tactical use alongside a squadron of C-47s, a couple of C-54s and some Beech D-18s. More Canberras were ordered in 1957, followed two years later by 41 Beech Mentors to replace the T-6 training force. A lull in this expansion was broken in 1967 when the first of 47 F-86K Sabres was delivered from surplus West German stocks.

Venezuela is a signatory of the Charter of the Organisation of American States, and also of the Tlatelolco Treaty which prohibits nuclear weapons in Latin America.

Below: Rockwell T-2D Broncos provide the FAV's intermediate stage of training before conversion to combat types.

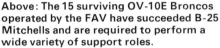

Above: The 15 surviving OV-10E Broncos operated by the FAV have succeeded B-25 Mitchells and are required to perform a wide variety of support roles.

Inventory

Venezuela's natural resources have enabled her to establish the FAV as one of the most efficient air forces in South America. The FAV has two combat groups, each comprising two or three squadrons. Venezuela followed other South American air arms when she ordered a batch of Mirages in 1971 to replace the F-86Ks in the interceptor role. Between 1952 and 1965 about 30 Canberras were purchased from the UK, and the aircraft have been periodically updated under servicing contracts with BAC. The first of the 16 Broncos arrived in Venezuela in May 1973, the type operating in a number of roles including bombing, reconnaissance and counter-insurgency, the unit previously flying B-25 Mitchell bombers supplied by the US between 1947 and 1952.

Fairchild Providers and Hercules give the air force the tactical and long-range heavy-lift capability so necessary in such a country. One of the more recent acquisitions is a single Boeing 737-200S fitted out for Presidential and VIP use.

Support

Pilot training in the FAV is administered by the *Comando Aéreo de Instruccion* and comprises the *Escuela de Aviacion Militar*, or School of Military Aviation, based near Maracay. Initial flying training is undertaken on Mentors. Future transport pilots then proceed to Queen Airs and C-47s for twin and instrument training before posting to the squadrons, while combat pilots go to the Palo Negro base for jet training on the Jet Provosts to Wings standard. Advanced tuition and weapon training are then given on the T-2Ds before conversion at squadron level on the two-seat Mirages, CF-5Bs and Canberras. The FAV Air Academy and Technical Training School is also situated at Maracay.

Navy

The air arm of the Venezuelan Navy – *Servicio de Aviacion Naval Venezolana* – is responsible for all ASW commitments in territorial waters and also conducts its own liaison and transport duties with a handful of aircraft. Like that of the air force, much of the Navy's equipment has been supplied by the United States, but a recent order was placed with Agusta for a couple of anti-submarine helicopters to supplement the half-dozen ex-US Navy Trackers.

Army

The *Aviacion del Ejercito Venezolana* has a modest helicopter force totalling almost 20 machines. Their use is generally confined to liaison and communications work.

VENEZUELA Navy and Army

Type	Role	Base	No
Navy			
HU-16B	SAR/mar pat	Caracas	4
S-2E	ASW/mar pat	Caracas	6
AB.212ASW	ASW	Caracas	10
Bell 47J	Liaison	Caracas	2
HS.748	Trans	Caracas	1
Army			
Alouette III	Liaison	Caracas	10
Bell 47G	Liaison	Caracas	6

VENEZUELA Air Force

Group	Unit	Type	Role	Base	No	Notes
Grupo 12	Esc/34	CF-5A	Int	Barquisimento	15	
	Esc/35	CF-5B/D	Train	Barquisimento	4	
		F-86K	Int	Barquisimento	15	
	Esc/35	Mirage IIIEV	Int	Barquisimento	10	
		Mirage 5V/5DV	Ga/train	Barquisimento	4/2	
Grupo 13	Esc/38	Canberra B.82	Bomb	Barcelona	18	
	Esc/39	Canberra B(I).88	Interdict	Barcelona	7	
		Canberra T.84	Train	Barcelona	2	
		Canberra PR.83	Recce	Barcelona	2	
	Esc/40	OV-10E	Coin	Barcelona	15	
Grupo de transporte	Esc/1	C-130H	Trans	Caracas	5	
		C-47	Trans	Caracas/Maracay	20	Some used for twin training
		C-123B	Trans	Caracas	12	
		Boeing 737-200S	VIP/trans	Caracas	1	
	Esc/2	HS.748	VIP/trans	Caracas	2	
		DC-9-15	VIP/trans	Caracas	1	
		Citation	VIP/trans	Caracas	1	
		UH-1N			2	
Grupo de reconcimento		UH-1D/H	Liaison	Maracay	12	
		Alouette III	Liaison	Caracas	15	
		Bell 206B	VIP	Caracas	6	
		Bell 206L	VIP	Caracas	1	
		UH-19	Liaison	Maracay	10	
		Queen Air	Liaison	Caracas/Maracay	9	Also used for twin conversion training
		Cessna 182N	Liaison	Caracas	12	
Escuela de Aviacion Militar		T-34	Train	Maracay	25	
		T-2D	Train/GA	Palo Negro	24	Two batches of 12 delivered; second batch fitted with weapons for the dual strike/trainer role
		MFI-15/17	Train	Palo Negro	20	Order reportedly placed in 1978

PERU

Area: 531,000 sq. miles (1·37 million sq. km.).
Population: 14·1 million.
Total armed forces: 89,000.
Estimated GNP: $13 billion (£6·5 billion).
Defence expenditure: $400 million (£200 million).

The *Fuerza Aérea del Peru* is one of the best equipped and trained air arms in Latin America. As well as the Air Force, both the Army and Navy have aviation elements and all benefit from the Government's policy of equipping the Services well. Peru steers a middle course between East and West and her armed forces' equipment reflects this, both the Army and Air Force having Soviet as well as American material in front-line use. The FAP has a combat force totalling some 150 aircraft of an inventory approaching 400 and manned by approximately 10,000 personnel. Headquarters are situated in the capital city, Lima. Organisation of the FAP is based on groups or *grupos*, each with up to three squadrons or *escuadrones*, the strength varying with the availability of aircraft.

A military air corps was first established in Peru in 1919 with the assistance of a French Air Mission and two dozen British and French aircraft. In 1924 the Peruvian Naval Air Service was formed and some Curtiss flying boats were bought. The two flying elements were combined into one force in 1929 to form the *Cuerpo de Aeronautica del Peru* and this amalgamation continued until the 1960s, when the Navy force was re-established as a separate arm.

Peru joined the Organisation of American States after the Second World War and began to receive United States aid, including Thunderbolt fighters, Mitchell bombers and some Harpoon reconnaissance aircraft. In 1950 the CAP became the *Fuerza Aérea del Peru*, and towards the middle of the decade received its first jet aircraft when F-86F Sabres arrived; then were followed by 16 Hunter F.52 fighters and eight Canberra bombers. More Canberras have since been purchased but the other two types are being phased out of use after more than 20 years service. Peru became the first Latin American country to initiate all-through jet training when it acquired a batch of Cessna T-37Bs in

Below: Like so many South American air arms, the FAP relies on the Cessna A-37B to back up its heavier attack types. Some three dozen are in service.

Above: The Canberra has served as a medium bomber in Peru for nearly a quarter of a century. It can be armed with AS.30 air-to-surface missiles for extra punch.

1961, following this with the first purchase by a South American air force of Mach 2 jet fighters when it announced in 1968 the receipt of the first French Mirage 5 for its fighter-bomber group.

Peru is a signatory of the Charter of the Organisation of American States, drawn up in 1948, and also of the Tlatelolco Treaty prohibiting nuclear weapons in Latin America. The United States continues to supply military and economic aid to Peru despite the country's recent acquisition of Soviet aid.

Inventory

The *Fuerza Aérea del Peru* created much controversy in 1976 when, after protracted negotiations with Western aircraft companies, it placed an order with the Soviet Union for 36 Sukhoi Su-22

fighter-bombers to replace older types then in front-line service. The order, worth about $250m, was a commercial deal between the two countries rather than a political one as some sources speculated. To assist in training Peruvian pilots a dozen ex-Cuban MiG-21s were assigned to the FAP, together with Cuban and Russian personnel. It is believed that the Su-22s have replaced the Sabres which have seen 20 years' service in Peru.

Currently flying in the strike role are two squadrons of a three-squadron bomber wing with 34 Canberras of varying marks, some dating back to 1956 when the FAP received its first batch of the type. Supporting these medium bombers are 36 Cessna A-37s flown in the light strike role.

Twenty-two Mirage 5s were ordered initially and repeat orders have increased the total to some 36 aircraft, although attrition has reduced this number. Canadian, American, Russian and Dutch types rub shoulders in Peru's transport force. To assist with developing the country's vast mineral resources, the FAP has a

fleet of Buffalo and Hercules transports for heavy-lift duties, while Russian Mi-8 helicopters and some An-26s provide support in the more remote regions of the country and fly light transport tasks. Aero-Peru is the result of a reorganisation of Satco, the airline element of the air force, and operates Fokker-VFW F.28s and F.27s on internal services. In this country of mountain and jungle the helicopter assumes a greater importance and the FAP has a fleet of nearly 60. Main FAP bases include Las Palmas, Talara, Iquitos, Pisco, Trujillo and Jorge Chavez.

Support

The Air Academy at Las Palmas is Peru's main training base, students undergoing primary tuition on Cessna T-41s, T-6s and T-34s and proceeding to jet-powered T-37Bs. Operational conversion is carried out at squadron level on two-seat combat trainers.

Navy

The Peruvian Naval Air Arm, *Servicio de Aviacion de la Armada Peruana*, has undergone rapid expansion in recent years and now boasts a modest ASW element supported by both fixed wing and helicopter types. The principal front-line aircraft is the Grumman S-2E Tracker, of

Above: Peruvian Hunter pilots in front of their mounts.

Below: One of two naval F.27M maritime patrollers, with prominent belly radome.

55

One of six Peruvian Navy AB.212ASWs operated on anti-submarine patrols comes aboard its parent frigate. The helicopters are sonar-equipped and can carry torpedoes.

aircraft in support of the ground forces. AOP and communications are the main roles, with casevac and light transport also conducted. The army also operates Russian SA-3 surface-to-air missiles.

PERU   Army

Type	Role	Base	No
Cessna 185	Liaison		5
Helio Courier	Liaison		5*
Alouette III	Liaison		4
Bell 47G	Liaison		8
SA-3	SAM		

*Two fitted with floats

PERU   Navy

Type	Role	Base	No
F.27M	Mar pat	Jorge Chavez	2
S-2A	ASW	Jorge Chavez	9
C-47	Trans	Callao	6
Aztec	Liaison		1
Alouette III	Liaison		2
UH-1D/H	Liaison		6
Bell 47G	Liaison		5
JetRanger	Liaison		10
AB.212ASW	ASW		6
SH-3D	ASW		4
Mi-8	Trans	Callao	42*
T-34A	Train	Callao	2
T-34C-1	Train	Callao	6

*Approx 36 in store

PERU   Air Force

Unit	Type	Role	Base	No	Notes
Grupo 12	Su-22	FGA		36	Includes four two-seaters
	MiG-21	Int		12	Ex-Cuban AF delivered 1977 for pilot training for Su-22 unit
Grupo 21	Canberra B.2/B.56/B(I).8	Bomb		22	
	Canberra B(I).68	Bomb		11	
	Canberra T.4	Train		2	
	A-37B	Att	Chiclayo	12	Total FAP inventory 36; also flown by Grupo 13
Grupo 13	Mirage 5P/5DP	Int/train	Chiclayo	36	
Grupo 31	HU-16B	SAR	Lima	4	
Grupo 41	L-100-20	Trans	Jorge Chavez	6	One sqn
	Buffalo	Trans	Jorge Chavez	16	One sqn
	An-26	Trans	Jorge Chavez	16	One sqn
	C-54	Trans	Jorge Chavez	4	One sqn
	DC-6	Trans	Jorge Chavez	5	One sqn
	An-26	Trans	Jorge Chavez	20	One sqn
Grupo 42	Twin Otter	Trans	Iquitos	7	Some with floats
AeroPeru	F.27-600	Trans	Lima	2	
	F.28-1000	Trans	Lima	4	One used by Presidential Flight
Grupo 8	Queen Air	Liaison		18	
	Turbo-Porter	Trans		12	
	Alouette III	Liaison		12	
	Bell 47G	Liaison		20	
	Bell 212	Liaison		17	
	Mi-6	Trans		5	
	Mi-8	Trans		6	
	Learjet 25B	Survey		2	
	T-34C	Train		6	
	T-34A	Train	Las Palmas	6	
	T-37B	Train	Las Palmas	26	
	T-41A	Train	Las Palmas	19	
	T-33A	Train	Las Palmas	8	
	T-6G	Train	Las Palmas	15	
	Pitts S-2A Special	Train	Las Palmas	6	

which the service has nine. Supplementing these are two Fokker F.27M maritime-patrol aircraft ordered in 1976.

Half-a-dozen AB.212ASW helicopters equipped with sonar and torpedoes for anti-submarine use are operated from frigates. The SAAP is Peru's largest operator of Mil Mi-8 transport helicopters, having a total of 42 in the inventory, although most remain in store. Reports have suggested that two more F.27MPAs are on option, and a more recent order has been placed for four SH-3D Sea Kings from Italy.

Army

Aviacion del Ejercito del Peru is the Army's small aviation element which operates nearly two-dozen fixed- and rotary-wing

BRAZIL

Area: 3·29 million sq. miles (8·52 million sq. km.).
Population: 107 million.
Total armed forces: 273,800.
Estimated GNP: $177 billion (£88·5 billion).
Defence expenditure: $2 billion (£1 billion).

The *Fôrca Aérea Brasileira*, or FAB, provides the air cover for the largest country in South America and the fifth-largest in the world. Some 600 aircraft make up the air force, personnel strength standing at approximately 35,000. Headquarters of the FAB is in the capital, Brasilia, and operational units come under a number of specialised commands as follows: *Comando de Defesa Aérea*, based at Anapolis, controlling the air-defence system; the *Comando Aérotático*, responsible for tactical elements; *Comando Costeiro*, controlling land-based maritime units; and the *Comando de Transportes Aéreos*, in charge of transport.

The country is divided into six regional air commands known as COMARs, each responsible for the FAB aircraft, equipment and personnel within its area. No 1 COMAR, with headquarters at Belém, covers the states of Amazonas, Pará and

Above: **The mainstay of the FAB's fighter force is the Northrop F-5E, which can carry Sidewinder air-to-air missiles.**

Acre, plus the Federal Territories of Roraima, Rondônia and Amapá; No 2 COMAR, with headquarters at Recife, covers the states of Piaui, Ceará, Rio Grande do Norte, Pernambuco, Alagoas, Sergipe, Bahia and part of Maranhão; No

3 COMAR, with headquarters at Rio de Janeiro, covers the states of Rio and Minas Gerais; No 4 COMAR, with headquarters at São Paulo, covers that state; No 5 COMAR, with headquarters at Porto Alegre, covers the states of Rio Grande do Sol, Paraná and Santa Catarina; and No 6 COMAR, with headquarters at Brasilia, covers the capital district, the state of Goiás and part of Maranhâo. Each of

these regional commands can operate independently if so desired.

A computer-based air-defence system is being installed by the French company Thomson-CSF. Known as DACTA I and II, the system will eventually embrace the whole of Brazil towards the end of the 1980s, possibly with the addition of SAMs.

Military aviation in Brazil began in 1913 with the formation of a Navy seaplane school equipped with an Italian Bossi seaplane. The Brazilian Army followed some time later, with the establishment of an Army Air Service founded under French supervision in October 1918 at Rio de Janeiro. Early equipment comprised a variety of French and British aircraft, with a number of American Boeing and Curtiss designs joining the Army before the beginning of the Second World War. Following Brazil's declaration of war on Germany and Italy in 1942, bases were put at the disposal of the Allies and further equipment was received in return from the US. Before this, on January 20, 1940, the army and navy air arms were merged to form the *Fòrca Aérea Brasileira* under the auspices of a newly established Air Ministry with headquarters in Rio de Janeiro. Over the next few years the FAB began laying the foundations of the air-defence network in a series of air zones embracing the whole of Brazil, and these continue in much refined form today. Also established in 1941 was the Air Force Academy for officer training at Campo dos Alfonsos.

After the war Brazil received quantities of Thunderbolts, Mitchells, Fortresses, Texans and some Neptunes, plus 60 Meteors from Britain in 1954. More jets followed in the late 1950s, this time from the US: Lockheed F-80C Shooting Star fighters and T-33 trainers, as well as some Fairchild C-119 Packet transports. In 1965 the FAB assumed control of all fixed-wing naval aircraft, including the ASW S-2 Trackers operating from Brazil's only aircraft carrier, *Minas Gerais*.

Brazil is a signatory of the Charter of the Organisation of American States (OAS) drawn up in 1948, and she has a bilateral military assistance agreement with the United States. The Tlatelolco Treaty, which prohibits nuclear weapons in Latin America, was signed by Brazil in February 1967. Brazil envisages no major threat to her sovereignty over the next decade.

Inventory

The *Fòrca Aérea Brasileira* has only one unit assigned the pure interceptor role, the 1st Air Defence Wing or *Ala de Defesa Aérea* (ALADA), equipped with 16 Dassault Mirage IIIs armed with MATRA R.530 missiles and two 30mm cannon. These were bought from France in 1970, though spares problems are reported to have reduced the effectiveness of the unit. No replacement for these aircraft has yet been decided upon, but an aircraft in the F-16 class is the likely choice.

To modernise the fighter force and increase its effectiveness, a $72.3m order for 42 Northrop F-5E Tiger IIs was placed in October 1974 and deliveries were completed by March 1976. Two *Esquadrãos*,

Above: The FAB was one of the earliest customers for the HS.748, one of many transport types needed to ferry personnel and equipment round one of the world's largest countries.

or squadrons, of the 1st *Grupo de Aviacão de Caca* (GAvCa) are equipped with this aircraft, one fewer than was originally planned as a result of economies. The F-5s are armed with Sidewinder air-to-air missiles and reports suggest further FAB orders for this aircraft. Also flying fighter duties are subsonic Embraer AT-26 Xavantes (licence-built Italian Aermacchi MB.326s). Within the FAB the Xavante is a true multi-role aircraft, performing ground-attack and training duties as well as equipping three fighter units. The FAB ordered the AT-26 in 1969 and deliveries from the Embraer factory in São Paulo began two years later. The initial order for 112 has been followed by two further contracts totalling 55 aircraft, with deliveries taking place throughout 1978–9. To further increase the versatility of this aircraft an Aerotec-manufactured starter pod can be fitted for deployment to airfields with no support facilities.

Reconnaissance and attack squadrons or *Esquadraões Mistos de Reconhecimento e Ataque* (EMRAs) total five; three have Xavantes, while the other two fly the indigenous armed version of the Neiva T-25 Universal. North American T-6s originally equipped these units and their phase-out has also coincided with the withdrawal of other older types, including Lockheed AT-33As, B-26Ks and Cessna T-37Cs. Complementing the attack units are three liaison and observation squadrons, or *Esquadrão de Ligacão e Observacão* (ELO), equipped with a variety of helicopters and light aircraft.

Comando Costeiro is the maritime branch of the air force. It has a unique position in that it supplies ASW aircraft for Brazil's only aircraft carrier *Minas Gerais* (ex-HMS *Vengeance*, purchased in 1957). These aircraft – S-2 Trackers supplied by the US and Canada – form the

Grupo de Aviacão Embarcada stationed at Santa Cruz, and six aircraft are normally assigned to the ship at any one time. For coastal patrol the 7th *Grupo de Aviacão* has 12 Embraer EMB-111 maritime-patrol derivatives of the Bandeirante light transport. The future MR force planned by the FAB envisages one long-range patrol squadron of Orion-type aircraft and four EMB-111 squadrons, but budget restrictions might force a revision of these plans. For SAR duties, two *Grupos* have a mixed force of RC-130E Hercules, Albatross amphibians and SH-1 helicopters.

A large transport component provides a valuable contribution to Brazil's widely spaced civil community as well as supporting the country's 170,000-strong army. Hercules and Buffaloes provide much of the airlift capacity, while Bandeirante transports carry out lighter duties. A total of 80 Bandeirantes are on order, 60 passenger-carrying machines and 20 freighters with strengthened floors and wide doors. The air force also has a tactical transport requirement which is expected to be met by the Embraer CX-2A, powered by two GE T64 turboprops.

The Brazilian Army operates nine batteries of Raytheon Hawk medium-range surface-to-air missiles, and in August 1977 it took delivery of the first of four fire units (plus 50 rounds) of the low-level Euromissile Roland 2.

Above: Brazilian Navy frigates operate the Westland/Aérospatiale Lynx on anti-submarine duties.

Above: Brazil's home-grown Neiva L-42 Regente is operated by the FAB in the liaison and observation roles, in which capacity it is supported by T-25 Universals and UH-1H helicopters.

The Air Force Academy at Pirassununga, São Paulo, provides officer training and flying tuition to wings standard (300–350hr) on T-23 Uirapuru primary trainers (50hr), T-25 Universals (150hr) and AT-26s (100hr). The other FAB training base is at Natal for reserve pilot instruction. Also at the base is a tactical weapons range for air-to-air and air-to-ground use. Twin training is performed on Bandeirantes and a helicopter training unit at Santos, São Paulo, has 34 Bell H-13Js. Technical ground training is conducted at Guaratingueta and São Jose dos Campos, with command and administrative training carried out in Rio de Janeiro.

A radio calibration unit operates two HS.125s and two EMB-110As.

Forca Aeronaval

Since 1965, when a Presidential decree ordered responsibility for all fixed-wing naval operations to be transferred to the air force, the Brazilian naval air arm has been a helicopter force. Tasks are wide-ranging and include ASW, ship-to-shore liaison and communications, support and training. The main shore base is the San Pedro de Aldeia Naval Air Station. Largest operational type is the SH-3 Sea King used for ASW from San Pedro and the aircraft carrier *Minas Gerais*. Following a tradition started with the Whirlwind and going on to the Wasp, the *Forca Aéronaval* has received the Westland Lynx anti-submarine helicopter. This British design equips the *Niteroi*-class frigates built in Britain and Brazil. The *Niterois* each carry two triple launchers for Short Seacat SAMs, this weapon also arming one destroyer.

BRAZIL   Air Force

Unit	Type	Role	Base	No	
1° ALADA	Mirage IIIEBR/DBR	Int	Anapolis	12/4	Ordered in 1970; known as F-103E and F-103D in FAB service
1° & 2° *Esquadrão*, 1 GAvCa	F-5E/B	Fight	Santa Cruz	36/6	$72.3m order placed in October 1974; delivered March 1975
1° & 2° *Esquadrão*, 4 GAvCa	AT-26 Xavante	Fight	Fortaleza		
1° *Esquadrão*, 14 GAvCa	AT-26 Xavante	Fight	Canoas		
Reconnaissance and attack squadrons (EMRAs)				total 167	
3° EMRA	AT-26 Xavante	Coin	Galeao		
4° EMRA	AT-26 Xavante	Coin	Cumbica		
5° EMRA	AT-26 Xavante	Coin	Canoas		
1° EMRA	T-25 Universal	Coin	Belem		Re-equipping 1977–78 from North American T-6s
2° EMRA	T-25 Universal	Coin	Recife		
Liaison and observation squadrons (ELOs)					
1° ELO	L-42 Regente		Campo dos Afoncos		Also flies Bell 206 JetRangers and Cessna 0-1s
2° ELO	T-25 Universal		San Pedro de Aldeia		
3° ELO	UH-1H/O-1E/L-42 Regente		Porto Alegre		
Maritime element (Comando Costeiro)					
2° *Esquadrão*, Grupo de Aviacão Embarcada	S-2A/S-2E Tracker	ASW	Santa Cruz	8/8	Ex-USN S-2Es; some S-2As now used as transports; six aircraft based on *Minas Gerais*
1° *Esquadrão*, 7° Grupo	EMB-111	ASW	Salvador	12	Designated P-95 in FAB service
6° *Grupo de Aviacão*	RC-130E Hercules	SAR	Recife	3	
2° *Esquadrão*, 10 Grupo	SA-16A Albatross	SAR	Florionapolis	13	
3° *Esquadrão*, 10 Grupo	SH-1D	SAR	Florionapolis	6	Plus a few Bell 47Gs
Transport Groups					
1° *Esquadrão*, I Grupo	C-130E/H/KC-130 Hercules	Trans/tanker	Galeao	7/3/2	KC-130s support F-5Es
1° *Esquadrão*, II Grupo	HS.748/EMB-110 Bandeirante	Trans	Galeao	6?	
2° *Esquadrão*, II Grupo	HS.748 Srs 2C	Trans	Galeao	6	
Parachute transport group (Grupo de Transporte de Tropas)					
2° *Esquadrão*	Buffalo	Trans	Campo dos Afoncos	21	Some with 5° & 6° *Esquadraos*
Special Transport Group (Grupo de Transporte Especial)					
	Boeing 737-200	VIP	Brasilia	2	
	BAC Viscount	VIP	Brasilia	1	
	HS.125 Series 3AR	VIP	Brasilia	8	
	Bell 206	VIP	Brasilia	6	
	EMB-110 Bandeirante	VIP	Brasilia		
	EMB-121 Xingu	VIP	Brasilia	5	
1°, 2°, 3°, 4° *Esquadrão*	EMB-110 Bandeirante	Trans	Belem/Recife/Galeao/Cumbica		Replaced C-47s
5° & 6° *Esquadrão*	Buffalo	Trans	Campo Grande, Manaus		
2° Grupo	PBY-5A Catalina	SAR/trans	Belem	6	Refurbished 1975–76; replacement required
Ala 435	UH-1D/Bell 206/OH-6A	Coin	Santos	6/4/4	
Miscellaneous					
	T-23 Uirapuru	Train	Pirassununga (Air Force Academy) and Natal	100	
	T-25 Universal	Train		132	
	AT-26	Train			
	EMB-110 Bandeirante	Train			
	H-13J	Train	Santos	34	
	HS.125	Radio calib		2	
	EMB-110 Bandeirante			4	

BRAZIL   Navy

Unit	Type	Role	Base	No	Notes
1° *Esquadrão ASW*	SH-3D Sea King	ASW	San Pedro de Aldeia	6	First delivered April 1970; also used aboard *Minas Gerais*
1° *Esquadrão General*	Bell 206B/Wasp/Whirlwind	Supp	San Pedro de Aldeia		
1° *Esquadrão de Instrucão*	Hughes 269/300	Train	San Pedro de Aldeia	10	Withdrawn 1977
	Lynx	ASW	San Pedro de Aldeia	9	£10m ($18m) order placed in 1975; for *Niteroi*-class frigates from 1978
	Seacat	SAM		13 launchers	Seven ships

BOLIVIA

Area: 415,000 sq. miles (1·07 million sq. km.).
Population: 4·7 million.
Total armed forces: 22,500.
Estimated GNP: $2·5 billion (£1·25 billion).
Defence expenditure: $90 million (£45 million).

The *Fuerza Aérea Boliviana* has counter-insurgency and transport as its chief duties. Headquarters are in the capital, La Paz. Before increased US military aid began in the early 1970s the FAB had suffered from internal unrest and political take-overs, but stability prompted a strengthening of the force. Defence expenditure for 1978 totalled 1820m Pesos ($90m).

The Bolivian Military Aviation Corps was formed in 1924 as part of the army, and in the first few years operated a miscellany of French and Dutch aircraft. In 1929 a small number of Vickers Vespa reconnaissance machines were obtained, followed by Curtiss Hawk fighters. An attempt in 1937 by an Italian Air Mission to reorganise the air arm foundered and it remained for an American Military Aviation Mission in 1941 finally to establish the force as a credible element of the Bolivian armed forces. New aircraft included C-47s, Stinson Voyagers and Beech AT-11s, and the country was divided into four air-defence zones for more efficient protection. Re-named *Fuerza Aérea Boli-*

viana, the force received further American aid towards the end of the 1940s in the shape of P-47 Thunderbolts, B-25J Mitchells and some T-6 trainers, and in 1958 six B-17G bombers arrived to serve as transports for a number of years.

Bolivia was a signatory of the Charter of the Organisation of American States, drawn up in 1948, and receives military and economic aid from the USA.

Inventory

Bolivian front-line combat strength lies mainly in an assortment of counter-insurgency aircraft obtained from the US (AT-6s, Hughes 500M gunship helicopters) and Canada (T-33A/N). Four F86F Sabres equip one fighter squadron and appear destined for replacement shortly. The transport role of the FAB is carried out by a military airline, *Transportes Aéreos Militares*, providing passenger and cargo services to outlying areas. Current equipment includes five of the six Israeli Aravas delivered in 1975–76, four Convair 580s and a force of seven C-47s. FAB personnel strength totals about 4000 and the main bases include La Paz, Puerto Suarez in the east, Santa Cruz, El Tejar, El Trompillo, La Florida and Charana.

BOLIVIA

Unit	Type	Role	Base	No	Notes
Air Pursuit Group	F-86F Sabre	Int	Santa Cruz	4	
Air Cover Group No 1	AT-6G	GA	Tarija	13	
Air Pursuit Group	Canadair T-33A/N	GA/train	La Paz	15	
Combined Air Group	UH-1H	Assault	Cochabamba	6	
	Lama	Rescue	Cochabamba	—	
Special Operations Group	Hughes 500M	Attack	Roboré	12	Delivered 1968; armed with single 7·62mm Minigun pod
	Electra	Trans	La Paz	1	
	DC-6	Trans	La Paz	2	
	Learjet 25B	Survey		1	
	Turbo-Porter	Trans		1	Delivered 1975
	Cessna 185	Util		15	
	Cessna Turbo-Centurion	Util		2	
	Cessna 414	Liaison		1	
	Cessna 421B	Liaison		1	Delivered 1976
	Super King Air 200	Liaison		1	Delivered 1976
	Aerotec T-23	Train	Santa Cruz	12	Basic training; 18 originally received
	T-6	Train	Santa Cruz	10	
	T-41D	Train	Santa Cruz	6	
	Cessna 310	Train	Santa Cruz		
	SF.260M	Train	Santa Cruz	6	Basic training
	PC-7 Turbo-Trainer	Train	Santa Cruz	16	
Transportes Aereos Militares					
	CV-580	Trans	La Paz	4	
	CV-440	Trans	La Paz	6	Ex-Aviaco aircraft bought in 1972
	C-130H	Trans	La Paz	3	
	C-47	Trans	La Paz	7	
	C-54	Trans	La Paz	2	
	Arava	Trans	La Paz	5	Six delivered; one lost in March 1976
	Sabreliner 60	VIP	La Paz	1	Presidential use

Below: This Learjet 25B is flown on survey missions by the FAB, being fitted with cameras housed in the bulge beneath the forward fuselage. The Learjet's speed makes it ideal for this role.

CHILE

Area: 290,000 sq. miles (751,000 sq. km.).
Population: 10 million.
Total armed forces: 85,000.
Estimated GNP: $9·8 billion (£4.9 billion).
Defence expenditure: $750 million (£375 million).

Both the Army and Navy have aviation elements, although by far the largest aircraft-operating Service is the *Fuerza Aérea de Chile* with some 11,000 personnel and more than 250 machines. Under the ruling military junta which seized power in 1973, all three Services come under the command of the Ministry of National Defence in the capital, Santiago. After suffering from political differences between Chile and some of her arms suppliers, the FAC began a re-equipment programme in the mid-1970s with the help of the United States and Brazil and currently stands as one of South America's major air arms. Its main roles differ little from those of other Latin American air forces, namely air defence, counter-insurgency, transport, liaison and training. Basic operational unit is the *Grupo* or Group, comprising a single *Escuadrilla* or squadron with a strength varying between 12 and 25 aircraft.

Military aviation in Chile began in 1913 with the establishment of a flying school equipped with Blériot monoplanes, followed two years later with the formation of two squadrons flying a total of ten aircraft. A Naval Aviation Service was founded in 1919 with a few seaplanes and by 1921 Chile boasted a small bomber force of D.H.4s acquired from Britain. The two air services were officially merged in 1930 to form the *Fuerza Aérea de Chile* flying a variety of British, German and American aircraft.

The FAC had six operational Groups by

Above: A-37B Dragonflies are flown by three FAC squadrons on Coin duties.

Above right: Three naval Bell 206s.

1941, when an American Aviation Mission arrived in the country and began re-organising the Service along similar lines to those of the USAAF. Curtiss P-40s and Thunderbolts arrived to replace old Hawk fighters, while Mitchell bombers and Catalina flying boats updated the bombing and reconnaissance elements. US aid continued in the 1950s, 50 Beech Mentor and D18S trainers being supplied together with some B-26 Invader light bombers. Canada supplied Beavers and Otters and, later that decade, Chile's first jets in the form of Vampire trainers. Twenty US-surplus F80Cs arrived in the 1960s and were followed by the first of more than 30 Hawker Hunters. Fifty Chilean-built Chincol primary trainers were delivered to the air force, joining the T-34 Mentors.

Before the overthrow of President Allende's Marxist Government in 1973, Russia offered Chile MiG-21s to replace her ageing equipment. Second-hand A-4 Skyhawks were proposed by the United States, but the final decision came down in favour of an order for F-5Es.

Chile has been a member of the Organisation of American States since its inception in 1948, and a military aid agreement exists with the United States although an arms embargo was announced in October 1976.

Inventory

Just before the 1973 military coup which overthrew the democratically elected Marxist government of President Allende, the *Fuerza Aérea de Chile* had accepted a letter of offer for a batch of Northrop F-5E/F interceptors to update its air-defence capability. Despite the take-over, the $40m order was honoured by the US and deliveries were made in 1976. Less smooth was Chile's effort to obtain spares, particularly Avon engines, for her Hunter fleet. Approaches to Britain met with hostile reaction and an embargo from workers in some of the engine overhaul facilities, while efforts to obtain spares from other sources, notably Jordan and India, met with little success. The Hunters continue in service with the FAC but they were supplemented in the ground-attack role in 1975 when deliveries began of the first of 34 Cessna A-37s ordered under the military aid agreement with the US. The last of these machines arrived in the

ARGENTINA

Area: 1·079 million sq. miles (2·79 million sq. km.).
Population: 23·36 million.
Total armed forces: 132,800.
Estimated GNP: $76·4 billion (£38·2 billion).
Defence expenditure: $1·66 billion (£833 million).

The second-largest South American republic, Argentina has suffered from recurring political instability and rapid inflation which have compounded defence problems in both procurement and operation. The country's armed forces have a manpower strength of some 132,800, plus 250,000 reservists, and all three of the Services have aviation elements. Chief among these is the *Fuerza Aérea Argentina* or Argentine Air Force. Responsible to the Ministry of Defence, with headquarters in Buenos Aires, the FAA has four executive commands: Air Operations Command, responsible for all air operations by FAA aircraft; Air Regions Command, respon-

sible for civil flying activities; Material Command, responsible for all equipment purchasing and distribution; and Personnel Command, responsible for all training and selection. Within Air Opera-

Above: The IA.58 Pucara is built by the FAA's own *Fabrica Militar de Aviones* of the *Area Materiel de Cordoba.* The type is replacing the IA.35 Huanquero on counter-insurgency duties.

country in 1977, replacing Second World War-vintage B-26s and AT-6s.

The territorial layout of Chile – in some places the distance between the sea and neighbouring Argentina is no more than 60 miles, although the country is some 2800 miles long – favours communications by aircraft; therefore of necessity the air force has a large transport group (*Grupo* 10), based at Los Cerrillos near the capital. Equipment includes two Lockheed Hercules, five Douglas C-118s and an assortment of twin-engined types; these are backed up by a helicopter element flying mainly American machines, but including an SAR section with six Aérospatiale Lamas purchased for high-altitude rescue work. Fixed-wing SAR and limited anti-submarine duties are performed by Albatross amphibians of *Grupo* 2, which is headquartered at Quintero and flies from a number of coastal bases down the country. Among the bases used by the FAC are Los Cerrillos, Los Condores, Puerto Montt, Punta Arenas, Temuco and Quintero.

Support

One of the main FAC training bases is the El Bosque flight academy near Santiago, where both primary and jet training is undertaken. Helicopter training is carried out at Temuco on UH-1s and Hillers, while navigation tuition is flown on Beech 99s at Quintero.

Navy

The *Armada de Chile* has a small naval air component flying a modest number of American-supplied aircraft plus some recently arrived Brazilian Bandeirante transports. Main roles are maritime patrol and air-sea rescue, for which the force uses Neptunes, Catalinas and Bandeirantes. For support duties the *Armada de Chile* relies to a large extent on helicopters, of which there are some 24 in use.

Army

The Chilean army air component, the *Ejercito de Chile*, operates in support of the 45,000-man ground force. Main roles are liaison and communications duties, with the added ability of air assault from a unit of Aérospatiale Puma helicopters. Helicopters predominate in the Army inventory and are backed up by a small fixed-wing element supported by half-a-dozen civilian light aircraft put at the disposal of the military.

CHILE

Unit	Type	Role	Base	No	Notes
Air Force					
Grupo 7	F-5E	Int	Antofagasta	15	Replacing Hunters
	F-5F	Train	Antofagasta	3	
Grupo 8	Hunter FGA.71	Att	Antofagasta	20	Fewer than 20 Hunters are
and 9	Hunter T.72	Train	Antofagasta	5	operational
Grupo 1	A-37B	Coin	Iquique	17	
Grupo 12	A-37B	Coin	Punta Arenas	17	
	C-130H	Trans	Los Cerrillos	2	
	DC-6A	Trans	Los Cerrillos	3	
Grupo 10	C-47	Trans	Los Cerrillos	8	
	King Air 100	Photo-survey	Los Cerrillos	1	
	Puma	VIP	Los Cerrillos	1	
	Twin Bonanza	Trans	Punta Arenas	5	
Grupo 6	Twin Otter	Trans/survey	Punta Arenas	2	
	Learjet	Trans/survey	Punta Arenas	2	
Grupo 5	Twin Otter	Trans	Puerto Montt	10	
	UH-1H	Trans/train	Temuco	10	
Grupo 3	UH-12E	Train/SAR	Temuco	2	
	S-55T	Liaison	Temuco	6	
	HU-16B	SAR/Mar pat	Quintero	12	
Grupo 2	T-34	Train	El Bosque	50	Primary trainers
	T-41D	Train	El Bosque	8	
Grupo 11	Beech 99	Train	Quintero	9	Nav trainers
	T-37B	Train	El Bosque	25	
Navy					
	EMB-111	Mar pat		6	
	SP-2E	ASW		4	
	Catalina	SAR		3	
	HU-16B	Mar pat	El Balloto	5	
	C.212 Aviocar	Trans		4	
	EMB-110C	Trans		3	
	Alouette III	SAR			
	Bell 47G	Train/liaison		14	
	Bell 206	Liaison		4	
	UH-1D	SAR		2	
	Navajo	Liaison		1	
	T-34	Train		6	
	Seacat	SAM	Destroyers and frigates	4 launchers	
Army					
	Puma	Assault		9	
	Lama	SAR		6	
	UH-1H	Liaison		3	
	Bell 206	Liaison		2	
	O-1	Obs		4	
	Cherokee Six	Util		2	
	Navajo	Util		4	
	T-25 Universal	Att/train		10	Fitted with weapon racks
	Cessna Hawk XP	Train	Santiago	18	
	C.212 Aviocar	Trans		6	

tions Command are five *Brigadés Aéreas* or Air Brigades; each is composed of three *grupos*, or squadrons, with a nominal establishment of three four-aircraft elements. The Air Force receives much of its equipment from the USA, but also buys from Europe – notably France, the Netherlands and Great Britain. In addition there is an industrial complex at Cordoba known as the *Area de Material Cordoba*, which comes under the control of Material Command and manufactures indigenous designs for transport, training and combat use. Some 17,000 officers and men make up the FAA, while the total national defence budget for 1978 amounted to $1.6bn (£833m), a figure affected considerably by high inflation.

The *Fuerza Aérea Argentina* has its origins in the establishment of a Military Aviation School, equipped with French training aircraft, at El Palomar in September 1912. Following the First World War an Italian Aviation Mission arrived in the country with six Ansaldo fighters, four

Caproni bombers and some support types. A year later, in 1920, a similar mission from France reorganised the Military Aviation Service and phased into use Dewoitine fighters and Breguet reconnaissance aircraft. A Military Aviation Factory was established in 1927 to produce aircraft under licence: initially British and French, and later American and German. A period of limited expansion followed for the Argentinian armed forces, which included the supply of a few DC-2 transports from the USA in 1944, while in the same year the MAS became the *Fuerza Aérea Argentina*. With the end of the Second World War the FAA began to replace its ageing equipment, 100 Fiat G.55 fighters being acquired from Italy, 20 Avro Lancasters and Lincolns from Britain and some C-47 and C-54 transports from the USA. These were followed by the Service's first jet equipment, 100 Gloster Meteor F.4s to equip the combat units, together with batches of Doves, Valettas and Bristol Freighters for the transport ele-

ments. In the late 1950s the country undertook licence production of 90 Beech Mentors and 48 Morane-Saulnier Paris trainers. Further modernisation followed

Above: Pucara can carry a wide range of external ordnance in addition to its built-in cannon and machine guns, including these rockets and 125kg bombs.

in 1960, when 28 F-86F Sabres supplemented the Meteors; A-4 Skyhawks were acquired in 1966.

Argentina is a member party of the Charter of the Organisation of American States drawn up in 1948 and as such became eligible to receive US military aid. She also purchases equipment from other sources. In 1967 Argentina signed the Tlatelolco Treaty, which prohibits nuclear weapons in Latin America. Argentina has no major enemies among neighbouring countries, but the armed forces have conducted anti-guerrilla operations against factions backed by communist movements in the northwest of the state.

Inventory

The combat strength of the *Fuerza Aérea Argentina* currently stands at some 140 aircraft, but the planned modernisation and expansion of the force has been hit by economic problems as well as production difficulties. This latter problem particularly applies to the IA.58 Pucara twin-turboprop light strike aircraft, designed and produced by *Fabrica Militar de Aviones* (FMA) at Cordoba. Flight trials of the Pucara began in 1969 to meet an FAA requirement for up to 100 aircraft, but money had been allocated for only an initial batch of 30 machines up to late-1977, plus appropriations for a further 15. The first unit to receive Pucaras is II *Escuadron de Exploration y Ataque*, based

Above: The Argentine Navy's air arm is one of several Latin American customers for the turboprop-powered Beech T-34C-1 trainer, which is based at Punta de Indio.

at Reconquista; the IA.58 is replacing an older Argentinian-designed type, the IA.35 Huanquero. Together with II EEA, the 1st *Escuadron de Bombardeo* with nine Canberras comes under the control of the Second Air Brigade (II *Brigada Aérea*), headquartered at General Urquiza air base, Entre Rios.

By far the largest combat force is the fleet of McDonnell Douglas A-4 Skyhawks operated by units under the control of the Fourth Air Brigade (IV *Brigada Aérea*). The A-4s, of which some 70 have been delivered, are designated A-4Ps in FAA service and have been modified from ex-US Navy A-4B/Cs; they have been fitted with Ferranti D126R Isis weapon-aiming sights for greater accuracy in the fighter-bomber role. For pure interception duties to defend the capital there is a squadron of Dassault Mirage IIIEs attached to the Seventh Air Brigade (VII *Brigada Aérea*).

When these aircraft were ordered from France in October 1970 they were intended to be the first of an eventual force of nearly 100. After delivery of the initial 14 aircraft, however, plans for a further 80 had to be shelved for economic reasons and instead 26 Israeli-built Atar-engined Mirage III variants, known as Daggers, were acquired in late-1978. Attached to the same Air Brigade is a counter-insurgency unit with 20 Bell and Hughes helicopters armed for the attack role.

There are two distinct transport elements within the FAA: the 1st Air Brigade (I *Brigada Aérea*) at El Palomar air base, Buenos Aires, which is equipped with Lockheed Hercules, more than 20 IA.50 Guarani IIs and some C-47s, plus a Presidential flight; and the state military airline LADE (*Lineas Aereas del Estado*). Founded in September 1940, LADE operates routes in isolated regions with limited traffic which would not sustain commercial operations, and includes a route to the Falkland Islands opened in 1972 in agreement with the British Government. There are a number of second-line units including a coastal patrol and rescue unit with Albatross amphibians and Lama helicopters, a liaison element operating Shrike Commanders and a special Antarctic unit with helicopters and fixed-wing aircraft used for logistic support of Argentina's bases in that area. *Fabrica Militar da Aviones*, having developed the Pucara to production standards, is designing a jet version for the COIN role and has designed a new basic trainer – designated IA.62 – intended to replace the T-34. To replace the training force of Paris jets, the FAA is considering a number of Western jet trainers including the British Aerospace Hawk, the Franco-German Alpha Jet and the Spanish CASA C.101.

Support

The main training organisation is the *Escuela de Aviacion Militar* at Cordoba, where pupils undergo primary and basic tuition on the fleet of Beech Mentors, followed by a course for pilots to wings standard on the Morane Paris. To replace the Mentors, the Argentine Government is considering an order for the FMA IA.62. Also at Cordoba is one of two regional workshops – the other being at Quilmes, Buenos Aires – which undertakes maintenance and overhaul work.

Navy

The *Comando de Aviacion Naval Argentina* has a history almost as long as that of the air force, being formed in 1919 with Italian assistance. A number of flying-boats of European origin were acquired during the 1920s, but by the beginning of the Second World War US landplanes, including some Douglas DB-8 attack-bombers and Corsair reconnaissance aircraft, had been procured. Little expansion occurred until after the war, when small numbers of US training and support aircraft were delivered together with some Catalinas, Mariners and Goose amphibians for long-range patrol and communication duties. Ten Corsair fighter-bombers entered service in 1956, followed in 1957 by some Lockheed Neptunes.

Below: The UH-1H Hueys operated by the FAA's VII *Brigada Aerea* at Moron can be equipped as gunships, although this example is carrying nothing more warlike than a powerful searchlight under its nose.

The aircraft carrier HMS *Warrior* was purchased from Britain in 1958, a year after Brazil obtained a similar vessel from the same source, and the ship was renamed *Independencia*. Grumman Trackers and the Corsairs operated from the carrier, providing the Navy with a wider area of operations and a new prestige. The beginning of the 1960s saw the purchase of 20 surplus Grumman Panthers from the United States, and these remained the backbone of the maritime attack element for some ten years until their retirement. A further light fleet carrier was obtained in 1969 from the Royal Netherlands Navy, being the ex-HMS *Venerable* before passing to Holland and renamed *Karel Doorman*. Arriving at Puerto Belgrano naval base in September, the new ship was named *25 de Mayo* and was a more modern vessel than *Independencia*, having an angled deck, mirror landing aid and other features – elements which helped to decide on the retirement of the older carrier a year later. More recently the ship has had a computer-assisted action information system installed for the control of aircraft away from the carrier.

Inventory

A separate command under Naval Operations, the *Comando de Aviacion Naval*, operates from three main shore bases – Punta de Indio Naval Air Base, Commandante Espora NAB and Ezeiza NAB. Each base accommodates a number of squadrons equipped in the main with US aircraft but also having a small number of types acquired from Europe. Following the scrapping of *Independencia* in 1971, *25 de Mayo* became the flagship of the Argentinian Navy and can execute most of the

Above: The Argentine President's Fokker-VFW F.28 poses over Buenos Aires. The *Dept de Aviones Presidenciales* has single examples of several VIP types ranging from an S-58T to a Boeing 707.

standard carrier roles; these include attack, for which detachments of A-4Q Skyhawks are deployed, and maritime reconnaissance flown by half-a-dozen S-2 Trackers. A total of 16 Skyhawks were acquired, of which some 15 remain in service, and associated with these machines is a detachment of eight Italian MB.326GB jet trainers for continuation training at the Naval Aviation School, Punta de Indio. Shore-based maritime reconnaissance is performed by six Lockheed Neptunes acquired from the US Navy from 1966, replacing a similar number of older aircraft originally procured from the RAF. These ageing machines are due to be replaced in the near future, and

reports have suggested that the Service is contemplating types including the British Aerospace Coastguarder and Fokker F.27 Maritime; any decision will depend largely on future budgets.

The naval transport fleet is composed of some 20 aircraft of seven different types, the most recent purchase being three Lockheed Electras bought from a US airline, refurbished and used for freight/passenger service. Based at Commandante Espora are two units for anti-submarine warfare, search-and-rescue and general communications work, one being equipped with Albatross amphibians and the other being a helicopter unit flying four Sea Kings, nine Alouette IIIs, S-55s and Bell 47Gs. Future helicopter procurement includes three Westland Lynx for ASW operations from newly constructed Type 42 destroyers.

A small coastguard force known as *Prefectura Naval Argentina* is equipped with five Short Skyvans for transport and SAR duties, plus six Hughes 500M helicopters. Main naval bases include Punta de Indio, Commandante Espora, Ezeiza, Rio Gallegos, Rio Grande, Ushiaia, Petrel, Puerto Belgrano and Mar del Plata.

Army

The modest *Commando de Aviacion del Ejercito* or Army Aviation Command became a separate military force late in 1959 and uses both fixed-wing aircraft and helicopters. Main roles are observation, liaison and transport, with aircraft detached to each Army Corps. Main base is Campo de Mayo near Buenos Aires, which is also the base for the *Gendarmeria Nacional* which operates border patrols with a variety of aircraft including Cessna 310s

ARGENTINA Air Force

Wing	Unit	Type	Role	Base	No	Notes
II Brigada Aerea	I Esc de Bombardeo	Canberra B.62	Bomb	General Urquiza	9	
		Canberra T.64	Train	General Urquiza	2	
	II Esc del Exploration y Ataque	Huanquero	Armed train	Reconquista	25	Being replaced by Pucaras
		Pucara	Coin	Reconquista	45	Total of 100 required by two squadrons
	I Esc Fotografico	Huanquero	Recce	Reconquista	20	
IV Brigada Aerea	I Esc de Caza Bombardeo	A-4P	Att	El Plumerillo	70	Total includes aircraft in IV and V Esc de Caza Bombardeo of V Brigada Aerea based at General Pringles AB
	II and III Esc de Caza Bombardeo	MS.760	Armed train	El Plumerillo	44	Total includes aircraft used by the Training School at Cordoba; a replacement for these is under consideration
		IAI Dagger	Int	El Plumerillo	26	Israeli-built Mirage III
VII Brigada Aerea	I Esc de Caza	Mirage IIIEA	Int	Moron	12	Delivered '72; further 80 required but order unlikely. Seven more Mirages ordered in 1977
		Mirage IIIDA	Combat/train	Moron	2	
	I Esc de Exploration y Ataque	UH-1H	Coin	Moron	6	Fitted as gunships
		Hughes 500M	Coin	Moron	14	
I Brigada Aerea	I Esc de Transporte	C-130E/H	Trans	El Palomar	3/4	A fifth C-130H was destroyed in '75
		Guarani II	Trans	El Palomar	22	
	Dept de Aviones Presidenciales	Boeing 707-320B	VIP	El Palomar	1	
		F.28 Mk 1000C	VIP	El Palomar	1	
		HS.748 Srs 2	VIP	El Palomar	1	
		Sabreliner	VIP	El Palomar	1	
		C-47	Trans	El Palomar	7	
		S-58T	Trans/VIP	El Palomar	2	
Lineas Aereas del Estado (LADE)		F.28 Mk 1000C	Trans	El Palomar	5	
		F.27 Mk 400/600	Trans	El Palomar	10	
		DC-6	Trans	El Palomar	3	
		Twin Otter	Trans	El Palomar	7	
		Shrike Commander 500U	Liaison		14	
	Esc de Busqueda y Salvamento	HU-16B	SAR		3	
		Lama	SAR		6	Used for high-altitude work
		Merlin IVA	Casevac		2	
Servicios Aereas de Estado Nacional (SADEN)		S-61R	Trans	Rio Gallegos	1	Also uses Marambio in Antarctica
		Beaver	Liaison	Rio Gallegos	3	Possibly only one in use
		Otter	Liaison	Rio Gallegos	3	
	I Esc Antartico	C-47	Trans	Rio Gallegos	1	
		S-61NR	Trans	Moron	2	
		UH-1D	Liaison		4	
		Bell 47G	Train		4	
		Bell 212	Liaison		8	Ordered in 1978
		UH-19	Liaison		6	
	Escuela de Aviacion Militar	T-34A	Train	Cordoba	35	Total of 90 received; 74 built by FMA

ARGENTINA Army

Type	Role	No	Notes
G.222	Trans	3	Replaced C-47s
King Air	Liaison	1	
Queen Air	Liaison	1	
T-41D	Train	5	
Cessna 207	Tactical obs	5	
Cessna 182J	Liaison	2+	Assembled by FMA
Twin Otter	Liaison	2	Total of three delivered
Navajo	Liaison	4	Total of five delivered
Turbo-Commander	Liaison	5	Bought from batch assembled in Argentina
Sabreliner 75	VIP	1	
Merlin IIIA	Liaison	4	
Bell 47G	Obs	2+	
Bell 212	Trans	2	
UH-1H	Trans	20	
JetRanger	Liaison	7	
FH-1100	Liaison	7	
Citation	Survey	1	Fitted with Wild cameras
Puma	Trans	9–12	

All based at Campo de Mayo

Above: The Argentine Navy's Twin Otter is fitted with skis for operations in the Antarctic.

and Piper Navajos. Army aircraft are maintained by the *Fuerza Aérea Argentina*'s VII *Brigada Aérea* at Moron. Virtually all the aircraft flown by the CAE are of American origin, although the Service took delivery in March 1977 of the first of three Aeritalia G.222 STOL transports from Italy; four Swearingen Merlins have also been received recently for liaison work. The sole jet in use is a Rockwell Sabreliner for VIP duties. As well as Army support work, the Service also conducts limited training for which a handful of Cessna T-41Ds are used.

ARGENTINA Navy

Unit	Type	Role	Base	No	Notes
I Esc de Ataque	A-4Q	FGA	Commandante Espora	15	Detachment deployed aboard carrier *25 de Mayo*
ASW *Esc*	S-2A	ASW	Commandante Espora	6	Detachment deployed aboard *25 de Mayo*. Part of Naval Air Wing No 2
	S-2E	ASW	Commandante Espora	4	Ex-US Navy delivered in 1978
Exploration *Esc*	SP-2H	Mar pat	Commandante Espora	10	Part of Naval Air Wing No 2
	HU-16B	SAR	Commandante Espora	3	Part of Naval Air Wing No 4
	Catalina	SAR	Commandante Espora	3	Possibly withdrawn from use
	C-54/DC-4	Trans	Ezeiza	3/2	Operated by
	L-188 Electra	Trans	Ezeiza	3	Naval Air Wing No 5
	C-47	Trans	Ezeiza	8	
	Guarani II	Trans	Ezeiza	1	
SAR *Esc*	HS.125-400	Calibration	Ezeiza	1	
	Twin Otter	Liaison	Punta de Indio	1	Used by *Grupo Aeronaval Antarctico*
	Beaver	Liaison	Commandante Espora	3	
	Queen Air	Liaison	Commandante Espora	4	
	Super King Air 200	Liaison	Ezeiza	2	
	Alouette III	Liaison	Commandante Espora	9	Used by Naval Air Wing No 4
	Bell 47G	Liaison	Commandante Espora	3	Used by Naval Air Wing No 4
	S-55	Liaison	Commandante Espora	5	Used by Naval Air Wing No 4
	S-61D	ASW	Commandante Espora	4	Used by Naval Air Wing No 4
	Lynx HAS.23	ASW		3	On order for ship use
	MB.326GB	Armed train	Punta de Indio	8	Used for cont training for A-4Q aircrew
Naval Av School	T-28	Train	Punta de Indio	28	Total of 45 Fennecs bought from France in 1966, modified for carrier use
	T-34C	Train	Punta de Indio	16	Possibly replacing T-28
	AT-6	Train	Punta de Indio	12	
	Seacat	SAM	Cruiser	1	
	Sea Dart	SAM	Type 42 destroyer	2	
Prefectura Naval Argentina	Hughes 500M	Liaison		6	
	Skyvan 3M	Trans		5	

URUGUAY

Area: 72,200 sq. miles (187,000 sq. km.).
Population: 2·76 million.
Total armed forces: 27,000.
Estimated GNP: $3·6 billion (£1·8 billion).
Defence expenditure: $72 million (£36 million).

Uruguay's air force, the *Fuerza Aérea Uruguaya*, operates a token combat element and has undergone little expansion over the past few years. It is responsible to the Ministry of National Defence and has its headquarters in Montevideo. With a total manpower of some 3000, the FAU is divided into two Brigades and three training schools: flying, technical and officer. In addition there is a naval air component, the *Aviacion Naval*, which has an ASW unit and an associated training and support element.

Military aviation in Uruguay was established in 1916 with the formation of the

Escuela Militar de Aéronautica equipped with two-seat Castaibert monoplanes. Under French and Italian influence a gradual expansion of the air arm, known as the *Aéronautica Militar*, took place through the 1920s. By 1931 Uruguay could boast a small naval air component (initially formed in 1925) with Italian Cant seaplanes and an air force with more than 50 aircraft. United States aid was forthcoming during the Second World War in return for the use of Uruguyan bases, and that aid continued on a somewhat reduced basis through the 1950s and 1960s. Replacing Mustang fighters and Mitchell bombers supplied in 1950, a force of Lockheed Shooting Stars arrived for FAU use in 1956 and 1960. On December 4, 1953, the air force was renamed *Fuerza Aérea Uruguaya*.

The country's naval air arm was expanded during the mid-1940s with the help of American aircraft, and by 1952 its inventory included Grumman Avengers

Above: No South American air arm is complete without a force of A-37B Dragonfly light attack aircraft, it seems. This is one of eight operated by the FAU.

and Hellcats, Vought Kingfisher floatplanes and North American SNJ-4 trainers. The force is now known as the *Aviacion Naval*.

PARAGUAY

Area: 157,000 sq. miles (406,000 sq. km.).
Population: 2·5 million.
Total armed forces: 17,000.
Estimated GNP: $2 billion (£1 billion).
Defence expenditure: $41 million (£20·5 million).

This small land-locked South American republic, bordered by Brazil, Bolivia and Argentina, has a relatively modest air arm, the *Fuerza Aérea del Paraguay*. It is tasked mainly with transport, internal policing and training duties and operates about 80 aircraft. The Ministry of Defence in Asuncion houses FAP headquarters, while the main air base is Campo Grande situated near the capital. Manpower totals some 2500 and the defence budget for 1978 amounted to 5190m Guaranies ($41m). Paraguay's small navy has a helicopter component, as does the Paraguayan army.

The country's air force was formed in the early 1930s as part of the army during the Paraguayan-Bolivian war. An Italian air mission helped to organise the handful of Potez XXV biplanes acquired and mercenary pilots were employed. In 1935 the *Fuerzas Aéreas Nacionales* was established as the country's air force; it was equipped with a number of Italian and American aircraft, the latter supplied in the 1940s.

Paraguay is a signatory of the Charter of the Organisation of American States and also the Treaty for the Prohibition of Nuclear Weapons in Latin America, signed in February 1967. A military assistance agreement exists with the United States.

Above: The Paraguyan Navy flies a pair of 1976-model Cessna 150Ms in the training role.

PARAGUAY

Type	Role	Base	No	Remarks
T-37B	Coin		6	
DC-6B	Trans		5	Ex-Varig Airlines, presented by Brazil
C-54	Trans		2	
Otter	Trans		1	
Twin Otter	Trans		1	
C-47/DC-3	Trans	Campo Grande, Asuncion	10	Operated by TAM for domestic flights
H-13	Liaison		14	⎤ For army use?
Hiller 12E	Liaison		3	⎦
T-23 Uirapuru	Train	Campo Grande, Asuncion	8	Replaced Boeing-Stearman PT-17As
Cessna 185	Liaison		5	
MS.870 Paris	Train		1	⎤ Ex-Brazilian
Fokker S-11	Train		8	⎦
Bell 47G	Liaison		2	⎤ Operated by Navy
Cessna 150M	Train		2	⎦

Inventory

Paraguay operates a variety of second-line aircraft procured over the past 20 years, the majority from Brazil and the US. For counter-insurgency and policing duties the FAP operates six A-37Bs, replacing a squadron of armed Texan trainers acquired from Brazil in 1960. Transport duties form the main task, and the force has a number of ex-Varig DC-6Bs in use together with some C-54s and C-47s. Economic reasons forced the cancellation of a $7m order for six Israeli Arava light transports early in 1977 but this is likely to be only a temporary measure. *Transporte Aéreo Militar* (TAM) is a branch of the air force and flies domestic passenger and cargo services with DC-3s.

Uruguay signed the Charter of the Organisation of American States in 1948 and the Treaty for the Prohibition of Nuclear Weapons in Latin America (the Tlatelolco Treaty) in 1967. A military aid agreement exists with the USA.

Inventory

Although the small combat force operated by the FAU has remained the same for some years, modern equipment has been phased into the transport element in the form of five EMB-110 Bandeirante light transports from Brazil for combined civil and military use, joining five F.27/FH-227s. Uruguay's economic situation has prevented the modernisation of the FAU's front-line force, but a squadron of Cessna A-37Bs was received in 1976. The FAU operates from bases including Carrasco near Montevideo, Melo, Durazno, Paysandu and Salto.

Aviacion Naval

The main base of the small naval air arm, Laguna del Sauce, houses three Grumman S-2A Trackers acquired for anti-submarine duties in 1965 plus helicopters and fixed-wing aircraft for training and support.

Basic flying training is carried out on T-6s and Beech AT-11s, jet conversion taking place on AT-33s at Carrasco. Technical trade training is based at Capitan Boiso Lanza near Montevideo.

URUGUAY

Brigade	Unit	Type	Role	Base	No	Notes
Air Force						
	⎡ Grupo 1	AT-33A	FGA	Carrasco	6	
		⎡ C-47	Trans		12	
Brigada 1	Grupo 3+4	F.27 Mk 100	Trans		2	
		FH-227	Trans		3	
		⎣ Queen Air 80	Liaison		2	
	Grupo 5	A-37B	Coin		8	Delivered 1976
	⎣ Grupo 6	Bandeirante	Trans		5	
Brigada 2	Grupo 1	T-6G Texan	Train	Durazno	10	
		U-17 Skywagon	Liaison		6	
		Beech AT-11	Train		10	
		UH-1H/B	Util		2/4	
		UH-12	SAR		2	
		T-41	Train		6	
		T-6 Texan	Train			
		C-45	Train		2	
Navy		S-2A Tracker	ASW	Laguna del Sauce	3	
		SH-34C		Laguna del Sauce	2	
		Bell 47G		Laguna del Sauce	2	
		SNJ-4		Laguna del Sauce	3	
		T-34B	Train	Laguna del Sauce	1	
		T-6 Texan		Laguna del Sauce	4	
		SNB-5		Laguna del Sauce	3	
		PA-12		Laguna del Sauce	2	

Western Europe

Western Europe contains a broad spectrum of states ranging from Iceland, which is a member of NATO but has no armed forces, through countries such as Sweden and Switzerland which maintain a position of heavily armed neutrality but are not part of the Atlantic alliance, to the major European powers such as Germany and Britain. France is no longer a member of NATO's military panel, yet she retains close ties with the alliance in some areas, such as early warning. The French arms industry also participates in collaborative development programmes involving NATO countries.

The main potential danger as far as most of Western Europe is concerned lies to the east, where the Warsaw Pact can deploy an awesomely powerful war machine. Other tensions can be more immediate, however, and in 1974 NATO was treated to the embarrassing spectacle of two of its members — Greece and Turkey — fighting each other over control of Cyprus. France, Holland and Britain retain a limited military presence overseas in former colonies and areas where long-standing ties remain. Portugal has shed most of her foreign possessions, though, and Western Europe is increasingly bound up in affairs near home rather than attempting to police large parts of the world.

Finland buys its weapons from both East and West, including Sweden, and Denmark has also purchased combat aircraft from its Scandinavian neighbour. The Swedes are finding it increasingly difficult to finance development of new airborne weapon systems unilaterally, however, and the doors to sales to NATO members are effectively barred.

Universal standardisation of equipment by NATO members seems as far away as ever, the problem being compounded by pinched defence budgets in Europe and the loss of potentially lucrative markets in Iran and elsewhere. The military activities of the continent will remain indivisibly bound up with those of the United States for the foreseeable future — unless the US does a sharp about-turn in her foreign policy — despite attempts to organise Western Europe into a cohesive economic and military bloc which can speak with one voice.

Although the wide variety of arms developed by Western European states does result in wasteful duplication of scarce resources, this is to some extent offset by the continent's ability to supply equipment which meets the needs of almost any given overseas customer. The French in particular are aggressive salesmen, putting their country in third position in the world league of arms exports, while Britain occupies fourth place.

IRELAND

Area: 26,600 sq. miles (68,900 sq. km.).
Population: 3 million.
Total armed forces: 14,500.
Estimated GNP: $9·2 billion (£4·6 billion).
Defence expenditure: $204 million (£102 million).

The Irish Air Corps was established as part of the army following independence from Britain in 1921 and has recently undergone a modest expansion. Three Alouette IIIs joined the corps between November 1963 and February 1964, becoming the Service's first rotary-wing aircraft, and this fleet has since been expanded to eight. The Alouettes are used on a wide variety of paramiliary and civil roles in addition to providing support for the army. The Aérospatiale Puma and Bell 212 have both been evaluated as troop transports and long-range search-and-rescue aircraft, but no order had been placed by the end of 1978. A wide variety of maritime-patrol and fishery-protection aircraft were also examined in 1977 but resulted in no purchases, a Beech King Air 200 entering service in April of that year on a three-year lease as an interim solution.

Eight Reims-Cessna FR172H Rockets were bought in 1972 to equip an army co-operation squadron. These aircraft can carry 37mm air-to-ground rockets in wing-mounted pods, and the six secondhand Aérospatiale CM 170-2 Magisters acquired to replace Vampires may also be armed, although their main role is advanced training. Also used mainly for training but with a secondary light attack role are ten SIAI-Marchetti SF.260W Warriors which have replaced Chipmunks and Provosts.

IRELAND				
Type	Role	Base	No	Notes
Magister	Train/light att	Casement (Baldonnel)	6	Replaced Vampires
SF.260WE	Train/light att	Casement (Baldonnel)	10	Replaced Chipmunk and Provost
Dove	Survey	Casement (Baldonnel)	3	
Alouette III	Army co-op/util	Casement (Baldonnel)	8	
FR172H Rocket	Army co-op/util	Gormanston	8	
Super King Air 200	Mar pat	Casement (Baldonnel)	2	

Right: As part of its updating process the Irish Air Corps has bought second-hand Magisters to replace ageing Vampires in the training and light attack roles.

UNITED KINGDOM

Area: 89,000 sq. miles (230,600 sq. km.).
Population: 54 million.
Total armed forces: 313,250.
Estimated GNP: $263·6 billion (£131·8 billion).
Defence expenditure: $13·84 billion (£6·92 billion).

The Royal Air Force, formed on April 1, 1918, by the amalgamation of the Royal Flying Corps and Royal Naval Air Service, is the world's oldest independent air arm. The Fleet Air Arm, Army Air Corps and Marines have their own aircraft, but the RAF is by far the largest contributor to British military aviation. A series of role changes since the mid-1960s has had dramatic effects on the structure and organisation of British military power. The nuclear deterrent, formerly based on Vulcan bombers dropping free-fall nuclear weapons and the Blue Steel stand-off missile, is now the responsibility of the Royal Navy's four nuclear-powered submarines carrying 16 Polaris A3 missiles each. Almost all RAF units based in the Far East and Arabian Gulf were withdrawn by the end of 1971, following the British Government's decision to abandon a permanent presence east of Suez, and the contraction is continuing with the relinquishing of bases in Cyprus and Malta (the latter was evacuated in October 1978).

The modern RAF is thus a European force tasked with supporting army operations, engaging hostile aircraft and attacking shipping and land targets such as marshalling areas and rear forces. Royal Air Force units in Britain are grouped into only two Commands – Strike and Support – following the latter's absorption of Training Command in mid-1977. A small number of helicopters are based in Cyprus and Hong Kong, but the only other RAF units based permanently overseas are those of RAF Germany.

Strike Command, with headquarters outside High Wycombe, comprises the former Bomber Command, now No 1 (Bomber) Group; 11 (Fighter) Group,

previously Fighter Command; 18 (Maritime) Group, ex-Coastal Command; and 38 Group, formerly part of Transport Command and Air Support Command. Support Command is run from a new headquarters at Brampton, the previous HQ at Andover being closed following the take-over of Training Command. RAF Germany has its headquarters at Rheindahlen, near Moenchengladbach.

Virtually all RAF combat and support aircraft are now assigned to NATO, following the creation of United Kingdom Air Forces – a NATO command – with the Air Officer Commanding-in-Chief Strike Command as its C-in-C. The exceptions are the tanker and airborne early-warning squadrons, which are retained under national command even though they are primarily engaged in NATO tasks in the United Kingdom Air Defence Region (UK ADR) and the Allied Command Atlantic area; and the squadron of Wessexes based in Hong Kong. Some of the Canberras and Nimrods operating from Malta were attached to the Central Treaty Organisation (CENTO) until their withdrawal in 1978. RAF personnel strength is just under 85,000.

Strike Command

Strike Command forces stationed in the United Kingdom are committed to the support of all three major NATO commanders: SACEUR (Europe), SACLANT (Atlantic) and CINCHAN (English Channel). The command's forces, operating under the direct control of four group headquarters, provide air defence for the United Kingdom together with air support for land and sea operations in the main NATO command regions; this latter task includes a range of deployment options throughout Allied Command Europe as part of SACEUR's Strategic Reserve (Air).

No 1 (Bomber) Group operates Vulcan heavy bombers, Victor tankers, Buccaneer heavy attack aircraft and several types – Nimrods, Canberras and Vulcans

Below: An RAF technician works on the twin Rolls-Royce/Turboméca Adour turbofans which power the British Aerospace/Dassault-Breguet Jaguar attack aircraft, developed jointly by Britain and France.

Bottom: A Jaguar of RAF Germany taxis out from its hardened shelter at Brüggen.

– for reconnaissance, electronic counter-measures (ECM), gathering of electronic intelligence (ELINT) and training.

The Vulcan B.2 bombers relinquished the role of nuclear deterrence to the Royal Navy's Polaris force in mid-1969 but still carry free-fall nuclear bombs for low-level strike, with conventional bombing as a secondary task. In the latter role the Vulcan can carry up to 21 1000lb para-chute-retarded bombs. The Vulcan force will convert to IDS (interdiction/strike) Tornados in the early 1980s. The Vulcan is also operated in the strategic recon-naissance role, this type having replaced Victor SR.2s. The Vulcan SR.2s fly high-level radar reconnaissance missions of up to 6hr and are a prime source of target information for No 1 Group's maritime-attack Buccaneers. The single squadron (No 39) of Canberra PR.9s and PR.7s em-ployed primarily on high-altitude recce was joined by aircraft from No 13 in October 1978, that unit having been with-drawn from Malta.

The RAF's tanker force now comprises two squadrons of Victor K.2s backed up by an operational conversion unit; all the earlier K.1s have been retired and the fleet has been reduced by one squadron. This force is to be augmented by ten con-verted VC20s: nine ex-airline, and one from RAE Bedford. No 1 Group also has the same number of units – two squadrons and an OCU – operating Buccaneer S.2s in the attack role. One of these, 12 Sqn, specialises in maritime operations and this part of the force was doubled in 1979 with the transfer of aircraft formerly used by the Royal Navy aboard HMS *Ark Royal*. The anti-ship Buccaneers are assigned a variety of overwater tasks, of which the most formidable is a concerted attack by up to eight aircraft simultaneously on a Soviet Navy surface action group, con-sisting of a guided-missile cruiser pro-tected by escorting vessels.

The Buccaneers normally carry either the British Aerospace/Matra Martel air-to-surface missile or 1000lb bombs, the latter being tossed at the target from a range of several miles; aircraft equipped to operate the missile are designated S.2B, the others in RAF service being S.2As (the Royal Navy equivalents were S.2C and S.2D). The S.2B can carry either the television-guided AJ168 Martel or the anti-radiation AS.37 variant; maximum load is three and four rounds respectively, the former version also requiring a pylon-mounted video/command pod. The tele-vision-guided Martels are to be supple-mented and replaced in the early 1980s by a turbojet-powered Martel development carrying a new active radar seeker de-signed by Marconi Space and Defence Systems. This weapon, known as P3T, was selected in mid:1977 in preference to the US Harpoon and will allow attacks on shipping to be launched from greater ranges than were possible previously.

The Buccaneers can also deliver nuclear weapons and may use 2in rockets against relatively soft targets. Lepus flares pro-vide illumination at night. The anti-ship Buccaneers are expected to be the last type to be replaced by Tornados, in the mid-1980s. The other No 1 Group Bucca-neers are assigned the overland strike/attack role, and all aircraft may also be called on to fly reconnaissance missions. Some of the Buccaneers are being fitted with Pave Spike designator pods, allowing them to deliver laser-guided weapons. This combination will also be carried by Tor-nado.

The tri-national Panavia Tornado, which will equip units in RAF Germany and Strike Command's 11 Group as well as replacing 1 Group's Vulcans, Canberras and Buccaneers, is planned to enter ser-vice progressively throughout the early and mid-1980s. First deliveries to the RAF were due to be made in mid-1979. Of the 385 required by the RAF, some 220 will be the IDS version – which is the initial variant – and the remainder will be ADV interceptors (see below). The Tornado, formerly known as MRCA (Multi-Role Combat Aircraft), will assume the roles of its predecessors and will be equipped with very much the same types of weapon. Those superseding Vulcans will be armed with nuclear and conventional free-fall bombs, as will the aircraft which take over from overland Buccaneers, and the anti-ship version will – like its forebear – be armed with P3T and bombs. A new recon-naissance pod is being developed for the variable-geometry type.

Strike Command's 11 Group, head-quartered at Bentley Priory, is charged with defending the UK ADR and patrol-ling NATO Early Warning Area 12. Between the retirement of HMS *Ark Royal* and the introduction of the Sea Harrier, 11 Group's Lightnings and more especially Phantoms are the only means of providing British fixed-wing air cover for the Royal

Above: The RAF's main combat aircraft in the 1980s will be the Panavia Tornado, which will perform the roles of both fighter and bomber.
Below: The British Aerospace Hawk, seen here threading its way through Welsh valleys near its home base on Anglesey, is replacing both Gnats and Hunters in the RAF's training fleet.

Navy and other NATO seaborne forces. The group's single squadron of Phantom FG.1 interceptors has been joined by four (plus an OCU) of FGR.2s following the introduction of Jaguars in the attack/ strike roles. The FGR.2s have largely replaced the Lightning force, although two squadrons are expected to remain operational until about 1984. These surviving F.3s and F.6s – which are armed with Firestreak and Red Top infra-red missiles and Adem 30mm gun packs – will be the first to be superseded by the ADV (Air Defence Variant) of Tornado, with the Phantoms following them into retirement.

The FGR.2s have been refurbished to extend their lives after spending several years operating at very low levels, and from the end of 1977 they began to carry the British Aerospace Sky Flash medium-range air-to-air missile in place of AIM-7E2 Sparrows. Sky Flash is based on the US Sparrow but carries a new Marconi Space and Defence Systems semi-active radar seeker which has greater resistance to countermeasures and which allows the weapon to snap down from a launching at medium altitudes to engage low-flying intruders in ground clutter. An EMI proximity fuze replaces the US unit in Sparrow. The FGR.2s can also be fitted with pod-mounted M61 Vulcan 20mm cannon for close-in work.

The Tornado ADV will also carry Sky Flash, illuminating targets with its MSDS Foxhunter long-range pulse-Doppler radar. For short-range engagements the aircraft will be armed with the AIM-9L variant of Sidewinder, to be built under licence by a European consortium, with the built-in Mauser 27mm cannon as a back-up. The ADV's fuselage is longer than that of the IDS version, allowing more fuel to be carried for long-endurance

standing air patrols with the wings swept forward. This should reduce the RAF's dependence on the tanker fleet on air-defence duties.

Since mid-1976 the force of manned interceptors in Britain has been joined by 85 Sqn's Bloodhound Mk 2 medium-range surface-to-air missiles. The original Bloodhound 1, of which 12 squadrons were operational at one time, entered RAF service in the late 1950s to protect the strategic-bomber bases in eastern Britain. The Mk 2, with greater resistance to ECM, an improved warhead and new ground-based radars, replaced the Mk 1 from 1964. The Type 87 target-illuminating radar used by 85 Sqn has a maximum range of some 250nm, and Bloodhound can engage targets at any altitude from 100ft or so up to more than 60,000ft. The missiles based in Germany are to be redeployed in Britain, allowing Bloodhound coverage to be extended up the east coast and providing a total of seven sites.

No 11 Group's interceptors – manned and unmanned – are operated in conjunction with the United Kingdom Air Defence Ground Environment (UKADGE) which is connected, via the Air Defence Data Centre at West Drayton, with six NADGE stations in Europe. Air-defence data received by digital link from NADGE are fed to Strike Command's Air Defence Operations Centre (ADOC) at High Wycombe and to the three Sector Operations Centres (SOCs) – at Buchan, Boulmer and Neatishead – which have replaced the old Master Radar Stations. Information is also available from the BMEWS (Ballistic Missile Early Warning System) site at Fylingdales.

The SOCs additionally receive inputs from airborne (see below) and surface-based radars in the United Kingdom. A wide variety of purely military and joint

military/civil radars are used, including the Decca Type 80 and Marconi Types 84, 85, 88 and 89. The Types 84 and 85, developed for use with the Linesman system intended to detect attacks and initiate a massive response under the tripwire philosophy of the 1960s, are large three-dimensional radars. The Types 88 and 89 were both developed for use with the British Army's Thunderbird surface-to-air missile, the former having back--to-back aerials and the latter being the height-finding element. The Plessy HF200 is also used for height-finding, and the Sperry TPS-34 and Marconi S259 are employed as back-up search radars. A number of Marconi Martello three-dimensional transportable long-range surveillance radars were ordered in early 1979 to replace earlier equipment.

The need to detect attacks from all directions has led to half a dozen air-traffic control radar sites being brought into the air-defence system. Radars at these sites which can feed data to the SOCs include the Plessey AR-5D and Cossor's SSR700 and SSR750. The UKADGE operates on a ring-main arrangement so that the chain of command and control is not broken if an individual link is severed. Information from the SOCs is fed by voice or video links to Strike Command headquarters, individual bases, airborne fighters and other aircraft, and to locations such as Bloodhound target-illuminating radars.

The SOC at Buchan, responsible for the northern sector, takes inputs from the High Power Reporting Point radar at Saxa Vord in the Faroes and from another radar at the Benbecula control and reporting post (CRP) on Benbecula in the Hebrides. Buchan handles NATO's Early Warning Region 12 and reports to the SHAPE operations centre. The Boulmer SOC is fed by CRPs at Bishop's Court

Left: The RAF's standard primary trainer is the British Aerospace Bulldog, which has taken over from the faithful Chipmunk.

Below right: A Victor tanker refuels a pair of Lightning interceptors, a further two squadrons of which are to be formed to bolster the RAF's defences. The bulk of the RAF's interceptor force now comprises Phantoms.

UNITED KINGDOM

Command	Group	Unit	Type	Role	Base	No	Notes
Air Force							
	1 (Bomber) Group	9 Sqn	Vulcan B.2	Bomb	Waddington		
		35 Sqn	Vulcan B.2	Bomb	Scampton		
		44 Sqn	Vulcan B.2	Bomb	Waddington		
		50 Sqn	Vulcan B.2	Bomb	Waddington	85	
		101 Sqn	Vulcan B.2	Bomb	Waddington		
		617 Sqn	Vulcan B.2	Bpmb	Scampton		
		230 OCU	Vulcan B.2	Train/bomb	Scampton		
		55 Sqn	Victor K.2	Tank	Marham		
		57 Sqn	Victor K.2	Tank	Marham	24	
		232 OCU	Victor K.2	Train/tank	Marham		
		27 Sqn	Vulcan SR.2	Strat recce	Scampton	4	
		39 Sqn	Canberra PR.9/PR.7	High-alt recce	Wyton	17	
		13 Sqn	Canberra PR.7	Recce	Wyton	14	
		51 Sqn	Nimrod R.1	Electronic surveillance	Wyton	3	
			Canberra B.6	Electronic surveillance	Wyton	4	
		360 Sqn	Canberra T.17/T.4	ECM	Wyton	19/1	Joint-Service unit
		231 OCU	Canberra B.2/T.4	Train	Marham		
		100 Sqn	Canberra B.2/T.4/ T.19/E.15	Target facilities Target facilities	Marham Marham	11/2 2/3	
		7 Sqn	Canberra TT.18/ T.4/T.19	Target facilities Target facilities	St Mawgan St Mawgan	11/2/2	
		12 Sqn	Buccaneer S.2	Mar att	Honington	45+16	Being augmented by ex-RN aircraft
		216 Sqn	Buccaneer S.2	Mar att	Honington		
		208 Sqn	Buccaneer S.2	Att	Honington		
		237 OCU	Buccaneer S.2	Train/att	Honington		
		115 Sqn	Andover E.3	Calib	Brize Norton	6	
	11 (Fighter) Group	5 Sqn	Lightning F.3/F.6/T.5	Int	Binbrook	nearly 40	
		11 Sqn	Lightning F.3/F.6/T.5	Int	Binbrook		
		23 Sqn	Phantom FGR.2	Int	Wattisham		
		29 Sqn	Phantom FGR.2	Int	Coningsby		
		56 Sqn	Phantom FGR.2	Int	Wattisham	80 approx	
		111 Sqn	Phantom FGR.2	Int	Leuchars		
		228 OCU (64 Sqn)	Phantom FGR.2	Train/Int	Coningsby		
		43 Sqn	Phantom FG.1	Int	Leuchars	17	
		8 Sqn	Shackleton AEW.2	AEW	Lossiemouth	12	To be replaced by AEW Nimrod
		85 Sqn	Bloodhound Mk 2	SAM	West Raynham (A flt) North Coates (B flt) Bawdsey (C flt) To be decided (D flt)	9 sections	Three other flights to use ex-RAFG missiles
	18 (Maritime) Group	42 Sqn	Nimrod MR.1/MR.2	ASW/Mar pat	St Mawgan		
		120 Sqn	Nimrod MR.1/MR.2	ASW/Mar pat	Kinloss		
		201 Sqn	Nimrod MR.1/MR.2	ASW/Mar pat	Kinloss	35	
		206 Sqn	Nimrod MR.1/MR.2	ASW/Mar pat	Kinloss		
		236 OCU	Nimrod MR.1/MR.2	Train/ASM/Mar	St Mawgan		
		22 Sqn	Whirlwind HAR.10	SAR	Chivenor (A flt)		
			Wessex HAR.2	SAR	Leuchars (B flt)		
			Wessex HAR.2	SAR	Valley (C flt)		
			Whirlwind HAR.10	SAR	Brawdy (D flt)		Being replaced by Sea King HAR.3
			Wessex HAR.2	SAR	Manston (E flt)		
		202 Sqn	Whirlwind HAR.10	SAR	Boulmer (A flt)	2	Being replaced by Sea King HAR.3
			Whirlwind HAR.10	SAR	Leconfield (B flt)	2	Being replaced by Sea King HAR.3 (to be based at Brawdy
			Whirlwind HAR.10	SAR	Coltishall (C flt)	2	Being replaced by Sea King HAR.3
			Whirlwind HAR.10	SAR	Lossiemouth (D flt)	2	
	38 Group	1 Sqn	Harrier GR.3/T.2/T.4	Army support	Wittering	30 approx	
		233 OCU	Harrier GR.3/T.2/T.4	Train/support	Wittering		
		6 Sqn	Jaguar GR.1/T.2	Army support	Coltishall		
		41 Sqn	Jaguar GR.1/T.2	Recce	Coltishall	45/3	
		54 Sqn	Jaguar GR.1/T.2	Army support	Coltishall		
		72 Sqn	Wessex HC.2	Assault/log	Odiham	20	
		33 Sqn	Puma HC.1	Assault/log	Odiham		
		320 Sqn	Puma HC.1	Assault/log	Odiham	36	
		240 OCU	Wessex HC.2/HAS.1/ Puma HC.1	Train/assault	Odiham		
		10 Sqn	VC10		Brize Norton	11	
		24 Sqn	Hercules C.1	Trans	Lyneham		
		30 Sqn	Hercules C.1	Trans	Lyneham		
		47 Sqn	Hercules C.1	Trans	Lyneham	45	30 to be lengthened
		70 Sqn	Hercules C.1	Trans	Lyneham		
		242 OCU	Hercules C.1	Train/trans	Lyneham		
		32 Sqn	HS.125–400 CC.1	Comms/light trans	Northolt	4	
			HS.125–600 CC.2	Comms/light trans	Northolt	2	
			Whirlwind HCC.12	Comms/light trans	Northolt	4	
			Andover CC2	Comms/light trans	Northolt	3	
			Gazelle HCC.4	Comms/light trans	Northolt	1	
		207 Sqn	Devon CC.2/2	Comms/light trans	Northolt (with flight at Wyton)	13	
		Queen's Flight	Andover CC.2	VIP	Benson	3	
			Wessex HCC.4	VIP	Benson	2	
		63 Sqn*	Hunter F.6/F.6A/	FGA	Brawdy	26 (6+	To be augmented
		79 Sqn*	T.7/FGA.9	FGA	Brawdy	6A/ 17/24	and replaced by Hawk
		234 Sqn*	Hawk T.1	FGA	Brawdy		Replaced Hunter 1978
		FAC Flight*	Jet Provost T.3A	FAC	Brawdy	2	
		(*No. 1 Tactical Weapons Unit)					
Strike Command		28 Sqn	Wessex HC.2	Util	Sek Kong (Hong Kong)		
		84 Sqn	Whirlwind 10	Util	Cyprus	4	Assigned to UNFICYP

in Ulster and Staxton Wold in Yorkshire. The ADDC at West Drayton produces its own recognised air picture display (RAPD), as do the SOCs, and additionally receives information from the radar at Burrington in Devon. The RAPD produced at Buchan can be supplied via West Drayton to the other centres.

No 11 Group has a single squadron of airborne early-warning aircraft which pass information on airborne and surface targets to the SOCs, fighters and other aircraft, including maritime-attack Buccaneers. The venerable Shackleton AEW.2, which carries an APS-120(F) radar fitted with AMTI (Airborne Moving-Target Indication), is scheduled to be replaced from 1982 by 11 AEW.3 Nimrods, full development of which was authorised in the spring of 1977. This aircraft, chosen in preference to the Boeing E-3A AWACS (Airborne Warning And Control System), will carry a new Marconi Space and Defence Systems radar with split aerials mounted in the Nimrod's nose and tail. The six-man tactical team – Tactical Air Control Officer, three Air Direction Offi-

UNITED KINGDOM

Command	Group	Unit	Type	Role	Base	No	Notes
		3 Sqn	Harrier GR.3	Army support/recce	Gütersloh	18	
		4 Sqn	Harrier GR.3	Army support/recce	Gütersloh	18	
		14 Sqn					
		17 Sqn					
		20 Sqn	Jaguar GR.1	Army support/strike	Brüggen	60-plus	
		31 Sqn					
RAF Germany		2 Sqn	Jaguar GR.1	Recce	Brüggen		
		15 Sqn	Buccaneer S.2	Att/strike	Laarbruch	approx 30	
		16 Sqn					
		19 Sqn	Phantom FGR.2	Int	Wildenrath	approx 30	
		92 Sqn					
		25 Sqn	Bloodhound Mk 2	SAM	Brüggen (A flt) / Wildenrath (B flt) / Laarbruch (C flt)	6 sections	To be redeployed in UK
		60 Sqn	Pembroke C.1	Comms	Wildenrath	12	
		18 Sqn	Wessex HC.2	Army support	Gütersloh	15	Supports 1st (British) Army Corps
		1 FTS	Jet Provost T.3A/T.5A	Train	Linton-on-Ouse	40/6	Direct-entry trainees
		2 FTS	Whirlwind HAR.10	Train	Shawbury	17	Helicopter training
		3 FTS	Jet Provost T.3A/T.5A	Train	Dishforth		Previously Leeming
			Bulldog T.1	Train	Dishforth		
			Gnat	Train	Valley	44	Being replaced by Hawk
		4 FTS	Hunter F.6/T.7	Train	Valley	20	Advanced training
Support Command			Hawk T.1	Train	Valley		Replacing Gnat
		6 FTS	Dominie T.1	Train	Finningley	13	
			Jet Provost T.4/T.5/T.5B	Train	Finningley	10/6/13	Navigator training
			Bulldog T.1	Train	Finningley		
		METS	Jetstream T.1	Train	Finningley	11	Trans from Leeming 1979
			Bulldog T.1	Train	Leeming		
		CFS	Jet Provost T.3A/T.5A	Train	Cranwell		RAF College
			Gazelle HT.3	Train	Shawbury	14	Helicopter instr. training
			Gnat	Train	Kemble		Red Arrows
			Gnat/Hawk	Train	Valley		
		College of Air Warfare	Dominie T.1	Train	Cranwell	6	
RAF Regiment							
	No 4 Wing	16 Sqn	Rapier	SAM	Wildenrath		
		26 Sqn	Rapier	SAM	Laarbruch		
		37 Sqn	Rapier	SAM	Brüggen	8 fire units each	
		63 Sqn	Rapier	SAM	Gütersloh		
		27 Sqn	Rapier	SAM	Leuchars		

Notes: OCU = Operational Conversion Unit; FTS = Flying Training School; METS = Multi-Engine Training Squadron; CFS = Central Flying School; UNFICYP = United Nations Forces in Cyprus.

Right: The British Army is receiving its first really potent attack helicopter in the form of the Westland/Aérospatiale Lynx, all of which are expected to be armed with TOW anti-tank missiles.

cers, Communications Control Officer and Electronic Support Measures (ESM) Operator – will monitor and control the flow of information between the on-board computers, which process data from the radar and ESM, and the 'customers': NADGE xtations, SOCs, air-defence ships, other AEW aircraft, tankers, fighters, maritime-patrol aircraft and attack types.

No 18 (Maritime) Group, with its headquarters at Northwood, operates mari-

UNITED KINGDOM

Command	Group	Unit	Type	Role	Base	No	Notes
Royal Navy			Polaris	SLBM	Submarines	64	
		702 Sqn	Lynx HAS.2	Train	Yeovilton	—	Formed 1978
		703 Sqn	Wasp HAS.1	Train	Portland		
		705 Sqn	Gazelle HT.2	Train	Culdrose	20	
		706 Sqn	Sea King HAS.1	Train	Culdrose	10	
		707 Sqn	Wessex HU.5	Train	Yeovilton	13	
		737 Sqn	Wessex HAS.3	ASW/train	Portland plus seven County-class destroyers	8	
		750 Sqn	Jetstream T.2	Train	Culdrose	16	Replacing Sea Prince
		771 Sqn	Wessex HAR.1	Train/SAR	Culdrose	11	
		772 Sqn	Wessex HU.5	Fleet support	Portland		Replaced Wessex HAS.1
		781 Sqn	Sea Heron C.1	Comms	Lee-on-Solent	4	
			Sea Devon C.20	Comms	Lee-on-Solent	8	
			Wessex HU.5	Comms/SAR	Lee-on-Solent	2	
		800 Sqn	Sea Harrier FRS.1	Fight/att/recce	Yeovilton/*Hermes*	8	
		801 Sqn	Sea Harrier FRS.1	Fight/att/recce	Yeovilton/*Invincible*	8	Formed late-1970s
		802 Sqn	Sea Harrier FRS.1	Fight/att/recce	Yeovilton/*Illustrious*	8	
		814 Sqn	Sea King HAS.2	ASW	*Hermes*/Prestwick	7	
		819 Sqn	Sea King HAS.2	ASW/SAR	Prestwick	6	One squadron also allocated to RFAs
		820 Sqn	Sea King HAS.2	ASW	*Blake*/Culdrose	4	
		824 Sqn	Sea King HAS.2	ASW	*Bulwark*	6	
		826 Sqn	Sea King HAS.2	ASW	*Tiger*/Culdrose	4	
		829 Sqn	Wasp HAS.1	ASW	Portland/frigates	42	
		845 Sqn	Wessex HU.5/	Assault	Yeovilton	24/15	
		846 Sqn	Sea King Mk 4				
		849 Sqn	Gannet AEW.3/COD.4/T.5	AEW/COD/train	Lossiemouth (HQ flt)	7/2/2	
		FRADU	Canberra T.22/TT.18 Hunter GA.11	Train (target tug	Yeovilton		
			Seaslug	SAM	County destroyers		
			Sea Dart	SAM	Type 82 and Type 42 destroyers		
			Seacat	SAM	See text		
			Seawolf	SAM	Type 22 and *Leander* frigates		
Army	No 1 Wing	No 1 Regiment 651 Sqn / 661 Sqn					
		No 2 Regiment 652 Sqn / 662 Sqn					
		No 3 Regiment 653 Sqn / 663 Sqn	Gazelle AH.1/ Scout AH.1/ Lynx AH.1	Obs/liaison/ anti-tank/ util	Detmold (HQ) plus Bünde, Münster, Herford, Soest, others	Eventually six Gazelle, six Lynx each	
		No 4 Regiment 654 Sqn / 664 Sqn					
		No 9 Regiment 659 Sqn / 669 Sqn					
	No 2 Wing	No 5, 6, 7, 8 Field Force 655 Sqn / 656 Sqn / 657 Sqn	Gazelle AH.1/ Scout AH.1		Netheravon		Formerly 666 Sqn / Formerly 664 Sqn / Formerly 665 Sqn
		No 7 Regiment 658 Sqn					
		8 Flt					
		11 Flt	Gazelle AH.1	Util	Sek Kong		Renumbered 1978 (ex-656 Sqn)
		16 Flt	Alouette II	Util	Dhekelia	8	
		UNFICYP	Alouette II	Util	Akrotiri		
		2 Flt	Gazelle AH.1	Util	Netheravon	8	Assigned to ACE Mobile Force
		6A Flt	Alouette II	Util	Netheravon	8	
		6B Flt	Beaver AL.1	Util	Netheravon	8	
		14 Flt	Gazelle AH.1/ Scout AH.1	Util	Netheravon		
		D+T Sqn		Train	Middle Wallop	7/3	
		Gazelle Conv Flt/ Advanced Rotary-Wing Flt	Gazelle AH.1/ Scout AH.1	Train	Middle Wallop		
				Train	Middle Wallop		
		Advanced Fixed-Wing Flt	Beaver AL.1/ Chipmunk T.10	Train	Middle Wallop	4/2	
		Intermediate Fixed-Wing Flt	Chipmunk T.10	Train	Middle Wallop	20	
			Rapier	SAM	various	72 launchers	
			Blowpipe	SAM	various		
Marines	3rd CBAS	Dieppe Flt/ 40 Commando	Gazelle AH.1	Util	Coypool	3	
		Salerno Flt/ 41 Commando	Gazelle AH.1	Util	Coypool	3	
		Kangaw Flt/ 42 Commando	Gazelle AH.1	Util	Coypool	3	
		Brunei Flt/ HQ Flt	Gazelle AH.1	Util	Coypool	3	
			Blowpipe	SAM	various		

Notes: FRADU = Fleet Requirement and Air Direction Unit; D+T Sqn = Demonstration and Trials Squadron; CBAS = Commando Brigade Air Squadron.

time-patrol Nimrods and a rescue helicopter fleet. The Commander-in-Chief is also Commander Maritime Air Eastern Atlantic and Commander Maritime Air Channel Command, with wartime responsibilities for reconnaissance, support of the Atlantic Striking Fleet, shipping control and anti-submarine warfare. A more recent addition to these tasks is patrol of North Sea oil and gas rigs as a peacetime measure against terrorist actions, with added significance in wartime. An additional Nimrod has been assigned to each of the four home-based squadrons for this role, code-named Operation Tapestry.

The Nimrods are in the process of being upgraded from MR.1 standard to MR.2. The updated version carries an EMI Searchwater radar in place of the earlier ASV-21; a Marconi Avionics AQS-901 acoustic processing and display system; and a new Central Tactical System based on the Marconi Avionics 920 ATC digital computer. Weapons carried by both versions include torpedoes, depth charges, mines, rockets and AS.12 wire-guided missiles. Eleven Nimrods originally built for the maritime-patrol role but rendered superfluous by the RAF's shrinking overseas commitments are being converted for AEW missions (see above), including three ex-203 Sqn aircraft withdrawn from Malta.

The search-and-rescue force of Whirlwinds and Wessexes is being upgraded by the introduction of 15 Westland Sea King HAR.3s, replacing most of the former type. The first two of the new helicopters were delivered to Lossiemouth in August 1978, forming D Flight of 202 Sqn. They will replace all 202's Whirlwinds, with four detached flights based along the coast.

No 38 Group encompasses a wide range of army-support, reconnaissance, logistic and other functions. The major combat element comprises a squadron of Harriers plus an operational conversion unit, two squadrons of attack Jaguars and a single unit operating the latter type in the reconnaissance role. Both types are also in service with RAF Germany; the Royal Air Force's total of 89 single-seat Harriers and 20 two-seaters is being augmented by an additional 24 aircraft ordered in the spring of 1977 to replace attrition losses. Deliveries of the RAF's 163 single-seat Jaguar GR.1s and 37 T.2 trainers has been completed, and various improvements – especially to the powerplant – are being studied. Both types carry a range of conventional weaponry, including iron bombs, cluster bombs and rockets (the last-named arms only Harrier, not Jaguar). The gun is the 30mm Aden, built-in in the case of the Jaguars and mounted in a pod for attachment to Harriers. Both are progressively being fitted with Ferranti LRMTS (Laser Rangers and Marked-Target Seekers) to improve weapon-delivery accuracy. The Harriers are assigned to NATO's ACE Mobile Force and the Jaguars to the United Kingdom Mobile Force. No 41's Jaguars are fitted with recce pods built by

British Aerospace, and containing HSD Type 401 infra-red linescan equipment in addition to optical cameras, for low-level operations over the central front and northern flank of the alliance.

A back-up combat force is available in the form of the Tactical Weapons Units, the Hunters of which would be available in an emergency. A second TWU to augment that at Brawdy is due to be established at Chivenor in early 1980, taking over the duties allocated to a temporary TWU at Lossiemouth from the autumn of 1978. The expansion has been necessitated by an increase in training and by the imminent introduction of Tornado.

The Jaguar began replacing Phantom FGR.2s on ground-attack duties only in 1974, but a successor for both this type and the Harrier is already being planned. By the end of 1978 the detailed specifications of the aircraft to meet Air Staff Target 403 had not been settled, although the likely solution is a type which can take off from a run of less than 1500ft with a heavy weapon load, land vertically and double as a fighter and ground-attack aircraft.

No 38's helicopter wing, based at Odi-

ham, operates Wessexes and Pumas assigned to the ACE Mobile Force and UK Mobile Force for assault and transport. The RAF received 40 of the Anglo-French Pumas, some of which are operated by 240 OCU. A medium-lift helicopter, the Boeing Vertol CH-47 Chinook, was finally ordered in 1978. Thirty-three aircraft will be delivered from August 1980. Each can carry 44 equipped troops, 24 stretcher cases or 28,000lb of underwing loads.

The group's fixed-wing transport fleet underwent a sharp contraction in the early and mid-1970s, with the retirement of all Belfasts, Britannias and Comets being accompanied by a reduction in strength and re-assignment of Hercules and Andovers. Thirty of the Hercules are being stretched – one by Lockheed and the remaining 29 by Marshall of Cambridge – over a four-year period. Two plugs totalling 15ft are inserted in the fuselage,

Below: A Royal Navy Sea King displays its weapon load of four torpedoes and the well for its Plessey dipping sonar. RN Sea Kings fly from shore bases, *Hermes, Bulwark,* the cruisers *Blake* and *Tiger* and will equip *Invincible*–class cruisers.

Above: An RAF Harrier looses off a salvo of air-to-ground rockets from its two underwing pods. Most other air forces have remained unconvinced of the Harrier's V/STOL advantages, but the type can operate more flexibly than any other.

increasing troop capacity by 36 to 128. Alternatively, 92 paratroops can be carried instead of 64, or 90 stretchers (rather than 84) together with six medical attendants.

RAF Germany
RAF Germany is run from Rheindahlen, near Moenchengladbach – the headquarters also of the British Army of the Rhine, the Northern Army Group and the 2nd Allied Tactical Air Force. The last-named comprises German, Belgian and Dutch forces in addition to the British element, and its Commander-in-Chief is also C-in-C of RAF Germany. The 2nd ATAF and its counterpart in southern Germany, the 4th ATAF (with German, US and Canadian units), comprise Allied Air Forces Central Europe (AAFCE), directly responsible to Allied Forces Central Europe (AFCENT).

The Commander of AAFCE is responsible in both peace and wartime to the Commander-in-Chief Central Europe for air defence of the central region, and RAF Germany's two squadrons of interceptors are therefore part of NATO's Command Forces. In peacetime the offensive squadrons remain under national control, although all RAF Germany's units are assigned to NATO and would be controlled by the Commander of the 2nd ATAF or of AAFCE itself in time of war.

RAF Germany has two squadrons of Phantom FGR.2 fighters, two of Harriers for army support, four of Jaguars and two of Buccaneers which may provide army support with conventional weapons or undertake nuclear strikes, one unit with reconnaissance Jaguars, a single squadron of Bloodhound surface-to-air missiles, a helicopter unit and a Pembroke-equipped communications unit.

The Phantoms, which have replaced Lightning F.2As, patrol the northern half of the Air Defence Identification Zone to prevent penetration. The Lightnings used in this role were based at Gütersloh, only 65 miles from the East German border and the sole RAF operational airfield east of the Rhine. The Phantoms, however, fly from Wildenrath near the Dutch border because their longer range allows them to be stationed further back, thus freeing Gütersloh for use by the shorter-legged Harrier force. The Phantoms are armed with Sparrow and Sidewinder air-to-air missiles and can carry pod-mounted M61 Vulcan 20mm cannon. They will be replaced in the mid-1980s by Tornado ADV.

The two Harrier units, Nos 3 and 4, have absorbed the aircraft formerly operated by No 20 and are now based at Gütersloh. The V/STOL aircraft also operate from dispersed sites and would be deployed to these in wartime. Weaponry includes free-fall and retarded 1000lb bombs, BL755 cluster bombs (maximum seven), Matra 68mm rocket pods (up to six with 19 projectiles each) or twin Aden 30mm cannon pods under the belly. RAF Germany's Harriers are being fitted with the Ferranti LRMTS, following its introduction with Strike Command's 1 Sqn, and No 4's aircraft can carry an EMI reconnaissance pod containing five cameras. All Harriers are also fitted with a single nose-mounted oblique camera.

As RAF Germany's Phantom FGR.2s have been transferred to air-defence duties their place has been taken by Jaguars, as was the case in Strike Command. The four squadrons at Brüggen can carry out nuclear strikes as well as conventional support missions, while 2 Sqn at Laarbruch operates primarily in the recce role with ground attack as a back-up. Two squadrons of strike/attack Buccaneers are also based in Germany and these units, in conjunction with those equipped with Harrier and Jaguar, operate along similar lines to their UK-based counterparts. One item of equipment which made its debut with RAF Germany, however, is the US-supplied ALQ-101 jamming pod. RAF Germany's Buccaneers are also being converted to carry the Pave Spike laser designator pod.

RAF Germany's Bloodhound SAMs are operated by 25 Sqn, with its headquarters at Brüggen. Flights are also deployed at the other Clutch airfields – Wildenrath and Laarbruch – to give overlapping coverage of the bases and surrounding areas. The Ferranti Type 86 target-illuminating radars are mounted on plinths to extend their range against low-flying targets; against which the missiles are primarily operated. The Bloodhounds in Germany are to be redeployed to Britain, with their former role being taken over by the Rapiers of the RAF Regiment (see below) as they convert to all-weather day-and-night operation with the delivery of Marconi Space and Defence Systems DN181 Blindfire radars.

The basing of low-level surface-to-air missiles in Germany is part of a co-ordinated programme intended to improve airfield defences. Other measures include the building of hardened aircraft shelters which will withstand anything up to a direct hit by a 1000lb bomb; hardening other facilities such as briefing rooms,

and Salalah have ceased and the island staging post of Gan has been closed. The Vulcans, Lightnings, Bloodhounds, Hercules and Argosies of the Near East Air Force have been withdrawn to Britain from their base of Akrotiri in Cyprus, and 1968 saw the disbandment of the Nimrod and Canberra squadrons formerly operating from Malta.

Support Command

Since mid-1977 Support Command has run the RAF's training programme. More than half the RAF's new pilots are university graduates, and the 16 University Air Squadrons are now equipped with Scottish Aviation Bulldogs in place of the earlier Chipmunks. Following some 70hr experience with the UASs, graduates undergo a further 75hr on Jet Provost T.5As at the RAF College, Cranwell. Direct-entry cadets are graded on Bulldogs for 15hr and then move to No 1 Flying Training School (FTS), Linton-on-Ouse, or Cranwell for 100hr on Jet Provost T.3As. One eight-aircraft squadron of Jet Provosts from No 1 FTS was due to move to Church Fenton in April 1979. These squadrons will be retained at Linton-on-Ouse, and a new FTS of 30 aircraft will gradually build up at Church Fenton.

Fast-jet pilots undergo a further 60hr on Jet Provost T.5As before transferring to No 4 FTS, Valley, for 70hr on Hawks and thence to the Tactical Weapons Unit for 50hr on Hunters or Hawks. The first of the 175 British Aerospace Hawks on order entered service at Valley in the autumn of 1976 and student training with the type began in July 1977. The first Hawk-equipped squadron in the TWU at Brawdy, No 234, completed the transition during 1978. Multi-engine pilots continue their instruction on Jetstreams of the multi-engine training squadron, also used for refresher flying. The Jetstreams have replaced Varsities formerly operated by No 5 FTS at Oakington in this role, and moved from Leeming to Finningley in the second quarter of 1979. Helicopter pilots transfer to 2 FTS at Shawbury, where they are trained on Whirlwind HAR.10s following the withdrawal of Bell 47 Sioux. Shawbury also houses the Central Flying School's Gazelle HT.3s, on which helicopter instructors are trained; the other elements of CFS are at Leeming (Bulldogs), Valley (Gnats) and Cranwell (Jet Provosts).

Non-pilot aircrew – Air Electronics Operators for the Nimrod squadrons and navigators for Nimrods, Victors, Vulcans, Phantoms and Buccaneers – are trained at No 6 FTS, Finningley. Initial familiarisation in the Bulldog is followed by instruction in the British Aerospace Dominie T.1. The low-level strike navigators also undergo 20hr in Jet Provosts.

RAF Regiment

The regiment, controlled by Strike Command's No 38 Group in the United Kingdom, defends RAF bases and equipment in the field. Five squadrons of British Aerospace Rapier low-level surface-to-air missiles are operational: one at each of the RAF's four active bases in Germany and the fifth at Leuchars in Scotland. A sixth squadron is planned to defend Lossie-

mouth. A squadron has eight fire units and will eventually also operate one Marconi Space and Defence Systems DN181 Blindfire radar for each launcher. The first Blindfires, which extent Rapier operations round the clock and allow engagements to be carried out in all weathers, entered service at the end of 1977.

The missiles have replaced many of the Bofors L40/70 guns, but these weapons are retained as a back-up and for use by units not equipped with Rapier. The regiment's field squadrons defend bases against attacks on the ground and are deployed alongside Harrier squadrons when they operate from dispersed sites.

Navy

Since 1969 the Royal Navy has operated Britain's nuclear deterrent in the form of four nuclear-powered submarines armed with 16 Lockheed Polaris A3 missiles each. The submarine-launched ballistic missiles, which have superseded Blue Steel stand-off weapons and free-fall nuclear bombs delivered against strategic targets by the RAF's Vulcans, each carry three re-entry vehicles with British warheads of some 200 kilotons yield. The missiles will not be replaced by Poseidon and are not, despite reports to the contrary, to be fitted with independently targeted re-entry vehicles.

At one time in the mid-1970s it appeared that the Fleet Air Arm would operate only helicopters in the 1980s, following the retirement of the RN's last attack carrier – HMS *Ark Royal* – at the end of 1978. Naval fixed-wing front-line operations will now continue, however, with the introduction of the British Aerospace Sea Harrier FRS.1 in 1979. Each of the three through-deck cruisers will carry five to eight Sea Harriers, and early experience is to be gained aboard the anti-submarine carrier HMS *Hermes* until the cruiser force builds up to its planned total. The V/STOL aircraft will be responsible for air-to-air and anti-ship operations in equal measures, each comprising some 45% of the total, and the remaining 10% will be devoted to reconnaissance. The Sea Harrier IFTU (Intensive Flying Trials Unit) was due to form at Yeovilton in June 1979, with the first operational squadron (No 800) forming at the end of the year. This unit was then planned to embark aboard *Hermes* in 1980.

The Sea Harrier carries a Ferranti Blue Fox radar and differs from the land-based version in a number of areas, but its performance is substantially similar. Weapons will include the British Aerospace P3T turbojet-powered Martel development for anti-ship operations, and the AIM-9L Sidewinder for air-to-air combat. The total Sea Harrier order is for 34 aircraft. The *Invincible*-class vessels and *Hermes* are being fitted with Ski-Jump ramps to improve the Sea Harrier's payload-range performance.

The single squadron each of Phantom FG.1s and Buccaneers in the *Ark Royal* air group were transferred to the RAF when the ship was retired, and the flight of airborne-early-warning Gannets was likewise handed over. The Gannets have

operations centres and vital storage areas; fitting filtration units so that the base can operate after an attack with nuclear, biological or chemical weapons; toning down structures with acid and paint so that they are difficult to detect from the air; training airmen and other personnel to deal with attacks by saboteurs; and strengthening gun defences by deploying Bofors L40/70s and general-purpose machine guns. Many of these improvements are also subsequently being carried out in the United Kingdom.

No 18 Sqn, equipped with Wessex HC.2s, supports 1 (BR) Corps which from its headquarters at Bielefeld is responsible, along with the other members of NORTHAG (German, Dutch and Belgian corps), for defending the northern section of the central front. The helicopters can each carry 15 fully equipped troops and may also transport underslung loads of up to 3000lb. The rotary-wing machines are not normally armed, although they can be fitted with a pair of sideways-firing 7.62mm machine guns.

RAF Germany's remaining unit is its sole communications squadron, equipped with the venerable Pembroke C.1 which has been resparred to extend its service life into the 1980s.

The RAF's other overseas commitments are limited to operating a Wessex squadron in Hong Kong and supplying four Whirlwinds to the United Nations Forces in Cyprus (UNFICYP). Hunters are rotated from Valley to Gibraltar, and other aircraft are deployed as needed to trouble-spots such as Belize, but the once-world-wide RAF presence has virtually disappeared.

On the periphery of the Indian Ocean, RAF operations from Singapore, Masirah

machine guns and 28 2in rockets, SS.11B1 missiles or the larger AS.12.

The RN's largest squadron operates 40-plus Wasps in the anti-submarine role with secondary tasks including anti-ship attacks. The helicopters are deployed aboard Type 42 destroyers and frigates in the Type 21, *Tribal*, *Leander* and *Rothesay* classes. The Wasps are in the process of being replaced by the Westland/Aérospatiale Lynx, which officially entered service in September 1976 with the formation of the intensive flying-trials unit. The first full training squadron, 702, was established at Yeovilton in January 1978, and the type will gradually be deployed aboard Type 42 destroyers, Type 21 and 22 frigates, and perhaps the through-deck cruisers on planeguard duties. The RN's ultimate requirement is for 100 Lynxes, although defence cuts have reduced the initial procurement to 88.

Lynx, which carries the Ferranti Sea-spray navigation/attack radar, is armed with either two torpedoes or up to four British Aerospace Sea Skua anti-ship missiles. The missiles are unlikely to enter service before the early 1980s, however, and AS.12 may be fitted as an interim weapon. The Lynx HAS.2 is being fitted with a passive sonobuoy processor and may later also be equipped with a dipping sonar.

Training of Sea Harrier pilots began in 1977 under RAF auspices, with operational flying training being carried out at Yeovilton. Royal Navy helicopter pilots receive their initial instruction on RAF Bulldogs and then transfer to the RN's 705 Sqn at Culdrose, where Gazelle HT.2s have replaced Hiller UH-12Es and Whirlwind HAS.7s. Pilots then convert to their operational type and other crew members undergo specialist training with other units. The Fleet Requirements and Air Direction Unit (FRADU), operated by Airwork, has three basic functions: providing aircraft to assist RN ships undergoing sea trials, acting as visual and radar targets, towing targets for live firings and participating in radio trials; providing aircraft for practice interceptions under the control of shipboard crews; and supplying aircraft for the Flying Standards Flight.

The Royal Navy is receiving 16 Scottish Aviation Jetstream T.2s to replace Sea Princes used in training observers, the first being handed over in October 1978. Communications types include Sea Devons and Sea Herons; one of the former type is used by each of the station flights at Prestwick, Yeovilton and Culdrose, while a Sea Heron acts as the 'Admiral's barge'.

The Royal Navy operates a variety of surface-to-air missiles: Seaslug, Seacat, Sea Dart and, from the late 1970s, Seawolf. The beam-riding medium-range Seaslug equips the seven County-class guided-missile destroyers. This ship/missile combination is now being replaced by the Type 42 armed with Sea Dart, a medium-range ramjet-powered weapon which was accepted for initial service in 1977. British Aerospace, Ferranti and Marconi have been awarded feasibility-study contracts to recommend improvements to Sea Dart,

Top: A Nimrod MR.1 of the RAF's 102 Sqn from Kinloss swoops over a Soviet Navy *Echo II* nuclear-powered submarine.

Above: A Westland/Aérospatiale Lynx anti-submarine helicopter, which is replacing Wasps, alights on the Type 42 destroyer HMS *Birmingham*.

recently been refurbished and are expected to remain in service into the 1980s.

The RN's standard large anti-submarine helicopter is the Westland Sea King, of which the navy has ordered 56 HAS.1s and 21 HAS.2s. The latter variant, which has uprated engines and other improvements, entered service at the end of 1976. The Sea Kings are deployed aboard two cruisers, *Blake* and *Tiger*; the anti-submarine carriers *Hermes* and *Bulwork*; the through-deck cruisers (when they enter service in the early 1980s); and at shore bases. The type is also augmenting the sole remaining sea-going anti-submarine Wessexes, which are deployed on the seven County-class guided-missile destroyers. Delivery of 15

Sea King Mk 4s for this role was due to begin in 1979. The aircraft have flotation gear, folding main rotors and tail boom, and improved avionics. They can carry up to 8000lb of slung load, allowing them to transport one-tonne Land Rovers and the Royal Marines' 105mm light gun.

The RN's Sea Kings are to be replaced in the 1980s by the Westland WG34, which may be built as a collaborative venture. It will carry sonobuoys rather than a dipping sonar and will additionally be equipped with MAD (magnetic anomaly detection) gear.

HMS *Hermes* retains a secondary assault role, but the specialised commando carrier *Bulwark* is to be converted for the ASW role. Similarly, only one of the two helicopter assault ships *Fearless* and *Intrepid* is kept at immediate operational readiness. The Wessexes can each carry ten marines over a radius of action of some 70 miles, and one helicopter in four is normally armed for use in the gunship role. Weapons include up to five 7.62mm

specifically in the area of electronic counter-countermeasures, for an improved version which is expected to enter service in the mid-1980s and remain operational beyond the year 2000. Sea Dart also arms the single Type 82 destroyer, HMS *Bristol*, and will be fitted to the through-deck cruisers.

The Short Seacat point-defence weapon is carried by *Hermes*, the assault ships *Intrepid* and *Fearless*, the cruisers *Tiger* and *Blake*, and frigates of the Type 21, *Leander*, *Rothesay*, *Tribal* and *Salisbury* classes. The most recent variant, GWS24, is installed in the Type 21s and can be operated completely from below decks.

Seacat's replacement, the British Aerospace Seawolf, is scheduled to enter service aboard Type 22 frigates at the end of the 1970s and will also be fitted to some of the *Leanders*. The weapon has consistently demonstrated the ability to intercept small, fast targets and is the West's only proven naval anti-missile missile.

Army
The Army Air Corps, formed in 1957 to succeed the Glider Pilot Regiment, has undergone a substantial re-organisation as the British Army itself has changed.

The corps has two wings: No 1, based at Detmold in Germany, controls five regiments with two squadrons each; and No 2, run from Wilton and part of the United Kingdom Land Forces. A further squadron is based in Hong Kong, and numerous flights and other units are operated.

The Westland/Aérospatiale Gazelle has replaced the Westland/Agusta-Bell AB.47 Sioux in the reconnaissance role, the turbine-powered helicopter being able to carry five men instead of three in addition to being faster, having a greater range and being capable of lifting heavier loads. The AAC has a total of 158 Gazelles on order. With the selection of the Hughes TOW in preference to the Euromissile HOT as the army's helicópter-mounted anti-tank missile it is increasingly likely that the Gazelles will carry squads armed with Milan missiles rather than being fitted with ATMs themselves.

The main missile-carrier, however, is the Westland/Aérospatiale Lynx, which is to be armed with TOW. One hundred are on order to replace Westland Scouts, with the first squadrons re-equipping in 1978 following the establishment of an intensive flying-trials unit in 1977. Each AAC regiment in Germany will eventually operate one squadron of missile-armed

Lynxes and one of liaison/observation/missile-squad Gazelles.

The British Army's only large surface-to-air missile, the BAC Thunderbird Mk 2, has been withdrawn from service largely because the 700 men of the 36th Heavy Regiment are needed for other tasks such as operating the Short Blowpipe shoulder-launched SAM. Blowpipe is deployed with three-man teams carried in Land Rovers or the Spartan armoured personnel carrier; the latter vehicle can transport ten rounds and an aiming unit.

The army is also equipping two light air-defence regiments – No 12 and No 22 – with the BAC Rapier low-level SAM. Each regiment has three battalions comprising three four-launcher troops. The British Army will thus eventually have 72 launchers, of which one-third will be operated in association with the Marconi Space and Defence Systems DN181 Blindfire radar; initial radar deliveries took place in 1978.

Marines
The Royal Marines operate their own Gazelles and Scouts in addition to using the Royal Navy's Wessexes for assault landings. The Marines are also receiving a small number of Short Blowpipe shoulder-launched surface-to-air missiles.

NORWAY

Area: 386,000 sq. miles (1 million sq. km.).
Population: 4·05 million.
Total armed forces: 39,000.
Estimated GNP: $36·2 billion (£18·1 billion).
Defence expenditure: $1·30 billion (£650 million)

Norway is a founder member of NATO and, along with Denmark, assigns her military resources to Allied Forces Northern Europe. *Kongelige Norske Luftforsvaret* (Royal Norwegian Air Force) is responsible for all military aircraft operations apart from army observation, which is carried out by the Field Artillery Observer Service. The RNoAF is divided into northern and southern commands, each with its own headquarters and control network, which are integrated into NATO's Nadge air-defence system.

The Norwegian army and navy began air operations in 1912, and the RNoAF was not formed until 1944, when these two services merged. The present personnel strength is some 10,000, of which 4000 are conscripts.

Inventory
The RNoAF is typical of several small NATO air forces, relying heavily on the Lockheed Starfighter and Northrop Freedom Fighter. A single interceptor unit flies F-104Gs supplied under a military assistance programme in 1963, and the purchase of ex-Canadian CF-104s has since allowed a further squadron to be

Right: Further P-3B Orion maritime-patrol aircraft may be bought to augment the RNoAF's present fleet of five, which have vast areas to patrol.

NORWAY

Unit	Type	Role	Base	No	Notes
Air Force					
330 Sqn	Sea King Mk 43	SAR	Bodø (HQ+A Flt) Banak (B Flt) Orland (C Flt) Sola (D Flt)	10	
331 Sqn	F-104G/TF-104G	Int/train	Bodø	15/3	
333 Sqn	P-3B Orion	ASW	Andøya	5	May receive additional three
334 Sqn	CF-104G/D	FGA/train	Bodø	19/3	
335 Sqn	C-130H Hercules	Trans	Gardermoen	6	
	Falcon 20	ECM/calib	Gardermoen	2	
336 Sqn	F-5A	Att	Rygge	16	
338 Sqn	F-5A	Att	Orland	16	
339 Sqn	UH-1B	Army supp	Bardufoss	15	
717 Sqn	RF-5A	Recce	Rygge	13	
718 Sqn	F-5B	Train/att	Sola	13	
719 Sqn	Twin Otter	SAR/liaison	Bodø	5	
	UH-1B	SAR/liaison	Bodø	8	
720 Sqn	UH-1B	Army supp/SAR	Rygge	9	
	Safir	Train	Vaernes	20	
	Nike Hercules	SAM	Oslo+eastern Norway	4 batteries	
Army					
	O-1E Bird Dog	Army obs		23	
	L-18C Super Cub	Army obs		9	Two squadrons
Navy					
	Sea Sparrow	SAM	5 frigates	5 launchers	
Coast Guard					
	Lynx	Mar pat	ships	4	Two more on option. 1981 delivery. Ordered by RNAF

equipped with Starfighters. The CF-104s have been modified to include F-104G equipment such as the 20mm M61 Vulcan cannon. The F-5s, of which 108 were supplied (including RF-5As and 14 two-seat F-5Bs), are used for light attack, reconnaissance and training in four squadrons. At least 17 have been lost. Both these types will be replaced from 1981 by the General Dynamics F-16, of which 72 (including a dozen F-16B two-seaters) have been ordered. The F-16s are expected to be armed with the McDonnell Douglas Harpoon anti-ship missile in addition to bombs and possibly guided air-to-surface weapons. They may also carry a medium-range air-to-air missile in the form of Sparrow or one of its derivatives.

Air defence is shared by four batteries of Nike Hercules long-range surface-to-air missiles deployed around the capital, Oslo, and in the eastern part of the country. Forty vehicles carrying the Euromissile Roland II short-range SAM were due to be introduced from 1979–80, but the order has been deferred and service introduction has slipped.

Maritime-patrol missions are carried out by five Lockheed P-3B Orions, and a further three are likely to be bought to improve coverage of Norway's extensive offshore fish and oil resources as well as monitoring Russian submarines operating in the Norwegian Sea. The Norwegian Coast Guard (*Kystvaksten*) was formed on the first day of 1977 and will operate the Westland/Aérospatiale Lynx helicopter, four of which were ordered in the autumn of 1978 with a further two on option. The helicopters will be delivered from 1981.

Initial training is carried out on Saab Safirs at Vaernes, pilots then transferring to the NATO course in Canada. The armament school is at Lista.

Army

The army's Field Artillery Observer Service operates Bird Dogs and Super Cubs which are flown by army pilots and maintained by the RNoAF. These observation types are due to be replaced, and whichever aircraft is chosen may also supersede the RNoAF's Safirs in the basic-training role.

Navy

The navy operates no aircraft but has five *Oslo*-class frigates being fitted with the NATO Seasparrow point-defence missile system.

Above: One of the RNoAF's ten search-and-rescue Westland Sea Kings winches aboard a survivor from his raft. The helicopters are based at four locations to cover Norway's coastline.

Below: RNoAF Twin Otters perform the dual roles of liaison and search-and-rescue – difficult tasks in such a mountainous country with a jagged coastline.

SWEDEN

Area: 173,400 sq. miles (449,100 sq. km.).
Population: 8·25 million.
Total armed forces: 65,500.
Estimated GNP: $83 billion (£41·5 billion).
Defence expenditure: $2·95 billion (£1·475 billion).

Sweden pursues a policy of armed neutrality, having traditionally relied on well equipped and trained armed forces to deter a potential aggressor. The country has not been involved in a major conflict since 1814, although units stationed overseas have seen action in a number of areas since then. Sweden measures 1500km from north to south, covering 15° of latitude and consequently experiencing a wide range of climatic conditions, yet its 450,000km² are occupied by only 8.25m people. These circumstances combine to make defence of the homeland a difficult task, leading to relatively heavy military spending, indigenous development of equipment requiring the minimum number of operators, unusually heavy emphasis (for Western Europe) on civil defence and reliance on rapidly mobilised reserves.

Sweden is not a member of NATO, and this fact has in the past reduced the potential market for the country's arms industry (although Bofors guns, for example, have been supplied worldwide and have been operated by opposing forces in many conflicts). The aerospace industry, in particular, relies on the home defence market for the great majority of its sales. The recent economic recession and widespread efforts to reduce military spending have thus cast doubts on the future of Sweden's arms industry, which in turn is bound to affect the country's defence policy. One of the great drawbacks of dependence on self-sufficiency in arms production is that the whole philosophy of non-alignment militarily and economically can be wrecked by a major increase in costs. In Sweden's case this could result in the Viggen being the last major combat aircraft to be built by Saab.

All three branches of the armed services operate aircraft, with the air force having the great majority. The Swedish Air Force (*Svenska Flygvapnet*, the *Kungl.* prefix for 'Royal' having been dropped in 1974), with headquarters in Stockholm, has about 20,000 personnel. The total defence budget for 1977–82 is some Kr50,000m ($21,000m) at early-1976 prices.

Sweden is divided into six regional military commands (*Milos*, for *Militärområde*): North Norrland, South Norrland, East, South, West and Bergslagen. They have unified tri-Service commands and each includes at least one air-defence sector, the commanders of which have responsibility for their own areas. *Flottiljer*, roughly equivalent to Wings, comprise two to four *divisioner* (squadrons) of 15–18 aircraft each and divided into three sections: flying, services and maintenance. An average *division* has some 20 officers (including pilots), 50 non-commissioned officers and 150 airmen. The services section typically comprises two NCOs, seven

technicians and 13 conscript airmen; corresponding strength of the maintenance section is three, 17 and 16 respectively.

Only the first four *Milos* listed above operate interceptors. The attack, reconnaissance and transport forces (apart from a light attack/recce squadron of F21 in North Norrland) come under the control of Attack Command (*Eskader* 1), headquartered at Gothenburg, rather than air-defence sector commanders. Responsibility for reconnaissance would be delegated to the *Milos* in time of war, but the attack squadrons would remain under *Eskader* 1 command.

The *Flygvapen* is in the process of reducing its 55 main bases (not all in permanent use) to 40, but a roughly equivalent number of road bases are available. Both the Draken and Viggen were designed for short-field operations, the Viggen being able to stop in 500m, with reverse thrust, following a 5m/sec descent on ILS using autothrottle and no flare. Major road bases have 2000m of strengthened highway with shelters or caves for the aircraft and hardened storage facilities for fuel, weapons and other stores in nearby woodland. Other auxiliary road bases consist of stretches of normal highway which have been slightly widened. At least eight aircraft in each combat squadron are kept permanently on alert and these would immediately be dispersed in times of tension. Facilities at permanent bases have also been hardened, and the Viggen's fin folds so that the aircraft can fit in low shelters. Four-colour camouflage on the upper surface of Viggens and Drakens aids concealment.

Sweden recognised the need for an integrated surveillance and defence network if the best use was to be made of limited facilities, and has developed STRIL 60 (*Stridsledning och Luftbevakning* 60), a counterpart of NATO's NADGE system. Surveillance over the complete country and warning of attack is provided by overlapping long- and short-range radars operating down to low level. New long-range three-dimensional radars for surveillance and ground-controlled interception, backed up by 25 sets of shorter-range

Above: An SK37 Viggen trainer of the Swedish Air Force is readied for its mission from a snow-covered dispersed base amongst the trees.

equipment, are scheduled to be operational by 1980. The radars are complemented by visual observations which pass via optical filter centres to radar group centres or sector operations centres. Each of the seven air-defence sectors has an SOC in which target information is processed automatically, presented to sector leaders and passed by radio link to other SOCs. Target data are supplied from the STRIL 60 network to fighter and attack units, missile and anti-aircraft artillery defences, civil-defence organisations and the national radio/television networks.

Swedish military aviation began in 1911, when the navy acquired a Blériot. The army followed a year later, and on July 1, 1926, these components were amalgamated to form the *Kungl. Svenska Flygvapen*. Most aircraft were imported, but in 1937 *Svenska Aeroplan Aktiebolaget* (Saab) was formed and began building foreign types under licence. The company later designed its own aircraft, such as the B18 twin-engined bomber and J21 fighter, which were supplied to the air force. The Swedish Air Board's own workshops, the FFVS at Bromma, also built aircraft following the German invasion of Denmark and Norway in 1940. By the end of the Second World War *Flygvapnet* had about 1000 aircraft in some 50 *divisioner*, making up 17 *Flygflottiljer*. Following a purchase of the de Havilland Vampire in 1946, the air force introduced Saab's J21R fighter (the first Swedish-built jet) in 1949 and the J29 (Europe's first post-war swept-wing type) two years later. With 661 constructed, the *Tunnan* (Barrel) – as it was affectionately known – was produced in larger numbers than any other Swedish aircraft.

In 1950 the *Flygvapen* had 34 fighter squadrons but this number has been shrinking ever since. Now it is down to 17, is planned to comprise ten in 1982 and will reach nine by 1987. More advanced types – the Lansen, Draken and Viggen (see below) – have succeeded the early jets, and quality has replaced quantity. On

October 1, 1966, the arrangement of four *Eskadrar* (Groups) was dissolved, the fighter force coming under the control of the *Milos* and *Eskader* 1 assuming responsibility for most other front-line types (see above). Since 1973, six *Flottiljer* – F2, F3, F8, F9, F12 and F18 – have been disbanded; F11 is expected to disappear in 1979–80, with F10 and either F16 or F20 following. The reconnaissance force is also to drop, and front-line combat-aircraft strength, which was 850 or so in the late 1950s and fewer than 450 in the mid-1970s, may well dip below 300 in the next decade.

Inventory

The *Flygvapen*'s fighter force has begun conversion from 17 squadrons of Saab 35 Drakens (Dragons), numbering some 300 aircraft plus reserves out of a total purchase of 560 (all marks), to nine squadrons of the same company's JA37 Viggen. The Draken's main armament comprises Rb27 and Rb28 air-to-air missiles: licence-built versions of the Hughes AIM-26B and AIM-4D Falcon respectively. Saab, Bofors, Svenska Radio and other companies collaborated on the licence-production programme, which turned out 800 Rb27s and 3000 Rb28s for the *Flygvapen* together with

others for the Swiss Air Force. Both the Draken and Viggen also carry the Rb24 Sidewinder, 2440 rounds of the AIM-9B variant having been bought. The latest version of the Draken is the J35F (J = *Jakt* – fighter), the type originally having entered service in 1960.

The interceptor Drakens were scheduled to be replaced from late 1978 by the JA37 fighter Viggen, the first production example of which flew in November 1977. The initial 30-aircraft batch was expanded in May 1978 by a contract covering 60 and 59 Viggens in the second and third batches. The deal is worth SwKr2000m for these latest 119 aircraft, although the Swedish parliament has until July 1980 to veto the third batch if it so desires. Between 160 and 200 JA37s were originally planned but the total is likely to stick at the 149 ordered so far, giving the *Flygvapen* a total of 329 Viggens (the other 180 being for attack, reconnaissance and training). The Dra-

Right: The Flygvapen is now receiving the JA 37 Viggen interceptor, seen here with instrumentation pod, two Sky Flash and three Sidewinders. Its 30mm gun has the same hitting power at 1.5km range as the Aden or DEFA at the muzzle, and the radar is a long-range pulse-doppler.

SWEDEN

Command	Wing	Unit	Type	Role	Base	No	Notes
Air Force							
Attack (*Eskader* 1)	F6 (Västgöta)		AJ37 Viggen	Att	Karlsborg	15–18	Last to convert from A32A
			AJ37 Viggen	Att	Karlsborg	15–18	
	F7 (Skaraborgs)		AJ37 Viggen	Att	Såtenäs/Tun	15–18	
			AJ37 Viggen	Att	Såtenäs/Tun	15–18	
			AJ37 Viggen	Att	Såtenäs/Tun	15–18	
			C-130E Hercules	Trans	Såtenäs/Tun	2	
			C-130H Hercules	Trans	Såtenäs/Tun	1	Second aircraft ordered
			Tp79 (C-47)	Trans	Såtenäs/Tun	7	
	F15 (Hälsinge)		AJ37 Viggen	Att	Söderhamn	15–18	
			Sk37 Viggen	Train	Söderhamn		Viggen OCU
	F11 (Södermanlands)		S32C Lansen	Recce	Nyköping/Skavstä		To be disbanded 1979–80?
			S35E Draken	Recce	Nyköping/Skavstä		
South Sweden	F10 (Skånska)		J35F Draken	Int	Angelholm/Barkakra	15–18	Likely to be disbanded
			J35F Draken	Int	Angelholm/Barkakra	15–18	
			J35F Draken	Int	Angelholm/Barkakra	15–18	
	F17 (Blekinge)		J35F Draken	Int	Ronneby/Kallinge	15–18	To be replaced by JA37 Viggen
			J35F Draken	Int	Ronneby/Kallinge	15–18	
			SH37 Viggen	Recce	Ronneby/Kallinge	15–18	
East Sweden	F1 (Västmanlands)		J35F Draken	Int	Västerås/Hässlö	15–18	To be replaced by JA37 Viggen
			J35F Draken	Int	Västerås/Hässlö	15–18	
	F13 (Bråvalla)		J35F Draken	Int	Norrköping/Bråvalla	15–18	
			J35F Draken	Int	Norrköping/Bråvalla	15–18	
			SH37/SF37 Viggen	Recce	Norrköping/Bråvalla	15–18	
	F16 (Upplands)		J35F/Sk35C Draken	Int/train	Uppsala	15–18	
			J35F/Sk35C Draken	Int/train	Uppsala	15–18	
South Norrland	F4 (Jämtlands)		J35D Draken	Int	Ostersund/Fröson	15–18	
			J35D Draken	Int	Ostersund/Fröson	15–18	
			J35D Draken	Int	Ostersund/Fröson	15–18	
North Norrland	F21 (Norrbottens)		J35D/F Draken	Int	Luleå/Kallax	15–18	
			SF37 Viggen	Recce	Luleå/Kallax	15–18	Replaced S35E Draken
			Sk60B/C	Att/recce	Luleå/Kallax		
	F5		Sk60A	Train	Ljungbyhed	75 approx	Secondary attack role
			Sk61 Bulldog	Train	Ljungbyhed	58	
	F13M		J32D Lansen	Target tug	Linköping/Malmslätt	12	Converted J32Bs
			J32E Lansen	ECM trials	Linköping/Malmslätt	12	
			Sk60B/C	Train	Linköping/Malmslätt		Weapon training
			Tp85 Caravelle III	Trans	Linköping/Malmslätt	2	
	F18		Sk60A	Liaison	Stockholm/Tullinge	6	Secondary attack role
			Sk50 Safir	Liaison	Stockholm/Tullinge	7	
	F20		Sk60B/C	Train	Uppsala		
			Hkp2 Alouette II	Util	Söderhamn	1	
			Hkp3 AB.204B	Util	Ronneby,	6	
			Hkp4 BV-107	Trans/SAR	Luleå (shared)	10	Under Navy control?
Navy		1st and 2nd helicopter sqns	Hkp2 Alouette II	Util		10 approx	
			Hkp4 Boeing/Kawasaki 107	ASW/util	Berga (1st sqn) and Säve (2nd sqn)	10	Seven Kawasaki, three Boeing
			Hkp6 AB.206B	Util		10	
			Seacat	SAM	Three ships	3 launchers	
Army			Hkp2 Alouette II	Util		6	
			Hkp3 AB.204B	Util		12	
			Hkp6 AB.206A	Util/obs		40	
			Sk61 Bulldog	Train/att		20	
			Fp153 Do27	Liaison/spotting		5	
			Fp151 L-21B Super Cub	Liaison/spotting		12	
			Rb67 Improved Hawk	SAM		3 batteries	
			Rb69 Redeye	SAM		440	
			RBS70	SAM			

kens will be retired by the early 1980s.

The JA37, developed from the AJ37 attack variant (see below), has an uprated Volvo Flygmotor RM8B turbofan (licence-built Pratt & Whitney JT8D), LM Ericsson PS-46/A pulse-Doppler search and fire-control radar, Singer-Kearfott/Saab SKC 2037 central digital computer, Smiths Industries head-up display and new weapons. The last-named include the Rb71 medium-range air-to-air missile (British Aerospace Sky Flash) – of which £60m-worth were ordered in December 1978 – and Oerlikon KCA cannon. Development of the Saab Rb72 AAM, using infra-red guidance, was frozen in the spring of 1978 as a result of cost increases. The cannon is the same calibre (30mm) as the Aden gun which can be pod-mounted on the attack Viggen, but the kinetic energy of a round at impact is claimed to be 6.5 times greater. Muzzle velocity is 1050m/sec and rate of fire 1350 rounds/min; the weapon has been fired at 7.6g in a centrifuge, indicating its usefulness in dogfights.

The medium attack force comprises six *divisioner* of AJ37 Viggens in three *Flottiljer*. The type entered service in June 1971, replacing A32A Lansens (Lances). The Viggen carries seven permanent weapon hardpoints – three under the fuselage and two beneath each wing – and an additional attachment point under each wing can be used if required. Standard weapons include the Saab Rb04E heavy anti-ship missile, developed from the Rb04D which equipped Lansens; the Saab Rb05A medium air-to-surface missile for use against ships, land targets and slow-moving aircraft such as helicopters; Bofors 135mm rockets in four 24-round pods; Aden 30mm guns mounted in pods; and up to 16 bombs. The Hughes Maverick air-to-surface missile – to be designated Rb75 – has been ordered to arm the Viggens in place of the cancelled Rb05B television-guided development of the command-guided Rb05A.

The AJ37's weapon system is based on an LM Ericsson PS-37 attack radar, Saab CK-37 digital computer and Marconi Avionics head-up display. Sidewinder (Rb24) and Falcon (Rb28) air-to-air missiles can be carried in the secondary interception role. A Viggen can be brought to combat readiness in about 15min by its ground crew comprising a technician and six conscript airmen.

The *Flygvapen* has one specialised light ground-attack *division*, part of F21. The squadron operates the Saab Sk60B (Sk = Skol – trainer); the very similar Sk60A used for training would be available for attack missions in wartime, however, and would expand the Sk60 tactical-support force to five *divisioner*. Weapons can include bombs, rockets and Rb05A missiles.

The Sk60s were due to have been replaced by the Saab B3LA, an attack aircraft with a secondary training role. In November 1978, however, the Swedish Defence Minister told parliament that this solution was no longer being considered. An alternative proposed by FMV is development of a simpler trainer designated Sk38, which could be followed later by an A38 attack variant. Yet another project being examined for the strike role is the A20 development of Viggen, powered by the RM8B turbofan and incorporating several design changes in both structure and equipment. Either type would probably carry the Saab B83 air-to-surface missile, which is planned to have infra-red guidance.

The multi-role Viggen has been developed in two reconnaissance versions in addition to the attack, fighter and trainer variants. The first SH37 (S = Spaning – reconnaissance) sea-surveillance Viggen was delivered in June 1975 and a further two *divisioner* have followed. The SH37 has replaced the unarmed day/night-recce S32C Lansen, of which two squadrons were originally operated. A belly-mounted pod contains a wide-angle camera, and air-to-air missiles are carried for self-defence. A second reconnaissance version of the Viggen, the SF37, is replacing S35E Drakens in the overland role. The eight recce squadrons will be reduced to six by 1982. The SF37 carries cameras in a modified nose, ousting the search/attack radar (which is retained in the SH37), and may also be fitted with pod-mounted night-photography equipment and infra-red sensors. Tactical reconnaissance can also be carried out by the Sk60C, a modification of the Sk60B with cameras in a long nose.

Transport is provided by a *division* of C-130 Hercules, designated TP84s (Tp = Transport), and C-47s (Tp79s). Caravelles operated by the National Defence Research Institute act as a back-up; the C-47 fleet has completed more than 30 years' service. All transports were operated by F8 at Stockholm before it was disbanded in 1974.

Training is carried out on the Scottish Aviation Bulldog (Sk61) and Saab Sk60, together with two-seat operational trainers. The Saab Safir (Sk50), having been supplanted by the Bulldog as basic trainer, is employed in the liaison role; each Wing has two or three for general communications duties.

Flygvapnet's helicopter fleet is understood to have been transferred to naval control following a trial period.

Navy

Swedish Naval Aviation operates helicopters on anti-submarine, minesweeping, search-and-rescue and other duties from two shore bases and the frigates *Visby* and *Sundsvall*. Most, if not all, the rotary-wing craft previously operated by the *Flygvapen* are thought to have been transferred to naval control. The Swedish Navy also has a single four-round launcher for Short Seacat surface-to-air missiles on each of the three *Östergötland*-class destroyers.

Army

Swedish Army Aviation operates its own helicopters and fixed-wing types for liaison, troop support and spotting. The Army additionally deploys Raytheon Hawk (Rb67) medium-range surface-to-air missiles, being updated to Improved Hawk standard, together with General Dynamics Redeye (Rb69) and Bofors RBS70 short-range SAMs. The Army is responsible for air-base defence as well as protecting its own forces, and the laser beam-riding RBS70 is being deployed partially in this role.

The *Flygvapen*'s only combat experience was gained during 1940, when F19 – a volunteer unit equipped with 12 Gloster Gladiators and four Hawker Harts – flew against Russian forces invading Finland. Twelve Russian aircraft were shot down for the loss of three F19 machines. Another special unit, F22, operated J29B fighters and S29C reconnaissance aircraft from Leopoldville and Kamina in support of the United Nations peacekeeping force in the Congo during 1962 and 1963.

All basic flying training is carried out at F5, the *Krigsflygskolan* (war flying school). Some 45 pilots are recruited annually, spending four years on flying training and ground studies. Initial instruction is carried out on the Scottish Aviation Bulldog (Sk61), the pupil going solo after about 18hr and building up to 45hr before converting to the Saab 105 (Sk60). Wings are awarded after 130–160hr on the jet, students then moving to F20, the *Flygvapen Krigsskola* (Air Force Academy), for a year of officer training and further flying on the Sk60. Commissioning follows after a total of three years' instruction. The officer spends his fourth year flying about 150hr

at an operational conversion unit: one of F16's *divisioner* for the Draken and the F15 OCU for the Viggen. Pilots for the Army, Navy and Police, as well as flying instructors and air-traffic controllers, are also trained at F5.

A career officer pilot, of which there are normally 250–300, is not considered fully qualified until he has logged some 500hr. The average annual flying time is 150hr per pilot. Nearly 500 short-service non-commissioned officers back up the regular pool; they serve six-year contracts and can extend the period or apply for a permanent appointment at any time. If a pilot elects to leave after his six years he is commissioned as a reserve pilot. Conscripts, used on ground duties only, undergo three months' training followed by nine months with an operational unit.

The Central Training Establishment at

Halmstad no longer operates Pembrokes on multi-engine training but still runs courses in aircraft maintenance, signals, logistics, base defence and other operations. Radar instruction is based at F18, Tullinge, which additionally trains fighter controllers, ground technicians and officer cadets. The F13M Wing provides weapon-training facilities and runs courses for squadron commanders.

Each air base is responsible for its own repairs, modifications and overhauls, in co-operation with the Test Materiel Centre. Central Maintenance Depots at Malmslätt and Arboga provide support facilities for powerplants, avionics, ejection seats and other equipment.

Below: The British Aerospace Bulldog, designated Sk61 by the Swedish armed forces, is operated by both the Army (seen here) and Air Force.

FINLAND

Area: 130,000 sq. miles (336,700 sq. km.).
Population: 4·75 million.
Total armed forces: 39,900.
Estimated GNP: $31·7 billion (£15·85 billion).
Defence expenditure: $454 million (£227 million).

The Finnish Air Force (*Suomen Ilmavoimat*) was formed in March 1918 as *Ilmailuvoimat* (the Aviation Force) and was involved in extensive fighting with the Soviet Union for the first quarter-century of its existence. The force distinguished itself in the 1939–40 Winter War and the following Continuation War of 1940–44, and its present shape was determined by the peace treaty signed with the Soviet Union and other countries in February 1947. Under the terms of this treaty the Finnish armed forces are limited to a total of 41,900 personnel, of whom only 3000 may be in *Ilmavoimat*. Nuclear weapons, bombers and, originally, guided missiles were prohibited, although the last-named of these restrictions was lifted in 1962 with the proviso that any guided weapons should be for defensive purposes only. The force is, however, still limited to a total of 60 combat aircraft.

Finland has the fifth-largest area of any

country in Europe but is sparsely populated, most of the 4.7 million inhabitants living in the industrialised south. To the east is the Soviet Union, with which Finland shares a border of more than 700 miles, while to the north and west are Norway (a NATO member) and neutral Sweden. The major role of *Ilmavoimat* is therefore to protect Finnish airspace and to co-operate as necessary with the other Services.

The Air Force has its headquarters at Tikkakoski, near Jyväskylä air base, and

Below: After leasing ex-Swedish Air Force Drakens, the Finnish Air Force bought them outright and ordered others built under licence by Valmet.

operates aircraft supplied by both the Soviet Union and the West in addition to those provided by the indigenous industry. Its commander reports to the commander-in-chief of the defence forces who is in turn responsible to the President, who acts as supreme commander. Defence of the country against an airborne attack is the responsibility of three air-defence wings named after Finnish provinces, each of these having the task of protecting its particular area and being further broken down into smaller military areas. The commanding officer of each wing or *Lennosto* is directly responsible to the C-in-C of *Ilmavoimat* and is required to co-operate with the other military heads within his assigned area. The northern half of the country is covered by the *Lapin Lennosto* or Lapland Wing, the south-east is the responsibility of the *Karjalan Lennosto* (Karelia Wing) and the west is allocated to the *Satakunnan Lennosto* (Satakunta Wing). Each wing is intended to have its own fighter squadron or *Hävittäjälaivue*, but only the first two have a combat component in fact, the unit assigned to the Satakunta Wing having a training role despite its fighter-squadron designation.

From 1963 until the mid-1970s *Ilmavoimat*'s main combat force comprised two squadrons of MiG-21F-12 day fighters supplied as part of a Finnish-Russian trade agreement negotiated in 1962, the foundations of which had been laid by the 1948 Treaty of Friendship, Co-operation and Mutual Assistance; this obliges Finland to resist any forces which might attempt to attack the USSR via that country, in addition to protecting its own homeland, and was extended by a further 20 years in 1970. The first ten MiG-21s were delivered in April 1963 and all 22 were on strength by late that November; the aircraft were armed with an integral 30mm cannon and could carry either UV-16-57 pods containing 16 55mm rockets each or, since the ban on such weapons was relaxed, AA-2 Atoll air-to-air missiles on a pair of underwing pylons.

The MiG-21s had been preceded in November 1962 by four MiG-15UTI advanced trainers and *Ilmavoimat* pilots also underwent instruction in the Soviet Union. A pair of MiG-21U trainers followed in March 1965 and were joined by the same number of MiG-21USs in June 1974; the three remaining MiG-15UTIs were retired in the mid-1970s and one of the original MiG-21Us has been lost, leaving three MiG-21U trainers remaining in service. The MiG-21 has been fully operational in Finland since 1965, and a replacement was sought from the mid-1970s under the designation MiG-X. This led to speculation that the new type might be the MiG-23MS Flogger E, but the selected aircraft turned out to be the MiG-21bis. The first two of what was expected to total 30 such all-weather fighters were delivered in October 1978 and the earlier model was expected to be phased out of service by the end of 1979. A two-seat version of the latest type is thought to have been supplied to *Ilmavoimat* in early 1978.

The MiG-21s are charged with defending south-east Finland, a task which is shouldered by Saab Drakens in the north. The first two of six Draken 35BSs – refurbished ex-Royal Swedish Air Force J35Bs with their all-weather avionics removed – leased from Sweden were delivered in May 1972 and were joined from April 1974 by the initial aircraft in a batch of 12 new Draken 35XSs which were assembled in Finland by Valmet (Valtion Metallitehtaat) under the terms of a contract signed with Saab in April 1970. All 12 were on strength by July 1975, allowing the Draken squadron to move from Jyväskylä/Luonet-

järvi to Rovaniemi. In April 1976 *Ilmavoimat* was authorised to buy a further 15 Drakens: the six leased aircraft, an additional half-dozen low-time ex-Swedish J35Fs (funding for which had been allocated the previous October) and three Sk35C trainers. These two-seaters, together with a flight simulator bought as part of the deal, allowed training of Draken pilots to be transferred from Sweden to Finland. The Sk35Cs were overhauled and modified in Finland, the first two becoming operational in June 1976.

The fighters operate within an air-defence system supplied by Plessey Radar and Elliott-Automation, which was completed in 1968. These long-range surveillance and height-finding radars are being augmented by Finnish-made low-level gap-fillers. The Finnish Army operates anti-aircraft guns, and these were joined in September 1978 by Russian-supplied SA-7 Grail shoulder-launched surface-to-air missiles. A further upgrading of the country's defences was due to begin at the end of 1979, when the first of an unknown number of SA-3 Goa medium-range SAMs was due to be emplaced near Helsinki. The missile purchases are being financed by a loan from the USSR, and the self-propelled SA-6 Gainful may also be introduced.

Another major re-equipment programme involves the British Aerospace Hawk light strike-trainer, 50 of which were ordered in December 1977, with the contract being ratified by the Finnish Government the following June. In a deal worth some £120m, BAe will supply four complete Hawk T.51s from October 1980 and Valmet will assemble the remaining 46 at Kuorevesi. The Finnish company will carry out about 10% by value of the work, manufacturing parts from materials supplied from Britain. Tooling was scheduled to be delivered to Finland in August 1979,

Above: Finland buys its aircraft impartially from East and West, this Russian Mi-8 operating alongside types of US, Swedish, French and British origin.

allowing production of some components to begin in October. The first Valmet-assembled Hawk is due to fly in February 1981 and deliveries will last until 1984–85, by which time up to 50 more Hawks may have been ordered.

The type which the Hawk will replace, the Fouga Magister, was also assembled under licence in Finland. Eighteen were supplied from the French production line, these being followed by 62 assembled by Valmet between 1960 and 1967. A modification programme was begun in 1972 to extend their effective life. Another venerable trainer scheduled for replacement is the Saab Safir, which is to be superseded by the Valmet Leko-70 Vinka (Blast). Thirty Vinkas were ordered in January 1977; the first was scheduled for delivery in April 1979, with the last being on strength by January 1981.

Training takes place at the air academy at Kuahava, using Safirs and Magisters until the Vinkas and Hawks are in service. Combat pilots then go to their squadrons for further instruction on the Sk35C Draken or MiG-21U. Other centres of instruction include a technical school at Kuorevesi and a communications school at Luonetjärvi.

The transport squadron, formed in March 1962, has a transport flight operating Dakotas (a mixture of ex-Finnair C-47s and C-53s) and a helicopter flight with a variety of rotary-wing craft. The Dakotas were planned to remain in service until 1985, but one crashed in October 1978 – killing all 15 people on board, including three members of parliament – and An-32s are now expected to replace these elderly machines.

FINLAND

Wing	Unit	Type	Role	Base	No	Notes
Lapin Lennosto	*HävLv* 11	35XS Draken	Int	Rovaniemi	12	Valmet-built
		35BS Draken	Int	Rovaniemi	6	Ex-Swedish AF
		J35F Draken	Int	Rovaniemi	6	Ex-Swedish AF
		SK35G Draken	Train	Rovaniemi	3	Ex-Swedish AF
		Magister	Train	Rovaniemi	some	
Satakunnan Lennosto	*HävLv* 21	Magister	Train/light att	Pori*, Tampere	60	Includes those with *HävLv* 11 and 31
Karjalan Lennosto	*HävLv* 31	MiG-21 bis	Int	Kuopio-Rissala	30	Replaced 18 MiG-21F
		MiG-21U/US	Train	Kuopio-Rissala	1/2	
		Magister	Train	Kuopio-Rissala	some	
	Transport Squadron	C-47/53	Trans	Utti	6	May be replaced by An-32
		Mi-8	Trans	Utti	6	
		Mi-4	Trans	Utti	3	To be replaced
		Hughes 500C	Trans	Utti	1	One other lost
	Liaison/Comms and Survey Flights	Cessna 402B	Liaison/comms/survey	Tikkakoski	2	Leased
		Cherokee Arrow	Liaison/comms/survey	Tikkakoski	5	Leased
	Air Academy	Safir	Train/comms	Kauhava, Luonet-järvi	24	Originally 36; to be replaced by Vinka
		SA-7	SAM	Helsinki and		Operated by Army
		SA-3	SAM	with troops		Operated by Army

*Expected to transfer to Tampere/Pirkkala, construction of which began in November 1977

DENMARK

Area: 17,000 sq. miles (44,000 sq. km.).
Population: 5·1 million.
Total armed forces: 34,000.
Estimated GNP: $43·8 billion (£21·9 billion).
Defence expenditure: $1·28 billion (£640 million).

The Royal Danish Air Force (*Kongelige Danske Flyvevåben*), with its headquarters at Vedbaek north of Copenhagen, is responsible for the great majority of military aviation, although both the army and navy have small air components. The three Services retain their individual identities but since 1970 they have been administered by a joint high command. The basic unit is an *eskadrille* (squadron), all of which come under *Flyvertaktisk Kommando* (Tactical Air Command) based at Kølvraa near Karup. This command is also responsible for an Air Defence Group of surface-to-air missiles. Under a 1973 defence act the air force's strength is laid down as 116 front-line aircraft in six squadrons; personnel strength is about 7100. Denmark is a member of NATO and her air force is assigned to Allied Air Forces Northern Europe.

Inventory

The six combat squadrons comprise two units each of Starfighters, Super Sabres and Drakens, with General Dynamics F-16s replacing the remaining F-100s from the end of 1979. The first of 55 ex-USAF F-100Ds were delivered in June 1959 and ten F-100F two-seaters followed in that year. The Super Sabres replaced F-84G Thunderjets and were upgraded by the installation of Saab BT9 bombing computers in the 1960s, but attrition and a shortage of fatigue life have combined to shrink the fleet. A modification programme to extend the useful life was begun in 1973, and 14 ex-US Air National Guard TF-100Fs were supplied during the following year, but by 1978 the operational force had been reduced to some 24 single-seaters and 14 two-seaters. At one time the F-100 fighter-bomber force totalled three squadrons, but this shrank to two units when *Esk* 725 converted to the Saab Draken. The remaining Super Sabres are to be replaced from late 1979 by the F-16, of which 36

single-seat F-16As and 12 F-16B trainers (with ten more aircraft on option) were ordered as part of the 'Sale of the Century' deal under which the type is also being acquired by Holland, Belgium and Norway. Danish industry is participating in the F-16 programme, and the first unit to operate the type will be *Esk* 724. This unit, which formerly operated Hunters from Skrydstrup until its disbandment in 1973, will replace *Esk* 730; the other remaining Super Sabre squadron, *Esk* 727, will also be disbanded when *Esk* 728 is operational as the second F-16 unit.

Although the F-16 was billed as a Starfighter replacement, the RDAF's F-104s will continue to form the interceptor element until 1985. The first of 25 Canadian-built F-104Gs was delivered in November 1964 and the batch was complete by the following June. Four Lockheed-built TF-104Gs were supplied at the same time and the force has since been bolstered by the acquisition of ex-Canadian aircraft – 15 CF-104Gs and seven CF-104Ds – which were ordered in September 1971. These second-hand aircraft were fitted with interceptor radars and fire control systems in Denmark and were additionally modified to take 20mm Vulcan cannon, an uprated engine and other improvements. Deliveries of the CF-104Ds – some of which are operated in the electronic-countermeasures role – began in November 1972, with the CF-104Gs arriving over an 11-month period from April 1973.

The newest RDAF combat type before the arrival of F-16s was the Saab Draken, which was selected in preference to the Dassault Mirage 5 and Northrop F-5. A contract placed in March 1968 covered 20 Draken A35XDs for ground attack and three Sk35XD trainers; these are designated F-35 and TF-35 respectively by the RDAF. The A35XDs incorporate a number of modifications specifically for Danish use, including Aden 30mm cannon mounted in the wing roots, four underwing stores, hardpoints and provision for large auxiliary fuel tanks beneath the belly. The main air-to-surface weapon is the AGM-12B Bullpup, 560 of which were built under licence by Konsberg Våpenfabrikk in Norway for the RDAF. The Drakens can also be fitted with the Sidewinder air-to-air missile, 650 AIM-9B FGW Mod

2s having been bought from Bodenseewerk. The force has since been expanded by the delivery of a further three two-seat TF-35s (Sk35XDs) and 20 RF-35s (S35XDs) which were ordered in June 1968. The RF-35s were originally equipped for day reconnaissance only, but Red Baron infra-red recce pods were ordered from Sweden in 1975. Finally, five additional two-seat Drakens were ordered in November 1973 and delivered from May 1976.

An Air Defence Group operates the long-range Nike Hercules surface-to-air missile and the medium-range Hawk. The former were taken over from the army in 1962 and were joined three years later by the Hawks, now being upgraded to Improved Hawk standard, at four fixed sites around Copenhagen.

The air force's sole large transport type is the Lockheed C-130 Hercules, all C-54s having been withdrawn. The three C-130s delivered in 1975 are expected to be supplemented by two more, and in late 1978 the RDAF was reported to have ordered two

DENMARK

Command		Unit	Type	Role	Base	No	Notes
Air Force							
Tactical Air Command (*Flyvertaktisk Kommando*)		*Esk* 723	F-104G/CF-104G/	Int/train/ECM	Aalborg	20/15	Figures approximate; some
		Esk 726	TF-104G/CF-104D			3/7	CF-104Ds detached for ECM
		Esk 725	F-35/TF-35	FGA/train	Karup	20/3	
		Esk 729	RF-35/TF-35	Recce/train	Karup	20/3	
		Esk 727	F-100D/F	FGA/train	Skrydstrup	24/14	To be replaced by F-16
		Esk 730					
		Esk 721	C-130H	Trans	Vaerløse	3	Replaced C-47/C-54
		Esk 722	S-61A	SAR	Vaerløse	8	Some detached
		Esk 724	F-16A/B	FGA/train	Skrydstrup	36/12	To replace F-100s from 1979–80
		Esk 728					
	Air Defence Group		Nike Hercules/Improved Hawk	SAM	various		
Army			T-17 Supporter	Train/util	Avnø	17	Six more allocated to station flights
			T-17 Supporter	Util	Vandel	9	Replaced L18C/L-21/KZ.VII
			Hughes 500M	Util	Vandel	15	
			Hamlet (Redeye)	SAM		200	
Navy			Alouette III	Util	frigates and fishery vessels	8	
			Lynx	Util	frigates and shore bases	7	
			Seasparrow	SAM	frigates and mine-layers		

Boeing 737s for VIP transport and government transport duties. The possibility of purchasing two smaller VIP transports such as Dassault Falcon 20s has also been mentioned.

The helicopter force is based on the Sikorsky S-61A, eight of which are operated. They form part of a search-and-rescue system directed from the RDAF's rescue co-ordination centre at Karup, which is connected to a similar centre run by the navy at Aarhus.

RDAF training begins on the Saab Supporter (designated T-17 in Denmark), students being graded on this type at Avnø. Denmark ordered 32 T-17s in January 1975, of which 23 are operated by the air force and the remainder by the army. Deliveries were completed in May 1977. Pupils then go to the United States for further instruction on T-41s, T-37s and T-38s before returning to Denmark for operational conversion on the two-seaters attached to combat squadrons. Helicopter training also takes place in the United States.

Army
Danish army aviation was re-established in July 1971 under the title *Haerens Flyvetjeneste* (Army Flying Service) and operates 15 Hughes 500M helicopters together with nine of the T-17 Supporters, the last-named replacing KZ.VII and Piper L-18C Super Cub liaison aircraft. The Danish Army has relinquished its large surface-to-air missiles to the RDAF, but the ground forces continue to operate 200 Hamlets, a developed version of the General Dynamics Redeye shoulder-launched SAM.

Navy
Naval crews fly eight Alouette III helicopters from frigates, fishery-protection vessels and shore bases, although the craft are maintained by the RDAF and share *Esk 722*'s facilities at Vaerløse. *Sovaernets Flyvevaesen* is being strengthened during 1979 by the delivery of seven Westland/Aérospatiale Lynxes which will fly from frigates and shore bases on reconnaissance and fishery-protection duties in Greenland, the Faeroes and Denmark itself.

The navy also operates surface-to-air missiles in the form of NATO Seasparrow, which arms frigates and minelayers.

WEST GERMANY

Area: 96,000 sq. miles (248,000 sq. km.).
Population: 61·3 million.
Total armed forces: 489,900.
Estimated GNP: $508·6 billion (£254·3 billion).
Defence expenditure: $17·26 billion (£8·63 billion).

Germany occupies a unique position, being sandwiched between the Soviet Bloc and the rest of western Europe. Her armed forces, all three of which operate aircraft, must therefore be able to blunt the initial thrust of an attack from the east to allow the rest of the NATO alliance time to reinforce the front line. The *Luftwaffe* is one of the largest and best equipped air forces in the world, and Germany spends more on defence than any other member of NATO apart from the United States.

Germany has always been well to the fore in military aviation. During the First World War the German army and navy operated both heavier-than-air machines and airships in tactical and strategic roles. Neither the high quality of her aircraft nor the flair of her pilots could save Germany from defeat, but the foundations had been laid for a resurrected air force. The *Luftwaffe* was founded as an independent Service in 1935, with the prime aim of supporting the army. Although strategic operations were again carried out, the force remained mainly a tactical arm and became increasingly devoted to home defence as the conflict turned against Germany.

Air Force
The present *Luftwaffe* was established in 1956, when the Federal Republic became a full member of NATO. In peacetime the force comes under the control of the Federal Defence Minister, on whose behalf the *Luftwaffe* Chief of Staff issues orders. The *Luftwaffe* CoS is a member of the Federal Armed Forces Defence Council and heads the Air Force Staff within the Defence Ministry, but he does not exercise operational control over his units. Control of the *Luftwaffe*'s combat forces is assigned to the Commanders of Allied Air Forces Central Europe and of the Baltic Approaches. The former controls two Allied Tactical Air Forces, both of which have *Luftwaffe* aircraft: the 2nd ATAF, which also includes Belgian, British and Dutch units, has its headquarters at Münchengladbach; and the 4th ATAF, with US and Canadian forces in addition to *Luftwaffe* squadrons, is run from Ramstein.

The *Luftwaffe*'s Nike Hercules and Improved Hawk surface-to-air missile batteries are fully assigned to NATO in peacetime, as are nine air-defence control and reporting centres and two low-altitude reporting battalions, the latter having 24 radars each, which are integrated with the Nadge early-warning and control network.

Four Hughes S-band three-dimensional air-defence radars were ordered in the autumn of 1978 to replace the same number of L-band two-dimensional FPS-7 sets. In wartime these forces would be joined by 24 *Staffeln* (squadrons) of manned interceptors, fighter-bombers and reconnaissance aircraft, making up about one-third of NATO's air forces in the central region.

The *Luftwaffe* Chief of Staff has three subordinate commands, all with headquarters at Porz-am-Rhein: Tactical Air Forces; Support; and General Air Force. Tactical Air Forces has four divisions: the 1st and 3rd, with headquarters at Lautingen and Munster respectively, are

Below: The *Luftwaffe*'s Pershing battlefield-support missiles complement the deep penetration capability which Tornado will confer.

WEST GERMANY

Command	Wing	Unit	Type	Role	Base	No	Notes
Air Force							
	JG 71 *Richthofen*		F-4F	Int	Wittmundhafen	30	
	JG 74 *Mölders*		F-4F	Int	Neuburg	30	Replaced F-104G
	JaboG 31 *Boelcke*		F-104G	FGA	Norvenich	36	To be replaced by
	JaboG 32		F-104G	FGA	Lechfeld	36	Tornado. Each *Geschwader* also
	JaboG 33		F-104G	FGA	Büchel	36	has six aircraft in reserve and
	JaboG 34		F-104G	FGA	Memmingen	36	eight in maintenance
	JaboG 35		F-4F	FGA	Pferdsfeld	36	Formerly LeKG 42 with G.91R
	JaboG 36		F-4F	FGA	Rheine-Hopsten	36	Replaced F-104G
	LeKG 41		G.91R/3	Light att	Husum	42	
	LeKG 43		G.91R/3	Light att	Oldenburg	42	To be replaced by Alpha Jet
	AG 51 *Immelmann*		RF-4E	Recce	Bremgarten	30	
	AG 52		RF-4E	Recce	Leck	30	Replaced RF-104G
	FKG 1		Pershing 1A	SSM	Landsberg	36 launchers	Four *Staffeln*
	FKG 2		Pershing 1A	SSM	Geilenkirchen	36 launchers	Four *Staffeln*
			Nike Hercules	SAM	various	216 launchers	Six battalions of four batteries with nine launchers each
			Improved Hawk	SAM	various	216 launchers	Nine battalions of four batteries with six launchers each
	LTG 61		C.160D Transall	Trans	Landsberg	36	
	LTG 63		C.160D Transall	Trans	Hohn	36	
	HTG 64		UH-1D	Trans/util	Ahlhorn	105	Four *Staffeln*
		WS 10	F-104G/TF-104G	Train	Jever		To be replaced by Tornado
		WS 50	G.91T	Train	Fürstenfeldbruck	55	To be replaced by Alpha Jet
		FFS-C	P.149D	Train	Utersen	40	
		FFS-S	C.160D Transall	Train	Wunstorf	14	
			Do28D-2	Train	Wunstorf	6	
		HFS	UH.1D	Train	Fassberg	25	
Luftwaffenaus-bildungskommando USA	3525th Pilot Training Wing (USAF)		T-37B	Train	Sheppard AFB	35	
	4510th Combat Crew Training Wing (USAF)		T-38A	Train	Sheppard AFB	41	
			F-104G/TF-104G	Train	Luke AFB	48/34	
			F-4E	Train	George AFB	10	
			Hansa Jet 320	ECM	Lechfeld	4	
		FBSS	Boeing 707-320C	VIP	Köln-Wahn	4	
			JetStar	VIP	Köln-Wahn	3	
			Hansa Jet 320	VIP	Köln-Wahn	4	
			VFW-Fokker 614	VIP	Köln-Wahn	3	Replaced CV-440
			UH-1D	VIP	Köln-Wahn	4	
			Do28D-2	VIP	Köln-Wahn	2	
			Do28D-2	Liaison/trans	various		Four assigned to each *Geschwader*
		Erprobungsstelle 61	Do28D-2	Train	Kaufbeuren	2	
			Do28D-2	Test	Manching	1	
			Hansa Jet 320	Test	Manching	2	
			Transall	Test	Manching	3	
			UH-1D	SAR	various		Three *Staffeln*
			Alouette II	SAR	various	10	
			Bell 212	SAR	Sardinia	3	
			Canberra	Misc	various	3	
			Pembroke	Misc	various	9	
			C-47	Misc	various	2	
			Noratlas	Misc	various	5	
Army							
HFK 1	LHFTR 10	HFVS 101	Alouette II	Liaison	Rheine-Bentlage	10	To be replaced by BO105
			UH-1D	Trans	Celle	20 20	
	MHFTR 15		CH-53G	Trans	Rheine-Bentlage	16 16	
HFK 2	LHFTR 20	HFVS 201	Alouette II	Liaison	Laupheim	10	To be replaced by BO105
			UH-1D	Trans	Roth	20 20	
	MHFTR 25		CH-53G	Trans	Laupheim?	16 16	
HFK 3	LHFTR 30	HFVS 301	Alouette II	Liaison	Niedermendig	10	To be replaced by BO105
			UH-1D	Trans	Fritzlar	20 20	
	MHFTR 35		CH-53G	Trans	Niedermendig	16 16	
		HFS 1			Hildesheim	10	
		HFS 2			Fritzlar	10	
		HFS 3			Rotenburg	10	
		HFS 4			Roth	10	
		HFS 5	Alouette II	Comms/obs/liaison	Niedermendig	10	To be replaced by BO105
		HFS 7			Celle	10	
		HFS 8			Oberschleissheim	10	
		HFS 10			Rheine-Bentlage	10	
		HFS 11			Celle	10	
		HFS 12			Niederstetten	10	
		HflWaS	CH-53G	Train	Buckeburg	15	
			UH-1D	Train	Buckeburg	30	
			Alouette II	Train	Buckeburg	40	
		HFB 6	UH-1D	Trans	Itzehoe-Hungriger Wolf	20	
			Alouette II	Liaison		?	
			Redeye	SAM	various	1000	
Navy	MFG 1		F-104G/RF-104G	FGA/recce	Schleswig	54/25	Three *Staffeln* of F-104G, one of RF-104G
	MFG 2		F-104G/RF-104G	FGA/recce	Eggebeck		
	MFG 3 *Graf Zeppelin*		Atlantic	ASW/MR/ECM	Nordholz	15	Two *Staffeln*
	MFG 5		Sea King Mk 41	SAR	Kiel-Holtenau plus detachments (Borkum, Helgoland, Sylt)	21	
			Do28D-2	Liaison	Kiel-Holtenau	20	
			H-34G	Liaison	Kiel-Holtenau	15	
			Standard	SAM	Destroyers	3 vessels	Replaced Tartar
			Seasparrow	SAM	Frigates	6 vessels	

Notes: JG = Jagdgeschwader; JaboG = Jagdbombergeschwader; LeKG = Leichte Kampfflugzeuggeschwader; AG = Aufklärungsgeschwader; FKG = Flugkörpergeschwader; LTG = Lufttransportgeschwader; HTG = Hubschraubergeschwader; WS = Waffenschule; FFS = Flugzeugführerschule; HFS = Hubschrauberführerschule; FBSS = Flugberietschaftstaffel; HFK = Heeresfliegerkommando; HFVS = Heeresfliege Verbindungsstaffel; L/M/HFTR = Leichte/Mittlere Heeresfliegertransportregiment; HFS = Heeresfliegerstaffel; HflWaS = Heeresfliegger Waffenschule; HFB = Heeresfliegerbatталion; MFG = Mannefliegergeschwader.

responsible for tactical offensive operations while the 2nd and 4th, run from Karlsruhe and Aurich, control air-defence units. The divisions' operations centres may be either Allied Tactical Operations Centres, which are combined German/NATO facilities, or the German section of an Allied Sector Operations Centre. *Luftwaffe* Support Command includes logistics units and the Air Material Office, and the General Office is responsible for air transport and training in addition to communications and other tasks. An anomaly is the allocation of Basic Training Command, USA, to Tactical Air Forces Command, however.

The *Luftwaffe* has some 110,000 personnel, including 39,000 conscripts, plus a further 100,000 in reserve.

Inventory
A high rate of spending on defence combined with a low inflation rate has allowed Germany to improve the equipment operated by her armed forces and to introduce new weapons. In the 1960s, under the doctrine of massive retaliation, the *Luftwaffe* emphasised its tactical nuclear forces. Five *Geschwader* (wings) – which in the case of combat forces each comprise two *Staffeln* – of F-104G Starfighters were assigned to the strike role, and until 1966–67 a quick-reaction force of 30 aircraft, each carrying a single one-megaton free-fall nuclear bomb, was kept on permanent alert. A nuclear capability is retained, but most of the Starfighter force has been re-allocated to conventional fighter-bomber operations under the doctrine of flexible response.

The theoretical strength of a Starfighter *Geschwader* is 50 aircraft, comprising two 18-strong *Staffeln*, six F-104s in reserve and eight undergoing maintenance. The high attrition rate suffered by the type, combined with a shortage of spares, has however reduced some *Geschwader* to 15 operational aircraft and the number in reserve varies from unit to unit. The four remaining fighter-bomber F-104 wings are due to re-equip with the Panavia Tornado multi-role combat aircraft from the end

of the 1970s; the *Luftwaffe* is scheduled to receive 212, with the last being delivered by September 1986. The Tornado will be allocated the roles of long-range and battlefield interdiction, air superiority and reconnaissance in addition to its naval missions. Weapons will include the AIM-9L Super Sidewinder air-to-air missile and probably the longer-range AIM-7F Sparrow, air-to-surface missiles and the MBB MW-1, which ejects submunitions from a dispenser. Tornado's terrain-following radar allows missions to be carried out in all weathers and at night, and the type is expected to carry advanced precision-guided munitions.

The Starfighter has already been replaced in the air-defence role by the F-4F Phantom. Two *Geshwader* each are assigned to interception and ground attack, in the latter case replacing G.91Rs of LeKG 42 as well as F-104Gs. The purchase of 175 slatted F-4s followed the acquisition of 88 RF-4Es, fitted with Goodyear sideways-looking airborne radar and a real-time data link for ground-processing of reconnaissance information. The recce Phantoms superseded RF-104Gs; 30 of the RF-4Es are operated by each of the two *Geschwader*, a further four are engaged in special duties and the remainder are held in reserve. The reconnaissance Phantoms

Above the MBB Kormoran anti-ship missile gives the *Marineflieger*'s F-104G Starfighters a new punch and will also arm Tornado.

are being fitted with advanced avionics and hardpoints by MBB so that they can carry out a secondary fighter-bomber role.

The light strike force had shrunk to two wings of Aeritalia G.91Rs by the mid-1970s, following the disbandment of LeKG 44 and the conversion of another to the F-4F as JaboG 35. The *Luftwaffe* received a total of 346 G.91s from Aeritalia (Fiat) and licence-production in Germany, but the type will be retired by the beginning of the 1980s. Its replacement is the Dassault/Dornier Alpha Jet 1A, the first production examples of which were delivered to the *Luftwaffe* in late 1978 following a maiden flight in April of that year. The initial two Alpha Jets were allocated to technical training of ground crews at Fassberg, with the next pair being based at Manching for extensive service trials. Pilot conversion and tactical training on the type were scheduled to begin at *Waffenschule* 50 in February 1979, the former G.91T unit being designated JaboG 49. Alpha Jets will then replace

Below: The *Luftwaffe* cut its teeth on RF-4E reconnaissance Phantoms such as this before deploying F-4Fs in the interception role.

G.91Rs at Decimomannu in Sardinia, where they are based for weapon training, with the combat wings LeKG 41 and 43 becoming operational with the type in 1980–81. These units will be redesignated JaboG 41 and 43 to reflect their fighter-bomber role. Dornier was scheduled to have completed eight production Alpha Jets by the end of 1978, and the rate is building up to six a month. The last of the *Luftwaffe*'s 200 Alpha Jets should be on strength by the spring of 1982.

The *Luftwaffe*'s Alpha Jets, unlike those ordered for the French *Armée de l'Air*, have a major combat role in addition to training. The type carries a Mauser 27mm cannon, as fitted to Tornado, and can be armed with 2000kg of weapons on four wing stations. Its roles include anti-helicopter operations and reconnaissance as well as close air support and training.

The deep strike role, at present allocated to F-104Gs and later to Tornados, is shared by Pershing 1A short-range ballistic missiles. The *Luftwaffe* has two wings, each of four *Staffeln*, equipped with a total of 72 launchers for the weapon. The force's surface-to-air missiles are being upgraded, with the Hawk batteries undergoing conversion to Improved Hawk standard. The *Luftwaffe* is also expected to order Euromissile Roland II low-level SAMs for airbase defence, the missiles supplementing radar-directed 20mm cannon. The Service will make use of the 18 Boeing E-3A Sentry AWACS (Airborne Warning And Control System) aircraft which have been ordered by NATO for basing in Germany.

The backbone of the transport fleet comprises the Dassault-Breguet Transall, the Noratlas having been retired. The *Luftwaffe* received 121 Dornier Do28D-2 Skyservants to replace Do27s and Pembrokes, four being allocated to each combat *Geschwader* for liaison and light transport. The air force also has 132 Bell UH-1Ds, scheduled for replacement in the mid-1980s.

Training
The *Luftwaffe* is virtually unique in operating only a handful of obsolescent types, and this applies to the training fleet as much as the combat forces. A replacement is being sought for the Piaggio P.149Ds used for initial grading, and the T-33 and Magister have been retired. The contenders for the P.149-replacement order were the Beech T-34C, Pilatus PC-7 and RFB Fantrainer, with an initial contract for 30 anticipated. Following their initial flights in the P.149Ds, students transfer to the United States for 130hr basic training on T-37s and a similar period of advanced instruction on T-38s. This part of the course lasts 13 months. An additional 125hr on F-104s leads to the award of a pilot's wings, and he then returns to Germany for Starfighter operational conversion with WS 10 or remains in the US for a similar course on F-4Es. The *Luftwaffe* will receive 47 dual-control Tornados, of which 22 will be allocated to the tri-national training base at RAF Cottesmore. Training of crews for multi-engine aircraft and helicopters takes place in Germany, at Wunstorf and Fassberg respectively.

The aircraft used for training in the United States carry USAF markings but are owned by the *Luftwaffe*. The cost of running these aircraft and the bases from which they operate is set against war reparations and maintaining US forces in Germany, where the USAF has nine bases. Between 2000 and 2500 *Luftwaffe* personnel are instructed in the United States every year; missile operators attend courses at Fort Sill and Fort Bliss, the latter being the headquarters of the German Training Command in the US.

Army
The *Heeresflieger* (army air corps) was formed in 1957, using helicopters bought from the United States and France together with Do27 liaison aircraft. Increased emphasis has been placed on assault helicopters following US experience in Vietnam, and the Hfl is charged with providing support for other NATO armies, notably those of the US and Britain. Since 1971 the army has been organised into three corps, each with its own *Heeresfliegerkommando* (army aviation command): No 1 HFK at Handorf; No 2 at Laupheim; and No 3 at Neidermendig. Each HFK comprises a headquarters company, a *Heeresflieger Verbindungsstaffel* (liaison squadron) and two regiments of two squadrons each for light transport (*Leichtes Heeresfliegertransportstaffeln*) and medium transport (*Mittleres* HFTS). The HFKs also have access to ten *Heeresfliegerstaffeln* which provide helicopters for communications and utility work for a corps, and observation and liaison at division level. The HFSs were designated battalions until 1972, and each had 15 helicopters compared with ten today. One battalion remains, however – HF *Battalion* 6, which supports Allied Forces Northern Europe. The Hfl has some 3000 personnel, including 750 pilots.

Inventory
The three MHFTRs are equipped with the Sikorsky CH-53G, of which 110 were built under licence by VFW-Fokker to replace CH-34s. Each MHFTR is intended to be capable of lifting a reinforced air-mobile battalion – comprising up to 1000 men and 120 tons of equipment – in a single assault wave on behalf of its parent corps. Twelve CH-53Gs can transport nearly 1000 troops a distance of 140km in 30min, compared with six hours with the Hfl's previous helicopters, and a fully equipped infantry company can be moved 250km by just three aircraft. Accommodation is provided for 38 fully equipped troops, or 66 in the high-density layout, and a pair of 2000lb cargo hoists are fitted.

The LHFTRs are equipped with Bell/Dornier UH-1Ds, of which 387 were built under licence in Germany. The Hueys are expected to be replaced in the 1980s by a new helicopter to be designed and built by a European consortium comprising some or all of MBB, Aérospatiale, Agusta and Westland. The major replacement programme in the late 1970s, however, involves the MBB BO105. The BO105M has been selected as the new VBH (*Verbindungs- und Beobachtungshubschrauber*), a communications and observation helicopter to succeed the Alouette IIs in the

HFVSs and HFSs. The first hundred VBHs are being delivered at a rate of four a month from September 1979, but the remaining 127 are to be converted from BO105Ps when these interim anti-tank helicopters are supplanted by purpose-designed gunships in the 1980s (see below).

Roles to be undertaken by the VBH BO105Ms include forward air control, transfer of commanders, artillery direction and casualty evacuation. An increased gross weight has necessitated strengthening of the transmission, and other improvements are being incorporated.

The armed BO105P, designated PAH-1 (*Panzer Abwehr Hubschrauber*), is scheduled to enter service in 1980 and all 212 should have been delivered by 1982. Each of the three army corps will have a 56-helicopter anti-tank regiment added to the present organisation, the PAH-1 carrying six long-range Euromissile HOT anti-tank missiles directed through an APX 397 sight. The type will also equip HFB 6 in Schleswig-Holstein, which is to have 21 PAH-1s.

The F-104G force has been brought up to nearly 100 by aircraft transferred from the air force, and some of the reconnaissance Starfighters are held in reserve.

One *Geschwader* is being armed with the MBB Kormoran anti-ship missile, replacing the Aérospatiale AS.30. The initial order is for 56 aircraft installations, each F-104G being equipped with two of the weapons. The Starfighters are scheduled to be replaced in the early 1980s by 113 Panavia Tornados, which will also carry Kormoran.

The single-*Staffel* MFG 3 operates 14 of the 21 surviving Dassault Atlantics, used for anti-submarine warfare and patrol in the Baltic. Five of the aircraft have been modified for electronic-countermeasures duties. The Atlantics were at one time to be replaced by 15 Lockheed S-3A Vikings, but in the autumn of 1978 the *Marineflieger* received approval to update 14 Atlantics rather than buying new aircraft. Dornier is main contractor for the work, which is aimed at improving the ECCM, ESM, radar and underwater detection equipment. Westland Sea Kings have replaced H-34s in the search-and-rescue role, and Do28 Skyservants have superseded Do27s and Pembrokes on liaison duties.

The navy also operates surface-to-air missiles in the form of the General Dynamics Standard, which has replaced the same manufacturer's Tartar in the three *Lutjens*-class destroyers. Germany has joined the NATO Seasparrow programme, and this weapon will equip the F122 frigates, of which the initial batch comprises six vessels. A point-defence missile based on the General Dynamics ASMD is also planned for these ships.

Left: Flame spurts from the twin 35mm barrels of a German Army Gepard anti-aircraft tank. Variants have also been ordered by Holland and Belgium.

Below: A *Heeresflieger* CH-53G disgorges a light vehicle armed with the Milan anti-tank missile.

Further in the future is the definitive PAH-2, able to attack at night as well as during daylight. Early design proposals such as the BO115 have been superseded, and the project is expected to be collaborative. Some of the PAH-1s will be redeployed as VBHs when they are replaced by PAH-2s (see above). About 200 PAH-2s are required from 1985.

The German army does not operate large surface-to-air missiles, these being the responsibility of the *Luftwaffe*, but it does have the Euromissile Roland II to defend armoured forces and also the infantry-operated General Dynamics Redeye. The army is scheduled to receive 340 Marder vehicles armed with Roland II.

Navy

The *Marineflieger* (fleet air arm) was formed in 1957 to provide shore-based attack, fighter, reconnaissance, anti-submarine and search-and-rescue aircraft. The two combat *Geschwader* operate three *Staffeln* of F-104G fighter-bombers and one of RF-104Gs received from the Luftwaffe, following their replacement by RF-4Es.

NETHERLANDS

Area: 13,500 sq. miles (35,000 sq. km.).
Population: 13·9 million.
Total armed forces: 109,700.
Estimated GNP: $104·1 billion (£52·05 billion).
Defence expenditure: $4·21 billion (£2·10 billion).

The Royal Netherlands Air Force, the KLu (*Koninklijke Luchtmacht*), operates nine squadrons of its own and a further three on behalf of the army. The navy has its own air arm with helicopters and fixed-wing types.

Dutch military aviation began with the establishment in July 1913 of the LVA (*Lucht Vaart Afdeling*) as part of the army. Holland was neutral during the First World War, and more than 100 aircraft of the belligerent powers were interned after they had landed on the wrong side of the border. Nearly half of these were impressed into LVA service. Between the wars the LVA was built up with Fokker and Koolhoven aircraft, and the air arm put up a spirited resistance to the invading German forces in 1940. Dutch squadrons were formed in the Royal Air Force after the fall of Holland, and in 1946 the air force was re-formed as the LSK (*Lucht StrijdKrachten*). The following year its title was altered yet again, to *Leger Luchtmacht Nederland*, and in 1953 the present name was adopted.

At the beginning of 1973 the KLu's four commands were reduced to the present two: CTL (*Commando Tactische Luchtstrijdkrachten*), the tactical air command responsible for all combat elements; and CLO (*Commando Logistiek en Opleidingen*), which runs logistics and training. The KLu has some 17,700 personnel, including 4100 conscripts, and about 18,300 reserves.

Inventory

The KLu operates only three combat types: the Lockheed F-104 Starfighter, General Dynamics F-16 and Northrop F-5 Freedom Fighter. Two of the Starfighter squadrons fly in the air-defence role in Sector 1 of the 2nd Allied Tactical Air Force (ATAF), which covers the Benelux countries and part of Germany. Holland is responsible for two radar stations in NATO's Nadge chain, at Nieuw Milligen and Den Helder, with the latter housing a GEC-Marconi Firebrigade air-defence centre for semi-automatic computer control of the interceptor Starfighters. Holland is contributing to the purchase and support of 18 Boeing E-3A Sentry AWACS (Airborne Warning And Control System) aircraft which will be operated by NATO in Europe from 1985. The 32nd Tactical Fighter Squadron of the United States Air Force, based at Soesterberg, is under joint KLu/USAF control. The unit recently re-equipped with F-15 Eagles in place of the previous F-4E Phantoms. Air-defence forces also include eight squadrons of Raytheon Hawk medium-range surface-to-air missiles and four squadrons of the longer-range Nike Hercules. Until the mid-1970s the KLu had five SAM groups or GGWs

(*Groepen Geleide Wapens*) in Germany: 1 GGW at Münster-Handorf; 2 GGW at Schöppingen; 3 GGW at Blomberg; 4 GGW at Hessisch; and 5 GGW at Stolzenau. The first two each had four eight-launcher squadrons of Nike Hercules, while the remaining three shared 11 squadrons with six Hawk launchers each. Economies have forced a reduction to four squadrons of the long-range weapon and eight of Hawk, however, and three of the latter units have been withdrawn to Holland to provide air-base defence. The Hawks are now in the process of being upgraded to Improved Hawk standard, and the Nike Hercules are due to be replaced in the period 1985–88.

A further two Starfighter squadrons are allocated an offensive role, and the fifth F-104 unit specialises in tactical reconnaissance, using the Orpheus pod developed by de Oude Delft. The pod, which weighs 400kg and is based on a US bomb container, carries five TA-8M cameras and infra-red linescan. About 118 Starfighters remain on KLu strength out of 120 F-104Gs and 18 two-seat TF-104Gs delivered between 1962 and 1965.

The Starfighters are being replaced by the General Dynamics F-16 air combat fighter. The first deliveries to 322 and 323 Sqns were planned for mid-1979, with 12 two-seat F-16Bs being accepted for conversion training. The KLu has 80 single-seat F-16As and 22 F-16Bs on firm order, including 16 and two respectively which were converted from options in June 1978. This additional order will allow the reconnaissance squadron to re-equip with the GD type, following conversion by the other four Starfighter units at the rate of one a year. All F-16s will be on strength by 1984. Original plans called for 112 new fighter-bombers, but this total was reduced by ten to save money. Further aircraft are expected to be ordered during the 1979–83 defence plan, however, to replace attrition losses.

The KLu's F-16s will be armed with AIM-9L Sidewinders and may carry a medium-range air-to-air missile such as Sparrow or Sky Flash, although the USAF has not specified such weaponry. Air-to-surface ordnance will initially comprise types already in service, such as the BL755 cluster bomb, and may be augmented by laser-guided bombs and/or air-to-surface missiles in the Maverick family.

Between 1969 and 1972 the KLu took

Above: Protection of offshore resources is one of many tasks for the Dutch Navy's Atlantics.

delivery of 75 Canadair-built NF-5As and 30 NF-5Bs. These equip three light attack squadrons plus an operational conversion unit which has a secondary combat role. The NF-5 will be replaced from the mid-1980s. A single transport squadron operates the F.27 and its military derivative, the Troopship. Larger and longer-range transports are expected to replace some of the F.27s.

KLu pilots undergo basic training in Canada since a joint programme with the *Force Aérienne Belge* expired and the remaining Fokker S.11s at Gilze Rijen were retired. Instruction begins on Beech Musketeers, students then converting to Canadair Tutors and T-33As. After two years in Canada the pilots return for operational conversion and tactical training with 313 Sqn, on NF-5s, or on TF-104s at Leeuwarden. The *Commando Logistiek en Opleidingen* operates a basic training school at Nijmegen, a school for officers at Gilze Rijen, a warrant officers school at Schaarsbergen, an electronics and technical school at Deelen and a flight-safety training and test centre at Soesterberg.

The KLu also operates the *Groep Lichte Vliegtuigen* (Light Aircraft Group) on behalf of the army. All the Piper L-21B Super Cubs have now been retired. The Alouette III still equips two of the three squadrons, however, while the other has converted to the MBB BO105C. Further BO105s armed with anti-tank missiles may

Below: The crew of a Dutch Navy Lynx rush an injured man to Den Helder hospital from the nearby air base. Other Lynxes are carried aboard ship.

be bought, and the Alouettes are scheduled to be replaced from 1983.

Navy

The *Marineluchtvaartdienst* (MLD), the Netherlands fleet air arm, was established in 1918 and has a current personnel strength of some 1900. The force has two squadrons of fixed-wing aircraft, one operating SP-2H Neptunes and the other Dassault-Breguet Atlantics (designated SP-13Hs by the MLD). The Neptunes are to be replaced by 13 Lockheed P-3C Orions ordered in December 1978. The first aircraft will be delivered in 1981, and the remainder at a rate of four a year. The P-3C Update II was selected in preference to the Dassault-Breguet ANG (Atlantic Nouvelle Génération).

Procurement of new rotary-wing equipment is going ahead in the form of the Westland/Aérospatiale Lynx. Six of the UH-14A version have replaced the same number of AB.204Bs in the search-and-rescue role; after joint training with the Royal Navy's 700L squadron in the winter of 1976–77, the first UH-14As were handed over to the MLD in the spring of 1977. These helicopters will be followed by ten anti-submarine SH-14B variants, replacing AH-12A Wasps in the shipboard role and also flying from new frigates as they are commissioned. The SH-14Bs, powered by the uprated Rolls-Royce Gem 4 engine, are cleared to operate at the higher gross weight of 10,500lb and are fitted with the CIT-Alcatel DUAV-4 dipping sonar as well as the Ferranti Seapsray radar. An additional eight Lynxes, designated SH-14C, were ordered in January 1978. These will be fitted with MAD (magnetic-anomaly detection) gear and may later also carry sonar. The Wasps will be retained in service until 1983–84, when they are expected to be replaced by a further 12 Lynxes.

MLD pilots undergo basic training on Cessna 150s and Safirs at the Dutch national airline school before transferring to the KLu's *Aanvullende Militaire Vlieg Opleidung* (Advanced Military Flying Training) course at Soesterberg, where they convert to Troopships. Rotary-wing pilots continue their instruction at Deelen on Alouette IIIs. Maritime-patrol crews undergo operational conversion with the MLD's 2 Sqn at Valkenburg, which operates no aircraft of its own but has access to the Neptunes and Atlantics which share the base. Observers and navigators are trained on three Troopships transferred from the Klu following the retirement of the MLD's Beech TC-45Js.

The navy (*Koninklijke marine*) operates surface-to-air missiles in the form of the Short Seacat from six *Van Speijk*-class

frigates; Standard from two *Tromp*-class destroyers; and Sea Sparrow from the *Tromps* and up to 12 Standard frigates planned.

Army

The Army's aircraft are operated by the Air Force, although the former force contributes to anti-aircraft defence. Weapons include 95 CA-1 Caesar tanks armed with twin Oerlikon 35mm cannon controlled by Signaal radars, and L70 40mm guns operated in conjunction with the Signaal Flycatcher radar. A low-level SAM such as Roland or Rapier is expected to be introduced in the 1980s.

Above: Dutch-operated Nike Hercules long-range surface-to-air missiles will remain in service until at least the mid-1980s.

NETHERLANDS Air Force

Command	Unit	Type	Role	Base	No	Notes
Commando Tactische Luchtstrijdkrachten	306 Sqn	RF-104G	Recce	Volkel (De Peel in wartime)	18	To be replaced by F-16
	811 Sqn	F-104G	Att	Volkel	18	Each also has two TF-104s
	312 Sqn	F-104G	Att	Volkel	18	
	313 Sqn	NF-5A/B	Train	Twenthe	4/10	OCU with secondary combat role
	314 Sqn	NF-5A/B	Att	Eindhoven		
	315 Sqn	NF-5A/B	Att	Twenthe	65/20 approx	Eindhoven base closing
	316 Sqn	NF-5A/B	Att	Gilze Rijen		
	322 Sqn	F-104G	Int	Leeuwarden	18	Plus six reserves each. Also seven TF-104G.
	323 Sqn	F-104G	Int	Leeuwarden	18	To be replaced by F-16
Commando Logistiek en Opleidingen	334 Sqn	F.27M Friendship	Trans	Soesterberg	9	Three Troopships assigned to Navy as navigation trainers
		F.27 Mk 100	Trans	Soesterberg	3	
		Nike Hercules	SAM	various	4 sqns	
		Hawk	SAM	various	8 sqns	Being upgraded to Improved Hawk

NETHERLANDS Army

Unit	Type	Role	Base	No
SAR Flight	Alouette III	SAR	Soesterberg	5
298 Sqn	Alouette III	Obs/liaison	Soesterberg	70
399 Sqn	Alouette III	Obs/liaison	Deelen	
300 Sqn	BO105C		Deelen	30

NETHERLANDS Navy

Unit	Type	Role	Base	No	Notes
320 Sqn	SP-2H Neptune	ASW	Valkenburg	15	Three Neptunes deployed to Hato, Curaçao (Neptunes to be replaced by 13 P-3C from 1981)
321 Sqn	SP-13H Atlantic	ASW	Valkenburg	7	
7 Sqn	UH-14A Lynx	SAR/VIP/liaison	De Kooy	6	Replaced AB.204Bs
860 Sqn	AH-12A Wasp/	ASW	Frigates/	12	Lynxes are replacing Wasps
	SH-14B Lynx	ASW	De Kooy	10	
	Seacat	SAM	6 frigates	12 launchers	
	Standard	SAM	2 destroyers	2 launchers	
	Sea Sparrow	SAM	2 destroyers, 12 frigates	14 launchers	

BELGIUM

Area: 11,800 sq. miles (30,500 sq. km.).
Population: 9·8 million.
Total armed forces: 87,100.
Estimated GNP: $73·4 billion (£36·7 billion).
Defence expenditure: $1·82 billion (£910 million).

The Belgium air force, *Force Aérienne Belge* in French and *Belgisch Luchtmacht* in Flemish, is responsible for virtually all fixed-wing military aviation. The army air arm does have a dozen Islanders, however, together with a medium-sized helicopter force. Rotary-wing types are also operated in small quantities by the gendarmerie and the navy.

Belgium is a member of NATO, and the majority of the FAéB is allocated to the 2nd Allied Tactical Air Force, which also includes aircraft from Holland, Britain and Germany. Combat units are administered by the *Commandement de la Force Aérienne Tactique* (Tactical Air Force Command) with headquarters at Evère. FAéB personnel strength is some 20,000, including nearly 5000 conscripts.

Inventory

Two intercepter squadrons are equipped with F-104G Starfighters and operate within NATO's Nadge air-defence system. Belgium is responsible for two Nadge ground radar stations, at Zemerzaeke and Glons, in the north of the country and in the Ardennes respectively. These were due to be replaced in early 1979 by a single General Electric GE592 solid-state phased-array three-dimensional radar – a land-based version of the naval TPS-59 – costing BFr200m and being installed at the former site.

The Starfighters are themselves scheduled to be superseded by General Dynamics F-16s, of which 102 have been ordered – 90 single-seat F-16As and 12 F-16B

trainers. Eighteen are due for delivery in 1979, to be followed in succeeding years by 12, 19, 22, 22 and nine aircraft respectively. A further 14 aircraft are on option. The first unit to re-equip is 350 Sqn, which should be fully operational by the second quarter of 1980. The F-16s are being assembled by Fairey and SABCA, Belgium being one of the four European NATO countries to announce a joint selection of

Below: The Belgian Air Force's first General Dynamics F-16 fighter takes off on its maiden flight in December 1978; a further 101 will follow.

Bottom: NATO countries use almost as many types for basic training as there are members of the alliance. The FAéB operates SF.260MBs in this role, students then progressing to Alpha Jets.

the type as a Starfighter replacement in the summer of 1975. The aircraft will carry conventional weapons such as the BL755 cluster bomb on air-to-surface missions and may be armed with a medium-range air-to-air missile in the Sparrow class. The Loral Rapport-2 electronic-warfare system, being installed in the FAéB's Mirage 5s, is also expected to be fitted to the F-16s. Two wings of Nike Hercules long-range surface-to-air missiles complete the air-defence forces.

A further two squadrons of F-104Gs fly in the nuclear strike and conventional ground-attack role, and these units will also re-equip with F-16s. The Starfighter force originally totalled 100 F-104Gs and 12 TF-104Gs, but only 82 remained in service by 1978. The remaining pair of

BELGIUM

Command	Wing	Unit	Type	Role	Base	No	Notes
Air Force	1 Wing	349 Sqn	F-104G/TF-104G	Int/train	Beauvechain	18/2	To be replaced by F-16
		350 Sqn	F-104G/TF-104G	Int/train	Beauvechain	18/2	
	2 Wing	2 Sqn	Mirage 5BA	FGA	Florennes	18	
		42 Sqn	Mirage 5BR	Recce	Florennes	18	
	3 Wing	1 Sqn	Mirage 5BA	FGA	Bierset	18	
		8 Sqn	Mirage 5BD/BA	Train/FGA	Bierset	12/6	Mirage OCU
	10 Wing	23 Sqn	F-104G/TF-104G	Strike/train	Kleine Brogel	18/2	To be replaced by F-16
		31 Sqn	F-104G/TF-104G	Strike/train	Kleine Brogel	18/2	To be replaced by F-16
	15 Wing	20 Sqn	C-130H	Trans	Melsbroek	12	Replaced C-119
		21 Sqn	Boeing 727-29QC	Trans	Melsbroek	2	Replaced DC-6B
			Merlin IIIA	Trans	Melsbroek	6	Replaced Pembroke
			Falcon 20	Trans	Melsbroek	2	
			HS.748-2A	Trans	Melsbroek	3	Replaced C-47
		7 Sqn	Magister	Train	Brustem	37	Being replaced by Alpha Jet
		9 Sqn	Magister	Train	Brustem		
		11 Sqn	T-33A	Train	Brustem	12	
			SF.260MB	Train	Goetsenhoven	33	
		40 Sqn	Sea King Mk 48	SAR	Coxyde	5	Replaced H-34
	9 Wing		Nike Hercules	SAM	Goch	4 sqns	Each squadron has
	13 Wing		Nike Hercules	SAM	Zulpich	4 sqns	16 launchers
Army		15 Sqn	Defender	Util	Brasschaat	7	Replaced Do27
			SA.330H Puma	Util	Brasschaat	5	
			Alouette II/Alouette-Astazou	Util	Brasschaat	24	
		16 Sqn	Alouette II	Util	Butzweilerhof	14	
			Defender	Util	Butzweilerhof	5	
		17 Sqn	Alouette II/SA.318C	Util	Werl	18	
		18 Sqn	Alouette-Astazou	Util	Aachen		
			Super Cub	Target tug		6	
		43 *Bataillon d'Artillerie*	Hawk	SAM	various	24 launchers each	
		62 *Bataillon d'Artillerie*					
Navy			Alouette III	Util	Support ships	3	
			Sea Sparrow	SAM	frigates		
Gendarmerie			Puma	Util	Brasschaat	3	
			Alouette II	Util	Brasschaat	5	

fighter-bomber squadrons operate the Dassault Mirage 5, built under licence by SABCA. A third unit equipped with the type specialises in the reconnaissance role, and the Mirage force is completed by an operational conversion unit with a secondary combat responsibility. The original force comprised 63 fighter-bombers, 27 reconnaissance aircraft and 16 two-seat trainers, but at least 18 of this total have been lost.

FAéB's training fleet is being upgraded with the introduction of 33 Dassault/Dornier Alpha Jet 1Bs, the first of which was delivered in December 1978. The elementary stage of instruction on SF.260MBs is being increased from 125hr to 150hr with the introduction of Alpha Jets, and 150hr on the new type replaces 125hr on Magisters and 100hr on T-33As. The T-33s were due to have been retired by June 1979 and the Magisters by the end of that year; the former type is being withdrawn from service, although about 30 of the latter type are to be retained for other roles. The Belgian Alpha Jets are fitted with four wing stations and could also be used in the close-support role; improved ECM equipment may be fitted.

Army

Aviation Légère de la Force Terrestre has a dozen Fairey Britten-Norman Defenders and a rotary-wing fleet of some 66 Alouette IIs and Alouette-Astazous, original deliveries having comprised 42 of each type. The present helicopters are due to be replaced by about 80 examples of a new multi-purpose type, the first batch being configured for anti-tank and observation duties and the second group for assault roles.

The army also operates two battalions of Hawk medium-range surface-to-air missiles and MBLE 43 Epervier battlefield reconnaissance drones; the 1er Corps d'Armée became operational with Epervier at the beginning of 1976.

Navy

Belgian naval aviation is confined to operation by the *Force Navale Belge* of three Alouette IIIs from the shore base at Coxyde and aboard two support ships. The new *Westhinder* class of frigate is equipped with an octuple launcher for the NATO Seasparrow point-defence missile system.

FRANCE

Area: 213,000 sq. miles (551,000 sq. km.).
Population: 52·6 million.
Total armed forces: 502,800.
Estimated GNP: $374·8 billion (£187·4 billion).
Defence expenditure: $17·52 billion (£8·76 billion).

France is one of the world's great military powers, with a large and varied arsenal of conventional and nuclear weapons. Defence and foreign policy are aimed at maintaining detente with the Communist Bloc, taking into account the growing number and importance of Third World states resulting from decolonisation, and working towards a unified economic and political organisation in Western Europe.

France's post-war defence policy has been moulded by a number of fundamental decisions made either voluntarily or as a result of defeat, whether political or military, in her colonies. As a result of one such decision made in 1956, all three armed forces now operate nuclear weapons developed in France: the *Armée de l'Air* (air force) is responsible for two nine-missile squadrons of SSBS (*Sol-Sol Balistique Stratégique*) intermediate-range ballistic missiles and has a force of Mirage IV supersonic bombers carrying free-fall strategic weapons, plus Mirage IIIE and Jaguar fighter-bombers carrying tactical nuclear bombs; the *Marine Nationale* (navy) has the MSBS (*Mer-Sol Balistique Stratégique*) equivalent of the air force's SSBS and the *Aéronavale* (fleet air arm) can equip its carrier-borne Super Etendard fighter-bombers with tactical nuclear weapons; and the *Armée de Terre* (army) operates the Pluton battlefield-support missile with a nuclear warhead.

Between 1960 and 1970, under the first and second defence build-up programmes, the major goal – establishing an independent nuclear deterrent – was reached. In the early 1960s the development of a strategic nuclear force was financed with money made available by the dramatic reduction in personnel costs following the end of the war in Algeria; in 1960, for example, defence accounted for 28.5% of the total central-government budget and 6% of gross national product. Later, however, money had to be drained from other programmes; despite the decision in 1962 to remodel and modernise the conventional forces, by 1970 the air force was 100 fighter aircraft short of its planned complement and all three Services had to postpone new developments as well as delay those in progress.

The third defence programme thereafter set a new goal of bringing the conventional forces up to the required level between 1971 and 1975, while at the same time improving and diversifying the strategic nuclear force and introducing tactical nuclear weapons. The amount of money made available would increase, although the percentage of gross national product would continue to drop. Subsequent events reduced the planned procurement, however, and led to the present pattern of late deliveries, smaller production batches and concentration on one major project. From 1972, inflation significantly affected the armed forces' purchasing power and procurement plans had to be trimmed. Also the proportion of the national budget allocated to defence continued to decline, reaching 16.9% in 1974. Thirdly, personnel costs increased greatly as a result of improvements in conditions; this came at a time when the rundown in manpower had halted, and the money had to be siphoned off an already stretched equipment budget and an operations allocation being hit by the 1974 rapid rise in fuel prices.

Despite the allocation of Fr7000m above the level foreseen in the third defence act, which had been intended to cover inflation, the yearly appropriations were adjusted upwards only in 1975. The proposed orders could then be placed, but the time needed to build the equipment combined with that required to overcome technical shortfalls has resulted in conventional forces remain-

Below: The *Armée de l'Air*'s standard interceptor is the Dassault Mirage F.1, which is replacing Mirage IIIs. From the mid-1980s it will be joined by the advanced-technology Mirage 2000.

ing below the planned levels, although the nuclear forces generally achieved their goals. Thus the equipment, manpower, preparedness for battle and future potential of France's armed forces have, like those of most Western nations, reached their present level as a result of financial strictures as much as military planning.

NATO member

France joined the North Atlantic Treaty Organisation from the outset and remains a full member, despite having withdrawn from the military organisation in 1966 in order to retain closer control over the deployment of her forces. Co-operation with other members of the alliance continues in a number of military areas, such as integration of the Strida early-warning and air-defence network with NATO's Nadge chain.

The President of the Republic is Commander-in-Chief of the French armed forces, with planning shared among the *Conseil Supérieur de la Défense Nationale*, the *Comité de Défense* and the *Comité de Défense Restreint*. The Prime Minister's responsibility for national defence is exercised through the *Sécrétaire General de la D″fense Nationale*, and the Prime Minister can deputise for the President in ordering a strategic nuclear response if France is attacked. Since 1962 the armed forces have been divided into the *Force Nucléaire Stratégique*, the operational forces of the three armed services, and home-defence forces comprising mainly army elements.

Air Force

France's *Aviation Militaire* played a major part in the First World War, and it was decided almost from the beginning that solely French aircraft and engines would equip the force. Assistance was given to Britain and later the United States, whose pilots flew French aircraft, and the *Aviation Militaire* ended the war as the world's largest air force. After the period of small military budgets which affected the development of most of the world's armed forces, the *Armée de l'Air* was established in April 1933 and became an independent Service in July of the following year.

The French aircraft industry, nationalised in 1936, designed a number of technically advanced types but production could not keep pace with the build-up of Germany's *Luftwaffe* and, despite a late

infusion of foreign aircraft and stubborn resistance by the squadrons, the *Armée de l'Air* was unable to prevent invasion in 1940. French-named units continued to operate from overseas bases throughout the war and from 1945 the home aircraft industry was re-established to feed French types to the re-formed *Armée de l'Air*. The policy of using indigenously designed and developed aircraft, latterly in collaboration with a close ally when practicable, continues to be pursued. French military aircraft are also sold throughout the world and bring in substantial foreign exchange.

Under the 1977–82 defence plan the non-strategic elements of the *Armée de l'Air* will receive just over 22% of the military budget, which is scheduled to rise from Fr58,000m in 1977 to Fr114,575m in 1982. During this period the defence budget as a proportion of government spending will rise from 17% to 20% and the percentage of gross national product is expected to increase from 3.0 to 3.6. The *Armée de l'Air* has some 80 squadrons, of which about 30 – totalling 450 aircraft – will be retained as a front-line combat force; this compares with a force of 650 combat aircraft which the Chief of the Air Staff would like to have seen in the 1980s. The present personnel strength is some 106,000, of which 39,000 are one-year conscripts.

Since 1964 the force has reorganised its command structure to fit its new roles and increasingly sophisticated equipment. Under the present arrangement large commands with specialised functions are supported by regional commands for joint logistics and support, with smaller units at the various air force bases.

Strategic forces

The Armée de l'Air is responsible for two arms of the strategic nuclear deterrent: the 1e] *Groupement de Missiles Stratégiques* (1e] GMS) and the *Forces Aériennes Strategiques* (FAS). The 1e] GMS occupies the Saint Christol base, covering 36,000ha of the Plateau d'Albion in the département of Haute Provence east of Avignon. It has two squadrons, each with nine SSBS (*Sol-Sol Balistique Stratégique*) S2 intermediate-range ballistic missiles in silos between 3km and 8km apart. The range of the missiles can be varied between 800km and 3300km and each carries a 150-kiloton nuclear warhead. The two underground *postes de conduite de tir*, PCT 1 at Rustrel (Vaucluse) and PCT 2 at Reilhannette

(Drôme), are normally responsible for one squadron each, but all 18 missiles can be launched from either control post if necessary. The weapons can be fired 3min 20sec after receipt of the firing order in normal peacetime conditions, or after 71sec if previously alerted.

From 1980 the S2s will be replaced by the shorter and lighter S3, carrying a 1.2-megaton thermonuclear warhead hardened to withstand the effects of exo-atmospheric nuclear explosions when intercepted by anti-ballistic missiles. The S3 will also carry penetration aids.

From 1964 to 1974 the *Forces Aériennes Strategiques*, headquartered at Taverny near Paris, operated three *escadres* (wings) of Dassault Mirage IV supersonic bombers dispersed to nine bases; one of the four aircraft in each of the nine *escadrons* (squadrons) was kept on permanent stand-by, able to scramble within 15 min. From 1974 to mid-1976 the quick-reaction force was reduced to six Mirage IVs, and on June 30 of that year one of the *escadres* was disbanded so that the bomber's service life could be extended to 1985. The 91e and 94e *Escadres* now each have 16 aircraft in three *escadrons* deployed at a total of six bases. The bases of the now-disbanded 93e *Escadre* at Istres, Cambrai and Creil retain the necessary support facilities and the *Depôt-Atelier de Munitions Speciales*, where the Mirage IV's AN-22 free-fall nuclear bombs of 60–70KT yield are assembled and stored, so that they can be used for dispersal in an emergency.

The last of 62 Mirage IVs was delivered in 1968, and the change of role from high-level bombing to low-altitude penetration has necessitated an airframe strengthening and modification programme. Twelve aircraft have been converted for long-range strategic reconnaissance, and these will be retained in this role after the majority of the force is retired in 1985. They will be equipped for all-weather reconnaissance and will have a new electronic countermeasures fit; French spending on ECM multiplied by ten times during the first half of the 1970s.

The FAS's 11 remaining (out of 12) C-135F tanker/transports are now based at Istres, having formerly been divided among the three Mirage IV *escadres*, each having its associated *Escadron de Revitaillement en Vol*. The C-135Fs can also back up the long-range transport fleet and are used to refuel tactical aircraft such as Jaguars and Mirage F.1s. They are expected to be re-engined with CFM56 turbofans.

The third element of the FAS is CIFAS 328 at Bordeaux, which operates four Mirage IVs, eight Mirage IIIB conversion trainers, ten Mirage IIIB-RVs for instruction in flight refuelling, five N.2501SNB (*Système Navigation-Bombardement*) versions of the Noratlas and some 17 T-33As.

Air defence

Air defence is the responsibility of CAFDA (*Commandement Air des Forces de Défense Aérienne*), with its force of fighters and surface-to-air missiles under the control of the Strida II early-warning and communications network. The mainstay of CAFDA's air arms is the Dassault Mirage F.1C, which entered service with the 30 *Escadre de Chasse Tous Temps* at Reims in 1974. This wing, which specialises in high-altitude, high-speed interceptions, was joined two years later by the 5 *Escadre de Chasse* at Orange, with specific responsibility for low-level operations over the Mediterranean. The F.1 replaced SO.4050N Vautours in the 30 *Escadre* and Mirage IIICs in the 5e; the latter aircraft were passed on to the 10e *Escadre* at Creil, augmenting those already operated by one of the two *escadrons* and superseding SMB.2 Super Mystères in the other. These Super Mystères in turn were allocated to the 12e *Escadre* at Cambrai to bolster that wing until it re-equipped with Mirage F.1s in 1976–77 (the first arrived in the autumn of 1976).

The normal complement of an F.1 *escadre* is 30 aircraft and one in reserve, with 40-plus pilots on strength. The new type has more thrust, an improved radar and better armament than the Mirage IIIC. The Cyrano IV radar can detect targets at ranges between 55km and 90km, depending on size, and the underwing mountings for the Matra R.530 air-to-air missile reduce the fuselage masking experienced with the Mirage IIIC's belly missile position. The R.530 will be replaced by the Super 530, which can snap up to intercept high-flying intruders. Matra's R.550 Magic dogfighting weapon has succeeded the Sidewinders which were carried by the Mirage IIICs, and additional range is conferred by the F.1's 4200lit internal fuel capacity compared with its predecessor's 3000lit. Three 1200lit external tanks may also be carried.

The 79th and succeeding F.1s are fitted with Thomson-CSF type BK passive radar warning to detect and classify hostile radar emissions, thereby giving warning of missile attack or interception by fighters; modifications which allow the warning equipment to be fitted retrospectively were incorporated from the 70th aircraft. The last 35 F.1s in the initial batch of 105 can also carry a removable but not retractable aerial-refuelling probe to extend their range. A further retrofit programme involves the *volet de combat*,

an all-speed automatic flap to enhance low-altitude manoeuvrability.

A further 63 Mirage F.1s out of a planned second batch of 109 have been ordered to add a fourth wing and maintain front-line strength in the early 1980s. The three wings already equipped with the type may also each be allocated a third squadron. A small batch of about six two-seat Mirage F.1Bs is expected to be bought for training in air refuelling, missile operations and systems instruction, with a secondary combat role. The aircraft are expected to form a third *escadron* in the 5 *Escadre de Chasse* from 1980.

The *Armée de l'Air*'s major new aircraft in the 1980s will be the Mirage 2000, which has absolute priority over other programmes to ensure that money and resources are not divided. The Mirage 2000 has a variable-camber delta wing with advanced high-lift devices, fly-by-wire flight controls, a new long-range pulse-Doppler radar and digital fire-control system. The initial requirement is for 127 single-seaters, with attack and reconnaissance versions (see below) later raising the total to 250–300. The first of four photoypes made its maiden flight in March 1978, and the initial aircraft is due to be delivered to CEAM (*Centre d'Expérience Aériennes Militaires*) at Mont-de-Marsan in 1982 for acceptance trials, establishing tactics, judging maintenance requirements and writing the operations manual. Ten aircraft are scheduled to have been delivered by the end of 1982 and the pro-

duction rate is expected to settle at 30–40 a year until 1990.

The Mirage 2000 will have a maximum speed exceeding Mach 2.2 at 60,000ft and will be as manoeuvrable as the F.1 at low level. The 80–100km range of the Thomson-CSF/Electronique Marcel Dassault Antilope radar (which is unlikely to be available for the first production aircraft) is double that of the Cyrano IV in the Mirage F.1. A typical interception would involve a climb to 60,000ft in 5min and launching a Matra Super 530 missile (of which two can be carried) against an intruder flying at up to 85,000ft and Mach 3.

Five Thomson-CSF Aladdin ground-based radars with a range of 50–100km against intruders flying at very low level are replacing older equipment. They will feed the Strida II (*Système de Transmission et de Représentation des Informations de Défense Aérienne*) network, which is also linked to NATO's Nadge chain. Command and control are exercised on three levels: at the top is CODA (*Centre d'Operation de Défense Aérienne*) at Taverny, with CDCSs (*Centres de Detection et de Contrôle de Secteur*) at the intermediate level and CDCs (*Centres de Detection et de Contrôle*) below them.

Batteries of Thomson-CSF/Matra Cro-

Above: The long-serving Max Holste Broussard soldiers on in a variety of hack roles with the *Armée de l'Air*, operating from bases all around the world. France's prolific aircraft industry has turned out many such utility types.

Below: The Noratlas will continue to form the major part of the *Armée de l'Air*'s transport fleet until it is succeeded by new Transalls in the 1980s.

Below: A reconnaissance Mirage III skims low over wooded hills on a typical mission. The type is to be replaced by a new version of the Mirage F.1 in the early 1980s.

tale low-level surface-to-air missiles are deployed to protect air bases and other CAFDA facilities. These are supplemented by 20mm cannon, and other defensive measures now being implemented include the provision of further hardened aircraft shelters.

Attack forces

Tactical nuclear strike, reconnaissance and ground attack are the responsibility of FATAC (*Force Aérienne Tactique*), which has some 300 aircraft deployed in six *escadres de chasse* and one *escadre de reconnaissance*, plus training units. Approximately 40 helicopters and communications aircraft are also placed permanently at FATAC's disposal by CoTAM (*Commandement du Transport Aérien Militaire*). A unique association of an operational command and a regional command links FATAC with the 1e e *Région Aérienne*, created at Metz in July 1965. The 1e eRA has responsibilities for 18 départements in north-east France, covering the regions of Alsace, Lorraine, Franche-Comté, Champagne and Bourgogne. Its 12 operational bases comprise the six occupied by FATAC's *escadres de chasse* (Dijon, Nancy, Luxeuil, St Dizier, Toul and Colmar), the reconnaissance base at Strasbourg, the fighter base at Reims (the home of CAFDA's Mirage F.1s of 30e

Escadre de Chasse Tous Temps), three *centres de détection et de contrôle* (Drachenbronn, Contrexéville and Romilly) and the headquarters at Metz. The 1e eRA is also responsible for Berlin-Tegel and Achern in Germany.

FATAC's missions in wartime are strike, army support, reconnaissance, destruction of enemy air power, secondary air defence, anti-ship operations, maintaining air and ground access to Berlin in collaboration with Britain and the United States, and overseas deployment. Air-to-air operations are controlled initially by a radar-equipped PDTA (*Poste de Direction Tactique Air*), which takes inputs from both the *Armée de l'Air* and the *Armée de Terre* and vectors aircraft on to intruders. 'Freelance' air-to-air operations in support of the Army are also carried out.

FATAC has two *Commandements*

Above: A base-defence soldier and his dog keep a watchful eye on a Jaguar attack aircraft, which can deliver nuclear as well as conventional weapons.

Aériens Tactiques, each with specific missions. The 1 CATAC, run from Metz, has some 13,500 personnel and exercises control over FATAC's force of Mirages and Jaguars. The 2 CATAC has its headquarters at Nancy and is tasked with providing the CAFI (*Composante 'Air' de la Force d'Intervention*), equivalent to NATO's ACE Mobile Force. *Escadres* from other commands would be allocated to 2 CATAC in an emergency.

FATAC has eight *escadrons* of Mirage IIIEs, two of Mirage 5Fs, two of Mirage IIIRs, one of Mirage IIIRDs and is building up to nine of Jaguars. The last F-100 Super Sabre *escadron*, EC 4/11 *Jura*, re-equipped with Jaguars in 1978. FATAC

FRANCE Air Force

Command	Wing	Unit	Type	Role	Base	No	Notes
	1er *Groupement de Missiles Stratègiques*		SSBS S2	IRBM	Saint Christol	18	Two squadrons of nine missiles each
	91e *Escadre de Bombardement*	EB 1/91 *Gascogne*	Mirage IVA	Strat bomb/recce	Mont-de-Marsan	4	
		EB 2/91 *Bretagne*	Mirage IVA	Strat bomb/recce	Cazaux	4	
		EB 3/91 *Beauvaisis*	Mirage IVA	Strat bomb/recce	Orange	4	
		ERV 4/91 *Landes*	C-135F	Tanker/trans	Istres	3/4	Escadres de Bombardement also have Mirage IIIB and T-33 trainers, plus Flamants for comms
Forces Aériennes Stratègiques		ERV 4/93 *Aunis*	C-135F	Tanker/trans	Istres	3/4	
	94e *Escadre de Bombardement*	EB 1/92 *Bourbonnais*	Mirage IVA	Strat bomb/recce	Avord	4	
		EB 2/94 *Marne*	Mirage IVA	Strat bomb/recce	St Dizier	4	
		EB 3/94 *Artois*	Mirage IVA	Strat bomb/recce	Luxeuil	4	
		ERV 4/94 *Sologne*	C-135F	Tanker/trans	Istres	3/4	
		CIFAS 328	Mirage IVA	Train	Bordeaux	4	
			Mirage IIIB			8	
			Mirage IIIB-RV			10	Refuelling trainers
			N.2501SNB			5	Mirage IVA nav-attack
			T-33			17	trainers
	5e *Escadre de Chasse*	EC 1/5 *Vendée*	Mirage F.1C	Low-alt int	Orange	31	
		EC 2/5 *Ile de France*					
	10e *Escadre de Chasse*	EC 1/10 *Valois*	Mirage IIIC	Int	Creil		To be replaced by 33 Mirage F.1s by 1980
		EC 2/10 *Seine*	Mirage IIIC	Int	Creil		
		EC 3/10 *Vexin*	Mirage IIIC	Int	Djibouti		
CAFDA	12e *Escadre de Chasse*	EC 1/12 *Cambrèsis*	Mirage F.1C	Int	Cambrai	31	
		EC 2/12 *Cornouaille*					
	30e *Escadre de Chasse Tous Temps*	ECTT 2/30 *Normandie-Niemen*	Mirage F.1C	High-alt int	Reims	31	
		ECTT 3/30 *Lorraine*	Mirage F.1C	High-alt int	Reims		
		EE 12/30 *Hautvilliers*	T-33	Train	Reims	3	
			Paris	Train	Reims	4	
			Flamant	Train	Reims	4	
			Crotale	SAM	various	10 batteries	For base defence
	2e *Escadre de Chasse*	EC 1/2 *Cigognes*	Mirage IIIE	Att	Dijon	20	
		ECT 2/2 *Côte d'Or*	Mirage IIIC	Train	Dijon	5	Mirage OCU, plus training for overseas pilots
			Mirage IIIB	Train	Dijon	9–12	
			Mirage IIIBE	Train	Dijon	11–16	
		EC 3/2 *Alsace*	Mirage IIIE	Att	Dijon	20	
	3e *Escadre de Chasse*	EC 1/3 *Navarre*	Mirage IIIE	Att	Nancy	20	Armed with AS.37 Martel; Jaguars replaced Mirage 5Fs
		EC 2/3 *Champagne et Brie*	Mirage IIIE	Att	Nancy	20	
		EC 3/3 *Ardennes*	Jaguar	Att	Nancy	15	
	3e *Escadre de Chasse*	EC 1/4 *Dauphinè*	Mirage IIIE	Strike	Luxeuil	20	Armed with AN-52 nuclear bomb
		EC 2/4 *La Fayette*	Mirage IIIE	Strike	Luxeuil	20	
		EC 1/7 *Provence*	Jaguar	Strike	St Dizier	15	Armed with AN-52 nuclear bomb
	7e *Escadre de Chasse*	EC 2/7 *Argonne*	Jaguar	Train	St Dizier	15	Jaguar OCU (ten Es, five As)
FATAC		EC 3/7 *Languedoc*	Jaguar	Strike	St Dizier	15	Armed with AN-52 nuclear bomb
	8e *Escadre de Transformation*	EC 1/8 *Saintonge*	Mystère IVA	Train/att	Cazaux	42	To be replaced by Alpha Jets from 1982
		EC 2/8 *Nice*	Mystère IVA	Train/att	Cazaux	42	
	11e *Escadre de Chasse*	EC 1/11 *Roussillon*	Jaguar	ECM	Toul	15	Armed with AS.37 Martel
		EC 2/11 *Vosges*	Jaguar	Att	Toul	15	
		EC 3/11 *Corse*	Jaguar	Att	Tout	15	
		EC 4/11 *Jura*	Jaguar	Att	Bordeaux	15	
	13e *Escadre de Chasse*	EC 1/13 *Artois*	Mirage IIIE	Att	Colmar	20	
		EC 2/13 *Alpes*	Mirage IIIE	Att	Colmar	20	
		EC 3/13 *Auvergne*	Mirage 5F	Att	Colmar	20	
	33e *Escadre de Reconnaissance*	ER 1/33 *Belfort*	Mirage IIIR	Recce	Strasbourg	20	
		ER 2/33 *Savoie*	Mirage IIIR	Recce	Strasbourg	20	
		ER 3/33 *Moselle*	Mirage IIIRD	Recce	Strasbourg	18	

also has a unit (ECT 2/2) responsible for training *Armée de l'Air* and overseas Mirage pilots and in wartime would control the Mystère IVs (Alpha Jets from 1982) of the 8e *Escadre de Transformation* at Cazaux, which normally provides weapon training. The remaining Mirage IIIs and 5s are now expected to be replaced by Mirage 2000s rather than by additional Jaguars.

FATAC's Mirage 5Fs, originally built for Israel and inherited by the *Armée de l'Air* when sales to that country were embargoed in 1967, are expected to remain in service until the early 1990s. Some of the Mirage IIIEs, however, were due to reach the end of their useful lives in 1978 and all will do so by 1985. These aircraft cannot be replaced on a one-for-one basis, or even by a single type. Some of their duties have been taken over by Jaguars,

and a proportion of the additional Mirage F.1s to be ordered may be allocated to the attack role. The Mirage 2000 is expected to be deployed in the attack and reconnaissance roles from the late 1980s after the fighter requirement has been met. On a ground-attack mission the Mirage 2000 can take off with full internal fuel and 5000kg of stores from a 1200m runway. Its radius of action will be 30% better than that of the Mirage III in the same configuration, and this can be extended by aerial refuelling. The ASMP (*Air-Sol Moyenne Portée*) nuclear stand-off missile is being developed to arm some of the strike Mirage 2000s, and the aircraft will have a total of nine hardpoints – five under the fuselage and two beneath each wing.

Two Mirage IIIE *escadrons* are equipped with Matra/British Aerospace AS.37 Martel anti-radiation missiles and a further

Above: Two pairs of Mirage F.1s patrol French airspace, on the alert for intruders. They can be armed with Super 530 and Magic air-to-air missiles.

FRANCE Air Force

Command	Wing	Unit	Type	Role	Base	No	Notes
CoTAM		ELA 41	Paris, Flamant,	Comms	Metz		
		ELA 43	Broussard,	Comms	Bordeaux		
		ELA 44 *Mistral*	Rallye 180GT	Comms	Aix-en-Provence		
		ELAS 1/44	Puma, Noratlas, Alouette II, Flamant, Broussard	Comms/SAR	Solenzara	2/2/1/?/?	
		GAM 50	Noratlas, Alouette II	Trans	St Denis, Reunion		
		ETOM 52	Puma	Trans	Noumea	3	
		ETOM 55	Noratlas	Trans	Dakar		
		GAM 56 *Vaucluse*	Noratlas, Puma	Trans	Evreux		
		ETAG 58	Aztec, Broussard, Alouette II	Trans	Pointe-à-Pitre		
	60e *Groupe de Transport*	ET 1/60 /GLAM)	Caravelle	VIP/trans	Villacoublay	3	Includes two ex-Air Zaire 11RS
			Falcon 20	VIP/trans	Villacoublay	5	
			Falcon 10	VIP/trans	Villacoublay	3	
			Puma	VIP/trans	Villacoublay	2	
			Alouette II	VIP/trans	Villacoublay	3–4	
			Broussard	VIP/trans	Villacoublay	?	
			Cessna 411	VIP/trans	Roissy	4	
		ET 3/60 *Estérel*	DC-8F	VIP/trans	Roissy	4	
	61e *Escadre de Transport*	ET 1/61 *Touraine*	Transall	Trans	Orleans	12	Total about 46
		ET 2/61 *Franche-Comté*	Transall	Trans	Orleans	12	
		ET 3/61 *Poitou*	Transall	Trans	Orleans	12	
	62e *Escadre de Transport*	ET 1/62 *Vercours*	Noratlas	Trans	Reims	15–18	Total about 120
		ET 2/62 *Anjou*	Noratlas	Trans	Reims	15–18	To be replaced with Transalls
	64e *Escadre de Transport*	ET 1/64 *Béarn*	Noratlas	Trans	Evreux	15–18	
		ET 2/64 *Bigorre*	Noratlas	Trans	Evreux	15–18	
	65e *Escadre de Transport*	ET 1/65 *Vendôme*	N.262D Fregate	Comms/train	Villacoublay	18	
		ET 2/65	Falcon 20	Comms/train	Villacoublay	3	
		Rambouillet	Paris, Broussard, Rallye	Comms/train	Villacoublay	24/12	
						1	
	67e *Escadre d'Hélicoptères*	EH 1/67 *Pyrénées*	Puma, Alouette II, Alouette III	Util/SAR/casevac	Cazaux		Each escadron has two Pumas, four to five Alouette IIs and four to five SA.319Bs; 4/67 supports the 1er *Groupement de Missiles Stratégiques*
		EH 2/67 *Valmy*			Metz/St Dizier		
		EH 3/67 *Parisis*			Villacoublay		
		EH 4/67			Apt		
		EH 5/67			Istres		
		GAM 82	Noratlas	Nuclear test support	Papeete	?	
			Falcon 20			1	
			Aztec			1	
			Alouette II			10	
		GOM 88	Noratlas	Util	Djibouti	3–6	
			Broussard			2	
			Puma			2	
			Alouette II			2–4	
		CIET	Transall	Train	Toulouse	?	
			N.262			6	
		CIEH	Alouette II	Train	Chambery	4	
			Alouette III			6	
CTAA		EE 54 *Dunkerque*	Noratlas	ECM	Metz	8	
		EC 57 *Commercy*	Noratlas	Radio calib	Villacoublay	4	
CEAA		EFIPN 307	CAP-10	Train	Clermont-Ferrand	26	Basic selection
		GI 312	Magister, Noratlas, Broussard, CAP-10, CAP-20	Train	Salon		Initial training
		GE 313 *Ecole des Moniteurs*	Magister, Flamant, Broussard	Train	Clermont-Ferrand		Used by overseas students
		GE 314 *Ecole de Chasse Ch. Martell*	T-33, Mystère IV, Broussard, Flamant	Train	Tours		Advanced jet training
		GE 315	Magister, Flamant, Broussard	Train	Cognac		Basic training for NCOs and naval pilots
		GE 316	Noratlas, Flamant	Train	Toulouse		Nav instruction
		GE 319	Flamant	Train	Avord	34	Twin conversion
		CEVSV 338	T-33	Train	Nancy		Blind-flying training
		CPIR 339	Falcon 20SNA	Train	Luxeuil	2	Mirage radar training

CAFDA = Commandement Air des Forces de Défense Aérienne; FATAC = Force Aérienne Tactique; CoTAM = Commandement du Transport Aérien Militaire; CTAA = Commandement des Transmissions de l'Armée de l'Air; CEAA = Commandement des Ecoles de l'Armée de l'Air; EB = Escadron de Bombardement; ERV = Escadron de Ravitaillement en Vol; CIFAS = Centre d'Instruction des Forces Aériennes Stratégiques; EC = Escadron de Chasse; ECTT = Escadron de Chasse Tous Temps; EE = Escadron d'Entrainement; ECT = Escadron de Chasse et de Transformation; ER = Escadron de Reconnaissance; ELA = Escadrille de Liaison Aériennes; ELAS = Escadrille de Liaison Aériennes et de Sauvetage; GAM = Groupe Aérien Mixte; ETOM = Escadrille de Transport d'Outre Mer; ETAG = Escadrille de Transport Antilles Guyane; ET = Escadron de Transport; GLAM = Groupe des Liaisons Aériennes Ministérielles; EH = Escadron d'Helicoptères; GOM = Groupe d'Outre Mer; CIET = Centre d'Instruction des Equipages de Transport; CIEH = Centre d'Instruction des Equipages d'Helicoptères; EE = Escadrille de Calibration; EC = Escadrille Electronique; EFIPN = Ecole de Formation Initiale du Personnel Navigant; GI = Groupement Instruction; GE = Groupement Ecole; CEVSV = Centre d'Entrainement en Vol Sans Visibilité; CPIR = Centre Prediction et Instruction Radar. Combat escadres also have their own SLVSVs (Sections de Liaison et de Vol Sans Visibilité) flying types such as Magisters, T-33s Broussards and Flamants.

two are armed with AN-52 free-fall tactical nuclear bombs. One of these, EC 2/4 *La Fayette*, received the weapon in 1972 to become the first *Armée de l'Air* unit with a tactical nuclear capability. Other Mirage IIIE *escadrons* are responsible for conventional air-to-surface operations, with air combat as a secondary role.

Reconnaissance

The two *escadrons* of Mirage IIIRs and one of IIIRDs form the 33e *Escadre de Reconnaissance*, the *Armée de l'Air*'s only tactical recce wing. In wartime the wing would work closely with the French 1st Army which, like the 33e ER, has its headquarters near Strasbourg. The Mirage IIIR, based on the IIIC, has five Omera 31 cameras in place of the nose radar. The IIIRD can also carry a belly-mounted SAT Cyclope infra-red pod or EMI sideways-looking radar in the same position. The IIIRs are due to be retired from 1982, but the IIIRDs are expected to last at least until 1986. Both variants are likely to be replaced by the Mirage 2000, for which the Super Cyclope pod and a new SLAR are being developed. The same type would also take over from the remaining strategic-reconnaissance Mirage IVs.

The major new type introduced by FATAC in the 1970s is the Anglo-French Dassault-Breguet/British Aerospace Jaguar fighter-bomber. The first of 200, comprising 160 single-seat A versions and 40 two-seat B trainers, arrived at St Dizier in June 1973 to begin replacing the Mystère IVs operated by EC 7, formerly based at Nancy. The type is planned to remain in service until 1995. Jaguar has a greater radius of action in the ground-attack and strike roles at low level than the Mirage IIIE and can be refuelled in the air, allowing it to patrol, for example, the whole of the western Mediterranean. The type made its combat debut in late 1977 when operating from Dakar against Polisario guerrillas in Mauretania.

The 7e *Escadre de Chasse* specialises in the strike role, EC 1/7 *Provence* having been declared fully operational with AN-52 nuclear bombs on September 1, 1974. This *escadron* has now been joined by EC 3/7 *Languedoc* while the *escadre*'s third *escadron*, EC 2/7 *Argonne*, acts as the Jaguar operational conversion unit. The second Jaguar wing, the 11e *Escadre de Chasse*, specialises in conventional ground attack; wartime roles include keeping open the corridors to Berlin and providing air support for the *Force d'Intervention*. All three *escadrons* are equipped with AS.37 Martel, one of which is carried in the central under-fuselage position. Other weapons include free-fall and retarded bombs of 125kg, 250kg and 400kg; 68mm and other rockets, of which some 65% hit within 20m of their target when fired at a range of 2000m; and air-to-air missiles. On a conventional ground-attack mission a typical load comprises bombs and a single 1200lit fuel tank, whereas for a penetration sortie the usual mixture is one AN-52 or AS.37 under the belly, two drop tanks on the inner wing pylons and electronic-countermeasures pods on the outer wing positions. One of 11e *Escadre*'s units, EC 1/11 *Roussillon*, specialises in

ECM work; 2/11 *Vosges* concentrates on attacks on ground forces; and 3/11 *Corse* is earmarked for overseas deployments on joint exercises and when a detachment is requested by states such as Ivory Coast and Senegal which have close links with France. Each pilot in 3/11 carries out two practice aerial refuellings per month, one in which fuel is transferred and another when contact is made and then broken. On overseas deployments the first refuelling is normally made at Flight Level 250 and 900km from base so that the pilot can return if necessary.

A Jaguar squadron, EC 3/3 *Ardennes*, also flies from Nancy and is equipped with AS.37 Martels. The type replaced Mirage 5Fs in that unit during 1977 and operates alongside Martel-equipped Mirage IIIE *escadrons* EC 1/3 *Navarre* and EC 2/3 *Champagne*. The 12 or so F-100 Super Sabres of EC 4/11 *Jura*, which formed an *escadrille*-strength *escadron* detached to Djibouti, were replaced by Mirage IIICs of EC 3/10 *Vexin* from January 1, 1979, on which date EC 4/11 re-formed with Jaguars at Bordeaux-Mérignac.

Jaguar units became operational with the Matra Magic air-to-air missile from the spring and summer of 1977, the weapon replacing the Sidewinders previously used in the self-defence and air-superiority roles. The last 80 aircraft to be delivered, starting in February 1977, are also being fitted with a laser range-finder to reduce the CEP (Circular Error Probable) by some 70%. Additional weapons and aiming aids being introduced over the next few years include the Matra Durandal anti-runway bomb and Beluga dispersion weapon, together with the Thomson-CSF/Martin ATLIS laser pod. An updated nav-attack system is expected to replace the present combination of Doppler and twin-gyro platform from 1985.

Transport

Transport duties are carried out by CoTAM *(Commandement du Transport Aérien Militaire)* with some 300 fixed-wing aircraft and 150 helicopters. CoTAM's wartime roles are support of the strategic nuclear forces; tactical support, including assault, aerial supply, general logistic duties and casevac; communications; some search and rescue; and assistance of civilian organisations.

CoTAM operates 47 Dassault-Breguet Transall transports flown by the three *escadrons* of the 61e *Escadre de Transport*

at Orléans, which has some 140 pilots, 60 navigators and 100 flight engineers out of CoTAM's total of approximately 800, 275 and 450 respectively. The Transall can air-drop 14 tonnes in a single pass, and the type's abilities were demonstrated during Operation Verveine in the spring of 1977, when 13 aircraft flew 1200hr between Morocco and Zaire to transport men and supplies. Considerable emphasis is also placed on paratroop operations, and training in this role accounts for some 25% of the Transall fleet's operations. These missions are often carried out at short notice and involve dropping troops at night from low level into rough areas without visual references. The present Transall force is expected to remain in service until at least 1995.

The N.2501 Noratlas still constitutes the largest part of CoTAM's fleet, equipping four regular transport *escadrons* and also being used for a variety of secondary roles; approximately 120 remain on strength. The aircraft are being refurbished and modernised to extend their lives to 1985 but, although they are structurally sound, they are becoming more and more difficult to operate as part of a mixed fleet. The engines and systems present increasing maintenance problems, fuel is scarce and operating costs are rising. CoTAM is therefore to receive 25 new Transalls from the re-established production line to replace the Noratlas from 1980. Some of the new transports may be permanently fitted out as tankers and the rest are likely to be equipped so that they can be refuelled in the air. This is increasingly important as potential staging posts such as Chad and Djibouti can no longer be relied on.

CoTAM's long-range transport fleet comprises the four DC-8Fs of ET 3/60 *Estérel* for Pacific reinforcement and VIP work, the latter also being the responsibility of GLAM (*Groupe de Liaisons Aériennes Ministerielles*) with Caravelles and a mixed bag of executive jets, helicopters and light transports. The former GAEL (*Groupe Aérien d'Entrainement et de Liaison*) was raised from *escadron* status (2/60) to *escadre* level in 1972,

Below: Marcel Dassault's new baby, the Mirage 2000, will serve the *Armée de l'Air* of the mid-1980s in the interception, attack and reconnaissance roles. By drawing together the threads of several new technologies, he has once again produced a Mirage to take on the rest of the world.

becoming the 65e *Escadron de Transport* with two *escadrons* of business jets and other transports. The 65e *Escadre* is responsible for training communications pilots and for providing the 2e *Region Aérienne* (Paris) with light transports, in addition to its general liaison duties. Other *regions aériennes* are served by the 41e, 43e and 44e *Escadrilles de Liaison Aériennes* at Metz (1e eRA), Bordeaux (3eRA) and Aix-en-Provence (4eRA) respectively. Each has a variety of transports and ELA 44 additionally operates a search-and-rescue flight in Corsica.

The *Centre d'Instruction des Equipages de Transport* has Transalls and Aérospatiale 262s for training, the latter having replaced Noratlases, and various transport units are stationed overseas. The most recent types to be introduced for foreign service are three ex-Air France Caravelle 11Rs, fitted with a cargo-loading system by UTA, which replaced DC-6s in the Pacific; and the Puma, deployed to New Caledonia in 1976.

Helicopters

The army is responsible for assault and gunship helicopters (see below), but the *Armée de l'Air* has a sizeable rotary-winged force in its own right. In addition to those operated by the *escadrilles de liaison aériennes* there are five *escadrons* which form the 67e *Escadre d'Helicoptères*. The CIEH (*Centre d'Instruction des Equipages d'Helicoptères*) trains pilots of other Services and government departments as well as for the *Armée de l'Air*. All Sikorsky H-34s have now been retired.

Most flying training is run by the CEAA (*Commandement des Ecoles de l'Armée de l'Air*). Initial selection takes place at the *Ecole de Formation Initiale du Personnel Navigant* (EFIPN), where students undergo 17–20hr in two-seat CAP-10s. About half fail at this stage. Until recently, officer candidates then moved on to the *Ecole de l'Air* (GI 312) at Salon for a further 145–150hr on Magisters, while other ranks carried out a similar course at GE 315, Cognac. In order to reduce costs, however, selection now takes place after 60–70hr on

Above: An *Aéronvale* Super Frelon dips its sonar transducer into a calm blue sea, although this French heavyweight can cope with much tougher conditions.

the Magister. The best pilots proceed on to fighter training, with more Magister flying followed by some 100hr on T-33s at GE 314, Tours; meanwhile the transport stream converts to the Flamant with GE 319. The *Armée de l'Air*'s 300-plus remaining Magisters are being refurbished to extend their lives at least until 1985. The Flamants are also being updated by SOGERMA and their scheduled retirement date of 1980 is very unlikely to be met.

CEAA's immediate priority is to replace its ageing T-33s, and this was due to begin with the delivery of the first of 20 Dassault/Dornier Alpha Jets in November 1978. The first Alpha Jet unit will be GE 314, which was due to have received the initial five of its planned 65 aircraft by May 1979, replacing 58 T-33s. A full 14-aircraft squadron was scheduled to be operational by September 1979, being followed by three more units before the winter of 1980–81. The *Patrouille de France* aerobatic team is re-equipping with Alpha Jets in 1979–80, and the Mystère IVAs of the 8e *Escadre de Transformation* will be succeeded by 30 of the new type in 1982. In addition, 14 will go to the *Centre d'Entrainement en Vol sans Visibilité* in 1981.

The transport stream, having completed some 100hr on Flamants at Avord, proceeds to CIET (*Centre d'Instruction des Equipages de Transport*) at Toulouse to obtain full transport qualification. Rotary-wing pilots go through a similar process with 3IEH at Chambery, leading to tactical qualifications. About 140 pilots are trained each year, of which half are officers and the rest non-commissioned officers. The breakdown is close to the ideal of 60 for combat aircraft, 60 for transports and 20 instructors.

Jaguar first-tourers, with some 350hr under their belts, join EC 2/7 for about 40hr conversion flying before moving on to their operational squadrons. There they complete an additional 130–160hr before they are regarded as being fully operational. A further 240hr is necessary before they reach deputy flight leader and a similar period before they become full flight leaders; by this time even the youngest has about 1000hr experience.

Young F-100 pilots with at least 1500hr experience can convert to Jaguars on a ten-mission short course, while older pilots and flight leaders undergo a 15-flight, one-month accelerated course. Mirage F.1 pilots may be on their first tour; they can be ex-Magister instructors with about 800hr, or they may be transferring from Vautours, Super Mystères

FRANCE Army

Regiment/Group/Unit	Type	Role	Base	No	Notes
Aviation Légère de l'Armée de Terre					
RHC (two attached to each of two army corps, two in reserve)	Gazelle	Recce			Two *escadrilles*
	Alouette III/ SA.342 Gazelle	Anti-tank			Three *escadrilles*
	Puma	Assault			Two *escadrilles*
GHL (one attached to each army corps)	Alouette II Alouette III Gazelle	Util		30	
GHL (one attached to each of military regions)	Alouette II Alouette III Gazelle	Util		20	
ESALAT	Bell 47/Alouette II	Train	Dax		Pilot training
EAALAT	Alouette II Alouette III Puma	Train	Le Luc	10	Instrument and tactical training
PMAH	Alouette II Alouette III Super Cub	Army co-op	Aix-les-Bains		Co-operates with 27 Brigade Alpine (mountain specialists)
EALAT	Bell 47 Alouette II Bird Dog	Comms	various		Attached to specialist Army schools
PD	Alouette Puma	Obs/assault	Djibouti	3 4	
PALTM	Tripacer	Obs/comms	Dakar	8	Attached to 1 *Regiment Inter-Armes d'Outre-Mer*
Armée de Terre					
	Hawk	SAM	various	54 launchers	
	Roland I/II	SAM	mobile	144/70 launchers	

RHC = Régiment d'Hélicoptères de Combat; GHL = Groupe d'Hélicoptères Légères; ESALAT = Ecole des Specialistes de l'ALAT; EAALAT = Ecole d'Application de l'ALAT; PMAH = Peloton Mixte Avions-Helicoptères; EALAT = Escadrilles d'ALAT; PD = Peloton Djibouti; PALTM = Peloton d'Aviation Légère des Troupes de Marine.

and Mirage IIICs or IIIEs. A new pilot has about ten familiarisation flights followed by 90–120hr over six months before he becomes operational. Ex-Mirage IIIE pilots undergo eight to ten days' ground familiarisation, some 15hr in the LMT simulator and about 15 conversion flights.

Communications

The RA 70 (*Réseau Air* 70) network covers three-quarters of France with direct and troposcatter communications, using four RAID (*Relais Automatiques de'Informations Digitales*) automatic processing and switching stations at Taverny, Metz, Bordeaux and Lyon Mont-Verdun; there are 23 other main stations and 56 secondary terminals at air bases. Six types of communication link are available: R1 (telephone) and RS (radio) together form the Tigre system; R3 is the Vestale troposcatter link; R4 is the navy's VLF link; R5 is an emergency network using commercial broadcasting stations; and R6 is a single-sideband back-up to Tigre.

'Poker' exercises are run some ten times a year to check the effectiveness of RA 70. During one such exercise the FAS had three C-135s and eight Mirage IVs airborne 11min after receiving the order to scramble. Five minutes later a total of 29 Mirage IVs from eight bases were in the air. The firing of six SSBS missiles had been simulated within 6min of receiving the launch order, and all 18 had been 'fired' after another 6min.

Navy

The *Marine Nationale* is responsible for the FOST (*Force Oceanique Stratégique*) sea-going arm of the *Force Nucléaire Stratégique* in the form of a fleet of nuclear-powered submarines carrying MSBS (*Mer-Sol Balistique Stratégique*) intermediate-range ballistic missiles. Surface ships also carry surface-to-air missiles, and the fleet air arm (*Aéronavale*, for *Aéronautique Navale*) has some 12,000 men and 400 aircraft. Of these, about 80 are embarked aboard two aircraft carriers and the remainder fly from shore bases or, in the case of helicopters, from ships of the *Marine Nationale*.

The fourth 16-missile SNLE (*Sous-Marin Nucléaire Lance-Engins*), *L'Indomptable*, put to sea for the first time in January 1977 and was to be joined in September 1979 by *Le Tonnant* with a sixth submarine, *L'Inflexible*, completing the fleet in 1983 if sufficient money is available. *L'Indomptable* is the first SNLE to carry the M20 version of MSBS, which has a single one-megaton MR-60 thermonuclear warhead in place of the 500-kiloton doped nuclear warhead fitted to the earlier M1 and M2. These versions, which have a range of 2500km, are being replaced in the first three submarines – *Le Redoutable*, *Le Terrible* and *Le Foudroyant* – by the 3000km M20.

The next stage is to fit the lighter MR-61 warhead, also with a yield of about a megaton, in place of the MR-60 so that accuracy can be improved. The heavier and fatter M4 development is due to enter service in the mid-1980s; this will carry six or seven re-entry vehicles of about 150 kilotons each. The SNLEs are based at the Ile Longue near Brest, and one of the major responsibilities of the *Aéronavale*'s Super Frelons is to patrol the exit and entry lanes.

The *Aéronavale*'s aircraft are operated by *flotilles* (squadrons), *escadrilles* (flights) and *sections*. The *flotilles* carry an F suffix and are numbered in the sequences 1 to 9 for shipboard ASW units, 11 to 19 for strike and fighter squadrons, 21 to 29 for maritime-patrol units and 31 to 39 for helicopters. *Escadrilles de servitude* are communications units often operating more than one type and carry an S suffix after the number: 1 to 19 for miscellaneous, 20 to 29 for helicopters and 51 to 59 for training units. Miscellaneous unnumbered *sections de liaison* are responsible for comms or training duties and may be former *escadrilles* which have been reduced in strength.

The *Aéronavale* provides the air groups for the 27,300-ton aircraft carriers *Foch* and *Clemenceau*, which embark both fixed-wing types and helicopters and which are scheduled to remain in service until 1992. The major new fixed-wing aircraft being introduced in the late 1970s is the Dassault Super Etendard strike fighter developed to replace the Etendard IVM, which has been in service since 1962. The Super Etendard is based extensively on its predecessor but is different in a number of areas. A new wing, with leading-edge slats and double-slotted flaps, considerably improves lift at low speed. The Snecma Atar 8K50 turbojet, a non-afterburning development of the 9K50 installed in the Mirage F.1, produces about 10% more static thrust than the Atar 80 in the Etendard IVM and has a lower specific fuel consumption. The thrust increase allows the Super Etendard to be catapulted at higher weights, thereby increasing maximum fuel load and payload/range.

A Thomson-CSF/Electronique Marcel Dassault Agave radar is used in both air-to-air and air-to-surface roles: fleet air defence, anti-ship attack and, to a lesser extent, overland operations. Two 30mm DEFA cannon are fitted and there are five attachment points for bombs, missiles or external fuel tanks. A reconnaissance pod is under development, although the specialised recce Etendard IVPs are expected to remain in service until at least 1985. On a typical attack mission, with four 400kg bombs (the Etendard IVM is limited to 250kg on its outboard pylons) and a centreline 600lit fuel tank – which its predecessor is again unable to carry in this position – the Super Etendard's radius of action is 250km with five minutes over the target. By carrying a 1100lit external tank in addition to 600lit under the belly, the radius can be extended to 620km on a high-low-high mission and 400km at low level.

The Super Etendard's normal anti-ship weapon is the AM.39 Exocet. Matra R.550 Magic air-to-air missiles can be carried for interception duties and self-defence, and the aircraft will also be armed with the AN-52 tactical nuclear bomb. The first Super Etendard squadron (11F) was scheduled to have received its 17 aircraft by January 1979 and was due to deploy aboard *Clemenceau* during the year. The second unit is to be 14F, which will donate its Crusaders to 12F. Seventy-one Super Etendards have been ordered.

The *Aéronavale*'s fighter component comprises two *flotilles* of Vought F-8E(FN) Crusaders, which are integrated with CAFDA units of the *Armée de l'Air* when they are shore-based. The Crusaders have undergone wing modifications to bring

Below: The Transall medium transport has been put back into production for the *Armée de l'Air* to replace Noratlases.

Unit	Type	Role	Base	No	Notes
Aeronavale					
4F	Alizé	ASW/SAR	Lann-Bihoué	12	⎤ 9F disbanded 1972 to provide Alizés for 4F and 6F
6F	Alizé	ASW/SAR/train	Nîmes-Garons	12	⎦
11F	Etendard IVM	Att	Landivisiau	18	Being replaced by Super Etendard
12F	⎤ F-8E(FN) Crusader	Fight	Landivisiau	16	⎤ Being re-organised as one squadron as 14F
14F	⎦			16	⎦ re-equipped with Super Etendard
16F	Etendard IVP	Recce	Landivisiau	14	
17F	Etendard IVM	Att	Hyères	18	
21F	Atlantic	ASW/recce/SAR	Nîmes-Garons	7	
22F	Atlantic	ASW/recce/SAR	Nîmes-Garons	7	
23F	Atlantic	ASW/recce/SAR	Lann-Bihoué	7	Also Pacific detachment
24F	Atlantic	ASW/recce/SAR	Lann-Bihoué	7	1 aircraft at Dakar for SAR; 1 SP-2H in Djibouti
25F	SP-2H Neptune	ASW/recce/SAR	Lann-Bihoué	10	
31F	SH-34J	ASW	St Mandrier	12	Provides detachments for carriers. Being replaced by Lynx
32F	Super Frelon	ASW	Lanvéoc-Poulmic	10	Provides shipborne detachments
33F	H-34	Assault	St Mandrier	16	Can carry SS.11 missiles
34F	Alouette III	ASW	Lanvéoc-Poulmic	12	Provides shipborne detachments
2S	⎡ Alizé	ASW	Lann-Bihoué	4	
	N.262	Liaison	Lann-Bihoué	4	Liaison for 1 and 2
	⎣ Navajo	Liaison	Lann-Bihoué	3	Maritime Regions
3S	⎡ N.262	Train	Hyères	4	Radar trainers
	⎣ Navajo	Liaison	Hyères	5	Liaison for 3 Maritime Region
10S	⎡ Alizé	Test	St Raphäel		Fixed-wing trials unit
	Navajo	Test	St Raphäel		
	N.2504 Noratlas	Test	St Raphäel	1	
	⎣ Rallye 100ST	Train	St Raphäel	6	Operated for ratings' school
12S	P-2H Neptune	Test/train	Papeete	3	
20S	⎡ Alouette II				
	Alouette III	⎤ Test	St Raphäel		Rotary-wing trials unit
	Super Frelon				
	⎣ H-34				
22S	⎡ Alouette II	⎤ Comms/SAR/train	Lanvéov-Poulmic		
	⎣ Alouette III				
23S	⎡ Alouette II	⎤ Comms/SAR	St Mandrier		
	⎣ Alouette III				
27S	⎡ Super Frelon	Trans	Hao	5	Supports nuclear test centre
	⎣ Alouette III	Comms	Hao	4	
55S	N.262	Train	Aspretto	4	Twin-engined conversion
56S	C-47	Train	Nîmes-Garons	12	Navigation training
59S	⎡ Zéphyr	Train	Hyères	12	⎤ Deck-landing training
	⎣ Etendard	Train	Hyères	12	⎦
Section Réacteur de Landivisiau	⎡ Paris	Comms	Landivisiau	9	
	⎣ Falcon 10MER	Train	Landivisiau	4	Super Etendard radar training
Section Jeanne d'Arc	⎡ Alouette III	⎤ Train	St Mandrier		Officer-cadet training
	⎣ H-34				
Section de Liaison de Dugny	⎡ C-54	VIP	Le Bourget/Dugny	1	
	N.262	VIP	Le Bourget/Dugny	2	
	⎣ Navajo	VIP	Le Bourget/Dugny	3	
Section de Liaison de Nouvelle Caledonie	⎡ C-54	Comms	Tantouta	1	
	⎣ C-47	Comms	Tantouta	1	Formerly 9S
Section de Liaison de Madagascar	⎤ C-47	Comms	Diego-Suarez		
Section de Vol Sportif Ecole Navale	⎡ Rallye 100ST	Train	Lanvéoc-Poulmic	10	Initial flying training
	⎣ MS.880B			10	
Marine Nationale Force Océanique Stratégique					
	MSBS	SLBM	Six submarines	96	
	Tartar/Standard	SAM	Destroyers, corvettes		
	Masurca	SAM	Cruiser, frigates		
	Marine Crotale	SAM	Frigates, corvettes, helicopter carriers		

them up to F-8J standard, and they are expected to remain in service until at least 1985.

The third carrier-based fixed-wing type is the Dassault-Breguet Alizé anti-submarine aircraft. As part of a modernisation programme 28 Alizés are being fitted with the frequency-agile Thomson-CSF Iguane radar in place of the previous DRAA 2A, together with new ESM equipment; these will be completed from 1980. Alizé armament includes AS.12 air-to-surface missiles in addition to anti-submarine weapons.

Maritime patrol

Shore-based maritime patrol and anti-submarine operations are the responsibility of the Dassault-Breguet Atlantic Mk 1, backed up by the remaining P-2H Neptunes. A 'New Generation' 2 version of the Atlantic, the ANG (Atlantic Nouvelle Génération), has been ordered as a successor to the Neptunes from about 1985 and all the Mk 1 Atlantics by 1990. The ANG, of which 42 have been ordered, will carry the Thomson-CSF Iguane radar and

Right: A posse of *Armée de Terre* Puma assault helicopters can deliver gap-plugging squads of fully equipped troops.

an all-digital tactical control system. A typical mission involves a 4hr patrol at 2500km from base while carrying up to four AM.39 Exocet anti-ship missiles.

The *Aéronavale* has 14 Aérospatiale Super Frelon helicopters for heavy-list and anti-submarine duties, backed up by smaller types. The Service is due to receive up to 40 Westland/Aérospatiale Lynxes for operations from ships and shore bases. The two primary missions are anti-submarine warfare and operations against fast patrol boats, hovercraft or hydrofoils carrying anti-ship missiles, using two Mk 46 torpedoes and four AS.12 missiles respectively. The helicopters are fitted with an Omera-Segid ORB-31/W radar and a CIT-Alcatel DUAV-4/HS-71 dipping sonar. The first of 26 ordered in the initial batch was handed over in September 1978.

The late delivery of Lynxes has led to an interim order for anti-submarine Alouette IIIs, the first of 12 being delivered to the new *Flottille* 34F at Lanvéoc-Poulmic in January 1977. The Alouette IIIs carry a Crouzet magnetic-anomaly detector and SFIM Type 146 autopilot and will be operated alongside the Lynx when it enters service; each unit is expected to comprise four Alouette IIIs and eight Lynxes.

Aéronavale fixed-wing aircrew are trained by the *Armée de l'Air*, which is also responsible for instruction on large helicopters following basic training given by ALAT (army aviation). *Escadrille* 22S carries out helicopter instrument instruction on its Alouettes. Other schools and test establishments are listed in the table.

The *Marine Nationale* operates several types of surface-to-air missile on its ships. The medium-range Ruelle Arsenal Masurca Mod 3 is carried by the cruiser *Colbert* and the frigates *Suffren* and *Duquesne*, the two last-named being responsible for carrier escort. The General Dynamics Standard is replacing the earlier

Tartar in four T47 destroyers and the three *Georges Leygues* corvettes so far authorised are expected to carry Standard 2. The point-defence Marine Crotale weapon is scheduled to arm up to 16 C70 corvettes, three F67 frigates, the helicopter carrier *Jeanne d'Arc* and the projected PA75.

Army

ALAT (*Aviation Légère de l'Armée de Terre*) was formed in 1954 to provide the army with an airborne observation and communications arm. The French Army is in the process of being reorganised into more compact, mobile fighting units, with one level in the chain of command being removed, and the ALAT is changing correspondingly. Before 1977 the force was organised into GALCAs (*Groupes d'Aviation Légère de Corps d'Armée*) and GALDivs (*Groupes d'Aviation Légère de Division*). One GALCA was allocated to each army corps and comprised ten Alouette IIs and Gazelles for light observation, 15 armed Alouette IIIs and 20 troop-transport Pumas. One 40-helicopter GALDiv was attached to each of the army's six divisions, being based near divisional headquarters; thus GALDiv 1 was at Tirer, 2 at Freiburg, 4 at Verdun, 7 at Mulhouse, 8 at Compiègne, and 11 at Pau and Dinan. A GALDiv comprised two *Pelotons d'Hélicoptères Légers*, each with ten Alouette IIs and Gazelles; a *Peloton d'Hélicoptères d'Attaque* with ten Alouette IIIs armed with SS.11 anti-tank missiles; and a *Peloton d'Hélicoptères de Manoeuvre* operating ten assault Pumas. GALDiv 11, which specialised in long-range intervention, had an additional Alouette II *peloton* attached to the amphibious operations headquarters at Dinan.

These traditional types of unit have been replaced by the *Régiment d'Hélicoptères de Combat* (RHC) and *Groupe d'Hélicoptères Légers* (GHL); three RHCs were formed in 1977, two more in 1978 and the last in 1979. Two RHCs are being attached to each of the two army corps, with their commands integrated with those of the corps headquarters, and the remaining two helicopter regiments are to be held in

reserve. Units can be detached to specific formations as necessary.

Each RHC has seven *escadrilles* (squadrons): two equipped with Gazelles for reconnaissance; three operating Alouette IIIs armed with SS.11 missiles and later SA.342 Gazelles carrying the second-generation HOT weapon; and two with Pumas. Each corps is additionally allocated a 30-helicopter GHL – comprising Alouette IIs and IIIs plus Gazelles – for observation, liaison and communications; a GHL of 20 machines is also attached to each of France's six military regions on the same lines as the GALDivs.

All 166 SA.341F Gazelles in the first batch have been delivered to replace Alouette IIs on observation and light transport duties. These are being followed by 160 uprated SA.342Ms armed with four Euromissile HOT anti-tank missiles; Alouette IIIs fitted with the first-generation SS.11 are being retained because the older weapon is complementary to HOT in some circumstances. The 84 or so remaining Alouette IIIs are expected to be replaced by the SA.360 Dauphin. A specialised attack helicopter is being studied for the 1980s. Resupply of the forward area, troop transport, logistics and emplacement of teams firing the Milan anti-tank missile are the responsibilities of 140 Pumas: 130 SA.330Bs and ten SA.330Hs. Emphasis is placed on blind flying, and the 75km-range Thomson-CSF Spartiate radar was introduced in 1974 for surveillance and control of ALAT helicopters in bad weather.

The French Army also has three regiments of Raytheon Hawk medium-range surface-to-air missiles, totalling 54 launchers. The missiles are being upgraded to Improved Hawk standard. The Euromissile Roland is being introduced for point defence of armoured columns and protection of important centres. The first production-standard units were delivered to the 54 *Régiment d'Artillerie* at Verdun towards the end of 1976, and the eventual total is expected to comprise 170 firing units for the clear-weather Roland I and 82 for the all-weather Roland II; of these, 144 and 70 respectively are likely to be fitted to AMX30 tank chassis, with the remainder being used as spares.

Training

The six-month basic flying training course at Dax includes 145hr instruction on Alouette IIs and Gazelles, following initial air experience gained in Broussards and L-19s. Basic training is followed by a three-week *pilotage tactique* course at Le Luc, comprising 18hr of handling and navigation at very low level. Missile operators are trained over a four-week period which includes simulator work and ten live firings. Once trained, operators are tested twice a year – once each on the simulator and with a live round. Unit commanders receive additional training.

Pilots from Dax convert to the Puma in a seven-week, 35hr course at Le Luc and can return there after two years with their squadrons for instrument-flight training, including 55hr simulator and flight time.

Below: A standard Etendard IV attack aircraft waits its turn as its successor, a Super Etendard, is about to be launched from one of the *Marine Nationale*'s two aircraft carriers.

SWITZERLAND

Area: 15,950 sq. miles (41,300 sq. km.).
Population: 6·35 million.
Total armed forces: 18,500.
Estimated GNP: $60·1 billion (£30·05 billion).
Defence expenditure: $1·55 billion (£775 million).

Switzerland has been neutral since 1815 and is committed to a policy of "total" or "general" defence, aimed at making invasion so unattractive for a potential aggressor that he is unlikely to waste time and effort on overrunning the country. This policy makes maximum use of the topography, which allows ground forces to be slowed down by destroying bridges and tunnels and blocking narrow passes. Aircraft operating from dispersal fields in steep-sided valleys can then play their part in attacking the immobilised enemy, secure in the knowledge that hardened facilities and a heavy concentration of anti-aircraft defences combine to protect their own bases.

The Swiss air force and anti-aircraft command (*Kommando der Flieger- und Fliegerabwehrtruppen*) constitutes one of five corps in the army and is divided into three brigades: Air Force Brigade 31, responsible for all flying units; Air Base Brigade 32, which oversees logistics; and Anti-aircraft Brigade 33, with surface-to-air missiles and cannon plus an associated command-and-control network. Air Force Brigade comprises three *Fliegerregimenten* (regiments), each containing six to eight *Staffeln* (squadrons). The regiments are subdivided into *Geschwader* or *escadres* (wings).

Only six squadrons are manned permanently. Five of these comprise the *Uberwachungsgeschwader* or *Escadre de Surveillance* (Surveillance Wing); three operate Hunters – which are being replaced by Northrop F-5Es – and the other two are equipped with Mirage IIISs. The air force has only 140 professional pilots, two-thirds of which are in the Surveillance Wing and the remainder allocated to training and helicopter units. The reconnaissance squadron, which operates Mirage IIIRSs and Venoms, is not officially part of the Surveillance Wing but has one-third professional pilots and two-thirds militia. The air force as a whole, including the anti-aircraft forces, has 3000 regular personnel, 6000 conscripts and 45,000 reservists. Pilots are split into 30% regulars, 40% ex-regulars and 30% militia.

Peacetime operations are concentrated at three bases – Dubendorf (the largest), Emmen and Payerne – but in war only one of these would remain operational. Most aircraft would disperse to the dozen emergency fields with hangars and other facilities carved out of the alpine valley walls. The organisation would also be split, with the squadrons in the Surveillance Wing joining their respective regiments.

Inventory

The interceptor role is shared by F-5Es, Mirages and Hunters; all can double as ground-attack aircraft, although the emphasis is on Hunters in the air-to-ground role. The Mirage IIISs, which entered service in 1966, are fitted with Hughes TARAN fire-control equipment and carry the same company's HM-55 and HM-58 air-to-air missiles, variants of the Falcon. The F-5Es, Mirages and Hunters may also be fitted with Sidewinder AAMs. The interceptors operate within the Florida air-defence system, also supplied by Hughes with some elements provided by Plessey and Ferranti.

Air defence is additionally provided by two battalions of BAC Bloodhound 2 medium-range surface-to-air missiles, each with 32 launchers. The missiles are spread over a wide area to provide defence in depth. The anti-aircraft element also has seven regiments of AAA, comprising 18 battalions with towed radar-directed Oerlikon 63 twin 35mm cannon and 19 battalions equipped with 20mm cannon. The 35mm weapons, grouped in units of 12 to

Above: After much debate and years of evaluations the Northrop F-5E has started to replace Swiss Air Force Hunters in the air-superiority role.

16, are linked in pairs to Contraves Fledermaus and Superfledermaus fire-control systems. Forty-five units of the same company's Skyguard FCS have been ordered to replace two-thirds of the earlier equipment, and further examples are expected to be procured to supersede the remainder. The effectiveness of the 20mm guns, which are deployed in batteries of four, is also being upgraded by the addition of Oerlikon Delta sights. Switzerland has a requirement for a low-level SAM and is expected to order the British Aerospace Rapier. The country may also order the Bofors RBS70, to which a share of the development costs has been contributed.

The Mirages can carry AS.30 air-to-surface missiles, but the ground-attack role is mainly the responsibility of Hawker Hunters. An initial batch of 100 bought in

SWITZERLAND

Wing	Unit	Type	Role	Base	No
Air Force Brigade 31					
	1 Sqn	Hunter F.58/T.68	FGA/train	Dubendorf	18
Uberwachungsgeschwader/Escadre de Surveillance	11 Sqn	Hunter F.58/T.68	FGA/train	Dubendorf	18
	18 Sqn	Hunter F.58/T.68	FGA/train	Dubendorf	18
	16 Sqn	Mirage IIIS/IIIBS	FGA/train	Emmen	15
	17 Sqn	Mirage IIIS/IIIBS	FGA/train	Payerne	15
	10 Sqn	Mirage IIIRS	Recce	Dubendorf	16
	10 Sqn	Venom	Recce	Dubendorf	8
	5 Sqn	Hunter F.58A/T.68	FGA/train	Meiringen	
	4 Sqn	Hunter F.58A/T.68	FGA/train		
	7 Sqn	Hunter F.58A/T.68	FGA/train		
	8 Sqn	Hunter F.58A/T.68	FGA/train		
	19 Sqn	Hunter F.58A/T.68	FGA/train		
	21 Sqn	Hunter F.58A/T.68	FGA/train		
	2 Sqn	Venom FB.50	Att	various	16-18 each
	3 Sqn	Venom FB.50	Att		
	6 Sqn	Venom FB.50	Att		
	9 Sqn	Venom FB.50	Att		
	13 Sqn	Venom FB.50	Att		
	15 Sqn	Venom FB.50	Att		
	20 Sqn	Venom FB.50	Att		
		Pilatus P-2	Train	Locarno, Sion, Emmen	50
		Pilatus P-3	Train	Locarno, Sion, Emmen	70
		Vampire T.55/FB.6	Train	Locarno, Sion, Emmen	16/20
		Venom FB.50	Train	Locarno, Sion, Emmen	
Zeilfliegerkorps		C-3605	Target tug	Sion	23
Transport fliegerkorps		Ju 52/3mG4E	Trans/paratroop	Dubendorf	3
		E50 Twin Bonanza		various	3
		Alouette II		various	26
Leichtefliegerstaffeln 1 to 7		Alouette III	Liaison/comms/SAR	various	70
		Do 27H-2		various	6
		PC-6A Porter		various	11
		PC-6B Turbo-Porter		various	24
Anti-aircraft Brigade 33					
		Bloodhound 2	SAM	various	64* launchers

*Two batteries

Northrop F-5E Tiger II entered service in early 1979. The type was selected after a long drawn-out evaluation which included the Harrier, A-7 and Mirage Milan; six two-seat F-5Fs and 13 F-5E single-seaters are being supplied direct by the US manufacturer and the remaining 53 F-5Es are being assembled by the Federal Aircraft Factory at Emmen. The radar-equipped Tiger II's will have an air-superiority role, using Sidewinders and built-in 20mm cannon, in addition to ground attack.

Two Hunter squadrons in the Surveillance Wing were due to convert to F-5Es in early 1979, their former aircraft then replacing two Venom units. Two further squadrons were later to convert to F-5Es. The majority of the Hunters are likely to be replaced by a different type; the F-16, F-18L and A-10 are among the contenders. The front-line element is completed by the reconnaissance squadron, equipped mainly with Mirage IIIRSs. These carry French cameras and use flares and photoflashes for night work; infra-red sensors may later be acquired if sufficient money is available.

The air force operates seven *Leichtefliegerstaffeln* (light support squadrons) on communications, rescue and other duties, one squadron being attached to each of the four army corps. Helicopter operations are based at Alpnach, with Dubendorf acting as the centre for rotary-wing SAR operations. Each base is allocated one or two Alouette IIIs for this role.

Training

Flying training for prospective air force pilots begins at the age of 18, with some 240 applicants undergoing a two-week aptitude-assessment course. The 160 who pass this stage go on to a second two-week flying course a year later, which whittles the number down to 80. These 80 20-year-old cadets are given four weeks' ground training at the Payerne recruit school and then move on to Locarno for the 17-week primary flying course – with nine weeks on Pilatus P-3s for the 30–40 who remain after the first month.

Students then move to Sion for basic flying training, being promoted to corporal after four weeks. Jet instruction begins with 17 weeks on two-seat and single-seat Vampires, followed by a similar period of advanced training on Venoms at Emmen. Graduates have about 300hr experience and are promoted to sergeant.

About one-third of the 20 or so newly qualified pilots join Swissair or take up a non-flying job, the rest being allocated to the Surveillance Wing. These latter professional air force pilots fly some 110hr a year, while the militia pilots notch up half this time in six single weeks throughout the year and a further ten days taken when appropriate. Surveillance Wing pilots selected for officer training undergo an additional 17 weeks' instruction before being promoted to lieutenant and then spend a similar period as instructors at the advanced flying school. Half the Surveillance Wing's pilots double as instructors, enabling them to keep up their hours. Mirage pilots join their squadron from a Hunter unit after about 1000hr on jets.

Top: Search and rescue is obviously vital in a country as mountainous as Switzerland. Here an Alouette III hoists a stretcher patient to safety.

Above: Bloodhound surface-to-air missiles complement manned interceptors.

Left: A pair of reconnaissance Mirage IIIRSs streak past one of the Swiss Air Force's surveillance radars in the high Alps.

the late 1950s has been followed by two further orders for refurbished aircraft, comprising 52 F.58As and eight two-seat T.68s. The 160th and last Hunter was delivered in July 1976; some 140 are operational, and the type is expected to remain in service throughout the 1980s. The Hunters, which are fitted with SAAB BT9 bombing computers, have replaced many of the obsolescent Venoms, although some 150 of the latter remain operational.

Both the Venom and the Hunter squadrons were due to be re-organised when the

AUSTRIA

Area: 32,300 sq. miles (83,600 sq. km.).
Population: 7·52 million.
Total armed forces: 37,000.
Estimated GNP: $47·7 billion (£23·85 billion).
Defence expenditure: $718 million (£359 million).

Austrian air forces took part in the First World War, Austria-Hungary comprising one of the Central Powers, and were merged with the *Luftwaffe* after Germany annexed Austria in 1938. The country was not formally reconstituted until 1955, and a treaty signed in that year limits the strength and equipment of the Austrian forces. Defence spending is small, and civil assistance is emphasised alongside the military roles.

The 4500-man *Osterreichische Luftstreitkräfte* (Austrian air forces) operate as part of the federal army, the Air Division being one of the army's four divisions. The *Luftabteilung* (Ministry of Defence) in Vienna is the highest planning authority for the OLk and supervises all activities, with day-to-day running devolved to the air headquarters and air operations centre in Vienna, from where aviation operations and training are directed. In an emergency, all activities would be directed from the underground air operations centre at Salzburg. Beneath this structure is the Air Division, based at Langenlebarn, which has a headquarters battalion, three air regiments, an air surveillance regiment, flying school and air material depot.

The *Fliegerstabsbataillon* (headquarters battalion) has its own HQ staff, four communications companies and a *Bildkompanie* responsible for aerial reconnaissance, support, photography and survey work. *Fliegerregiment 1* (Air Regiment No 1) at Langenlebarn has Helicopter Wing No 1 (which includes one squadron of assorted fixed-wing aircraft), Air Defence Battalion No 1 with two companies operating 20mm and twin 35mm anti-aircraft guns, and an aircraft maintenance depot. *Fliegerregiment 2* at Zeltweg includes the *Überwachungsgeschwader* (Surveillance Wing), Helicopter Wing No 2 (with its own maintenance base), Air Defence Battalion No 2, and another aircraft maintenance depot. *Fliegerregiment 3* at Hörsching incorporates the *Jagdbombergeschwader* (Fighter-bomber Wing), Helicopter Wing No 3, Air Defence Battalion No 3 (four companies operating 20mm and twin 35mm weapons) and a further maintenance depot. The *Flugmelderegiment* (Air Surveillance Regiment) at Salzburg has two sites for the Selenia RAT-31S three-dimensional surveillance radar, at Kolomannsberg and Koralpe, together with mobile radars, ground-to-air communication sites, and support and training units. Surveillance data are transmitted to the air operations centre in real time over an automatic data-processing system, the information being

Below: An S-65Oe banks over two rows of Saab 105Oes, used for all combat roles.

Bottom: The OLK's two heavy-lift S-65Oes are widely used on civilian construction and SAR duties in the Alps, in addition to military work.

Top: The OLk's two squadrons of Alouette III fly utility duties from Aigen.

Above: Saab Safirs provide fledgling OLk pilots with their first taste of aircraft handling.

co-ordinated with that from civilian radars and air-traffic control. The organisation is completed by the flying school at Zeltweg and the *Luftzeulager* (Air Depot) at Hörsching.

The OLk's primary missions are air defence, tactical air support, reconnais-sance, transport and utility operations, and radar surveillance. The combat force comprises 34 Saab 1050es remaining from the total of 40 bought in two equal batches, the first of which was ordered in July 1967. The 1050e, deliveries of which began in July 1970, is operated on interception, ground-attack and reconnaissance duties, carrying two gun pods, 12.5cm rockets, two 250kg bombs or up to six 125kg bombs. The first batch replaced Vampires and Magisters, while the second was intended as an interim successor to the Saab J29

until a more effective combat aircraft could be procured. This process had still not been completed by the end of 1978, however, at which time the Dassault Mirage F.1, Israel Aircraft Industries Kfir-C2, Saab Viggen and Northrop F-5E were under consideration, with the last two as favourites. The official requirement is for 24–30 new interceptors but budget restrictions may prevent this total being met, at least initially.

A large part of the OLk's fleet is made up of helicopters. Two dozen AB.212s were ordered in July 1978 to replace the same number of AB.204Bs; these are being delivered in 1979–80 and are equipped for duties including casualty evacuation, search and rescue, and firefighting in addition to transport. The dozen OH-58Bs are fitted with M-27 Minigun pods containing 7.62mm machine guns for armed escort in addition to the other roles of reconnaissance, target observation and marking, liaison and search and rescue. The largest rotary-wing type is the S-65, a pair of which were acquired in 1970 for use primarily on rescue and civilian heavy-lift duties in the Alps. The S-65s carry a hoist as fitted to the HH-53B/C and have provision for auxiliary fuel tanks in addition to accommodation for up to 38 passengers. Alouette IIIs are operated on a variety of tasks and a dozen AB.206As are employed for training and secondary roles.

Initial training takes place on the Saab 91D Safirs operated by the OLk's flying school. After about 40hr on this type, combat pilots proceed to the Saab 1050e for an additional 100hr before joining an operational squadron. Helicopter pilots progress from the Safir after a total of 80hr, completing their training with about 150hr on AB.206s. The OLk has a requirement for 24 new primary trainers to be introduced in 1981–83.

The *Goldhaube* (Golden Hood) programme aimed at providing the ground-based element of an air-defence network, within which the new intercepters will operate, is based on the Selenia RAT-31S three-dimensional search radar (in both fixed and mobile installations) combined with the same company's MRCS-403 mobile automatic reporting and control stations. In addition to the three OLk air-defence battalions which defend air bases and headquarters installations, the Austrian Army has its own air-defence battalions operating twin M-42 40mm guns, twin Oerlikon 35mm cannon controlled by the Contraves Skyguard fire director, and 20mm guns. Each Austrian Army combat battalion additionally has its own 20mm anti-aircraft artillery.

AUSTRIA

Regiment	Wing	Unit	Type	Role	Base	No	Notes
Fliegerregiment 1	Helicopter Wing No 1		AB.212	Trans/casevac/SAR/util	Langenlebarn	12	One sqn; replaced AB.204B
			AB.206A	Train/SAR		12	One sqn
			OH-58B	Obs/liaison/recce		12	One sqn
			Skyvan	Trans		2	
			PL-6B Turbo-Porter	Trans		12	One sqn
			L-19 Bird Dog	Trans		16	
Fliegerregiment 2	Surveillance Wing		Saab 1050e	Int	Zeltweg Graz	17	Two sqns
			Saab 1050e	Int			
	Helicopter Wing No 2		Alouette III	Util	Aigen	24	Two sqns
Fliegerregiment 3	Fighter-bomber Wing		Saab 1050e	FGA		17	
			Saab 1050e	Recce			
	Helicopter Wing No 3		AB.212	Trans/casevac/SAR/util	Hörsching	12	One sqn; replaced AB.204B
			S-650e	SAR/heavy-lift		2	
		Flying school	Safir	Train	Zeltweg	19	

106

ITALY

Area: 131,000 sq. miles (339,000 sq. km.).
Population: 56 million.
Total armed forces: 362,000.
Estimated GNP: $193·7 billion (£96·85 billion).
Defence expenditure: $5·06 billion (£2·53 billion).

The Italian air force, *Aeronautica Militare Italiana* (AMI), forms the major part of NATO's 5th ATAF (Allied Tactical Air Force) and has overall responsibility for operations carried out by the air arms of the navy (*Marinavia*, for *Aviazione per la Marina Militare*) and army (ALE, for *Aviazione Leggera dell'Esercito*). A tri-Service secretariat in the defence ministry supervises military operations and the AMI has three commands (air defence, air transport and rescue, and training) together with the same number of inspectorates (responsible for logistics, naval air operations and communications/air-traffic control).

Italy fought on the Allied side during the First World War, the *Corpo Aeronautico Militare* ending the conflict with nearly 1800 aircraft on strength. The *Regia Aeronautica*, an independent air arm with the same status as the army and navy, was formed in 1923 and saw extensive action in Abyssinia and the Spanish Civil War before the outbreak of the Second World War.

The present AMI has a personnel strength of about 70,000, including some 26,000 conscripts. Italy's military spending has been falling steadily in the past few years, but the AMI is receiving more than

Right: Macchi's MB.326 has been an export moneyspinner and is being followed by the MB.339.

Below: The F-104 will remain the AMI's main combat type until Tornado enters service.

half of an additional L2365 billion allocated for a ten-year equipment build-up. Most of this will be made available in the first half of the 1980s.

Italy is divided into three air regions: the 1a *Regione Aerea*, headquartered at Montevenda outside Milan; and the 3a *Regione Aerea*, with headquarters at Martina Franca, near Taranto. The third regional headquarters, of the 2a *Regione Aerea*, has been disbanded. Until 1967 the AMI was organised into *aerobrigate* (air brigades), based on the USAF wing, but most have since been split into smaller *stormi* (regiments). The *aerobrigate* comprised up to three *gruppi* (squadrons), each having 16 to 25 aircraft plus reserves and being deployed to its own base. The present *gruppi* have only 12 to 18 aircraft each and this reduction in size, combined with the high cost of operating the bases, means

that the policy of having only one *gruppo* per base has been dropped.

The AMI has 19 aircraft *gruppi* assigned to NATO (six of interceptors, five of fighter-bombers, three reconnaissance, three light strike/recce and two anti-submarine) and eight *reparti* (reduced from 12) of surface-to-air missiles. In time of war the rest of the AMI would come under 5th ATAF control.

Inventory

The AMI is receiving 205 Lockheed/Aeritalia F-104S Starfighters to succeed the F-104G variant and the interceptor and ground-attack roles. The F-104S is the only Starfighter version to carry the Raytheon AIM-7 Sparrow medium-range air-to-air missile, which is also built under licence in Italy. Selenia has developed the Aspide multi-role missile, based on Sparrow, to replace the US-designed weapon in the

Brigade	Wing	Unit	Type	Role	Base	No	Notes
1 Aerobrigata Intercettori Teleguidati	7° Reparto		Nike Hercules	SAM	Istrana (Traviso)	96 launchers	Deployed in eight gruppi (reduced from 12)
	16° Reparto		Nike Hercules	SAM	Istrana (Treviso)		
	17° Reparto		Nike Hercules	SAM	Istrana (Treviso)		
	2° Stormo CBR Mario d'Agostini	14° Gruppo	G.91R	Light att/recce	Treviso-San Angelo		
		103° Gruppo	G.91R	Light att/recce	Treviso-San Angelo		
3 Aerobrigata Ricognitori Tattici		28° Gruppo	RF-104G/F-104S	Recce	Verona-Villafranca	18	Approx 30 RF-104G plus some F-104S
		132° Gruppo	RF-104G/F-104S	Recce	Verona-Villafranca	18	
	4° Stormo CI Amedeo d'Aosta	9° Gruppo	F-104S	Int	Grosseto	12	
	5° Stormo CB/CI	23° Gruppo	F-104S	Int	Rimini-Miramare		
		102° Gruppo	F-104S	FGA	Rimini-Miramare		
	6° Stormo CB Alfredo Fusco	154° Gruppo	F-104S	FGA	Ghedi (Brescia)		
	8° Stormo CB Gino Priolo	101° Gruppo	G.91Y	Light att	Cervia-San Giorgio		
	9° Stormo CI Francesco Baracca	10° Gruppo	F-104S	Int	Grazzanise (Caserta)	12	
	14° Stormo GE	8° Gruppo	PD.808	ECM	Pratica di Mare	6	
			EC-130H	ECM	Pratica di Mare	1	
			EC-119	ECM	Pratica di Mare	3	To be replaced by G.222
		71° Gruppo	EC-47	ECM	Pratica di Mare	2	
			F.27	Calib	Pratica di Mare	2	
			C-45	Util	Pratica di Mare		
			T-33	Util	Pratica di Mare	3	
	15° Stormo SA Stefano Cagna	84° Gruppo	Hu-16A	SAR	Ciampino	12	Being replaced by HH-3F
		85° Gruppo	AB.204	SAR	Ciampino	14	Being replaced by HH-3F
		85° Gruppo	AB.47J	SAR	Ciampino	7	
		20° Gruppo AO	TF-104G	Train	Grosseto	28	
	30° Stormo AS	86° Gruppo	Atlantic	ASW	Cagliari Elmas	9	
		92° Gruppo	AS-61TS	VIP	Ciampino	2	
	31° Stormo TS	306° Gruppo	DC-9-30	Trans/VIP	Ciampino	2	
			DC-6B	Trans/VIP	Ciampino	2	
			PD.808	Trans/VIP	Ciampino		
	32° Stormo CBR Armando Boetto	13° Gruppo	G.91Y	Light att/recce	Brindisi		
	36° Stormo CI/CB	12° Gruppo	F-104S	Int	Gioia del Colle	12	
		156° Gruppo	F-104S	FGA	Gioia del Colle		
		636° Gruppo	P.166M	Util	Gioia del Colle		
	Athos Ammannato	88° Gruppo	Atlantic	ASW	Catania-Fontanarossa	9	
46 Aerobrigata Trasporti Medi		2° Gruppo	C-119G/J	Trans	Pisa-San Giusto	22	To be replaced by G.222
		98° Gruppo	G.222	Trans	Pisa-San Giusto	15	
		50° Gruppo	C-130H	Trans	Pisa-San Giusto	13	
	51° Stormo CI/CB Ferruccio Serafini	22° Gruppo	F-104S	Int	Istrana (Treviso)		
		155° Gruppo	F-104S	FGA	Istrana (Treviso)		
	53° Stormo CI Guglielmo Chiarini	21° Gruppo	F-104S	Int	Cameri	12	
		653° Gruppo	Various	Util	Cameri		
		201°, 204° and 205° Gruppi SVBAA	G.91T	Train	Foggia-Amendola	100 approx	To be replaced by MB.339?
			G.91T	Train	Foggia-Amendola		
		207° Gruppo	P.166M	Train	Latina		
			SF.260AM	Train	Latina	20	Replaced P.148
		206° Gruppo SE	AB.47G/J.	Train	Frosinone		
		209° Gruppo SE	AB.204B	Train	Frosinone		
		211° Gruppo SVBIE T-6			Alghero		
		212°, 213° and 214° Gruppi SVBIA	MB.326	Train	Lecce-Galatine		To be replaced by MB.339
	2 Regione Aerea Reparto Volo	303° Gruppo	T-6G	Util	Guidonia		
		307° Gruppo	P.166M	Util	Guidonia		
			S.208M	Util	Guidonia		
			C-47	Util	Guidonia		
		311° Gruppo RSV	Various	Test	Pratica di Mare		
		313° Gruppo PAN	G.91R-1A/G.91PAN	Demo	Rivolto (Udine)		Frecce Tricolori aerobatic team
		Scuola Militare	P.148	Train	Guidonia		
		Volo a Vela	S.208M	Train	Guidonia		

air-to-air and surface-to-air roles. The Italian air-defence system, SIDA (*Sistema Integrao Difesa Aerea*), forms part of NATO's Nadge chain. In addition to its manned interceptors the AMI has 96 launchers for Nike Hercules surface-to-air missiles in the north of the country. A major part of the money to be spent on updating AMI equipment is allocated to improving air defences; purchases will include at least four Selenia Argos 10 long-range early-warning radars, five Argos 12 height-finders and up to 20 batteries of the same company's Spada surface-to-air missile system, which also uses the Aspide. The Spada batteries are due to enter service between 1981 and 1987.

The major new type scheduled to be introduced in the early 1980s is the Panavia Tornado, of which Italy is planning to buy 100 (88 of the interdictor-strike variant and 12 dual-control trainers). The front-line force at any given time will comprise the 12 trainers and 54 others organised

Left: Technicians attend to licence-built SH-3D anti-submarine helicopters at the Italian Navy's Luni base.

Brigade	Wing	Unit	Type	Role	Base	No	Notes
Navy			SH-3D	ASW	⎫ Shore plus cruisers,	24	Two squadrons
			AB.212ASW	ASW	⎬ destroyers, frigates	28	Replacing AB.204ASs
			AB.204AS	ASW	⎭	30	Two squadrons
			AB.47	Util		12	
			Terrier	SAM	Cruisers		
			Tartar/Standard	SAM	Destroyers		
		NATO Seasparrow/Albatros/Aspide	SAM	Frigates		
Army			CH-47C	Logistics		26	
			AB.205B	Util		138	
			AB.204B	Util		50	
			AB.206	Scout		142	
			AB.47G/J	Scout		100	
			A.109 Hirundo	Util/gunship trials		5	Two carry TOW anti-tank missile
			SM.1019E	Liaison/obs		80	
			L-18/L-21 Super	Liaison/obs		50	Being replaced by
			Cub	Liaison/obs		50	SM.1019E
			O-1E Bird Dog	Liaison/obs		60	
			Hawk	SAM		68 launchers	Being upgraded to Improved Hawk

CBR = Caccia Bombardieri Ricognotori; CI = Caccia Intercettori; CB = Caccia Bombardieri; GE = Guerra Elettronica; SA = Soccorso Aerea;
AS = Antisommergilibi; TS = Trasporti Speciali; SCIV = Scuola Centrale Istruttori di Volo; SVBAA = Scuola Volo Basico Avanzato Aviogetti; SE = Scuola Elicotteri;
SVBIE = Scuola Volo Basico Iniziale Elica; SVBIA = Scuola Volo Basico Iniziale Aviogetti; RSV = Reparto Sperimentale Volo; PAN = Pattuglia Acrobatica Nazionale
(the Frecce Tricolori display team).

Above: Students training as AMI combat pilots in the 1980s will gain their first jet experience in the MB.339.

Above: Troops embark on a CH-47 Chinook medium-lift helicopter built by the Agusta group for the *Esercito*.

either into three fighter-bomber *gruppi* with 14 aircraft each and a reconnaissance unit with 12 Tornados, or three 18-aircraft multi-role *gruppi*. Tornado is scheduled to become operational in late 1980 or early 1981, all 100 being in service by 1985. The 34 reserves will be used to make up attrition losses. Refurbishing of the F-104Ss will begin in 1980 and the type is scheduled to remain in service until 1990.

The F-104S also equips fighter-bomber *gruppi*, three of which operate alongside interceptor squadrons; the fourth, the single-*gruppo* 6° *Stormo*, converted from the earlier F-104G in 1976. Some of the F-104Gs retired from ground-attack duties as they were replaced by the newer variant have been converted to reconnaissance RF-104Gs, equipping three 18-aircraft *gruppi*. These aircraft, together with recce F-104Ss, carry the Dutch Orpheus pod. The RF-104Gs are due to be retired from 1980 but may be required to soldier on beyond that date. Some two dozen two-seat

TF-104Gs are also in service. The Aeritalia G.91Y is operated in the light attack/reconnaissance role by two *gruppi*, and a further two are equipped with the earlier G.91R. The G.91Rs may be replaced by Tornadoes, with both the F-104Gs and the G.91Ys being succeeded by the planned Aeritalia/Macchi AM-X light attack aircraft. The Italian Air Force's schedule calls for the first of six prototypes to fly in late 1982, with deliveries to the Service starting in 1985. The AMI requires 187 AM-Xs, all of which should be in service by 1991. AM-X will be powered by the Rolls-Royce Spey turbofan and is to be used for interdiction, close air support, reconnaissance, low-level air combat and anti-ship operations. It will be armed with two air-to-air missiles and a gun, carrying up to 6000lb of external stores.

The front-line strength also includes two maritime-patrol and anti-submarine *gruppi* operating the Dassault-Breguet Atlantic. These are scheduled to be replaced from the mid-1980s, but a shortage of money is likely to result in them remaining operational longer than had been planned. Aeritalia G.222s are replacing

Fairchild C-119s on transport duties, some being equipped as electronic-countermeasures aircraft.

Basic flying training is on T-6 Texans and SIAI-Marchetti S.208Ms and SF.260s; the latter have replaced Piaggio P.148s, although this type is retained as a glider tug. Combat pilots then proceed to the MB.326 (to be replaced by the MB.339) and thence to the G.91T. The latter may also be replaced by the MB.339, 100 of which are on order, although retirement is not planned before 1985. Transport pilots move from the T-6 to the Piaggio P.166M. The AMI also provides helicopter instruction for all three Services.

Navy

The AMI is responsible for fixed-wing anti-submarine operations, leaving *Marinavia* to carry out helicopter duties. The 24 Agusta-built Sikorsky SH-3D Sea Kings are flown mainly from shore bases; they are to be fitted with the Sistel/SMA Marte

Above: The G.91 has served the AMI well in the light attack and recce roles. It will be replaced by a combination of Tornado and AM-X.

anti-ship missile system as well as operating in the ASW role. Twenty-eight AB. 212ASWs, in batches of 13 and 15, are being delivered for operations initially from the navy's larger ships, with AB.204s deployed aboard smaller vessels until they are replaced by AB.212s. The navy additionally operates surface-to-air missiles: Terrier on cruisers, Tartar (being replaced by Standard) on destroyers and the NATO Seasparrow on frigates. The Selenia Albatros SAM/gun system, which will use the Aspide multi-role missile, is expected to replace Seasparrow aboard the frigates.

Army

ALE (*Aviazione Leggera dell'Esercito*) has a comparatively large force of helicopters and fixed-wing aircraft for medium-lift, assault, anti-tank, observation and communication duties. Each RAL (*Reparto Aviazione Leggera*) – of which six are assigned to corps or divisions and eight to brigades, with provision for allocating one RAL to each of five infantry brigades – comprises one or two sections of fixed-wing aircraft and one of reconnaissance helicopters, with a further section of utility helicopters. Other helicopters are assigned to headquarters units. The ALE has evaluated five Agusta A.109 utility helicopters, and about 100 are likely to be supplied for a variety of military and paramilitary applications. Two of the A.109s were fitted with the Hughes TOW anti-tank missile system, but the Italian Army's definitive gunship helicopter is expected to be the A.129. The first of three prototypes is scheduled to fly in 1982 and the ALE is likely to buy at least 60 production two-seaters armed with eight TOWs each.

The army also operates 68 launchers for the Raytheon Hawk medium-range surface-to-air missile and is in the process of upgrading these batteries to Improved Hawk standard. The Sistel Indigo-MEI SAM is scheduled to enter service in 1982 to defend fixed installations and armoured formations.

SPAIN

Area: 196,700 sq. miles (509,500 sq. km.).
Population: 36·2 million.
Total armed forces: 315,500.
Estimated GNP: $123·6 billion (£61·8 billion).
Defence expenditure: $2·36 billion (£1·18 billion).

All three branches of the Spanish armed forces operate both fixed-wing aircraft and helicopters. The *Ejército del Aire* became an independent Service in 1939, following the Spanish Civil War, and operated several German-designed types built under licence by the national aircraft company, CASA. Formal ties with the United States date back to the signature in 1953 of the Spanish-American Defence Treaty, which resulted in a modernisation of Spain's military inventory and organisation. The latest renewal of this treaty, in 1976, gives US forces continued use of the air bases at Torrejon, Zaragoza and Moron, plus access to the naval facilities at Rota. Spain is not a member of NATO, but may join in the future.

The mainland is divided into three air regions: No 1, with headquarters in Madrid; No 2, centred on Seville; and No 3, run from Zaragoza. A separate air zone, with headquarters in Las Palmas, covers the Canaries. The EdA, with a personnel strength of some 35,700 including 8400 conscripts, operates three commands – air defence (*Mando de la Defensa Aerea*), tactical (*Mando Aviacion Tactica*) and

transport (*Mando de la Aviacion de Transporte*) – plus a training organisation (*Aviacion de Entremaniento*) and an air-sea rescue service. The basic operating unit is the *escuadron* (squadron), one or more of which may form an *ala* (wing) or *grupo* (group).

Inventory

Air defence is the responsibility of five squadrons, two with Mirage IIIEEs, two with F-4C(S) Phantoms and one operating Mirage F.1CEs. The Phantoms are expected to be replaced and supplemented in the 1980s by four squadrons (72 aircraft) of General Dynamics F-16s; when renegotiating the defence agreement the US offered to replace the F-4Cs with F-4Es as an interim measure until delivery of the F-16s, but this offer was declined by Spain and the F-4C force may be built up by the supply of an additional four F-4Cs and the same number of RF-4Cs. Only 15 Mirage F.1s were originally accepted, as a result of budgetary cut-backs, although a further ten were ordered in January 1977 to allow a second squadron, No 142, to be formed. Up to 48 more F.1s may be acquired. The interceptors operate within the Combat Grande air-defence system, based on seven

Right: The T-6G Texan (E-16) is due for replacement but has done trojan service in EdA colours. This example is operated by the *Academia General del Aire*.

Above: Spanish Navy Matadors are AV-8As ordered initially via the USMC for operation from the aircraft carrier *Dedalo*. Sea Harriers may follow.

Right: The HA.200D, designated E-14B in its training role, provides conversion training to jet types.

Below, right: The EdA has declined an offer of F-4Es but retains its F-4Cs.

long-range radars, which is being up-graded and modernised by Hughes Aircraft.

The EdA's major attack type is the F-5 Freedom Fighter, 62 of which were assembled by CASA after the direct supply of eight aircraft by Northrop. The aircraft are designated SF-5A, SF-5B and SRF-5A in the attack, training and reconnaissance roles respectively. They also carry a Spanish designation (see table), in common with all other aircraft operated by the armed forces. Tactical forces also include a variety of transports, tankers, light-attack and miscellaneous types, plus an anti-submarine force which comprises obsolescent Grumman Albatrosses (being replaced by F.27Ms) and ex-USN P-3A Orions. The three KC-130H tankers, formerly allocated to the Phantom wing, are now operated alongside the transport Hercules.

The introduction of the CASA C.212 Aviocar into the transport fleet has allowed many older types to be retired. The EdA is receiving at least 61 Aviocars for use as transports, navigation trainers and survey aircraft. The training and light attack forces are additionally due to be modernised by the introduction of 60 CASA C.101s, the first of which flew in the summer of 1977. An initial production batch of 60 aircraft was ordered in March 1978, with deliveries taking place from October 1979 to 1981. The first training squadron is expected to be operational by the end of 1979.

Training

Initial instruction on Bonanzas and Mentors is followed by advanced training on HA.200s and SF-5Bs. Multi-engine instruction is given in Aviocars, King Airs and Barons. Helicopter pilots for all three Services are taught on a variety of types at Cuatro Vientos. The BO105 has been selected in preference to the SA.342 Gazelle for utility use by all three Services; 50 will be assembled by CASA. The EdA also has a requirement for about 130 piston-engined primary trainers to replace T-6s and T-34s; CASA is offering its C.102 design. The syllabus would then comprise 70–80hr on the new aircraft, followed by 150hr of advanced training on the C.101 and 30–40hr weapon training with the same type, leading to conversion on to the F-5B.

Army

FAMET (*Fuerzas Aero Moviles del Ejército de Tierra*) operates an expanding army air arm in the logistic, liaison and armed support roles. The medium-lift element comprises nine CH-47C Chinooks, the last three of which are cleared to carry 24,000lb externally. They also have an automatic flight-control system, greater contingency power, improved avionics and a crash-resistant fuel system. Some 64 UH-1Hs will be in service in the assault role when the most recent order for eight is

SPAIN Army

Unit	Type	Role	Base	No
Unidad I	OH-58A/Bell+AB.205/Alouette II	Util	Los Remedios	3/16/2
Unidad II	OH-58A/UH-1H	Support	Virgen del Camino	3/12
Unidad III	OH-58A/UH-1H	Support	Agoncillo	3/12
Unidad IV	OH-58A/UH-1H	Support	El Copero	3/12
Unidad V	CH-47C	Trans	Los Remedios	9

SPAIN Navy

Unit	Type	Role	Base	No	Notes
Esc 001	Bell 47/AB.47	Train	Rota	12	
Esc 003	AB.212ASW/AB.204B	ASW/assault	Rota	3/4	Replaced AB.204AS
Esc 004	Comanche/Twin Comanche	Liaison	Rota	2/2	
Esc 005	SH-3D	ASW	Rota/*Dedalo*	23	
Esc 006	Hughes 500M	ASW/SAR	Rota/destroyers	17	
Esc 007	AH-1G	Anti-ship	Rota	20	
Esc 008	AV-8A/TAV-8A Matador	FGA/train	Rota/*Dedalo*	5/1	Five more ordered
	Sea Sparrow	SAM	Frigates		
	Standard ER	SAM	Frigates		

SPAIN Air Force

Wing/Group	Unit	Type	Role	Base	No	Notes
Air Defence Command (*Mando de la Defensa Aerea*)						
Ala 11	*Esc* 111	Mirage IIIEE/DE (C-11/CE-11)	FGA/train	Manises	11/3	
Ala 11	*Esc* 112	Mirage IIIEE/DE (C-11/CE-11)	FGA/train	Manises	11/3	
Ala 12	*Esc* 121	F-4CR(S) (C-12)	Int	Torrejon	33	
Ala 12	*Esc* 122	F-4CR(S) (C-12)	Int	Torrejon		
Ala 14	*Esc* 141	Mirage F.1CE (C-14)	Int	Los Llanos	14	Replaced F-86F
Tactical Air Command (*Mando Aviacion Tactica*)						
Ala 21	*Esc* 211	SF-5A/B/SRF-5A (C-9/CE-9/CR-9)	FGA/train/recce	Moron	18	
Ala 21	*Esc* 214	HA.220 (C-10C)	Light att	Moron	23	
	Esc 221	P-3A (P-3)	ASW	La Parra	2	Being supplemented by four P-3Cs
	Esc 301	C-130H (T-10)	Trans	Zaragoza	4	
	Esc 301	KC-130H (TK-10)	Tank	Zaragoza	3	
	Esc 407	O-1E (L-12)	Liaison/obs	Tablada	7	
	Esc 407	Do2*i* (L-9)	Liaison/obs	Tablada	12	
Ala 46	*Esc* 461	C.212 (T-12B)	Trans	Las Palmas	12	Replaced C-47
Ala 46	*Esc* 462	HA.200D (C-10B)	Att/train	Las Palmas	15	
Ala 46	*Esc* 463	T-6D (E-16)	Att/train	Las Palmas	20	
Ala 46	*Esc* 464	SF-5A/B/SRF-5A (C-9/CE-9/CR-9)	FGA/train/recce	Las Palmas	9/2/8	Formerly with *Esc* 212
Air Transport Command (*Mando de la Aviacion de Transporte*)						
Ala 35	*Esc* 351	C.212 (T-12B)	Trans	Getafe	9	
Ala 35	*Esc* 352	C.212 (T-12B)	Trans	Getafe	9	
Ala 37	*Esc* 372	Caribou (T-9)	Trans	Villanubla	12	
Grupo 91	*Esc* 911	C.212AVI/C.207 Azor (T-7)	Trans/VIP	Getafe	4/8	
Grupo 91	*Esc* 912	Navajo (E-18)/Baron B55/Aztec E	Trans/VIP	Getafe	1/5/6	
	Esc 401	T-6 (E-16)		Getafe	26+	
		DC-8-52	Trans/VIP	Getafe	1	
Training Command (*Aviacion de Entremaniento*)						
	Esc 791	T-34A (E-17)/F33C (E-24)	Train	Murcia	25/29	Basic training
	Esc 792	C.212EI (TE-12)	Train	Murcia	5	Crew training
	Esc 792	AISA I.115	Train	Murcia	8	Basic training
	Esc 793	T-6D/T-6G (E-16)/HA.200A (C-10A)/	Train	Murcia	30/37	Conversion training
		HA.200D (C-10B)			62	
	Esc 731	SF-5B (CE-9)	Train	Talavera/	23	
	Esc 732	SF-5B (CE-9)	Train	La Real		
	Esc 741	C.212 (TE-12)	Train	Matacan	8	
	Esc 742	F33A (E-24)/T-42A	Train	Matacan	24/	
		Baron (E-20)/ King			18/6/2	
		Air C90/A100 (E-22/23)				
	Esc 751	AB.47G-2A/3B (Z-7/7B)/UH-1H (Z-10B)	Train	Cuatro Vientos	24/3	
	Esc 752	Hughes 269C	Train	Cuatro Vientos	19	
	Esc 721	C.212 (TE-12)	Train	Alcanterilla	2	
	Esc 801/802/803	SA-16A (AD-1)/AB.205 (Z-10)/AB.47J (Z-11)/ Do27 (L-9)	SAR/VIP	San Son Juan/ Las Palmas/ Getafe	9/ 14/3/ 2	SA-16As being replaced by three F.27M
	Esc 403	Do27 (L-9)/C.212B (TR-12A)	Survey	Cuatro Vientos	5/5	
	Esc 404	CL-215 (UD-13)	Water-bomb	Torrejon	7	

Left: This gaudy CL-215 (UD-13) water-bomber is one of seven operated by the EdA on firefighting missions in support of the civil fire brigade.

completed; some are armed with 7.62mm machine guns for army support. Other helicopters and fixed-wing types fly on liaison, communications and spotting duties. The OH-58 fleet is being expanded to 29 aircraft by the purchase of a further 18. Training is carried out initially by the EdA before completion with FAMET. The army also has a battalion of Raytheon Hawk medium-range surface-to-air missiles and a second unit with the uprated Improved Hawk.

Navy

Arma Aerea de la Armada was formed in mid-1954 with Bell 47s and received its first anti-submarine equipment in the form of S-55s the following year. Ten years later the Service took delivery of four AB.204ASs for the anti-submarine role and these were joined in 1966 by the first SH-3Ds. The AB.204s have since been replaced by AB.212ASWs and have been converted to carry out land-based support roles. A dozen additional AB.212ASWs were ordered in 1978 to supplement the three remaining aircraft from the original four. The main attack helicopter is the AH-1G HueyCobra, and the Hughes 500M also operates from ship platforms.

In 1967 the *Armada Espanola* acquired the helicopter carrier *Dedalo*, and this vessel also carries AV-8A Matadors, making the Spanish navy the third operator of the V/STOL Harrier. Up to 24 Matadors may eventually be ordered. The navy also has surface-to-air missiles in the form of Sea Sparrow and extended-range Standard.

PORTUGAL

Area: 34,000 sq. miles (88,000 sq. km.).
Population: 9·3 million.
Total armed forces: 63,500.
Estimated GNP: $16·4 billion (£8·2 billion).
Defence expenditure: $533 million (£266·5 million).

Portugal is a member of NATO but her air force (*Forca Aérea Portuguesa*) – which is the only Service to operate aircraft – contributes no squadrons to the alliance since the single squadron of maritime-patrol Neptunes was withdrawn at the end of 1977. These may be replaced by ex-USN P-3A Orions. Until the mid-1970s the air force had a major overseas role in Africa, but since the independence of Angola, Mozambique and Guinea-Bissau the Service has assumed a metropolitan role. The withdrawal from colonial operations has allowed large numbers of obsolescent aircraft to be retired, and Germany has continued her traditional support of Portugal by supplying surplus ex-*Luftwaffe* equipment. The FAP's personnel strength is some 8000.

Inventory

The air-defence organisation is based on an obsolescent early-warning and control network, the *Grupo de Deteccao, Alerta e*

Above: The FAP operates half a dozen ex-USAF T-38A Talon trainers and may receive its more warlike derivative, the F-5E. Although a member of NATO, Portugal makes little contribution.

Conduta de Intercepcao, with radar sites at Monsanto, Montejunto, Pacos de Ferreira and Sao Romao. The interceptor element is limited to a squadron of F-86 Sabres, long overdue for replacement. Recent donations by the *Luftwaffe* include a further 14 Aeritalia G.91R light strike aircraft to add to the 16 or so remaining from the 45 supplied previously. Half a

dozen G.91T trainers have also been transferred, although one crashed during delivery, and F-104Gs may follow to replace the Sabres. The FAP is reported to have requested a squadron of Northrop F-5Es to replace the G.91s, possibly with another squadron taking over from the F-86s.

The delivery of five C-130H Hercules, the first two arriving in the autumn of 1977, has allowed older types such as the C-54 and Noratlas to be retired. The Hercules have modifications including an auxiliary power unit which can be operated while airborne, a new weather radar and improved air-conditioning. They have been responsible for search-and-rescue and fishery-protection duties since the Neptunes were retired. The C-45, C-47 and F-84G have been withdrawn also, the smaller transport types being replaced by Casa C.212 Aviocars. Budgetary limitations forced the order to be reduced from 28 aircraft to 24, and the options on an

PORTUGAL

Type	Role	Base	No	Notes
F-86F	Int	Monte Real (BA-5)	20	
G.91R-4	Att	Montijo (BA-6)	18	
G.91T-3	Train	Montijo (BA-6)	6	
Puma	Mar pat	Lajes (BA-4)	4	
Puma	Assault/trans	Montijo (BA-6)	8	
Alouette III	Assault/trans	Montijo (BA-6)	24	
C-130H	Trans	Lisbon	5	Replaced DC-6 and Noratlas
C.212A1	Trans	Tancos (BA-3)	14?	Replaced C-47
		Lajes (BA-4)	6?	
C.212B2	Survey	Sintra (BA-1)	4	
F337	Coin/recce	Ota (BA-2)	22	Replaced T-6 and some Do27
F337	Coin/recce	Sao Jacinto (BA-7)	10	
T-38A	Train	Monte Real (BA-5)	6	
T-33A	Train	Monte Real (BA-5)/Sintra (BA-1)	12	
T-37C	Train	Sintra (BA-1)	15	
Chipmunk	Train	Ota (BA-2)	30	
Alouette III	Train	Tancos (BA-3)	10	
Do27A-4	Liaison	Tancos (BA-3)	12	

Note: the FAP does not use squadron numbers. Units are described by the base to which they are allocated—BA (*Base Area*) main bases and AB (*Aerodromo Base*) secondary fields

additional 12 are thought to have been dropped. Only a few of the 153 Do27s acquired in the 1960s remain operational, and the fleet of 252 T-6s has been whittled down to fewer than 20. Both types have been superseded in the liaison and counter-insurgency roles by Reims-Cessna 337s, including 16 FTB.337G Midiroles and eight equipped for photo-reconnaissance. At least 179 Alouette IIIs were supplied to the FAP, but fewer than 40 remain operational.

Training begins on de Havilland/OGMA Chipmunks, students then progressing to T-37s and T-33s before conversion to the T-38A, six of which have been supplied from USAF stocks. Transport crews are trained on Aviocars, and helicopter instruction is given in Alouette IIIs.

MALTA

Area: 94·9 sq. miles (245·8 sq. km.).
Population: 320.000.
Total armed forces: Not known.
Estimated GNP: $524 million (£262 million).
Defence expenditure: $11·03 million (£5·515 million).

Malta established a small air arm when it became independent from Britain on September 21, 1964. Its first aircraft comprised four Bell/Agusta-Bell 47G helicopters donated by West Germany in

1971 and an AB.206 presented by Libya in 1973. Fixed-wing aircraft in the shape of three surplus ex-German Do27s were expected to follow, but in the event Malta has strengthened her ties with Libya and is receiving aircraft from the Arab country. The Bell/AB.47s have been replaced by the same number of SA.319 Alouette IIIs operating from the ex-Royal Navy helicopter base at Hal Far, which has been re-established with Libyan assistance. The Alouettes are reported to be equipped for missile launching but are

thought to be employed mainly on search-and-rescue and coastal-patrol duties. Maltese crew have been trained on two Libyan Super Frelons, however, and these were expected to be transferred to the Armed Forces of Malta.

MALTA

Type	Role	Base	No	Notes
Alouette III	SAR/patrol	Hal Far	4	Ex-Libyan
Super Frelon	ASW		2	May be transferred from Libya

GREECE

Area: 51,000 sq. miles (132,000 sq. km.).
Population: 8·8 million.
Total armed forces: 190,100.
Estimated GNP: $26·3 billion (£13·15 billion).
Defence expenditure: $1·52 billion (£760 million).

The Hellenic Air Force (*Elliniki Aéroporia*) formed part of NATO's 6th Allied Tactical Air Force until Greece's withdrawal from active participation in the alliance in 1974. The rift resulted from the Turkish invasion of Cyprus and was continued when relationships between the two countries became even more strained over rights to offshore resources in the Aegean. Both Greece and Turkey have built up their forces rapidly in a mutual arms race, and it is ironic that if Greece were to rejoin NATO's military activities as a full member her contribution would be greatly strengthened as a result of this re-equipment process. Relations with the United States were also strained, but in July 1977 a new defence agreement was signed by the two countries. The US is allowed the use of some Greek facilities in exchange for $410m-worth of military aid over four years. The air force has 22,000 men, including 14,000 conscripts, plus about 20,000 reserves.

The HAF is organised into *mire* (squadrons) grouped in *pterighe* (wings) under the control of specialised commands. The army operates its own air arm, and the navy has a small helicopter force.

Inventory
The mainstay of the HAF from the mid-1960s to the mid-1970s was, as is the case in many smaller NATO air forces, a mixed fleet of Starfighters and Freedom Fighters. Original deliveries of Starfighters comprised 35 F-104Gs and four TF-104G two-seaters; this force was augmented in turn by five F-104Gs from storage in the United States, a further five F-104Gs and two TF-104Gs from the same source, and finally an additional nine single-seaters and a two-seater from Spain. One Starfighter *mira* remains operational.

The HAF received an original batch of 52 Northrop F-5As and 16 reconnaissance RF-5As, plus nine two-seat F-5Bs. A further 12 ex-Iranian Freedom Fighters have since entered service, and the type now equips three *mire* of day interceptors armed with Sidewinder air-to-air missiles in addition to the pair of built-in 20mm M39 cannon, two *mire* on reconnaissance duties, and a training unit.

An embargo on arms sales to Greece by the United States precipitated an order for 40 Dassault Mirage F.1CGs to replace

GREECE   Air Force

Command	Wing	Unit	Type	Role	Base	No	Notes
28th Tactical Air Command	110 *Pterighe*	345 *Mira*	A-7H	Att	Larisa	18	Replaced F-84F
		348 *Mira*	RF-5A	Recce	Larisa	14	
		349 *Mira*	RF-5A	Recce	Larisa	14	
	111 *Pterighe*	337 *Mira*	F-5A	Day int	Nea Ankhialos	15	
		341 *Mira*	F-5A	Day int	Nea Ankhialos	15	
		343 *Mira*	F-5A	Day int	Nea Ankhialos	15	
	114 *Pterighe*	336 *Mira*	Mirage F.1CG	Int	Tanagra	20	Replaced F-102
		342 *Mira*	Mirage F.1CG	Int	Tanagra	20	Replaced F-102
	115 *Pterighe*	338 *Mira*	A-7H	Att	Souda	20	
		340 *Mira*	A-7H	Att	Souda	20	
			F-104G/TF-104G	FGA/train	Araxos	30	
	117 *Pterighe*	339 *Mira*	F-4E	FGA	Andravidha	18	
		? *Mira*	F-4E	FGA	Andravidha	20	
		? *Mira*	RF-4E	Recce		24	
		363 *Mira*	HU-16B	SAR/ASW	Elefsis	8	Ex-RNoAF
Air Material Command		355 *Mira*+356 *Mira*	Noratlas/C-130H/C-47/Gulfstream I	Trans/VIP	Elefsis	40+/12/30/1	
		357 *Mira*	Bell 47G	Util	Elefsis	10	
		359 *Mira*	Sikorsky H-19D	Trans	Elefsis	12	
		362 *Mira*	AB.205A	Trans	Elefsis	14	Some detached
Training Command		360 *Mira*	T-2E	Train	Elefsis	40	Replaced T-33A
		361 *Mira*	T-37B	Train	Elefsis	18	
		National Air Academy	T-41D	Train	Dhekelia	20	
			Nike Hercules	SAM	various	1 battalion	

GREECE   Army

Type	Role	Base	No
L-21	Util/liaison	Megara	15
Cessna U-17A	Util/liaison	Megara	25
Commander 680FL	Util/liaison	Megara	2
AB.204/205	Trans	Megara	47
UH-1D	Trans	Megara	10
Bell 47G	Util	Megara	5
Hawk	SAM	various	12 launchers

GREECE   Navy

Type	Role	Base	No
Alouette III	Anti-ship/util		4
Albatros/Aspide	SAM	4 destroyers	4 launchers

the 20 Convair F-102As (plus five or six two-seat TF-102As) in the interception role. The Mirages are armed with Matra R.550 Magic close-range air-to-air missiles. A growing force of McDonnell Douglas F-4 Phantoms operates in the fighter/ground-attack role, following delivery of an initial batch of 38 F-4Es. An additional two aircraft were supplied soon afterwards to replace attrition losses, and these were in turn succeeded by eight reconnaissance RF-4Es, costing $91m with spares and support. President Carter approved the sale of 18 more F-4Es to Greece in early 1977, bringing the total to 64 (plus the two replacements).

Above: The HAF retains only one operational Starfighter squadron, but – as with so many NATO air forces – the type bridged the gap from early jet equipment to modern combat aircraft.
Below: The HAF was the first export customer for the Vought A-7 attack aircraft and has since placed a follow-on order for two-seat trainers.

The HAF was the first export customer for the Vought A-7 attack aircraft, the last of 60 A-7Hs being delivered in the summer of 1977. Greece is interested in fitting these aircraft, as well as the Phantoms and Starfighters, with US electronic countermeasures. The arms deal involving the Corsairs also included 40 Rockwell T-2 Buckeye jet trainers; the HAF's T-2E variant doubles as a light strike aircraft, with armour-plated fuel tanks to protect against small-arms fire and a total weapon load of nearly 1600kg on six hardpoints. The third element of the deal comprised an extra eight Lockheed C-130 Hercules

Below: The HAF flies six Canadair CL-215 water-bombers on behalf of the civil authorities. The aircraft can alight on the sea or a lake and scoop up a load of water in a few seconds.

worth more than $71m with spares and support. Some of these are planned to be fitted with ELINT (electronic intelligence) equipment. In January 1978 the HAF ordered six two-seat A-7s, two each for 338, 340 and 345 *Mire*. Five are new-production TA-7Hs and the other is a converted Greek single-seater. Deliveries begin in April 1980.

The HAF is responsible for a single battalion of Nike Hercules long-range surface-to-air missiles. The transport and training fleets have been strengthened by aircraft received from NATO members and more recently from Israel.

Basic flying training is carried out at the National Air Academy (*Ethniki Aéroporiki Acadymia*) at Dhekelia. Students then progress to 361 *Mira* at Elefsis for 125hr on T-37Bs before moving to 360 *Mira* for some 75hr of advanced tuition on T-33s

and T-2s. Operational conversion is then carried out on two-seat variants of the F-104 and F-5.

A maintenance and limited manufacturing facility has been established by Lockheed at Tanagra, some 50 miles north of Athens. Some 3000 personnel were to be employed at the facility, which will undertake depot-level maintenance for the HAF.

The HAF was involved in sporadic combat with the Turkish Air Force during and after the 1974 invasion of Cyprus.

Army

The army has a completely independent air arm, Aéroporia Stratou, with headquarters at Megara outside Athens. The force was substantially strengthened in the mid-1970s, receiving an additional 40 AB.204/205s and ten UH-1Ds during 1975 with later deliveries to include 35 utility helicopters – probably further UH-1s – from the United States. The fixed-wing force is being reduced as the helicopter fleet builds up.

The Greek Army also operates one battalion – 12 launchers – of Raytheon Hawk medium-range surface-to-air missiles.

The HAF operates the eight HU-18 Albatross maritime-patrol and search-and-rescue aircraft on behalf of the navy. In 1976 the navy took delivery of its own patrol force in the form of four Alouette III helicopters armed with AS.12 missiles.

The Greek Navy has ordered Selenia Albatros point-defence missile systems, using the Aspide weapon, to arm its four *Themistokles*-class destroyers.

TURKEY

Area: 301,000 sq. miles (780,500 sq. km.).
Population: 42·1 million.
Total armed forces: 485,000.
Estimated GNP: $46·6 billion (£23·3 billion).
Defence expenditure: $1·7 billion (£850 million).

Turkey's air force (*Turk Hava Kuvvetleri*) is almost totally committed to the 6th Allied Tactical Air Force of NATO Allied Air Forces Southern Europe. Turkey is also a member of the Central Treaty Organisation (CENTO). The country shares a common border with the Soviet Union and has long been a major recipient of modern defence equipment from the United States and, to a lesser extent, from other NATO members. Relationships with the other partner in 6th ATAF, Greece, were however soured by Turkey's invasion of Cyprus in the summer of 1974. This led to Greece's withdrawal from NATO's military operations and precipitated an embargo on US arms sales to Turkey. The latter, country retaliated by evicting US personnel from bases on its soil, and ordered new equipment in the form of F-104S Starfighters from Italy. The United States lifted its arms embargo in the autumn of 1978 and the THK – which had been reduced to about half its normal strength – was expected to be operating normally again by March 1979. A dozen

ex-US Air National Guard F-100Fs are thought to have been supplied to make up for attrition losses, and new contracts for both Sparrow and Sidewinder air-to-air missiles have been signed.

All three armed services operate their own aircraft. The air force has some 45,000 personnel, including 25,000 conscripts, and is run by Air Forces Central Command which is responsible to the Turkish General Staff. All combat aircraft are grouped into the 1st Tactical Air Force, with headquarters at Eskisehir, and the 2nd Tactical Air Force centred on Diyarbakir. Other commands are responsible for air transport and training. Aircraft are organised into *filos* (squadrons) with an average strength of 15–20.

The ten-year REMO defence re-organisation and modernisation programme, launched in 1973, was contracted to six years after the Cyprus invasion to allow re-equipment to proceed more quickly.

Inventory

In recent years only two squadrons have been allocated solely to air defence, although this force was increased to three units in 1975 with the formation of an F-104S squadron in the interception role. The THK received 36 ex-USAF F-102As and three two-seat TF-102A trainers in 1969 to equip two interceptor squadrons; these have now been replaced by Starfighters. Air defence is also shared by two

battalions of Nike Hercules surface-to-air missiles, organised into six squadrons of 12 launchers each.

The mainstay of the attack element from the late 1950s was the North American F-100 Super Sabre, deliveries of an initial batch of 260 ex-USAF aircraft beginning in 1958. Additional Super Sabres were supplied in the early 1970s and the type saw extensive action during the 1974 invasion of Cyprus. Five F-100 squadrons remain operational and a large number of aircraft are in storage. The THK's strike force comprises two squadrons of F-104Gs and F-104Ss. Italy supplied 40 Aeritalia/Lockheed F-104Ss in two batches of 18 aircraft each and one of four; a further 40 were placed on option. Of those covered by the original contracts, half are allocated to interception and the remainder primarily to ground attack. All can be armed with the Sparrow air-to-air missile built under licence by Selenia, however, for counter-air operations.

The THK received 95 Northrop F-5A fighter-bombers and RF-5A reconnaissance aircraft, together with 13 two-seat F-5B trainers, from 1965 and has since received additional aircraft from Iran and Libya. Although primarily employed in the ground-attack role, the Freedom Fighters can carry Sidewinder air-to-air missiles for self-defence and for fighter missions. A further dual-role type is the McDonnell Douglas F-4 Phantom, which

entered service in 1974. Initial deliveries comprised 40 F-4Es in two batches of 20 each, these being followed from 1977 by an additional 32 F-4Es and eight reconnaissance RF-4Es. The Phantoms are likely to be armed with the Hughes Maverick air-to-surface missile; the earlier Bullpup arms the F-100 and F-104.

The THK has a requirement for up to 200 General Dynamics F-16s to replace the F-104Gs and F-102s from 1980–81, and negotiations for some 56 Dassault/Dornier Alpha Jet light strike/trainers have been held. Plans to establish an aerospace manufacturing industry known as TUSAS have slipped, but the Macchi MB.339 has been selected for eventual licence-production. The deal would cover 60 aircraft over a 64-month programme, with Turkish involvement building up to 90%.

The THK has comparatively large transport and training fleets with a wide variety of types. All C-54s have now been withdrawn but many aircraft of similar vintage soldier on. Basic flying training is carried out on the T-34 and T-41, with some T-6s remaining in the inventory. Jet conversion takes place on the T-37, followed by

advanced instruction on the T-33 and operational conversion on TF-104s and F-5Bs, together with two-seat variants of the older combat types. Twin-engined training is the responsibility of the remaining C-45s and AT-11s, together with some C-47s.

The THK has 12 major bases, plus a further 17 or so which can be used for dispersal. Most main bases have their own station flights and rescue units operating T-33s, C-45s, UH-1D/Hs and other miscellaneous types. Dispersal fields also frequently have full jet facilities and may be allocated their own T-33/C-47 station flights.

Army

Turkish army aviation (*Kara Ordusu Havaciligi*) originated with the supply of 200 Piper L-18B/C Super Cubs for observation and liaison; many of these remain in service, and a substantial rotary-wing force has since been built up; recent orders include 56 AB.205s in two batches of 28 for commando units. KOH is responsible to Turkish Ground Forces Command and is controlled by the Central Army Aviation Establishment at Güvercinlik, near Ankara. Field armies, corps and divisions all maintain flying units at their own airfields, of which two of the largest are Sinop and Konya. Some of the AB.205s are armed, and the helicopter fleet was employed extensively in the 1974 invasion of Cyprus. Major maintenance and repair facilities are located at Eskisehir and Erkilet/Kayseri.

Training begins on the Citabria 150S, followed by the Do28 and T-42 for fixed-wing pilots and the AB.47 for helicopter specialists.

Turkey has conducted negotiations with Germany for the supply of Euromissile Roland II mobile surface-to-air missile systems, and the Army wants at least 100 helicopters such as the Hughes 500M-D Defender, armed with the TOW anti-tank missile, to be built in Turkey.

Navy

Naval aviation (*Donanma Havaciligi*) operates one squadron each of fixed-wing maritime-patrol aircraft and anti-submarine helicopters.

TURKEY

Command	Unit	Type	Role	Base	No
Air Force					
1st Tactical Air Force (*Birinci Taktik Hava Kuvveti*)	111 Filo	F-100D/F	FGA	Eskisehir (1st Jet Air Base)	16
	112 Filo	RF-5A/F-5B	Recce		18
	113 Filo	F-4E	FGA		18
	131 Filo	F-100D/C/F	FGA	Konya (3rd Jet Air Base)	16
	132 Filo	F-100D/C/F	FGA		16
	141 Filo	F-104G/TF-104G	Strike	Mürted (4th Jet Air Base)	16/2
	142 Filo	F-104S/TF-104G	Int		18
	161 Filo	F-5A/B	FGA	Bandirma (6th Jet Air Base)	
	162 Filo	F-4E	FGA		18
	191 Filo	F-104G/TF-104G	Strike	Balikesir (9th Jet Air Base)	16/2
	192 Filo	F-5A/B	FGA		
2nd Tactical Air Force	151 Filo	F-5A/B	FGA	Merzifon (5th Jet Air Base)	
	152 Filo	F-5A/B	FGA		
	171 Filo	F-100C/D/F	FGA	Erhac-Malatya (7th Jet Air Base)	
	172 Filo	F-100C/D/F	FGA		
	181 Filo	F-5A/B	FGA	Diyarbakir (8th Jet Air Base)	
	183 Filo	RF-5A/F-5B	Recce		18
	182 Filo	F-104S/TF-10AG	FGA?	Mürted (4th Jet Air Base)	18
Air Transport Command (*Hava Nakil Kuvveti*)	222 Filo	C-130E	Trans	Erkilet/Kayseri (12th Air Transport Base)	7
	221 Filo	Transall	Trans		20
	VIP Flight	Cessna 421C	VIP	Etimesgut/Ankara	3
		UH-1H	VIP		
		C-47	VIP		1 or 2
	VIP Transport Sqn	Viscount 794	VIP	Yesilkoy/Istanbul	3
		UH-19	VIP	Yesilkoy/Istanbul	5+
		UH-1H	VIP	Yesilkoy/Istanbul	6
Air Training Command (*Hava Egitim Komütanligi*)	123 Filo	T-41D	Train	Gaziemir	19
		T-34A	Train	Gaziemir	24
		T-6	Train	Gaziemir	30–40
	122 Filo	T-33A	Train	Cigli/Izmir	30
	121 Filo	T-37B/C	Train	Konya	23
		Islander	Survey		3
		Nike Hercules	SAM	various	6 sqns
Army		Cessna 421B	Trans		3
		UH-1B	Trans		20
		AB.205	Trans		100
		AB.206	Obs/liaison		15
		U-17	Obs/liaison	Various, including Güvercinlik/ Ankara (HQ), Sinop and Konya	Up to 100
		Beaver	Obs/liaison		8
		O-1E Bird Dog	Obs/liaison		65
		L-18 Super Cub	Obs/liaison		76
		Do27	Obs/liaison		15
		Cessna 206	Obs/liaison		8
		Citabria 150S	Train/util		40
		AB.47G-3B	Train/util		50
		TH-BT	Train/util		10
		Do28B-1	Train/trans	Various, mainly Güverncinlik/ Ankara	5
		Do28D-1	Train/trans		9
		T-42A Baron	Train/trans		5
Navy		S-2A Tracker	Mar recce	Karamürsel (HQ), detachments at Sinop, Izmir, Antalya	8
		S-2E Tracker	Mar recce		12
		TS-2A	Train	Karamürsel	2
		AB.205AS	ASW	Karamürsel	3
		AB.212ASW	ASW	Karamürsel	6

Eastern Europe

Since the end of World War II, Eastern Europe has been under the dominance of the Soviet Union to a greater or lesser extent. East Germany, Poland, Hungary, Czechoslovakia, Romania and Bulgaria are members, along with the USSR, of the Warsaw Pact, set up in 1955. Albania was an original signatory but left in 1968 and has allied itself with China, while Jugoslavia has never been a member and treads a wary path between the Soviet Bloc and the West.

Members of the Warsaw Pact are very much under the influence of the Soviet Union as far as their military activities are concerned, and all except Romania have Russian forces based within their borders. Any attempt to deviate from the prescribed line is punished immediately, as happened in Hungary in 1956, Czechoslovakia in 1968 and to a lesser extent on a number of occasions in Poland.

The Soviet Union has its own tactical aircraft based in East Germany, Poland, Hungary and Czechoslovakia to augment those flown by the national air forces, and in time of war both these elements would come under the overall control of Soviet commanders. Warsaw Pact countries have a distinct advantage over NATO in that they operate much the same types of equipment, most of which is developed and produced in the Soviet Union. Other members of the Warsaw Pact do contribute a limited number of designs, however, such as jet trainers from Czechoslovakia and Poland.

Romania and Jugoslavia have collaborated on development of the Orao fighter-bomber, but this is unusual. Jugoslavia does buy weapons from the West as well as the East, however, and it has its own arms industry. Any instability following the death or retirement of President Tito could well result in significant changes in Jugoslavia's military stance, though, despite the country's ties with the West.

The degree of support which Warsaw Pact forces would give the Russians in any full-scale war is an interesting subject for debate, bearing in mind that little love is lost between several members of the alliance and their Soviet masters. NATO would be unwise to bank on any such dissension reducing the weight of an assault, however, since some of its own members appear less than totally enthusiastic about their commitment to resist such an onslaught.

EAST GERMANY

Area: 41,800 sq. miles (108,200 sq. km.).
Population: 16·8 million.
Total armed forces: 157,000.
Estimated GNP: $54·6 billion (£27·3 billion).
Defence expenditure: $3·15 billion (£1·575 billion).

East Germany's air force, the *Luftstreikraefte* (LSK), forms – together with the Soviet Union's 16th Air Army – the first line of the Warsaw Pact's European air strength. Apart from a small force of East German Navy helicopters responsible for search and rescue in the Baltic Sea, it has a monopoly of East German air power. Its commander-in-chief is usually a first deputy in the Ministry of National Defence.

The Soviet Union started to build up the nucleus of the LSK before either of the Germanies was permitted to have an air force under international treaty. Initially known as the VP-L (*Volkspolizei-Luft*, or Air Police) the force was formed in the early 1950s and in late 1956 became the LSK. Single-seat and two-seat MiG-15s were delivered in the following year, and the LSK has now built up to become one of the largest forces in the Warsaw Pact.

Uniquely among Warsaw Pact nations, East Germany has placed its armed forces under the peacetime command of the Soviet Union. (Other non-Soviet Pact forces only come under Soviet command in case of a collective threat to Pact security, and revert to national control when the threat recedes.) Powerful as the LSK is, it is only half the size of the 16th Air Army, which controls it. The LSK is in fact virtually an extension of the 16th Air Army, using different bases (only Parchim seems to be shared) and German personnel.

The Mikoyan MiG-21 Fishbed in its various forms is the most numerous type in the LSK inventory. About 250 are in service in the fighter-bomber and reconnaissance roles, and the remaining MiG-21F clear-weather fighters are likely to be replaced by newer types such as the MiG-21SMT Fishbed K used by the 16th Air Army. Despite reports that about 90 Sukhoi Su-7 Fitter As equip two regiments

in the close-support role, none have been supplied to East Germany by the Soviet Union. The proportion of obsolescent types in the LSK is the lowest in the Pact outside the Soviet Union; some 40 MiG-17 Frescos remained in service at the end of 1978, but the force was reported to be dwindling.

The LSK transport force operates mainly in support of the fighter and strike units, as well as running a few aircraft in the service of the government. The most numerous type is the Ilyushin Il-14; about 20 are in service and, judging from the pattern in other forces, they are likely to be replaced by Antonov An-26s or the new An-32. A few An-2 and An-14 light transports operate in the communications role. In early 1977 it was reported that some of the Il-14s were in service as electronic intelligence (Elint) platforms.

The East German Government fleet – now diminished by the sale of an Antonov An-24 to Vietnam – may be an extension of the national airline Interflug rather than

GERMANY (EAST)

Division	Regiment	Type	Base	Number
3rd Air Defence (Neubrandenburg)	2nd Fighter	MiG-21	Neubrandenburg	36
	9th Fighter		Peenemunde	
	13th SAM	SA-2	Parchim	12 batteries
	17th SAM	SA-3	Uhlenkrug	12 batteries
	18th SAM		Sanitz	12 batteries
1st Air Defence (Cottbus)	1st Fighter	MiG-21	Cottbus	36
	3rd Fighter		Preschen	
	7th Fighter	MiG-21	Drewitz	36
	8th Fighter	MiG-21	Marxwalde	36
	14th SAM	SA-2	Steuergraebchen	12 batteries
	16th SAM	SA-3	Ladeburg	12 batteries
	31st Helicopter		Brandenburg-Briesen	
	27th Transport		Dresden-Klotzsche	

an LSK operation. Two Il-18s, three Tupolev Tu-124s and a pair of Tu-134s are operated, but the Tu-124s are probably due for retirement.

The LSK has about 80 transport and training helicopters, with WSK Mi-2s and Mil Mi-4s now being replaced by the Mi-8. The East German Navy flies about eight

Below: A MiG-21 FL taxis past a runway light during night-flying practice.

Above: The LSK is one of several Warsaw Pact air forces to operate the Czech-built Aero L-29 Delfin trainer. The aircraft is carrying external fuel tanks in place of weapons.

Mi-4s from Baltic bases for search and rescue duties.

A variety of training aircraft are operated. Aero L-39 combat trainers are being delivered, and Aero L-29 Delfins and MiG-15UTIs are operated, together with MiG-21Us attached to operational units for conversion training. The LSK has 20-plus MiG-21Us.

In structure and deployment the LSK follows the pattern of a Soviet Air Army, with combat units grouped in divisions and separate support regiments; fighter units, for instance, are grouped in divisions, as are the SAM units which are integrated with the Soviet PVO defence network. LSK headquarters is at Strausberg-Eggersdorf. The SAM units operate some 60 batteries of SA-2 Guideline and SA-3 Goa missiles.

The training structure of the LSK appears to be mostly independent of the Soviet Air Forces now, although Frontal Aviation still exerts a greater influence over LSK training than it does over that of the other Warsaw Pact air forces. Like its neighbours, East Germany appears to be moving from a three-tier training system to a two-level structure. The present three training aircraft – Yak-11, L-29 and MiG-15UTI – seem likely to be replaced by a more modern basic trainer and the Aero L-39 Albatros; no direct advanced-trainer replacement for the MiG-15UTI seems to be emerging.

119

POLAND

Area: 121,000 sq. miles (313,000 sq. km.).
Population: 35 million.
Total armed forces: 306,500.
Estimated GNP: $86·1 billion (£43 billion).
Defence expenditure: $2·55 billion (£1·275 billion).

Above: Poland's aerospace industry provides its own air force and those of other countries with a variety of aircraft, including the PZL Wilgas which are operated by the PWL in the utility role.

Poland's is the largest air force in the Warsaw Pact outside the Soviet Union, forming the second line of defence in Europe together with the Soviet 37th Air Army. Interceptor, SAM and fighter-bomber divisions are responsible to the Command of Aviation Forces, which in turn is subordinate to the Ministry of Defence in peacetime. Poland is also the only Warsaw Pact member outside the Soviet Union to have a fixed-wing naval air force. The Polish Air Force (Polish initials PWL) is the only Warsaw Pact air force to be larger than the Soviet Air Army stationed in its home country, and in recent years has emerged as the most technically advanced of the non-Soviet forces.

The PWL originated in the Soviet 6th Air Army, formed with Polish and Polish-Russian personnel in 1944 to take part in the offensive into Germany. The 6th Air Army was completely modelled on the Soviet pattern in concept, organisation and structure; following the March 1946 Russo-Polish Air Agreement it started to expand, but under very close Soviet control. Mikoyan MiG-15s were delivered in the early 1950s and Ilyushin Il-28 light bombers followed in 1955. The MiG-15 and MiG-17 were placed in production in Poland under the designations LiM-2 and LiM-5. After the events in Hungary of 1956 the Polish Air Force was granted rather more independence from deeply resented Soviet control, but at the same time its expansion was slowed down and later

aircraft in the MiG series were not licence-built. During the early 1960s the Soviet Union attempted to discourage the manufacture of military aircraft in Poland and gave priority to the Czech air force and industry. The emphasis has reversed since the Czech attempt at independence and the invasion of 1968.

Poland is a signatory of the Warsaw Pact, and in time of threat to the Pact its forces would fall under the supreme command of the Soviet Union. In that event the PWL's fighter-bomber divisions would be assigned to the 37th Air Army of Soviet Frontal Aviation, with its headquarters at Legnica. The interceptor divisions are in any case integrated into the Soviet PVO network of air-defence zones. The 37th Air Army is part of the Northern Group of Soviet Forces.

Poland was the first non-Soviet Warsaw

Pact force to take delivery of aircraft more modern than the MiG-21/Su-7 generation. (The Soviet Union has been surprisingly niggardly about supplying new-type aircraft to the Warsaw Pact, particularly in comparison with the massive deliveries of the MiG-23/27 family to the Middle East.) At least one regiment of 35 swing-wing Su-20 Fitter Cs was delivered in 1974 to replace LiM-5 fighter-bombers, and deliveries of this quasi-new-generation variant of the standard Su-7 are continuing, probably to replace the second LiM-5 regiment. Ilyushin Il-28 light bombers still equipped one regiment in 1976.

The MiG-21 Fishbed and the standard fixed-wing Su-7 Fitter A still form the greatest part of the PWL front-line strength. Upwards of 300 MiG-21s of all marks equip nine fighter regiments forming three divisions; these include some 30

Below: The PWL was the first Warsaw Pact air force outside the Soviet Union itself to receive variable-geometry Su-20 Fitter Cs, which have replaced LiM-5s (licence-built MiG-17s).

MiG-21U Mongol conversion trainers, and in addition the PWL has a few MiG-21RF Fishbed H reconnaissance fighters. All MiG-21s carry the AA-2 Atoll and AA-2-2 Advanced Atoll AAMs. About 160 Sukhoi Su-7 Fitter As equip five fighter-bomber regiments, with a proportion (15–20 aircraft) of Su-7U Moujik trainers. At the time of writing one regiment still operated 35–40 LiM-5s (Polish-built MiG-17s), but these have been replaced by about 40 Su-20s.

Other combat aircraft equip Poland's Naval Air Division, which is still responsible to the Command of Aviation Forces and is thus part of the PWL. The LiM-5 Fresco remains in service at regiment strength, but there is just one squadron of 12 MiG-21F Fishbed C day fighters. A few Ilyushin Il-28 Beagles remain in service for coastal strike.

The training fleet is largely equipped with Polish-designed aircraft. These include the WSK-Mielec TS-11 Iskra jet trainer, developed in the early 1960s in competition with the Czech Aero L-29, and the WSK TS-8 Bies primary trainer.

Above: Although bearing the designation of the Mil design bureau, the Mi-2 was built in Poland by WSK-Swidnik and not in the USSR. The helicopter equips the PWL and several other air forces.

Some Yak-18s are operated and the ubiquitous MiG-15UTI is used for advanced training. WSK has not developed an aircraft in the class of the Aero L-39 Albatros and the new Czech trainer will probably be acquired to replace the TS-11 and MiG-15UTI.

Tactical transport and logistic support of combat units is provided by a mixed transport regiment of about 45 aircraft, including Antonov An-12s and An-26s. A few An-24s and An-2 biplanes are also operated, as are 15 Ilyushin Il-14s. A single Tu-134 is operated by state airline LOT on behalf of the Polish Government, and a military VIP element operates an Il-18, numerous Yak-40s and four Mil Mi-8s. Some Polish PZL Wilgas serve in the utility/liaison role.

Helicopter forces include a substantial number of WSK Mi-2s and improved Mi-

2Ms; the Mil-designed Mi-2 was never placed in production in the Soviet Union and WSK has delivered aircraft to the USSR as well as other nations. Mi-4s and Mi-8s, the latter armed with unguided rockets for fire suppression, form the transport force. The Naval Air Division has a similar but smaller helicopter transport force, with some Mi-2s and Mi-4s serving in the air-sea rescue role.

PWL missile forces, integrated into the Soviet PVO network, comprise 30 SA-2 Guideline sites protecting strategic targets against medium-altitude intruders. The Polish Navy is the only non-Soviet Warsaw Pact navy to possess SAMs; a single SAM Kotlin destroyer is fitted with the SA-N-1 missile system, using the Goa missile.

Details of PWL deployment are not available, but it is believed that there are some 35–40 military airfields active in Poland. Some of these are used by Soviet Frontal Aviation units. In addition to permanent bases there is a considerable number of dispersal strips in the forward area, often using hardened stretches of road and equipped to receive control and support facilities.

Training was the responsibility of the Aviation Inspector's Office until 1967, when the AIO was merged into the Command of Aviation Forces. Primary training is carried out on piston-engined aircraft, with basic training on the TS-11. Pilots proceed to the squadrons after advanced training on the MiG-15UTI. Normally, one squadron in each regiment is assigned the training role, four of its complement of 12 aircraft being two-seaters.

Below: Yet another Polish-built type which has been deployed by the PWL and other air forces is the WSK-Mielec TS-11 Iskra (spark) trainer. It is expected to be replaced by the Czech L-39 Albatros.

CZECHOSLOVAKIA

Area: 49,400 sq. miles (128,000 sq. km.).
Population: 15 million.
Total armed forces: 186,000.
Estimated GNP: $49·6 billion (£24·8 billion).
Defence expenditure: $1·82 billion (£910 million).

The Czechoslovak air arm (*Ceskoslovenske Letectvo*) is responsible for air defence and the support of Czechoslovak and allied ground forces. It is believed to be divided into two Air Armies, one of which is responsible for air defence and the other for support of ground forces. The air-defence force, the 7th Air Army, is independent of Army command and responsible directly to the Ministry of Defence, while the 10th Air Army carries out tactical support under the direction of the local front commander.

Unlike other Warsaw Pact air forces the Czech air arm is not entirely a Soviet creation. The air force revived immediately after the 1939–45 war, when the country was attempting to maintain its independence from the Soviet Union. Before the coup d'etat of 1948 aligned Czechoslovakia closely with the Soviet Union, the Czech aircraft industry had equipped its air force with Avia S-199 fighters derived from the Messerschmitt Bf 109 and had exported some of those aircraft to Israel; a programme was under way to put the Messerschmitt Me 262 jet fighter into production under the designation S-92 and new jet fighters were under development. Following the 1948 coup,

however, the air force expanded under Soviet control to become the strongest of the Warsaw Pact air arms by the second half of the 1950s. Although its equipment was Soviet-designed, many aircraft were Czech-built and some Czech-produced aircraft were exported. Licence-production covered the Mikoyan MiG-15, the MiG-17, the MiG-19 and the Ilyushin Il-28 in the 1950s, and in the early 1960s the Czech industry built MiG-21F-13s for the air force.

Czechoslovakia's position in the Warsaw Pact is somewhat ambiguous following the events of 1968, when an increasing trend towards 'liberal Socialism' was abruptly stopped by Soviet occupation, with the assistance of other Warsaw Pact forces. The Czech air arm's pre-eminence among the Warsaw Pact forces has diminished, and it is Poland rather than Czechoslovakia which is at the forefront of modernisation. Under the Warsaw Pact the Czech forces are taken under Soviet command if a collective threat faces the members of the Pact; in the case of the Czech air arm this would mean that the 10th Air Army would be attached to the Soviet Air Army based in Czechoslovakia (headquartered at Milovice, north-east of Prague) under the command of the Central Group of Soviet Forces. The 7th Air Army would fall under Soviet PVO control, and is almost certainly integrated into the PVO structure in any case.

Czechoslovakia has now become the first non-Soviet Warsaw Pact signatory to take delivery of the Mikoyan MiG-23 Flogger. One regiment of the type was

reported to be operational at Pardubice in mid-1978, and further re-equipment of the CL with this type is to be expected. Backbone of the air force is the fleet of 340 Mikoyan MiG-21s, although early Czech-built clear-weather MiG-21Fs must still form a large proportion of the 220 MiG-21s assigned to the intercept role. Eighty MiG-21s operate in the ground attack role and 40 MiG-21RF Fishbed Hs operate in the fighter-reconnaissance role with cameras and electronic countermeasures (ECM) equipment. Some 40 Sukhoi Su-7 Fitter As provide close-support strength and are being supplemented with swing-wing MiG-23s. A single MiG-15bis squadron remains tasked with patrolling the Czech-West German border, each aircraft carrying a pilot and an observer. Aero L-39s will replace these aircraft when demand for export of the type permits. The specialised L-39D strike version of the trainer is armed with a GSh-23 twin cannon of 23mm calibre and can be armed with AA-2 Atoll AAMs.

Aero L-39s are beginning to replace L-29 basic and MiG-15UTI advanced trainers, and the Zlin Z-226 – an indigenous product like the L-39 – has supplanted the Yak-11 primary trainer. About ten Sukhoi Su-7U Moujik and 30-plus Mikoyan MiG-21U Mongol conversion trainers are included in the total operational fleet.

About 40 transport aicraft include a

Below: The Aero L-39 Albatros is replacing CL L-29 Delfins and MiG-15UTIs. The light strike trainer is additionally being supplied to other Warsaw Pact air forces.

Above: The S-104 (Czech-built MiG-17F) continues to serve the CL in the attack and ground-support roles.

Right: The MiG-21U two-seat trainer was not built under licence in Czechoslovakia, having been supplied to the CL from Russian production lines.

single VIP Tupolev Tu-134A, Antonov An-24s and older Ilyushin Il-14s. Mil Mi-4 and Mi-8 transport helicopters are operated, with a total force of about 100. Ilyushin Il-76s are reported to have been requested.

It is likely that more modern aircraft will be supplied to both Poland and Czechoslovakia now that the 'front-line' units in East Germany have been largely re-equipped. The only air-launched missile in Czech serice remains the AA-2 Atoll AAM which arms the MiG-21. The 7th Air Army is believed to command about 60 SAM batteries, probably equipped with SA-2 Guidelines and SA-3 Goas.

The 7th Air Army comprises two regional divisions. Each of these includes three interceptor regiments with about 36 MiG-21s each, together with support aircraft, and five SAM regiments, normally of 12 batteries. One of the divisions is presumably centred on Prague. The 10th Air Army has four fighter-bomber regiments and the MiG-21RFs equip two reconnaissance regiments, but the detailed allocation of bases is not known.

The Czech forces conduct their own training with the L-29, L-39 and primary trainers, and conversion trainers attached to operational units. Pardubice, Pilsen and Bratislava are among training bases.

HUNGARY

Area: 36,000 sq. miles (93,200 sq. km.).
Population: 10·6 million.
Total armed forces: 114,000.
Estimated GNP: $25·2 billion (£12·6 billion).
Defence expenditure: $658 million (£329 million).

The smallest of the Warsaw Pact air arms, the Hungarian Air Force is subordinate to the command of the Hungarian Army. Its main role is tactical support of the army, but its interceptor and SAM squadrons form part of the Warsaw Pact air-defence system and defend major targets in Hungary itself. Apart from the air force itself, the only paramilitary air activity consists of a few helicopters attached to the Danube river patrol force.

Like most of the Warsaw Pact forces, the Hungarian air arm is a post-war Soviet creation, the wartime force having been totally defeated and dismantled by the Soviet Union after the 1939–45 war. The final 1947 peace treaty between Hungary and the Soviet Union set a technical limit to the strength of the air force, but this limit has been disregarded since the air force expanded under Soviet control. By

Below: A crew member leaps from one of the Hungarian Air Force's small number of Mi-8 assault helicopters. This example also carries weapon attachments on each side of the fuselage.

the mid-1950s Hungary had the strongest air force in the Balkans.

Some Hungarian air force units took part in the 1956 rebellion against Soviet occupation forces, and following the suppression of the revolt the air force was demobilised. Its restoration started in 1957, but since that time the Soviet Union has maintained a close control over the Hungarian armed forces.

Under the Warsaw Pact the Hungarian Government retains nominal control over its armed forces in peacetime. However, control would revert to the Soviet Union

Above: The L-29 Delfin, designed and built in Czechoslovakia, is widely used by Warsaw Pact air forces, including that of Hungary. The more modern L-39 Albatros is also in Hungarian service.

in the event of a collective threat to the Pact. The Hungarian air force is probably outnumbered by the Soviet Air Army based in Hungary, which is believed to have some 200 combat types in addition to support aircraft, and which is taking delivery of MiG-27s. The Soviet Air Army,

with its headquarters at Tokol south of Budapest, would take over control of the Hungarian Air Force in wartime and both forces would operate under a Soviet front commander. As in Czechoslovakia the Soviet forces have the dual role of allies and occupying forces.

The Mikoyan MiG-21 Fishbed is the most important type in the inventory, with a total force of about 80 aircraft including perhaps eight MiG-21U Mongol two-seaters. A single regiment operates 35 Sukhoi Su-7s in the close-support role.

Antonov An-26 Curl freighters have recently been delivered to replace some of the older aircraft – Lisunov Li-2s and Ilyushin Il-14s – in service with the single transport regiment. A total of 25–30 transports include An-2s and An-24s as well as the types previously mentioned. State airline Malev operates VIP TU-134s for the Government. A small number of Mil Mi-8s have been delivered to supplement the older Mi-4s.

Training types include Yakovlev Yak-11s and Yak-18s, as well as the standard Aero L-29 Delfin and L-39 Albatros. The MiG-21Us are presumably attached to operational units. Some MiG-17UTIs may be retained as intermediate trainers.

Hungary's military aircraft constitute a single Air Division, with two fighter regiments, the Su-7s in a fighter-bomber regiment, and a single transport regiment. The other component of the Hungarian air forces is an Air Defence Division, which commands two battalions of SA-2 Guideline missiles.

Above: The Hungarian Air Force's single transport regiment has been supplied with An-26s, of which three are seen here, to replace some of the long-serving Li-2s and Il-14s. The older types remain in limited service, however.

JUGOSLAVIA

Area: 98,700 sq. miles (255,800 sq. km.).
Population: 21·3 million.
Total armed forces: 267,000.
Estimated GNP: $37·8 billion (£18·9 billion).
Defence expenditure: $2·33 billion (£1·165 billion).

Jugoslavia's air force is highly important to the country's policy of independence from both East and West and typifies this policy with its mixture of Western, Soviet and domestically developed and manufactured aircraft. The Jugoslav Air Force is independent of the army and is devoted more or less equally to the two roles of home air defence and tactical support of the army, with a current move-ment of emphasis in favour of the latter role. In addition, the Jugoslav Navy operates a fleet of ASW and communications helicopters.

The Jugoslav Air Force has been built up since the end of the 1939–45 war, initially with Western aid. Two hundred Republic F-84G Thunderjets were acquired under MAP in the early 1950s and Jugoslav pilots were trained by the US Air Force in West Germany. A major change occurred in the early 1960s, when the first Soviet first-line aircraft and missiles were delivered. Towards the end of the decade, however, the first home-grown combat types began to enter service.

Jugoslavia has stood firmly independent of the Warsaw Pact since it was formed in 1955. It is however associated with the Council for Mutual Economic Assistance (Comecon), although it is not a full member, and maintains friendly trading relations with other East European nations, particularly Romania. It is associated with Eastern nations by numerous pacts of assistance and co-operation, and during the 1973 Middle East crisis allowed its air base at Mostar to be used as a staging post for the Soviet VTA airlift of extra arms and supplies to Egypt and other Arab countries.

Below: The Soko Galeb (gull) armed trainer, one of several indigenous types operated by or being developed for the Jugoslav Air Force, is equipped with a pair of 0.50in machine guns and wing pylons.

Above: The Orao (eagle) attack aircraft is being developed jointly by Jugoslavia and Romania, using equipment and systems from both Warsaw Pact and Western sources. It will replace several older types.

At the time of writing it is not clear what shifts in Jugoslav policy will take place when Marshal Tito finally leaves the political arena. It is felt that Tito's personal power, based on his standing in the eyes of his people, has been instrumental in holding Jugoslavia aloof from the Warsaw Pact, and that Jugoslav foreign policy will move eastward with his departure. Clearly the Soviet Union is prepared to use all means to persuade Jugoslavia to join the Warsaw Pact.

The main strength of the Jugoslav Air Force, its 110 Mikoyan MiG-21 interceptors, is Soviet-supplied, and no attempts have been made to seek Western replacements. The interceptor force has been kept up to date by expansion and replacements, and includes MiG-21PF Fishbed D/F all-weather fighters and third-generation MiG-21MF Fishbed J fighter-bombers. All are armed with AA-2 Atoll infra-red homing AAMs. North American F-86D and F-86L Sabre all-weather fighters provided an important link in the Jugoslav defensive chain in the early 1960s, before MiG-21PFs were delivered, but there are fewer than 50 aircraft of the type still in service and they must be due for retirement.

The main future planned purchases will be of the new Orao light strike fighter, developed by Soko of Jugoslavia in collaboration with the Romanian aircraft industry. Some 200 Oraos will replace older combat aircraft, such as the 15 Thunderjets remaining in service and the 15 Lockheed RT-33A Shooting Stars still

used in the reconnaissance role. MiG-21s will probably be replaced by Oraos in the fighter-bomber role, and will in turn supersede the Jugoslav Sabres. The 150 Soko Jastreb light strike aircraft – single-seat versions of the Galeb trainer – will be replaced eventually by Oraos.

The future of the Orao might be threatened if Jugoslavia joined the Warsaw Pact, because it uses considerable amounts of Western components and equipment including the Rolls-Royce Viper engine. Deliveries of the Orao, originally due to start about the beginning of 1977, have been delayed by weight and cost problems.

Also in service in the strike role is the little piston-engined Soko Kraguj counter-insurgency aircraft. Other domestic products in service include 30 RJ-1 reconnaissance versions of the Jastreb, 60 G-1 Galeb jet trainers and 30 TJ-1 two-seat weapons trainers with the more powerful engine and the beefed-up structure of the Jastreb.

A mixed transport force is now being strengthened with the arrival of Antonov An-26 Curl tactical transports to replace Douglas C-47s, about fifteen of which are

in service. Twleve An-12s form the heavy transport force. Four DC-6s are on hand, as are single examples of the Ilyushin Il-18, the Il-14 and the Caravelle 6, although the last-named may not be operational. Government/VIP aircraft for official use include a Boeing 727 – probably operated by the national airline Jugoslovenski Aerotransport (JAT) rather than the air force – and three Yakovlev Yak-40s. A substantial number of Jugoslav UTVA-66s fly in the liaison/utility role.

Air Force helicopters are a mixture of East and West. About 50 of the 132 Aérospatiale Gazelle helicopters being built under licence by Soko are destined for the air force; 20 Alouette IIIs are already operational. Five Agusta AB.205s are in service, and 12 Mil Mi-8s provide a medium transport force.

The Air Force is also responsible for

Below: The Soko Jastreb is a single-seat light-strike derivative of the Galeb. The aircraft is fitted with three 0.50in machine guns in the nose and can carry a variety of underwing armament.

missile defences. Eight batteries of medium-range SA-2 Guideline missiles are backed up by shorter-range SA-3 Goa SAMs and share the defence of strategic points and major cities with the interceptor force.

The Jugoslav Navy operates Mil Mi-8 Hips for transport duties and the Kamov Ka-25 Hormone for ASW, although it is not known how close the latter is to the Soviet Navy standard. Soko-built Gazelles are being delivered for liaison duties.

The Jugoslav Air Force has a Training Corps with its headquarters at Mostar and bases at Batajnica, Titograd, Niksic, Zalusani and Pola. Pupils progress from the Aero 2/3 basic trainer – due for replacement by the UTVA-75 – to the Galeb before undergoing weapons training on the TJ-1. Conversion training to the MiG-21 is carried out at regiment level.

Right: Jugoslav troops practise casualty evacuation with the aid of an Mi-8.

Below: Jugoslav Air Force MiG-21s stored in a hardened underground shelter.

ALBANIA

Area: 10,700 sq. miles (27,700 sq. km.).
Population: 2·458 million.
Total armed forces: 41,000.
Estimated GNP: $1·1 billion (£550 million).
Defence expenditure: $154 million (£77 million).

The Albanian People's Army Air Force is one of the smallest in Eastern Europe. It forms an integral part of the Albanian People's Army and operates mainly in the close-support role, under direct Army command.

Apart from an abortive experiment at the beginning of the 1914–18 war, there was no military aviation in Albania until the country came under Soviet influence in 1947. Initially formed with wartime Yakovlev Yak-3 fighters and general-

purpose Polikarpov Po-2s, the Albanian air arm was modernised with Soviet assistance after Albania signed the Warsaw Pact in 1955.

From 1962, however, the Albanian Government ceased to take an active role in Pact affairs, and it formally broke with the Pact in 1968 after the invasion of Czechoslovakia. It now has few links with other European countries, but maintains close relations with the People's Republic of China. Since the early 1960s the Chinese Shenyang State Aircraft Factory has supported the Albanian air arm and supplied it with new aircraft, all Shenyang-built versions of Soviet designs. Albania's main strategic importance to the Soviet Union and the Warsaw Pact is as a convenient staging-post on the route to the Middle East – a role filled by Jugoslavia during the 1973 Middle East crisis –

and as a Mediterranean base, a role now taken over by Libya.

The most numerous type in the inventory of the APAAF is the Shenyang F-6 (Mikoyan MiG-19), still a useful fighter despite its age. About 36 are reported to be in service, equipping three squadrons. Rather fewer Shenyang F-7s (Mikoyan MiG-21Fs) equip one air-defence squadron; it would not be surprising if these were armed with a copy of the AA-2 Atoll IR-homing AAM. Up to 20 F-7s may be in service. Some 50 F-2s and F-4s (MiG-15s and MiG-17s) are in service in the fighter-bomber role. Shenyang-built Il-28s (B-5s) have also now been supplied.

Shenyang is likely to be the source of any future acquisitions. The F-6 bis all-weather fighter derived from the F-6 may be supplied. The Chinese are said to be now working on a new fighter powered by the licence-built Rolls-Royce RB.168 Spey.

Before the break with the Warsaw Pact the APAAF took delivery of about 20 helicopters – Mil Mi-1s and Mi-4s – and some of these continue in service. Similarly, the training fleet of Yakovlev Yak-11s, Yak-18s and Mikoyan MiG-15UTIs is Soviet-supplied, and these aircraft cannot have much life left. A few Antonov An-2s, Ilyushin Il-14s and Il-18s also fly in the transport role, and are Shenyang-supported.

Full details of deployment are not available, but it is likely that the F-7s are based at Shijak near Tirana and that one of the F-6 units is deployed at Kucove to defend the southern part of the country. Other fighter-bomber units are based at Durazzo and Valona.

In the apparent absence of conversion trainers based on the F-6 and F-7, training relies heavily on the MiG-15UTI advanced trainer, pupils progressing to this aircraft from piston-engined trainers and proceeding to the squadrons.

Right: Albanian pilots study a map in front of a line of F-6 (Chinese-built MiG-19) fighters, of which three squadrons are thought to be operational. The APAAF flies several Chinese-built types.

ROMANIA

Area: 91,700 sq. miles (237,500 sq. km.).
Population: 21·6 million.
Total armed forces: 180,500.
Estimated GNP: $51·4 billion (£25·7 billion).
Defence expenditure: $923 million (£461·5 million).

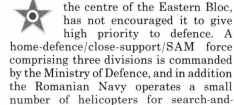

Romania's position, isolated in the centre of the Eastern Bloc, has not encouraged it to give high priority to defence. A home-defence/close-support/SAM force comprising three divisions is commanded by the Ministry of Defence, and in addition the Romanian Navy operates a small number of helicopters for search-and-rescue duties in the Black Sea.

The Romanian Air Force is a post-war Soviet creation, having started as a Romanian-manned division of the Soviet air force. Modernisation and increasing independence followed in the mid-1950s when the first MiG-15 jet fighters were delivered.

Romania is a member of the Warsaw Pact, and is pledged to put its forces under Soviet control in the event of a collective threat to the Pact nations. There are no Soviet forces stationed in Romania; in wartime the Romanian air divisions could either be attached to the 15th Air Army in the Odessa military district or to the Southern Group of Soviet Forces based in Hungary.

In recent years Romania has moved to reduce its dependence on the Soviet Union, developing close links outside the Warsaw Pact with Jugoslavia, and producing an increasing amount of military equipment.

Typifying and leading this trend is the air force, which is expected to take delivery of some 80 IAR.93 Orao strike fighters, developed in collaboration with Soko in Jugoslavia and incorporating a considerable amount of Western equipment, including Rolls-Royce Viper engines. The Orao will replace 80 obsolescent MiG-17s now in service with two fighter-bomber regiments. Orao development appears to have been subject to delay, however.

Two regiments fly about 80 MiG-21F Fishbed Cs and MiG-21PF Fishbed Ds, the latter having limited all-weather capability, in the interception role. A single air-defence division backs up the fighters with SA-2 Guideline medium-range SAMs. The fighters themselves carry AA-2 Atoll IR-homing missiles. A single regiment of Sukhoi Su-7 Fitter As (about 35 aircraft) provides close support.

The combat aircraft form two mixed air divisions, the small size of the air force ruling out the usual practice of assigning specific roles to each division. The two air divisions probably correspond to the two Romanian military districts, centred on Cluj and Bucharest.

The transport force operates some 30 Il-14s, ten Antonov An-2s and an Ilyushin Il-18 in a single independent regiment. In early 1979, Romania was negotiating a deal to build British Aerospace One-Elevens under licence. Some of these would be pure freighters, and could be used by the air force. Fifty Aérospatiale Alouette IIIs have been built under licence for the helicopter force, which also operates

Below: The MiG-21F remains in front-line service with the Romanian Air Force, along with a number of more recent MiG-21PF interceptors which have a limited bad-weather capability.

WSK Mi-2s, Aérospatiale/Westland Pumas and a smaller number of Mil Mi-4s and Mi-8s. The Romanian Navy SAR fleet comprises four Mi-4s. Some Czech L-200 Morava light twins are flown for liaison.

Training follows the standard Warsaw Pact pattern, and the Romanian Air Force is also likely to move from a three-level to a two-level structure. Primary training is currently carried out on piston-engined Yak-18s, with basic training on L-29 Delfins and advanced instruction on MiG-15UTIs before conversion. The more capable L-39 Albatros is however expected to replace the L-29 and the MiG-15UTI, with the Jugoslav UTVA-75 being the most likely replacement for the Yak-18.

Right: Romanian troops run from a newly landed Mi-4 assault helicopter.

BULGARIA

Area: 43,000 sq. miles (111,000 sq. km.).
Population: 8·7 million.
Total armed forces: 150,000.
Estimated GNP: $18·6 billion (£9·3 billion).
Defence expenditure: $432 million (£216 million).

Most of Bulgaria's military air forces operate under direct army command, with the exception of an air-defence division – equipped with SAMs – and a few helicopters operated by the Navy for air-sea rescue and patrol in the Black Sea. The air-defence division is under the direct command of the Minister of Defence, who is the commander-in-chief of the Bulgarian armed forces.

Like many other East European air forces, the Bulgarian air arm was formed under Soviet tutelage shortly after the 1939–45 war. Although the strength of the Bulgarian air force is limited technically to 90 aircraft by the terms of the 1944 peace treaty with the Soviet Union, it was allowed to grow well beyond that figure after Bulgaria came under Soviet influence. The air arm has not been greatly expanded or modernised since the early 1960s, however.

Under the terms of the Warsaw Pact the Bulgarian armed forces are independent until the Pact is faced by a collective threat. In that event the forces come under the control of the Soviet High Command. Air units would probably be attached to the Soviet Union's 15th Air Army, normally based in the Odessa military district. Missile units are presumably integrated into the Soviet PVO defence network in any case. Bulgaria is in the first line of defence between Greece and Turkey and the Romanian oilfields, but is well away from the European central region.

Bulgaria is reported to be a Mikoyan MiG-23 recipient, with 20 aircraft, and up to 100 Mikoyan Mig-17 Frescos still form the greatest part of the air force, but the balance is likely to shift in favour of the MiG-21 Fishbed, about 50 of which are now in service in intercepter, fighter-bomber and fighter-reconnaissance roles. In 1979

Above: A Bulgarian Air Force MiG-19PM all-weather fighter, which is unusual in having no pylons for AA-1 Alkali air-to-air missiles. MiG-21s are gradually succeeding the MiG-17s.

a handful of Ilyushin Il-28s were reported to be still flying with a reconnaissance squadron. Training units operate the Yakovlev Yak-11, the Yak-18 and the Czech Aero L-29 and L-39, with the MiG-15UTI for advanced training. Transport aircraft include half-a-dozen Antonov An-2s, ten Ilyushin Il-14s and a few Antonov An-24/30s. Two Tupolev Tu-134 VIP transports are operated. Army transport helicopters include 40 Mi-4s and a handful of Mi-8 helicopters. The Navy operates about six Mil Mi-4s and some WSK Mi-2s for search and rescue.

Bulgaria will probably not receive the latest generation of Soviet combat aircraft for some time to come, but will take de-livery of newer versions of the MiG-21 to make good attrition in its current fleet.

Details of deployment are not available, although it is known that the headquarters of the tactical air forces is located at Plovdiv. The tactical aircraft form a single composite division, in contrast to the practice in larger Warsaw Pact forces where each division is assigned to a specific role. The division comprised three large regiments, two of which operate in the fighter-bomber role – with an intercept capability in the case of the MiG-21s, which are likely to be based near Sofia – and one is responsible for reconnaissance.

The air-defence division operates SA-2 Guideline SAMs, in three zones centred on Sofia, Plovdiv and Yambol.

Training is carried out on Yak-18 piston-engined trainers. Pupils proceed via the L-29 and the MiG-15UTI to operational squadrons, the only operational conversion trainers being MiG-21Us.

Soviet Union

The Soviet Union is one of the world's two super-powers, a status achieved by virtue of its great size, population and technical might. Other countries such as China, India and Brazil have one or both of the first two attributes but they do not possess Russia's political will to play a major role on the world stage nor the military muscle to make such a stance viable.

The Soviet Union emerged from World War II with immense armed forces but limited technology. The work carried out in Germany during the war proved a vital windfall, however, in the form of both captured equipment and design personnel. When combined with development already under way in Russia, this allowed the USSR to proceed rapidly with building new military aircraft, guided missiles, jet engines and other areas of advanced technology. A crash programme to develop nuclear weapons was also inaugurated, so that the time during which the United States was the sole possessor of weapons of mass destruction would be as short as possible.

Soviet forces have traditionally been well equipped in terms of numbers, although the sophistication of individual weapon systems has usually been inferior to that of their Western equivalents. This is now changing, and the new breed of Russian armaments are in many cases at least the equal of those deployed by NATO. The numerical superiority remains, so the USSR is becoming ever stronger militarily.

The Soviet Union has its own 'Triad' of land-based strategic ballistic missiles, bombers and submarine-launched ballistic missiles. The Soviet Navy is being strengthened all the time, allowing it to exert influence in areas such as the Indian Ocean which were once the preserve of the West. The Mikoyan and Sukhoi design bureaux continue to turn out new combat aircraft to cause sleepless nights for NATO planners, while similar advances take place in land armaments.

The Soviet Union exerts pressure in many areas around the world, but its regular forces have not been involved in any major conflict since the end of World War II. Skirmishes on the border with China continue and other wars are fought by proxy — friendly powers, such as Cuba and Vietnam, are supplied with lavish amounts of equipment backed by training for their nationals in the Soviet Union, the provision of Russian 'advisers' on the spot and, when appropriate, the presence of Soviet warships nearby.

Area: 8·62 million sq. miles (22·3 million sq. km.).
Population: 257·9 million.
Total armed forces: 3,638,000.
Estimated GNP: $688 billion (£344 billion).
Defence expenditure: $70 to $124 billion (£35 to £62 billion); estimates vary.

The giant size, strength and attitude of the Soviet Union distorts the Warsaw Pact, in that it is hardly a free collaborative grouping of sovereign nations enjoying collective defence but a master and what from the start have often been called the 'Satellite countries'. Though the United States is by far the largest and most powerful member of NATO, it has neither the authority nor the wish to give commands to the other partners, whose total freedom to take their own decisions is jealously guarded to the point where it frequently and dangerously damages military effectiveness. The WP Satellites, on the other hand, have twice been kept in line at the point of the gun (Hungary in 1956, Czechoslovakia in 1968).

As spelt out in the companion book *The Soviet War Machine*, the Soviet Union finds it difficult, or impossible, to live with the rest of the world in a state of impartial objectivity. From the bloody formation of the country almost 60 years ago (one cannot take November 1917 as a starting date because for years the outcome of the civil war was far from obvious) the Soviet Union has believed that what it calls 'the Capitalist Nations' or 'the Imperialists'

wish to attack it and see it overthrown. The violent aggression by Nazi Germany in June 1941 is held to reinforce this view. Soviet history books do not mention the fact that, to throw back the Nazi invader, the capitalist Allies in 1941–45 provided all the help they could muster to the Soviets until the Allied forces met on the Elbe. The school textbooks also omit to mention that, after World War II when the USA alone possessed nuclear weapons and the means to deliver them, this gigantic and unrepeatable imbalance of power

Above: Despite criticism by some Western observers the Tupolev Tu-126 'Moss' has been in combat service for at least eight years and appears to be a satisfactory 'AWACS' type radar surveillance platform.

was not used to destroy the Soviet Union in the way that the Soviet leaders would have their population believe is forever sought by the Western world.

There has to be a threat before massive defence funding can be justified, and – though any citizen of any NATO country

knows all too well that the idea is ludicrous – the Soviet Government continues to build up enormous forces by land, sea, air and space on the basic premise that these forces are needed to resist attack by the Capitalists. So successful has the unrelenting campaign against NATO been that, in the Soviet Union especially, the armed forces are among the most important and popular sectors of society. One has only to imagine the impact (or lack of it) in Western media of photographs captioned 'Our Heroes' or 'On guard' to recognise the gulf between the Communist and Capitalist society. The giant paradox is, of course, that the Soviet Union does have the capability and, if need be, the motivation, to take over by force the whole of Europe, while the supposed aggressive European NATO countries lack the means and the motivation to invade anyone, let alone the Soviet Union!

Unlike those of NATO, or any other Western-style society, all WP troops are subjected to continual political indoctrination. This is all part of the centrally controlled political and economic infrastructure that pervades every aspect of Soviet life, and of which military affairs are an integral and central part. Without any overt wish to oppress, the fact remains that the entire WP population is subjected to totally managed media, centralised ideology and an overall pressure for standardisation. This is manifest in such

end-results as suppression of minorities, harsh treatment of 'dissidents', and universal teaching of the Russian language, Communist Party beliefs and ideology (especially among the military) and Soviet military organisation and doctrine, down to the finest points of detail. Of course, this makes for maximum military efficiency, which is tested rigorously in large-scale exercises involving several WP countries at a time.

Of all the WP countries, the Soviet Union is the only one to possess strategic weapons, major offensive weapon-systems, large naval units and, in particular, nuclear weapons. Only the Soviet Union has the industrial and financial strength to create the full range of modern weapon-systems. The WP satellites have extremely limited potential to create weapons, and in the field of air power it is limited to helicopters in the small to medium size (Poland), light strike trainers (Czechoslovakia and Poland), light general-aviation aircraft such as utility transports and agricultural machines (Poland, Romania and, to a lesser degree, Czechoslovakia) and sailplanes (Poland, Czechoslovakia and Romania). Special mention must be made of the unique Jugoslav/Romanian Orao twin-jet tactical attack aircraft designed and built jointly by Romania and a non-member of the WP, Jugoslavia. This is discussed in the entry on Romania.

While Soviet aircrew eventually graduate to a position of trust, flying long transport, reconnaissance and electronic-warfare missions thousands of kilometres from Communist territory, other WP aircrew are rigorously prohibited from crossing

their own frontiers. No other WP country but the Soviet Union has any counterpart to the Strategic Rocket Troops (the élite of all Soviet armed forces) nor the Zenith Rocket Troops, the SAM element of the PVO, the air defence of the homeland. Soviet fighter and interceptor squadrons of the IA-PVO (air defence) and FA (frontal aviation, ie tactical multi-role) have since 1971 been almost wholly re-equipped with new and extremely advanced types of aircraft, a high proportion of them with variable-geometry 'swing wings'.

The aviation and missile forces of the Soviet Union are numerically the largest in the world, not only meeting the needs of one of the world's two superpowers but also giving the Soviet Union military domination over an organisation of secon-

Above: One of the vast force of Soviet silo-launched ballistic missiles. Today the same silo probably contains a larger ICBM using the 'cold launch' technique.

dary industrial powers in the Warsaw Pact. In recent years the rapid expansion and modernisation of Soviet military aviation has given rise to increasing concern in the West.

There is no single 'Soviet Air Force', comprising all military aviation under one command. The armed forces of the Soviet Union are integrated beneath a single General Staff. Three main aviation contingents report directly to the General Staff and another is under naval control.

Of the three main elements one – the Strategic Rocket Forces (Russian initials RVSN) – has no combat aircraft at all. The Air Defence Forces (designated the PVO, from their Russian initials) command extensive surface-to-air and anti-ballistic missile defences (ZA-PVO) as well as what is by far the most powerful manned interceptor force in the world (IA-PVO). Other aviation activities are the responsibility of the Air Force (VVS) in its three operating sections: these are Frontal Aviation (FA), which provides tactical support to the Ground Forces; Long-Range Aviation (DA), which constitutes the Soviet Union's strategic bomber force, and Military Transport Aviation (VTA). The VTA fleet is backed up by the civil freighters of the Soviet Union's state airline, Aeroflot.

The fourth aviation element of the Soviet forces is Naval Aviation (AV-MF). It is subordinate to the Soviet naval command rather than directly to the General Staff as are the RVSN, VVS and PVO. Its historical role can be loosely described as 'support of the fleet' and its importance has expanded dramatically in the last two decades with the growth of the Soviet Fleet.

In the Soviet Union it is difficult to say where civilian government ends and military command begins. Communist Party Secretary Leonid Brezhnev chairs the Council of Defence, composed of high-level military and political officials including Defence Minister Dmitry Ustinov. In wartime the Council of Defence would probably assume the government of the Soviet Union: in times of peace it ensures the country's industrial and political readiness for war. Between the Council of Defence and the General Staff is the Main Military Council, which in peacetime is responsible for the strategic direction of the armed forces. It is thought to be chaired by Defence Minister Ustinov. Ustinov and Brezhnev were awarded high honorary military rank in 1976.

Both the General Staff and the Main Military Council have responsibilities reaching higher and lower than those of Western military commands, which delegate more of their work to departments of the civilian government. The Soviet high command is responsible for long-range planning and the analysis of both military and political situations, as well as the administration of training and support services.

The close links between military and civil administration have in the past led to sweeping political decisions which have had profound effects on the air services. The strategic bomber force was regarded as an élite in the 1930s; it declined during World War II, but came back into the lime-

Above: This Tupolev Tu-20 (Tu-95) is of the basic sub-type called 'Bear-D' by NATO, with extensive electronic-warfare, missile-guidance and navigation equipment. It was photographed by the pilot of an RAF Lightning.

light once nuclear weapons were available. Under Krushchev in the late 1950s and early 1960s, however, it lost all priority in favour of the RVSN, and is only now showing signs of catching up in terms of importance with the USAF Strategic Air Command. Another decision with long-term consequences was Kruschchev's dictum in the mid-1950s that there should be no more specialised ground-attack aircraft; it was to be 20 years before Frontal Aviation could field a specialised strike fighter in the shape of the MiG-27.

RVSN is the youngest of the services. Soviet experiments with long-range military rockets started after 1945 with caprured German equipment and engineers, and by the second half of the 1950s had reached the stage where Krushchev could found the RVSN and order the development of an intercontinental ballistic missile (ICBM) force. This proved more difficult than anticipated and it was not until 1962 that an effective ICBM entered service. In the same year an attempt to locate medium-range missiles in Cuba, within range of the United States, led to the Cuba missile crisis. The RVSN is reported to be responsible for anti-satellite systems, and its activities are now covered by the Vladivostok agreement of November 1974 on control of strategic weapons.

VVS has its origins in the Revolution, when some units of the former Imperial Russian forces took part in the Civil War. Post-revolutionary development was hampered by the destruction of the Civil War and the loss of some leading designers who emigrated to the West. By the 1930s the VVS had been built up into a large and powerful force, but the battles between VVS detachments and the German Condor Legion in the Spanish Civil War showed

that technologically it was in the second rank at best.

Stalin's purges swept away many VVS leaders and effectively prevented the force from modernising itself after the Spanish Civil War. The VVS was heavily defeated in the early days of the German invasion in the second half of 1941 and early 1942, and the force which emerged from battle experience was very different from the pre-war VVS. By 1945 the VVS had become a higher command level as far as the tactical combat forces were concerned, and the next few years saw the same development in the strategic bombing and transport arms.

The tactical combat force, FA, dates its status as the 'first among equals' of the three VVS commands back to 1942. In that year the Soviet tactical forces were formed into Air Armies charged with support of the front line and subordinated to the commander-in-chief for a particular military front. By the war's end there were some 17 Air Armies. Armies were assigned to a specific area; the next largest unit was the division, normally assigned to a specific role. The division comprised three or four regiments, each operating some 50 aircraft of a single type. This pattern of organisation remains in force.

FA fielded 12,000 combat aircraft in the early 1950s, but many of these were wartime types or their developments. Priority was given to strategic arms – bombers and interceptors rather than strike aircraft –

132

and right up to the late 1960s FA was equipped with aircraft not well suited to the close-support role. But the end of the Krushchev era, and the replacement of the 'nuclear tripwire' strategy with 'flexible response', meant that the FA had to be re-armed for its new role. Now an increasing number of FA's 5000-plus aircraft are the new types which were designed and developed in the late 1960s and early 1970s.

The strategic bomber component of the VVS – designated DA since 1946 – has suffered more than most from the fluctuations in Soviet policy. Russia pioneered the heavy bomber in the 1914–18 war, with Sikorsky's *Ilya Mourometz* class of bombers. Some fought briefly for the Bolsheviks in the Civil War, but the heavy bomber element disappeared by 1923.

The VVS formed a new bomber force in the early 1930s, equipped with the TB-1 and the then advanced TB-3. They had been mainly replaced by more modern types by the outbreak of war in 1941, but the force was virtually annihilated during suicidal unescorted daylight raids on tactical targets. Reformed in March 1942 under the designation ADD, the bombers, mainly Pe-8s and Il-4s, carried out a number of attacking raids on German targets before being placed under direct VVS control as the 18th Air Army in December 1944.

Two events led to the reconstitution of the independent strategic bombing force as the DA in 1946. The first was the internment in 1944 of three USAAF Boeing B-29 Superfortresses which force-landed in Soviet territory after raids on Japan. The second was the development of nuclear weapons. Copied in every detail as the Tupolev Tu-4, the B-29 became the carrier of the Soviet fission bomb.

The new DA was responsible for divert-ing a great deal of US military effort into strategic defence after the US 'bomber scare' of the 1950s. But by the time the first Soviet intercontinental bombers entered service it was realised that the subsonic medium-altitude bomber was a sitting duck. The supersonic Myasishchev Mya-52 was the DA's last hope to hold its own against the rising RVSN, but it failed dismally to achieve its design range. Krushchev was disgusted and the programme was cancelled in 1960. Since that time DA has been predominantly a medium-range force, and only now, with the Backfire bomber and a new tanker fleet, is it showing signs of recovering its

Above: This 'Backfire-B' was photographed in 1978 by an aircraft of the Swedish air force during NATO Exercise Northern Wedding.

Below: A Tu-22 Blinder-B of Soviet AV-MF (naval aviation), with engines in full afterburning thrust. The main gears are retracting into their wing pods, which also house damage-assessment cameras.

strength. A new, larger bomber with fully strategic range is expected in service by the early 1980s, indicating that the Soviet Union intends continued reliance on the Triad concept of bombers, submarine-launched missiles and land-based ICBMs.

Other developments affecting the DA include the evolution of long-range cruise missiles and, possibly, a specialised subsonic missile carrier.

The youngest VVS command is the VTA freighter fleet, with its origins in the late 1920s. Closely associated with paratroop forces up to the mid-1950s, the VTA began to increase in size and effectiveness with the delivery of specialised Antonov freighters from 1957. The big Mil Mi-6 helicopter also flew in that year. In the late 1960s VTA took delivery of the very large and extremely long-range Antonov An-22 Antei, giving it a new global airlift mobility. The smaller Antonovs are now being replaced by the Ilyushin Il-76, which can carry twice the payload twice as far.

Represented on the General Staff with the VVS and RVSN is the air defence organisation, the PVO. Created in early 1941, as the alliance between the Soviet Union and Germany became more strained, the PVO started life as an organisation of district commanders who co-ordinated interceptor and anti-aircraft artillery forces on a regional basis. Its autonomy within the VVS increased throughout the war, and it became wholly independent in the immediate post-war years.

The PVO forces and particularly the independent Moscow PVO District were

Above: A line of Ash air-to-air missiles goes aboard Tupolev Tu-28P 'Fiddler' long-range interceptors of the IA-PVO; infra-red and radar missiles are mixed.

Below: Tu-28P 'Fiddler' interceptors on the flight line at a Soviet airbase equipped with acres of concrete. This extremely large interceptor led to today's 'Backfire'.

the priority units in the 1950s as the bomber threat grew, and most research and development efforts were biased to its needs. With the advent of the surface-to-air missile in the late 1950s the PVO forces split into two subdivisions: the IA-PVO manned interceptor force and the ZA-PVO in command of strategic SAMs (not including the tactical SAMs attached to Ground Forces).

PVO planning in the late 1950s and early 1960s assumed that the bomber threat would continue to expand, but the expansion and modernisation of the interceptor force were slowed down slightly when Western planners seemed to abandon the manned bomber. In the 1970s IA-PVO has started a new re-equipment programme,

and ZA-PVO has acquired a new role with the development of the Soviet anti-ballistic missile (ABM) system.

In recent years the AV-MF has grown as fast as the surface fleet it supports. Its force of flying boats and fighters for coastal defence expanded rapidly during the war, but in 1953 the fighter squadrons were taken over by the local PVO commands and Air Armies as part of a general downgrading of the surface fleet. The AV-MF was a small but enterprising force, armed with twin-jet Tu-14 torpedo-bombers and pioneering the use of shore-based Mi-4 helicopters for anti-submarine warfare (ASW).

AV-MF's current growth followed Admiral Gorshkov's revival of the surface fleet. With the need to support a worldwide fleet, AV-MF acquired long-range aircraft surplus to the requirements of the unfavoured DA, and built up a powerful maritime strike, reconnaissance and electronic intelligence (Elint) force. In 1962 AV-MF deployed its first shipboard helicopters for ASW. In 1967 the specialised helicopter carrier *Moskva* was commissioned. Towards the end of the same decade the first specialised Soviet ASW aeroplane, the Ilyushin Il-38, entered service. In 1976 the AV-MF was sharing first deliveries of the swing-wing Backfire with DA and, even more significantly, carried out open-sea trials with Forger vertical-take-off-and-landing (VTOL) light attack aircraft on the carrier *Kiev*, adding a new dimension to the fastest-expanding of all the Soviet air arms.

Above: A frame from a propaganda film of PVO air defence operations, showing a group takeoff by MiG-25 'Foxbat' interceptors. Note landing lights on main-gear doors.

The Soviet Union's main alliance is the Warsaw Pact, on which its security against attack from the West is based. The other Pact countries are Poland, Czechoslovakia, Hungary, East Germany, Romania and Bulgaria. The armed forces of these countries are under Soviet command and their role is mainly defensive. Their aircraft industries are confined to the development of smaller types.

Signature of the Pact in May 1955 confirmed the Soviet Union's hegemony over Eastern Europe, established in fact between 1945 and 1950. The Soviet Union has been ready in the past to use force if any of the Pact countries steps out of line.

Three Eastern European nations are members neither of the Warsaw Pact nor of NATO. One, Albania, has no links at all with the Soviet Union. Finland and Jugoslavia, however, have a looser relationship with the Soviet Union than the other Warsaw Pact countries. Neither state abuts on to the central NATO area, so they do not present a severe security risk. The air forces of both rely on the Soviet Union for first-line aircraft, but Finland buys some equipment in the West and Jugoslavia is developing its own aircraft. Close co-operation between Jugoslavia and Romania has led to the development of a joint combat aircraft, the Orao.

Outside Eastern Europe, the Soviet Union buys influence through military and civil aid programmes. Beneficiaries of 'MiG diplomacy' in the Middle East have included Libya, Syria and Iraq. That there is another side to the coin is demonstrated by the example of Egypt, where the supply of spares for Soviet-built aircraft has fluctuated with changes in international relationships. Cuba is an important outpost and ally in Central and South America, having now taken delivery of aircraft in the MiG-23/27 series, while Peru has now received Sukhoi Su-22s. However, the rudimentary equipment standard of the latter has come as an unpleasant surprise to the Peruvian Air Force.

The Soviet Union's two main potential enemies are NATO and China. Conscious efforts have been made to reduce the tension between the Soviet Union and its allies and NATO. The Chinese Government appears convinced that war with the Soviet Union is highly likely. As yet the Chinese armed forces, particularly the air arm, are technologically and operationally decades behind those of the Soviet Union, but China is putting massive effort into military expansion and modernisation. The proportion of Soviet defence spending assigned to Chinese-aligned forces will inevitably have to increase as these measures – such as the acquisition of ICBMs and SLBMs, the development of a supersonic interceptor and the procurement of Harrier V/Stol fighters – take effect over the next five years.

The first round of Strategic Arms Limitation Talks between the Soviet Union and the USA reached an accord at the end of 1974, and the Salt II agreement was signed in June 1979. So far efforts to achieve a similar agreement on tactical arms in the European theatre (Mutual and Balanced Force Reduction, or MBFR) have been to no avail, mainly because of Western disquiet at the rapidly increasing strength of the Soviet conventional forces which would operate in the European theatre.

Inventory

The Soviet forces are expanding at an unprecedented rate, and older weapon systems which were clearly inferior to their Western contemporaries are being replaced by new weapons which match the best Western products.

The RVSN (Strategic Rocket Forces) typifies this trend, sharing as it does the task of strategic deterrence with the sea-launched ballistic missiles of the Soviet Navy and, to a lesser extent, the bombers and air-launched cruise missiles of DA. The total number of intercontinental delivery vehicles is limited by international agreement, but current second-generation missiles are now being replaced by new, more accurate weapons and fourth-generation weapons are now being introduced. The RVSN operates a considerable force of intermediate-range ballistic missiles (IRBMs) and first-generation weapons have now been almost entirely replaced by land-mobile systems such as SS-12 Scaleboard, SS-14 Scamp/Scapegoat and SS-15 Scrooge, carried on cross-country transporter/erectors. These systems – not controlled by Salt agreements, as they are not counted as strategic weapons – are themselves being replaced by SS-20, comprising the first two stages of the SS-16 ICBM. The SS-20 carries three independently targeted warheads, with a circular error probable (CEP) of only 750m under optimum conditions. Alternatively, it is feared, the three MIRVs could be replaced by a single warhead and a third booster stage, producing an ICBM outside the control of Salt.

The only first-generation Soviet ICBM to remain in service in late 1978 was the SS-9 Scarp, armed with a single warhead of 25MT yield. The liquid-fuel SS-11 Sego, armed with multiple or a single warhead,

Below: A formidable aircraft in the class of the F-111 and Tornado, the Su-19 'Fencer' is still little known in the West. This is the only released photograph.

135

Above: One of the later sub-types of MiG-21 in Soviet service is the model called 'Fishbed-L' by NATO (its true identity is unknown); related aircraft have visited Western airbases and are built in India.

is still the most numerous missile, but its replacement by later weapons has already started. The solid-fuel SS-16 is replacing the SS-13 (generally assumed to have been less than a complete success, and the first Soviet solid-fuel ICBM) and forms the basis for the SS-20 IRBM.

New missiles like SS-17 (60 in service by mid-1978) and SS-19 (200 by mid-1978) are replacing the SS-11 force. They are equipped with MIRVs and are cold-launched, which in theory could allow silos to be quickly refurbished and re-loaded after firing. More powerful than either SS-17 or SS-19 is the SS-18, of which more than 100 have been deployed to replace SS-9s. Throw weight is nearly 30% greater than that of SS-9, and missiles on test have demonstrated alarmingly small CEPs which suggest that the Soviet Union may acquire the capability to launch a first strike against the US ICBM force and survive the weakened retalia-tion. The Mod 1 and Mod 3 carry single re-entry vehicles, the Mod 3 having much improved accuracy, while the Mod 2 is armed with eight to ten retargettable vehicles (RVs).

Soviet missile developments are con-tinuing, with such programmes as the MARV (manoeuvring re-entry vehicle) and fourth-generation weapons for ser-vice in the mid-1980s. The RVSN may also be closely involved in the development of a Soviet re-usable space vehicle (aero-dynamic test vehicles are reported to have flown) and the so-called 'charged-particle

weapon' which was the subject of much discussion in 1978.

Numerically the most important SLBM in Soviet service is the SS-N-6 Sawfly, which started to close the technological gap between US and Soviet systems in 1967. More than 1000 missiles now arm 34 *Yankee*-class boats, each with 16 launch tubes, and the SS-N-6 Mod 3 with three MRVs is replacing earlier versions.

The Soviet Union has now taken the lead in SLBM development with the entry into service of the 8000km-range SS-N-8, fitted as standard with MRVs and arming Delta submarines. Eight twelve-round Delta Is are being joined by a rapidly expanding force of 16-round Delta IIs and 20-round Delta IIIs. New SLBMs under development include the SS-N-17, a solid-fuel weapon which will be refitted to SS-N-6-armed *Yankee* submarines, and the liquid-fuelled SS-N-18, arming Delta IIIs and the so-called Typhoon class and replacing the SS-N-8.

Of the three combat elements in the VVS it is the tactical FA which has been most modernised and re-equipped over the past ten years. FA is estimated to take more than half the 1000 high-performance com-bat aircraft which Soviet aircraft factories produce every year. Moreover, these new aircraft are much more suited to the tactical support role than the MiG-21 and Su-7 which formed the backbone of the FA throughout the 1960s after being deve-loped to meet air-defence-orientated re-quirements. The 1970s have also seen the Soviet Union attempting to catch up with the West in development of guided air-to-surface missiles (ASMs) and weapon-aiming systems, with new weapons begin-ning to reach front-line regiments in 1977.

Exact estimates of the composition of

the FA are hard to come by, particularly in view of the rapid introduction of the latest types of aircraft and the difficulty of telling whether older aircraft have been retired or placed in reserve. But it is currently estimated that FA deploys some 5000 first-line combat aircraft, split among three groups: new-technology aircraft introduced since 1971; relatively modern types; and aircraft which by Western standards are obsolescent. The last group is already being taken out of service in front-line areas, although the Soviet Union tends to keep aircraft in service for some years after their Western contem-poraries have been declared obsolete. A corollary of this policy is that Soviet aircraft fly fewer hours per year, extending their airframe lives.

A Frontal Aviation Army's largest sub-unit is the division, which is assigned to a specific role. (All units are, however, trained to some extent in the delivery of tactical nuclear free-fall weapons.) The main equipment of fighter divisions is still the Mikoyan MiG-21 Fishbed series, flown in the early 1950s in prototype form and in service since the later years of the decade. More than 1000 of the 1500-plus MiG-21s in first-line Soviet service fly with FA fighter, fighter-bomber and reconnais-sance units, and an increasing proportion are the latest third-generation versions (Fishbed H/J/K/L/N) with increased pay-load and an internal gun. They are re-garded as true multi-role aircraft, albeit with a limited range. Later MiG-21s can carry a mix of up to four AA-2-2 Advanced Atoll radar homing and Atoll infra-red homing air-to-air missiles (AAMs), or bombs and unguided rockets for ground attack. Frontal Aviation intends to use the MiG-23S as a top-cover fighter, with

Above: A squadron of specially simplified MiG-23 fighters of the Soviet Frontal Aviation visited Finland in 1978, creating a very good impression.

the new MiG-21bis operating at lower levels. The MiG-21bis features greatly expanded avionics compared with earlier versions, together with the bulged dorsal spine seen previously only on the MiG-21SMT Fishbed K.

Since 1971, however, FA fighter divisions have been phasing in the much more capable MiG-23MS Flogger B variable-sweep air-superiority fighter. Its useful combat radius is far greater than that of the MiG-21 – some 500nm compared with 150nm for the smaller fighter – and it

carries a heavier armament, comprising a pair of medium-range AA-7 Apex AAMs (usually, one IR-homing and one semi-active homing) and two IR dogfight missiles known as AA-8 Aphid. Like the later MiG-21s it has an internal GSh-23 twin-barrel cannon of 23mm calibre. More than 800 MiG-23MS fighters are in service, and production is estimated to be running at almost 200 aircraft a year, for the Soviet Union and for export. Significant improvements have been incorporated on the latest MiG-23s. Fire-control appears to have been improved, and the US Department of Defence now credits the aircraft with some look-down/shoot-down capability. Recent aircraft are fitted with a more powerful engine and provision for

extra fuel tanks, and may be slightly more manoeuvrable than initial aircraft.

The MiG-23MS has a secondary ground-attack capability superior to that of the MiG-21 or even the Sukhoi Su-7, the standard Soviet strike fighter of the 1960s. The roles of the fighter and fighter-bomber divisions are, however, diverging. The considerable expansion of the Ground Forces' surface-to-air missile formations has reduced the need for fighters over the

Below: The MiG-27, a ground-attack version of the MiG-23, is extremely well equipped and is in service in very large numbers. Its main deficiency is a modest weapon load, carried on three body and two wing pylons.

battelfield, so aircraft like the MiG-23MS would be used to carry the air war outside SAM range. Fighter-bomber units similarly are equipping with more-specialised and longer-range aircraft.

Probably the most important of the new ground-attack types is the MiG-27 Flogger D, derived from the MiG-23MS but with a new nose section, a modified engine and completely different armament and weapon-aiming systems. The high commonality of the MiG-27 with the MiG-23 allowed a rapid build-up of production and more than 400 have been deployed since early 1974. Under development to arm the MiG-27 are television-guided laser-guided and anti-radiation ASMs and 'smart bombs' and some of these at least are already operational. The new weapons include the 6nm-range AS-7, thought to be command-guided, and the AS-X-10 of similar range but with TV guidance. Cluster bombs and fuel-air munitions are under development for the MiG-27 and other strike aircraft, which for the first time feature multiple stores racks. Another new weapon is a 240mm unguided missile for use against aircraft shelters.

The MiG-27 could probably attack targets in Eastern England at full load, but is more suited to operations within continental Europe. FA is introducing a deep-penetration strike aircraft, however, in the shape of the Sukhoi Su-19 Fencer, and deployment of this aircraft has considerably reduced the dependence of forward commanders on the long-range strike aircraft of the DA.

Fencer flew in 1970 and was reported operational in regiment strength towards the end of 1976, production having taken some time to build up. It is a two-seater, far heavier than the MiG-27, and can reach most important European targets from East German bases. Later versions may have a multi-mode attack radar, but early aircraft have the same avionics fit as the MiG-27. Armament includes the GSh-23 rather than the six-barrel 30mm Gatling gun of the MiG-27, but the Su-19 will add a pair of 50nm-range AS-X-9 anti-radar missiles to the array of ASMs carried by the smaller aircraft. The Su-19 has not appeared in service as widely as expected, and contrary to earlier reports it has not yet been deployed to East Germany. This has led to speculation that the development of the aircraft or the multi-mode air-to-surface attack radar vital to its low-level all-weather strike role may have been delayed. It has been suggested that the Su-19 has been deliberately kept away from areas where it might come within range of unfriendly cameras, but it is hard to see how a good-quality photograph of the type could do as much harm to the type's operational readiness as such a restriction on deployment.

The third basic type of variable-sweep combat aircraft in FA service is the Sukhoi Su-17 Fitter C, D/E, derived from the Su-7 via what was intended to be a purely experimental variable-sweep Fitter B test-bed flown in 1967. It offered a considerable improvement in field/payload/range performance over the Fitter A, however, and could be placed in production fairly rapidly. Compared with the experimental

Fitter B the Su-17 has an uprated engine, and it has eight stores pylons as against six on later Fitter As. It has not been seen with the sensors associated with the new generation of Soviet ASMs, however, and is probably an interim type. Over 350 Su-17s have been delivered to FA.

The new strike aircraft are replacing some of the older FA types such as the 1951-vintage MiG-17 – only withdrawn from FA units in East Germany in late 1976 – and the Yak-28 Brewer light bomber. About 1000 of these types remain in service. The 500-plus Sukhoi Su-7s are likely to stay in service for some time, despite their limited range and weapon load. A new specialised air-combat fighter is likely to be adopted as a replacement for the MiG-21, and a number of prototypes from different design bureaux have been tested over the past few years. The Sukhoi bureau is credited with the design of a specialised ground-attack aircraft, replacing the Su-7 and eventually the Su-17 in the close-support role, and this type may be closer to service than the new fighter.

Tactical reconnaissance for FA is mainly provided by MiG-21RF Fishbed Hs, with wingtip electronic counter-measures pods and a reconnaissance pack replacing the GSh-23 cannon. Perhaps 300 are in service. Since 1971 FA has operated the Mach 3 MiG-25R Foxbat B for optical and electronic reconnaissance, but so far this type has been used only in small numbers and it is doubtful whether there are as many as 100 aircraft of this type attached to FA. The two-seater MiG-23U Flogger C, basically a conversion trainer variant of the MiG-23MS, has also been reported in the electronic countermeasures (ECM) role, with external ECM pods replacing missiles. But Frontal Aviation is also converting Yakovlev Yak-28 Brewer light bombers to the Brewer E ECM configuration, because its internal weapons bay offers more space for electronics than is available in the MiG-23 airframe.

FA acquired a new capability in 1973 with the introduction of its first 'gunship' helicopters. Several hundred Mil Mi-24 Hind helicopters are in service, and progressive improvements in armament and weapon aiming have now led to the definitive Hind D variant, with sophisticated sighting systems and a multi-barrel machine gun in a nose turret. Unguided rockets and AT-6 anti-tank missiles are carried on stub wings. The new missile may be TV-command guided with IR terminal homing. Some sources say that there is a completely new anti-tank helicopter under development, but these reports may have arisen from early accounts of Hind D.

The Mi-24 is a squad carrier as well as a gunship, with a cabin for 8–12 troops. The primary FA transport helicopter, however, is the Mil Mi-8 Hip, a twin-turbine type with 35 seats. More than 1000 Mi-8s are in service, and most can be armed with unguided rockets for fire suppression in combat. Sources differ on the division of the VVS helicopter fleet between FA and VTA, but it is not likely that the larger Mi-6 Hook and Mi-10K Harke would be used by FA in forward areas because of their size and vulnerability to small SAMs. There are persistent reports of a new large helicopter under development, possibly a heavy-lift type or alternatively an updated aircraft closer in size to the Mi-8.

Support for FA divisions is organic, each having a transport squadron and a communications flight. The twin-turbo-plus-jet-booster Antonov An-26 Curl is used by FA, as are the An-14 and the An-2 biplane. The Czech Aero 45 is used for liaison.

The strategic bomber arm of the VVS, the DA, fielded a relatively modest force throughout the 1960s and early 1970s. Its intercontinental strike force now comprises about 50 subsonic Tupolev Tu-95 Bear Bs, armed with turbojet-powered

AS-3 Kangaroo cruise missiles. For strike and reconnaissance within the European theatre DA continues to operate some 170 medium-range Tu-22 Blinder As and Bs, the latter being in the majority and armed with the 750km-range AS-4 Kitchen missile. AS-3 and AS-4 are both large weapons, carried singly. Tanker support for the DA has up to now been provided by some 85 Myasishchev Mya-4 Bison As, converted from the bomber role after the type failed to achieve its design range.

DA is however now being re-equipped with the Tupolev Backfire swing-wing bomber, first flown in 1969 and reported as fully operational, after a fairly protracted development, in 1977. Western intelligence sources estimate that some 450 of the definitive version of Backfire, the refined and improved Backfire B, will enter DA service.

Backfire's main purpose is to replace the Tu-22 fleet in the intermediate-range strike role. Compared with the Tu-22 it can fly a much greater proportion of its mission at low level, substantially enhancing DA's ability to survive Nato air defences.

The flexibility of the turbofan-powered, variable-sweep Backfire allows it to present the same sort of long-range threat to the USA as is now posed by the ageing Tu-95s, because it can cover a considerable proportion of the USA on high-altitude, subsonic missions with flight refuelling.

Efforts have been made by the USA to have Backfire included among the weapons controlled by the Salt agreements, but the Soviet Union has insisted that it should be excluded, as a medium-range system. The balance of argument may shift if DA puts into service a tanker version of the Ilyushin Il-86 to support the Backfire fleet, as the US Department of Defense expects.

The development of two completely new bombers was reported by the US journal Aviation Week in early 1979. Although the information is not believed to have been accurate in all respects, the report referred to a new, supersonic swing-wing bomber considerably larger than Backfire and equivalent to the cancelled Rockwell B-1, and a large subsonic replacement for the Tu-95, intended for the strategic missile-launching and maritime roles. However, the DoD posture statement for 1979 made reference only to the first type, which was then believed to be approaching the roll-out stage.

VTA, the transport and airlift command

Above: Two Su-11 interceptors, each armed with an infra-red and a radar missile; this type has now been replaced by the Su-15 and probably other interceptors.

of the VVS, is a respectably sized force in its own right, but is backed up by the considerable resources of Aeroflot, particularly for strategic airlift and the movement of troops within the Soviet Union. Some 40 of the Soviet Union's 50 Antonov An-22 freighters are Aeroflot aircraft, and all of them appear to be equipped to full military standard with mapping radar and paradropping sights. VTA has some 600 Antonov An-12 tactical transports which differ from Aeroflot's 250 aircraft only in possession of a tail gun turret with two 23mm guns. Reports indicate that a new turbofan-powered transport, the An-40, is being designed to replace the An-22.

The An-12s are being replaced by the new Ilyushin Il-76. This four-jet freighter flew in March 1971 and joined service test units in late 1974. It is designed to have twice the payload and twice the range of the An-12 while using the same airfields. VTA aircraft have a tail turret similar to that of the An-12.

Smaller fixed-wing transports include the Antonov An-26 and the older An-2 and

Below: Still in Warsaw Pact service in large numbers, the Su-7B is seen here on a rocket-firing run. It also has powerful cannon, but the bomb load and range are poor.

Below right: One of the Sukhoi Su-17 swing-wing tactical aircraft developed from the much less effective Su-7B.

An-14. Some An-8s – the twin-engine, unpressurised ancestor of the An-12 – may remain in service, probably in the training role. Some 500 Mil Mi-6 heavy-lift helicopters are probably under VTA command. Smaller transport helicopters are likely to be under FA control.

VVS training bases are estimated to have nearly 2000 Czech-designed Aero L-29s in service. These aircraft and the remaining MiG-15UTI advanced trainers are being replaced by another Czech design, the Aero L-39 all-through trainer, which was being phased into VVS service in 1975. All combat units have two-seater aircraft attached for conversion training; typically, one squadron per regiment has four two-seaters out of 12 aircraft. The MiG-23U, MiG-25U and the trainer versions of the third-generation MiG-21s lack the full avionic equipment of the single-seaters; only the Su-70 Moujik of FA types is a fully operational combat aircraft. Another important trainer is the Mil Mi-2 helicopter, a Soviet design built in Poland and which is the main VVS rotary-wing trainer.

The thawing of the Cold War and the recession of the US bomber threat led to a slowdown in the modernisation of the air-defence forces of the PVO. Re-equipment is now gathering momentum, however, to meet the threat posed by US cruise missiles and the increasing numbers of long-range strike aircraft (F-111 and Tornado) in the NATO inventory.

The Sukhoi Su-15 Flagon is probably the most numerous interceptor in the IA-PVO, more than 700 being in service. Some of these are early-production Flagon As, with the same Skip Spin radar and AA-3-2 Advanced Anab missiles as the 1961-vintage Yakovlev Yak-28P Firebar (some 200 in IA-PVO service), but an increasing number are of the later Flagon E and F variants. Flagon E – which may well have a different bureau designation from the earlier aircraft – has a modified wing, more powerful engines and better avionics than Flagon A.

There are only some 280 MiG-25 Foxbat A interceptors in PVO service because the high-flying supersonic bomber (US XB-70 Valkyrie) which the type was designed to intercept failed to materialise. Just as the PVO was re-equipped to meet the threat posed by Mach 3 bombers after it was clear that these were not going to become operational, so it is now being equipped with missile and aircraft systems designed to counter the cancelled Rockwell B-1. Presumably the Soviet leaders feel that a more hawkish US Government might well restart development of the B-1 or order a substitute like the General Dynamics FB-111H.

The PVO's current standard long-range interceptor is the Tupolev Tu-28P Fiddler. Some 200 of these very large two-seat fighters, carrying four AA-5 Ash AAMs, remain in service to defend vast areas of the Soviet periphery where the SAM screen is thin.

Older, less effective types are deployed in the interior air-defence zones. They include some 600 Sukhoi Su-9s and Su-11s, similar to the FA's Su-7 but of tailed-delta layout. They carry the same armament

as Firebar and Flagon A. Other units operate a total of around 250 older interceptors with limited all-weather capability – probably MiG-17s. The Soviet Union's new interception system is reported to be based on the Super MiG-25, carrying up to six AA-X-9 missiles. The new aircraft has a crew of two, and will presumably operate with considerably more autonomy than the present MiG-25.

It is debatable whether the new aircraft will really be a MiG-25. The existing MiG-25 is highly limited in its range and loiter performance, and Mach 3 maximum speed is unlikely to be of great importance to an aircraft designed to intercept subsonic, low-level targets. A possible explanation for references to a Super MiG-25 is that the new aircraft shares the basic layout of the MiG-25 and F-15 Eagle. Until the new system enters service, the PVO has acquired some 200 MiG-23S Flogger Bs to provide a limited defence against low-flying targets.

One very important type in the IA-PVO inventory is the Tu-126 Moss airborne warning and control system, derived from the Tu-114 airliner with a mushroom radar on top. A small force of these aircraft – reported at no more than 20 – provides early warning and control of interceptors outside the range of ground controllers. Unlike the USAF Boeing E-3A, the Tu-126 is an overwater-only system, but it was in service eight years ahead of the E-3 and is expected to be replaced in the 1980s by a more advanced system which would presumably match the E-3's operational capability.

Training in the IA-PVO is separate from VVS but follows a similar pattern, with L-29s being replaced by L-39s for basic and intermediate training, pilots converting via the Su-11U Maiden or Su-15U Flagon C. IA-PVO units have An-2s and An-26s for support purposes.

The main strategic SAM in ZA-PVO is

Above: By far the most numerous heavy airlift transport in Soviet VTA service is the Antonov An-12, a line of which are seen here with ASU-85 assault guns just flown in. Coming in to land is a larger An-22.

the SA-5 Gammon, a high-altitude, long-range weapon which started to replace the SA-2 Guideline in the mid-1960s. Some 1100 SA-5s are reported to be in service,

Below: First seen in 1976, the 40,000-ton Kiev is the world's most heavily armed, and in many ways most versatile, warship. It can operate against almost every kind of sea or strategic targets.

SA-2, each battery containing six launchers, a loading vehicle and a Fan Song van-mounted fire-control radar. The Army also has SA-3 Goa SAMs, together with SA-7 Grail shoulder-launched weapons, the SA-9 Gaskin development using uprated missiles fired from four-round launchers on BRDM wheeled vehicles, and the new SA-8 Gecko. The last-named is roughly equivalent to the West's Euromissile Roland and carries a four-round launcher, surveillance radar, tracking radar, low-light television camera and command links on one six-wheeled amphibious vehicle.

The Naval Air Arm

Naval SAMs are closely related to land-based weapons. Widely used on earlier ship classes is the SA-N-1 system, which uses the same missile as the land-based SA-3 Goa. Standard area-defence system on the more modern large ships is the SA-N-3, which is believed to use the rocket/ramjet missile of the SA-6 system, on automatic twin launchers fed from below decks. The close-range SA-N-4, discreetly concealed in a flush 'silo' installation, is thought to use the SA-8 missile. The shipboard Guideline, the SA-N-2, has on the other hand not been successful, and only one installation has been commissioned.

The naval air arm, AV-MF, is perhaps the most diverse of all the Soviet air forces and one of the fastest expanding. As the Soviet surface fleet has grown and expanded its area of influence the AV-MF has followed it with a fleet of missile carriers. Elint, ECM and maritime reconnaissance aircraft, and now VTOL aircraft have gone to sea.

The bulk of the AV-MF anti-shipping strike force is still composed of 290 elderly Tupolev Tu-16 Badgers delivered in the early 1960s as the DA phased the type out of its inventory. Most of the Tu-16s are Badger Gs, carrying a pair of 320km-range AS-5 Kelt ASMs, but there are still some Badger Cs in service with a single 200km-range AS-2 Kipper. About 50 Tu-22 Blinder Cs serve in the reconnaissance/Elint role with a free-fall strike capability.

These aircraft and missiles are likely to be replaced in the anti-surface-shipping role by the vastly more capable Tupolev Backfire B now completing work-up with AV-MF units, and the 220km-range AS-6 Kingfish missile. Some Tu-16s have been seen with a single AS-6 Kingfish missile under the port wing. Backfire versions are likely to be configured for Elint, ECM and

and a general upgrading of the SAM system is in prospect.

Now under development to meet the low-level threat is the hypersonic (Mach 6) SA-10 system. Associated with continuous-wave tracking radars, the SA-10 missile is believed to have a maximum range of around 50km and operates at altitudes between 300m and 5000m. Deployment of such a weapon will have to take place on a massive scale, because of its limited range, if complete defensive coverage is to be achieved. The system is not intended to defeat the USAF's new cruise missiles, as seemed likely at one stage, but is designed to deal with low-level aircraft.

Remotely piloted vehicles used for SAM training and development include conversions of obsolete combat types such as the Yakovlev Yak-25RD Mandrake strategic reconnaissance aircraft and the MiG-19 interceptor. Purpose-built drones include the Lavochkin La-17 and a high-altitude, supersonic-cruise RPV with a single Tumansky R-266 engine.

ZA-PVO is also responsible for the anti-ballistic-missile (ABM) system, based on the Galosh and Improved Galosh interceptor missiles. There are four 16-missile sites around Moscow, with Hen House radars on the Soviet periphery for target acquisition and tracking. A new ABM, designated SH-4, and a high-acceleration missile similar to the US Martin Orlando Sprint are reported under development.

The development of mobile surface-to-air missiles has been very important to Soviet tactics, effectively freeing Frontal Aviation for the strike and longer-range air-superiority roles. Each Soviet Army Group – of which there are five permanently deployed in East Germany, for example, with a sixth acting as back-up in time of war – is responsible for an area measuring 50km along the battle front and 100km deep. Every Group has five batteries of SA-6 Gainful medium-range SAMs powered by integral rocket/ramjets, nine batteries of the longer-range SA-4 Ganef and three of the SA-2 Guideline.

An SA-6 battery comprises a Straight Flush radar vehicle, using the same PT-76 light-tank chassis as the missile vehicles; three of the latter, carrying three rounds each; and a reloading vehicle. Three of the batteries travel with the ground forces about 5km back from the forward edge of the battle area and the remaining two are deployed some 10km further back in the gaps between the forward systems. An SA-4 battery consists of an H-band Pat Hand acquisition and fire-control radar, three twin-launcher tracked vehicles and a reloader. Surveillance and initial target acquisition for both the SA-4 and SA-6 are the responsibility of E-band Long Track radars. The leading three SA-4 batteries travel some 10km behind the Army's forward forces with the other six moving in a belt 25km back from the front.

Defence of rear areas is provided by the

Right: A close-up of one of the Yak-36 'Forger-A' jet V/STOL single-seaters carried by Kiev on her first ocean cruise. This is judged to be an interim aircraft.

Right: A close-up of one of the Yak-36 'Forger-A' jet V/STOL single-seaters carried by Kiev on her first ocean cruise. This is judged to be an interim aircraft.

141

reconnaissance, and an eventual AV-MF force of 100 mooted in the West may be on the low side. The 75 Badgers serving in the tanker role with AV-MF may be replaced by the Ilyushin Il-76 tanker.

AV-MF's costal-defence strike units still retain about 20 Ilyushin Il-28T Beagles armed with torpedoes. According to some sources the Il-28Ts are being replaced by Yak-28 Brewers retired from FA, but it is now reported that Sukhoi Su-17 Fitter C variable-sweep strike fighters are being delivered.

For wide-ranging maritime reconnaissance the AV-MF relies on the Tu-95 Bear now that the M-4 Bison C is being retired. There are rather more than 50 Tu-95s in AV-MF service, although some have been lost, and their long range and endurance are invaluable. Bear Cs and Fs are in service in the maritime reconnaissance role and Bear Es carry cameras for photo-reconnaissance and mapping. Some Bear Ds, with a ventral radar for mid-course guidance of SS-N-3 Shaddock surface-to-surface missiles, may have been converted to Bear F configuration. There are also about 50 maritime reconnaissance/ECM/Elint Badger Ds and Fs in service, and these have recently been joined by Antonov An-12 Cub C ECM aircraft converted from VTA or Aeroflot freighters.

The Elint/ECM force has recently been augmented by the Il-18 Coot A, equipped with a wide assortment of radomes, and the An-12 Cub B.

AV-MF has always flown flying boats for anti-submarine warfare (ASW) and currently has some 75 Beriev Be-12 Mail amphibians armed with rockets and free-fall weapons. Since 1970, however, it has been phasing in its first specialised land-based ASW system, the Ilyushin Il-38 May, and more than 55 are in service. Like the very similar Lockheed Orion the Il-38 was adapted from an airliner, the Il-18.

Above: There are numerous sub-types of Ka-25 'Hormone' helicopter, but all are thought to be used from Soviet warships. This one has a ventral weapons bay, and is basically an ASW 'Hormone-A'.

The AV-MF shore-based ASW helicopter units are also being upgraded with the introduction of the Mil Mi-14 Haze, an amphibious ASW helicopter based on the Mi-8 Hip (a few of which equip specialised AV-MF minesweeping units). First flown in 1973, the Mi-14 started to replace Mi-4 Hounds in AV-MF service in early 1977.

Shipboard helicopter units are mainly equipped with the Kamov Ka-25 Hormone, a less than elegant but very compact twin-turbine co-axial ASW helicopter. Some 160 have entered service since the mid-1960s, including some Hormone Bs, which provide over-the-horizon missile guidance associated with the SS-N-3 Shaddock, and a few transport versions. Ka-25s are carried on the helicopter carriers *Moskva* and *Leningrad* as well as by the aircraft carrier *Kiev*. *Kiev* carries both Hormone sub-types, indicating that the new SS-N-12 needs over-the-horizon guidance, like the older SS-N-3.

Some sources expect the Ka-25 to be replaced by a new helicopter on the larger ships, possibly related to a new gunship for FA, and it is possible that the *Kiev* class will deploy Mi-24s in support of

amphibious operations or that the Mi-14 will go to sea.

Perhaps the biggest single advance in AV-MF equipment, however, is the deployment of the Forger VTOL light attack aircraft, believed to be a product of the Yakovlev design team. As few as 30 Forgers may be in service and the aircraft on *Kiev* form a trials unit of perhaps a dozen single-seat Forger As and a pair of Forger B two-seat conversion trainers. Forger can carry AA-2 Atoll or AA-8 Aphid missiles, but has not yet been seen with advanced ASMs or associated avionics. This and the fact that Forger is a purely VTOL type, which cannot make rolling take-offs and take advantage of the angled decks of the new carriers, support the contention that Forger is mainly an experimental aircraft and that a truly V/STOL follow-on will appear.

Details of the deployment of the Soviet forces are even more scarce than information about the total inventory. The strict 'need-to-know' security policy operated by the Soviet Union, under which virtually all information is classified to some degree, combines with the natural fluidity of military deployments to produce a near-impenetrable security blanket.

Soviet sea-launched strategic missiles are of course highly mobile, most of the Soviet missile submarines ranging widely from their bases with the Northern Fleet in the Murmansk region and the Pacific

Below: One of the best photographs of the AV-MF Il-38 'May' ocean patrol aircraft. These Orion-like platforms have been encountered by Western aircraft from the Pacific to the Indian Ocean, Atlantic and Arctic.

Above: Originally thought to be a version of Mi-24, the 'Hind-D' large multi-role tactical helicopter is now known to be the Mi-28. It appears to have night and all-weather equipment.

Fleet at Vladivostok. Land-based ICBMs of the RVSN are siloed in a belt 500km long and 400km wide, stretching from a point east of Moscow to the Lake Baikal region. Mobile and fixed IRBMs are deployed in European and Asian regions, threatening China and Europe.

Rather more may be said about the deployment of the VVS forces. Frontal Aviation is divided into at least 15 Air Armies, all except four being attached to the Military Districts within the Soviet Union. The four exceptions are attached to the Groups of Soviet Forces in the Warsaw Pact countries: the mighty 16th in East Germany; the 37th, headquartered at Legnice in Poland; one Air Army centred on Milovice in Czechoslovakia; and a fourth headquartered at Tokol, near Budapest in Hungary.

The Air Army's largest sub-unit is the division, assigned to a specific role. A division will usually operate a number of different types in three regiments. Each regiment operates a single combat type from a single base, although units of a division will normally be grouped closely together within the Air Army's district. The 16th is fairly typical in structure, although it fields more than 1000 combat aircraft compared with 250–300 in other Air Armies.

Transport and reconnaissance are the responsibility of individual regiments attached directly to the Army, rather than being organised into divisions. This system has flexibility in some cases: the MiG-25R regiment deployed with the 37th Air Army in Poland has detached squadrons to East Germany under 16th Air Army command.

Regiments comprise three squadrons of 12 aircraft each, and one of these usually operates about four two-seater aircraft for conversion training.

The long-range bomber force of the DA is not numerically larger than an FA Army, and the largest sub-unit of DA is the division. There are three DA divisions, one in the Far East district, based on the coast of the Bering Sea; one in the Murmansk area; and one in European Russia. Each comprises three regiments of three ten-aircraft squadrons. The bulk of the Tu-22 medium-range bomber force is European-based, although some aircraft face China in the Far East, and the new Backfire started to join the European units first.

Below: One of the original 'Hind' models with Mi-24 designation; these assault helicopters have a wide flight deck, but the same weapon carrying wing as the Mi-28.

The long-range Tu-95s are concentrated in the Northern and Siberian Divisions, the latter now receiving Backfires, and all units include M-4 Bison tankers.

VTA is so closely associated with Aeroflot, presumably using the same training and support facilities, that it is virtually impossible to separate VTA deployment from Aeroflot's vast network.

The manned interceptors of the IA-PVO are deployed in two types of organisation. First there are two PVO Districts, Baku and Moscow, which between them protect the central industrial regions of the Soviet Union. About half the IA-PVO interceptors, predominantly the more modern types, are assigned to the PVO districts, the division being the largest unit. The other half of the PVO divisions are attached to the other Soviet Union Military Districts, and some are reported to be under the command of the Groups of Soviet Forces in Germany, Czechoslovakia and Hungary.

The PVO divisions attached to the Military Districts mainly operate the Su-15 and older, less capable types of aircraft. Exceptions are the Tupolev Tu-28P long-range fighters based in Northern and Eastern Siberia to counter over-the-pole strikes by USAF B-52s, and the MiG-25s and Su-15s now being deployed in the Far East against a growing Chinese bomber threat.

ZA-PVO missiles are deployed in belts around the main industrial areas of the Soviet Union, as well as in target-defence installations protecting major cities such

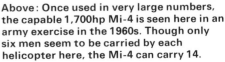

Below: First seen in 1957, the Mi-6 'Hook' is still much larger than any helicopter in use with any NATO country. Despite its ungainly appearance it has the high speed of up to 186mph (300km/h).

as Moscow and Sverdlovsk. So far only Moscow is defended by an ABM system.

AV-MF deployment follows the organisation of the surface fleet, in four main groups: the Baltic fleet headquartered at Kaliningrad; the Northern fleet based around Murmansk; the Black Sea fleet at Sevastopol and the Pacific fleet at Vladivostok.

The Baltic AV-MF units operate mainly the shorter-range types, including probably the greater part of the Tu-22s, Be-12 amphibians and Su-17 strike fighters. The Mi-4P anti-submarine helicopters are being replaced by the new Mi-14 Haze.

Atlantic surveillance and strike is the responsibility of the Northern AV-MF units, ranging all the way down to the Azores after flying around the North Cape of Norway. Some of the AV-MF Backfires,

Above: Once used in very large numbers, the capable 1,700hp Mi-4 is seen here in an army exercise in the 1960s. Though only six men seem to be carried by each helicopter here, the Mi-4 can carry 14.

Tu-95s and Tu-16 missile carriers are based in the Northern region, together with An-12 Cub C Elint aircraft and Ilyushin Il-38s. Tu-16 tankers provide support.

Support for the Soviet fleet in the Mediterranean and the Indian Ocean is provided by AV-MF aircraft attached to the Black Sea fleet: it is reported that Tu-95s have overflown Iran on their way down to the Indian Ocean, and Mozambique is reported to be allowing the construction of an AV-MF base on the island of Bazaroto. Il-38s and Cub Cs have been reported in the Indian Ocean as well as in the Mediterranean, and have operated in Egyptian markings from Egyptian bases. Be-12s mainly confine their activities to the Black Sea region itself, but Tu-16s range over the entire Mediterranean in the reconnaissance role. Lastly, the Pacific fleet uses the Tu-95 predominantly, because of its great range and endurance.

All the Soviet fleets presumably possess land bases to support the helicopters (and now fixed-wing VTOL aircraft) deployed aboard surface ships (see table).

The Soviet air forces have had no direct combat experience since the 1941–45 war, although the Soviet Union was closely involved in the Vietnam war. Where Soviet material has seen action, as in Vietnam and the Middle East, the capture of Soviet pilots or rocket crews would have been highly embarrassing and they have not directly engaged in hostilities. However, MiG-25Rs based at Cairo West made some overflights of Israeli-held territory in late 1971.

Training in the Soviet aerospace forces is extremely basic and very formal, giving little encouragement to the development of individual initiative. The VVS and PVO operate separate basic training systems; only the PVO has an advanced school, at Armavir. The Soviet air forces rely on regiment training; pilots stay with their regiments much longer than is usual in Western forces, and each regiment has a

group of two-seater aircraft with training as their primary role. Even the Tu-22 exists in a conversion-trainer version, assigned to DA regiments.

Engineering support for the Soviet air forces is presumably the responsibility of the Ministry of Aircraft Production plants which manufacture all combat aircraft. Soviet aircraft are designed on the assumption that a defective item can always be removed, replaced from stock and returned to the factory for overhaul, this philosophy extending to engines as well as to avionics. The main function of the organic light transport groups within Frontal Aviation is probably spares supply and the ferrying of components for overhaul.

SOVIET UNION   AV-MF shipboard power

Class (number + in build)	Aircraft	No	SAMs	Notes
Kiev (2+1)	Forger	12	SA-N-3	SS-N-12 cruise missiles may have strategic role.
	Ka-25	20	SA-N-4	New V/STOL aircraft, Mi-14 Haze and new light helicopter may be embarked on later ships
Moskva (2)	Ka-25	18	SA-N-3	Moskva has ferried Mi-8 minesweepers. Used for Forger sea trials
Sverdlov (1)			SA-N-2	One conversion only
Sverdlov (2)	Ka-25	2	SA-N-4	Two command ships
Kara (4+1)	Ka-25	2	SA-N-3	Most modern Soviet cruiser class
			SA-N-4	
Kresta I (4)	Ka-25	2	SA-N-1	Hormone B for missile guidance
Kresta II (9+)	Ka-25	2	SA-N-3	Hormone A
Krivak (11)			SA-N-4	
Kynda (4)			SA-N-1	
Kashin (19)	Ka-25	2	SA-N-4?	In process of modernisation
Kanin (7)	Ka-25	2	SA-N-1	
Kotlin SAM (8)			SA-N-1	
Grisha (21+)			SA-N-4	Corvette
Nanuchka (16+)			SA-N-4	Corvette
Ugra (9)	Ka-25K	2		Submarine support
Don Helo (2)	Ka-25K	2		

WARSAW PACT/NATO TACTICAL AIR POWER BALANCE

STRIKE AIRCRAFT

	Introduced	Number	Tactical radius (nm) hi-lo-hi	All-weather capability	Low-level performance	Avionics
Mikoyan MiG-27	1974	600, rising	500	no	good	Doppler, laser, EO
Sukhoi Su-19	1977	150, rising	1,000	yes	good	Multi-mode radar? Doppler, laser, EO
Tupolev Tu-26	1976	130, rising	1,400	yes	good	Bombing radar, Doppler, inertial?
BAe Buccaneer	1968	80 in service	700	limited	medium	Doppler, attack radar
BAe/AMD Jaguar	1972	150 in service	600	no	good	Laser, inertial
General Dynamics F-111	1967	500 in service	1,500	yes	good	Multi-mode radar, inertial
Grumman A-6 Intruder	1964	350 in service	600–750	yes	medium	Multi-mode radar, inertial, infra-red, laser
Panavia Tornado IDS	1981	650 on order	700–900	yes	good	Multi-mode radar, Doppler/inertial, laser
Vought A-7 Corsair	1968	600 in service	600	limited	medium	Laser, attack radar

Includes strike aircraft with radius of action over 500 nm. Soviet forces include Su-7 strike aircraft; NATO still uses considerable numbers of CF-104/F-104G of comparable or higher effectiveness, and dual-role F-4s.

CLOSE-SUPPORT AIRCRAFT

	Introduced	Number	Armour	Runway requirement	Avionics	Gun armament	Stores load
Sukhoi Su-17/20	1971	600, rising	no	medium	Laser? EO	2×20mm	6,500lb
New close-support type	1980?		yes	short	Laser, EO	1 multi-barrel	
BAe Harrier	1968	100 in service	no	V/Stol	Laser, inertial	2×30mm	4,000lb
BAe Hawk	1974	175+total	no	medium		1×30mm	5,000lb
AMD-BA/Dornier Alpha Jet	1980	230 on order	no	medium		1×30mm	5,000lb
Fairchild A-10	1976	730 on order	yes	short	Laser	1×30mm multi-barrel	10,000lb

AIR SUPERIORITY AND DEFENCE

	Introduced	Number	Dogfight capability	MR AAM	SR AAM	Gun
Mikoyan MiG-21 late model	1968	1,500, rising	good	none	4	yes
Mikoyan MiG-23S Flogger B	1973	800, rising	medium	2	2	yes
New air combat fighter	1982?		good	2	2	yes
Dassault Mirage F.1	1972	40	medium	2	2	yes
General Dynamics F-16	1978	1,700 on order	good	none	4	yes
Lockheed F-104S Starfighter	1970	240 in service	low	2	2	no
MDC F-4M/K Phantom	1968	170 in service	low	4	2	no
MDC F-4F Phantom (slatted)	1974	250 in service	medium	4	2	yes
MDC F-15	1976	730 on order	good	4	4	yes
MDC/Northrop F-18 Hornet	1982	800 on order	good	2	2	yes

This table excludes aircraft committed to strategic or worldwide roles, such as USN F-14s, RAF Tornado F.2s and aircraft reserved for the defence of the continental Soviet Union.

Middle East

The Middle East is the world's largest source of oil and as such has become increasingly important to the rest of an energy-hungry world since the end of World War II. This need to protect oil supplies has been one dominant factor in the area, with the traditionally hostile relations between Israel and her Arab neighbours as the other.

The state of Israel was established in 1948 and immediately became involved in war with the Arabs. In 1956 Israel, aided by France and Britain, attacked Egypt in an attempt to secure continuing use of the Suez Canal and to expand her borders. On that occasion the United States intervened, forcing an end to the action, but in June 1967 the Israelis launched a pre-emptive attack on Egypt, Syria and Jordan in a war which gave them control of Sinai, the Gaza strip, the Golan Heights and the west bank of the Jordan in only six days.

The Arabs regained much of their military pride in October 1973 and inflicted a series of defeats on the Israelis until the US launched a massive airlift of supplies. Since then the US has catalysed a willingness by President Sadat of Egypt and Israel's Prime Minister Begin to renounce warfare, leading to a peace treaty being signed in the presence of President Jimmy Carter in the spring of 1979.

Other Arab countries, particularly Syria, have refused to be party to such dealings, however, and the Arab Organisation for Industrialisation – backed by money from Saudi Arabia, Qatar and the United Arab Emirates as well as from Egypt itself – was disbanded as a protest by Egypt's partners against the treaty. At the time of writing it was too early to judge the long-term effect of this move.

Egypt has moved gradually into the US sphere of influence since the 1973 war, but Libya, Iraq and to a certain extent Syria remain firmly in the Russian camp. A period of consolidation in the Arabian Gulf seemed likely after the war in south-western Oman came to an end, but this hope was destroyed in late 1978 and early 1979 when the Shah of Iran was deposed and the Ayatollah Khomeini declared the country an Islamic republic. Iran's massive arms-buying spree is now well and truly over, although at the time of writing the Soviet Union's influence over Iran was not as great as had been feared by the West and other Middle Eastern states such as Saudi Arabia.

With the Israeli-Egyptian peace treaty taking the heat off that area, the West is looking increasingly to Saudi Arabia as a stabilising influence in the rest of the Middle East.

LIBYA

Area: 810,000 sq. miles (2·09 million sq. km.).
Population: 2·9 million.
Total armed forces: 37,000.
Estimated GNP: $18·5 billion (£9·25 billion).
Defence expenditure: $448 million (£224 million).

The Libyan Republic Air Force was formed as the Royal Libyan Air Force in 1959, using two ex-Egyptian Gomhouria primary trainers and Austers donated by the Royal Air Force. From 1963 the United States supplied T-33s and C-47s, but in 1967 Libya requested both Britain and the

146

United States to withdraw from the bases which they had been using as staging posts and for training. The US continued to supply aircraft in the form of Northrop F-5s, however, and an air-defence system comprising missiles, radars and command-and-control facilities was ordered from Britain.

In 1969 Libya was declared a republic, following a military coup, and early the next year the air-defence contract was cancelled. In its place the restyled LARAF (Libyan Arab Republic Air Force) ordered Mirages, Magisters, helicopters and missiles from France. Training was carried out by French, Egyptian and Pakistani personnel, and crews from the two last-named countries flew most of the combat aircraft. Relations between Libya and Egypt deteriorated after the war with Israel in October 1973, and in the following year Russian equipment and personnel were supplied. Further French and Italian arms were procured in the mid-1970s but the US embargoed sales and has refused to deliver eight C-130H Hercules on order. Relations between Libya and Egypt reached a new low in mid-1977 when open warfare broke out for several days.

The LARAF has a personnel strength of some 5000 and the army also has a small air arm.

Inventory

The LARAF's new combat aircraft are being supplied by France and the Soviet Union in roughly equal numbers, with 38 Mirage F.1s being delivered from 1978 to supplement the large force of Mirage 5s and the MiG-23 Floggers supplied by the USSR. A further 50 Mirage F.1s are on option, and additional supplies of Soviet equipment are expected. The manned interceptors are backed up by a missile-based air-defence network comprising SA-2s, SA-3s, SA-6s and Crotales. Many of the Mirage 5s are non-operational because of a lack of crews, and some were seconded to Uganda to replace MiGs destroyed in the Israeli raid on Entebbe to rescue airline passengers being held captive after their aircraft was hijacked.

The active combat element is thought to comprise about 50 MiG-23 Floggers (including trainers) divided into one squadron each for interception and ground attack, two fighter-bomber squadrons and a training unit with Mirage 5Ds and 5DDs, two further squadrons with Mirage 5DE interceptors and flights of reconnaissance 5DRs, and two bomber units operating Tu-16s and Tu-22s. From 1978 the force was augmented by the 38 Mirage F.1s. Most aircraft continue to be flown by Pakistani and Russian crews, and Libyan control over the Tu-22s is unlikely to be more than nominal. The Soviet Air Force has free

Above left: Libya was one of the first export customers for the MiG-23 Flogger, which is used for both interception and ground attack. The Flogger E variant operated by the LARAF is different in several aspects from Soviet Air Force versions.

Left: A Libyan Tu-22 Blinder bomber is escorted by a US Navy F-4 Phantom over the Mediterranean.

Above: The LARAF uses its Super Frelons for search-and-rescue work as well as for anti-submarine operations. A small number of the helicopters have been based in Malta and may be handed over to Maltese forces for future use.

access to Libyan bases as staging posts and since early 1977 has operated a squadron of five reconnaissance MiG-25 Foxbats from Libyan soil, one of which crashed in November 1978; these aircraft were formerly based in Syria.

Other sources of new equipment include Jugoslavia, which re-opened the Galeb production line to fulfil an order for at least 50, and Italy. Italian-built CH-47Cs have been supplied, and a Libyan assembly line for SF-260s has been established. The US embargoed the supply of T64 turboprops for Aeritalia G.222 transports, so the 20 G.222Ls ordered for delivery in 1980–81 will be powered by Rolls-Royce Tynes.

Some 38 Libyan Mirages were flown by

Pakistani pilots from the Egyptian bases at Almaza, El Faiyum and Beni Suef during the 1973 October War, carrying out about 400 missions against Israel. Libyan and Egyptian forces themselves clashed in July 1977, and Egypt admitted the loss of two Su-20s; Libya claimed to have shot down more than 20 Egyptian aircraft in the four-day conflict.

LARAF training centres on the Air Force Academy, opened in 1975 and staffed by Pakistani and Jugoslav personnel. Initial instruction is carried out abroad, students then spending two years at the academy. The Galeb is in the process of replacing refurbished Magisters as the major training aircraft. Twin conversion is carried out on C-47s, and operational conversion is the responsibility of the combat squadrons, using Mirages and Floggers.

The army has its own fleet of helicopters and some fixed-wing aircraft, but most support tasks are carried out by LARAF types.

LIBYA

Type	Role	Base	No	Notes
Air Force				
Mirage F.1ED	Int		16	
Mirage F.1AD	Att		16	
MiG-23	Int/Att	El Adem	50	Includes two-seat trainers
Mirage 5DE	Int	Okba Ben Nafi?	32	
Mirage 5D	Att	Okba Ben Nafi?	58	
Mirage 5DR	Recce	El Adem	10	
Tu-16	Bomb	Okba Ben Nafi	10	
Tu-22	Bomb	Okba Ben Nafi	12	
C-130H	Trans		8	
C-47	Trans		9	
CH-47C	Trans		20	
Mi-8	Trans		12	
Boeing 707	VIP		1	
JetStar	VIP		1	
AS-61A-4	VIP		1	
Mirage F.1BD	Train		6	
MiG-23U	Train		5	
Mirage 5DD	Train		10	
Falcon 20ST	Train		2	
CM.170-2	Train	Zawia	12	
T-33A	Train	Zawia	3	
G2-A-E Galeb	Train	Zawia	50	Replacing Magister
SF.260	Train		230	Approx 110 from SIAI-Marchetti, others licence-built
SA.321M	ASW/SAR		9	
SA.316B	Comms/util		10	
Alouette II	Comms/util		3	
Bell 47G	Comms/util		3	
Agusta-Bell 212	Comms/util		2	
SA-2	SAM			
SA-3	SAM	Okba Ben Nafi,		
SA-6	SAM	Tripoli, others		
Crotale	SAM			
Army				
O-1E	Liaison/obs		10	
Alouette III	Liaison/obs		4	
AB.47G	Liaison/obs		6	
AB.206A	Liaison/obs		5	
SA-7	SAM			

EGYPT

Area: 385,000 sq. miles (1 million sq. km.).
Population: 38 million.
Total armed forces: 395,000.
Estimated GNP: $13·3 billion (£6·65 billion).
Defence expenditure: $2·81 billion (£1·405 billion).

The Egyptian Air Force, now known as the Air Force of the Arab Republic of Egypt, was formed as the Egyptian Army Air Force in 1932 and is thus the second-oldest Arab air arm. The EAF fared badly in the Palestine War of 1948, the invasion of Suez by British, French and Israeli forces in 1956 and the pre-emptive strike with which Israel began the 1967 Six-Day War, but in the October War of 1973 the tide was turned. Since 1973 Egypt's relations with the Soviet Union, which had supplied jet combat types before Suez, have rapidly deteriorated and in March 1976 the friendship treaty between the two countries was renounced by Egypt. China has since provided a limited amount of assistance, but President Sadat has increasingly turned towards the West as the main source of arms.

All military aircraft are operated by the air force, although some are assigned to the Air Defence Command; this acts as virtually an independent fourth Service operating gun and missile defences as well as controlling interceptors. The air force has 30,000 personnel and a further 75,000 members of the army serve with the Air Defence Command.

Inventory

The Egyptian Air Force is divided into regiments along Russian lines, each regiment operating one type and being assigned to a single specific role. The interceptor force comprises three regiments with nine squadrons operating MiG-21MFs, each squadron having about a dozen aircraft. Despite the cool relations between Egypt and the USSR, and statements by President Sadat that Russian-supplied equipment would soon be no better than scrap iron, it is evident that Soviet combat aircraft are still very much operational. Up to 150 MiG-21s have been overhauled and re-engined in Russia, and these are expected to remain operational well into the 1980s.

Although the question of re-equipment has yet to be settled, the Air Defence Command already operates a formidable interlaced network of missiles and guns in addition to the MiG-21 interceptors. The command was born in 1968, during the War of Attrition with Israel which followed the Six Day War. The command has about 100 battalions, each with 200–250 men and operating a single type of missile or guns of similar calibres. Battalions not equipped with weapons are allocated specific tasks such as radar surveillance or logistics.

The command's basic operational unit is the brigade, each comprising four to eight battalions and assigned to the defence of a specific area. For the protection of Cairo, however, several brigades are grouped together to form a division tasked with air defence. The air-defence network is run from joint command posts at brigade level, with an air force colonel and air-defence brigade commander sitting side-by-side in a hardened underground bunker and controlling eight missile battalions together with the manned interceptors.

The joint command post receives inputs from long-range surveillance radars such as the Russian P-35 Barlock (with its associated Thin Skin height-finder) and Spoon Rest. Other data are supplied by visual observers, acquisition radars at missile sites, air-traffic control radars, IFF (identification, friend or foe) posts and other sources. The brigade commander assigns one or more missile battalions to intercept incoming aircraft, directs the operations of manned interceptors and assigns appropriate countermeasures to deal with a specific threat. The overlying divisional organisation can assume general command if a joint centre at brigade level is inoperative, and individual missile-battalion commanders can engage targets on their own initiative if necessary.

The missile force comprises some 80 single launchers for SA-2 surface-to-air missiles, 60 twin launchers for the shorter-range SA-3 and about the same number of vehicles carrying triple launchers for the ram-rocket SA-6. Other battalions operate several hundred radar-directed and manually laid guns with calibres of 20mm, 23mm, 37mm, 57mm, 85mm and 100mm. The Air Defence Command additionally supplies the army with ZSU-23-4 Shilka tracked vehicles carrying quadruple 23mm cannon for operation alongside the SA-6s. The SA-2s and SA-3s can be transported from site to site but are normally emplaced semi-permanently.

Above left: The Egyptian Air Force is in the middle of transition from operating predominantly Russian-supplied aircraft to relying more on the West, but it still deploys modern types supplied by the USSR. These Su-20 attack aircraft operate alongside the earlier Su-7 on which they are based and are complemented by MiG-23 Floggers.

Left: Westland has supplied Egypt's Navy with six Sea King ASW helicopters as well as the Army with Commandos.

Missile bases operated by SA-2 or SA-3 battalions are located 5km to 11km apart and typically comprise six missile ramps separated by 50–100m, a Spoon Rest surveillance radar, a Flat Face surveillance radar (at SA-3 sites), a Squint Eye acquisition radar, the missile fire-control radar (Fan Song or Low Blow for SA-2 and SA-3 respectively) and a command post. A missile battalion has four companies responsible for operating the radars, firing

the missiles, general support including reloading, and site defence with machine guns.

The battalion command post is contained in a van normally located in a hardened concrete shelter some 35ft underground. The post contains the battalion commander, who is also the missile-guidance officer, and three other officers. Each of the subordinate officers handles the interception in one dimension – azimuth,

Above: Egyptian Air Force Tu-16 Badger bombers launched a substantial number of AS-5 Kelt missiles (seen here beneath each wing) against Israeli targets in the 1973 October war.

elevation or range – and the commander's display shows all relevant information for an engagement. A battalion can launch three missiles at different targets simultaneously, with the launchers being trained from the command post itself or by the individual crews. Reloading a single SA-2 ramp or both rails of a twin SA-3 launcher takes less than a minute.

The SA-6 vehicles, together with their vehicle-mounted Straight Flush radars, are not deployed at fixed sites and move with the field army, as do the Shilkas.

The Egyptian Air Force has a medium-bomber regiment operating about 25 Tu-16 Badger Gs, of which perhaps a dozen are equipped to carry two AS-5 Kelt air-to-surface missiles each. Some 20 Il-28 light bombers may still be operational. The MiG-21 is a mainstay of the fighter-bomber force as well as the interceptor element, equipping two regiments in the former role. The aircraft are being updated by the installation of Ferranti inertial-navigation platforms and Smiths Industries head-up displays, and other British companies are assisting in keeping the fleet operational. A further regiment operates ageing MiG-17s in the fighter-bomber role, with low-level interception as a secondary task, and some MiG-19s remain as a second-line combat element.

Other types assigned to the fighter-bomber task, with emphasis on counter-air operations, are about 24 MiG-23 Flogger Bs and the single-seaters from the force of 52 Mirage IIIEEs and two-seat IIID trainers; the first 38 of these were ordered by Saudi Arabia on Egypt's behalf, and Egypt herself signed for a further 14 in 1977.

EGYPT

Type	Role	No	Notes
Air Defence Command			
MiG-21MF	Int	108	Nine squadrons
SA-2	SAM	80 launchers	
SA-3	SAM	65 launchers	
SA-6	SAM	60 launchers	
Air Force/Army/Navy			
Tu-16	Bomb	25	Some carry AS-5 Kelt ASM
IL-28	Bomb	20	May not be operational
MiG-23 Flogger B	Int	24	
Mirage IIIEE	Int	52	Total includes some two-seaters
MiG-19	Int	60?	Second-line only
MiG-21PFM	FGA	150	
MiG-21F	FGA	100	
MiG-17	FGA	125	
MiG-23 Flogger F	Att	18	
Su-7	Att	120	Perhaps only 80 operational
Su-20	Att	48	
L-29	Train/light att	100+	
EC-130H	ECM	2	
Sea King Mk 47	ASW	6	Operated by Navy
SA.342 Gazelle	Anti-tank/obs/util	54	Most operated by Army; some carry HOT ATGW. Navy has one squadron
Mi-8	Assault/log	70	Most operated by Army
Commando Mk 1	Assault/log	5	Operated by Army
Commando Mk 2	Assault/log	23	Operated by Army
C-130H	Trans	3	One other lost
An-12	Trans	16	Likely to be replaced by C-130
IL-14	Trans	40	
Mi-4	Trans	20	May be operated by Army
Mi-6	Trans	12	In storage?
Fournier RF-4	ELINT/obs	6	
Wilga	Liaison		
Boeing 707	VIP	1	
Boeing 737	VIP	1	
Falcon 20	VIP	1	
Commando Mk 2B	VIP	2	
Gomhouria	Train	200	
Al Kahira	Train		
Yak-11	Train		
MiG-23U	Train	6	
MiG-21U	Train		
MiG-15UTI	Train	50?	
Su-7U	Train		
SA-7	SAM		Operated by Army
SA-9	SAM		Operated by Army

149

In 1978 the United States agreed to supply Egypt with 50 Northrop Tiger IIs – 42 F-5E single-seat fighters and eight F-5F two-seat trainers – and further examples are likely to follow; President Sadat originally requested 250 in all, but 120 is a more realistic eventual total.

The Arab Organisation for Industrialisation, financed by Saudi Arabia, Qatar and the United Arab Emirates, was due to build British and French aircraft and engines in Egypt. This plan fell through in the spring of 1979, however, when Saudi Arabia withdrew its money following the Egyptian–Israeli peace treaty.

The Egyptian Air Force's specialist ground-attack fleet operates three regiments of Su-7s together with some 48 of the Su-20 Fitter C variable-sweep development and about 18 MiG-23 Flogger Fs. The L-29 Delfins operated mainly as trainers can also undertake the light strike role to support front-line forces. The licence-built Alpha Jets will undertake similar tasks.

The Egyptian Air Force has taken delivery of six Lockheed Hercules, following the relaxation by the United States of its ban on the supply of military equipment; four are transport C-130Hs (one of which has been lost) and the other two are electronic-countermeasures EC-130Hs. At least a further 14 Hercules are expected to be acquired to replace the 16 or so remaining An-12s from an original fleet of 25–30. The aged Il-14 serves on in the transport and training roles. Other possible purchases include Transalls, G.222s or HS.748s.

Helicopters are operated largely on behalf of the army and navy. Transport/assault types include the Westland Commando and the Russian Mi-8, some of which may be deployed to transport two-man army teams carrying AT-3 Sagger anti-tank missiles. At least some of the 54 SA.342 Gazelles are fitted with Euromissile HOT anti-tank missiles. Anti-submarine helicopters were included in President Sadat's list of weapons requested from the United States in return for a relaxation of hostility towards Israel.

Aircrew are trained mainly at the air force academy, Bilbeis. After five years at a special school followed by six months on a tri-Service basic military course, cadets attend the academy for two and a half years. Basic training is carried out on the Gomhouria, a licence-built derivative of the Bücker Bestmann, with most advanced instruction taking place on L-29 Delfins. Newly graduated pilots then spend a year converting to their operational type. Navigators for the transport and bomber fleets are trained on Il-14s. Several types,

Above: The Egyptian Air Defence Command operates an interlaced belt of medium-range SA-3 Goa surface-to-air missiles (seen here on their twin launcher) and the long-range SA-2 (visible behind).

Below: Flogger F attack aircraft – which are apparently designated MiG-23 rather than MiG-27 – fly over Cairo to commemorate the October war. Egypt also operates MiG-23S fighters.

Combat experience

The Egyptian Air Force is one of the world's most experienced in recent combat, having emerged from virtual destruction in the Six Day War of June 1967 to cut new teeth in the War of Attrition from 1968 to 1971 and finally to emerge from the shadows in the October War of 1973. The missile umbrella operated by Air Defence Command acquitted itself particularly well in 1973, the Israeli Air Force and its US advisers being taken by surprise at the effectiveness of the SA-6 and ZSU-23-4SP Shilka.

Egyptian aircraft losses were substantial, however, and the inferiority of the MiG-21/Atoll combination compared with Mirages and Mirage derivatives carrying Shafrir air-to-air missiles was overwhelming. Manned interceptors were for the large part held in reserve while the SAM belt was performing so well, and the lessons of 1967 had been well learnt in the areas of base defence and hardening of vital points. Egypt and Libya fought a

short but sharp engagement in July 1977, with extensive air activity. The EAF carried out strikes against El Adem and other Libyan bases, losing at least two Su-20s in the process.

Army

The Egyptian Army contributes men and equipment directly to Air Defence Command, which supplies mobile SA-6 and Shilkas for defence of armoured forces. The army also has its own SAMs in the form of the shoulder-launched SA-7 and its vehicle-mounted SA-9 derivative. Both weapons performed well in the October War, but Egypt has expressed interest in acquiring Western low-level missiles in the Thomson-CSF Crotale class. The army operates an extensive fleet of helicopters.

Navy

The Egyptian Navy operates a small force of helicopters, including Sea Kings and Gazelles.

Above: The Egyptian Army was the first customer for the Westland Commando assault/logistics helicopter, based on the Sea King ASW machine.

Right: The first Egyptian C-130 Hercules poses in front of the Pyramids. The Lockheed aircraft could well form the backbone of the EAF's future transport fleet and is also used for ECM duties.

including the Scottish Aviation Bulldog, SIAI-Marchetti SF.260 and FFA AS.202 Bravo, have been examined as possible replacements for the Gomhourias.

ISRAEL

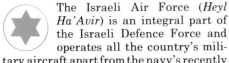

Area: 8,000 sq. miles (20,700 sq. km.).
Population: 3·65 million.
Total armed forces: 164,000.
Estimated GNP: $14·2 billion (£7·1 billion).
Defence expenditure: $3·31 billion (£1·655 billion).

The Israeli Air Force (*Heyl Ha'Avir*) is an integral part of the Israeli Defence Force and operates all the country's military aircraft apart from the navy's recently acquired maritime-patrol IAI 1124 West-

winds. The air force came into being with the founding of Israel in 1948 and was operating combat aircraft by the middle of that year. Continued conflict with the neighbouring Arab states has led to Israel building up one of the world's best equipped and trained air forces, with extensive combat experience. After the 1967 Six Day War France embargoed further arms supplies to Israel and the United States has since become by far the major supplier of *Heyl Ha'Avir* equipment. Israel has the highest per capita military spending of any state and relies heavily on US aid.

Inventory

The Israeli air-defence network is based on ten batteries of Raytheon Improved Hawk medium-range surface-to-air missiles combined with a force of manned interceptors: the surviving Dassault Mirage IIIs, Israel Aircraft Industries Neshers and Kfirs, and McDonnell Douglas F-15s, soon to be joined by General Dynamics F-16s. The *Heyl Ha'Avir* received 72 Mirage IIICJs armed with Matra R.530 air-to-air missiles and twin built-in DEFA 30mm cannon between 1963 and 1966 for use as interceptors and fighter-bombers, and they fought with distinction in these roles during the 1967 Six Day War with the Arab states.

The succeeding French embargo prevented the air force from receiving the 50 Mirage 5Js ordered for ground-attack work, and McDonnell Douglas A-4 Skyhawks were ordered in their place. Israel Aircraft Industries also set about copying the Mirage, however, and the resulting Nesher (Eagle) made its maiden flight in

Left: The distinctive canards immediately show that this is a Kfir-C2, and the *Heyl Ha'Avir*'s older -C1s are likely to be upgraded to the same standard.

September 1969. The type, powered by a SNECMA Atar 9C turbojet, entered service in 1972 and about 40 saw action during the October War in the following year. Deliveries are thought to have totalled about 50 Neshers, with the survivors remaining in service.

IAI then progressed from this interim aircraft to the Kfir (Lion Cub), superficially resembling a Mirage but with Israeli avionics and other differences, of which the major one is the use of a General Electric J79 turbojet – as fitted to the air force's F-4 Phantoms – in place of the Atar. The second variant of the Kfir to enter service is the C2, with foreplanes fitted to improve manoeuvrability and with other detailed improvements, which was revealed in July 1976. Between 150 and 200 Kfirs in all are being built as interceptors and fighter-bombers for the Israeli Air Force, and IAI could increase the production rate to six aircraft a month if export sales of the type are concluded. Both early proposed contracts, with Ecuador and Taiwan, have however run into difficulties since the US can prevent such deals by embargoing the sale of their engines.

The *Heyl Ha'Avir* received its first four McDonnell Douglas F-15s in December 1976, these refurbished pre-production aircraft being joined by 21 new machines to form an initial batch of 25 – a pair of two-seat F-15B trainers and 23 single-seat F-15As. A further 30 F-15A/Bs are expected

Top: The US is perturbed that Israeli F-15 Eagles took part in a dogfight with Syrian air force fighters over South Lebanon on June 27, 1979, during which five Syrian MiG 21s were reportedly destroyed. The F-15s were protecting other Israeli aircraft which attacked Palestinian targets along the Lebanese coast; at least one Kfir also took part in the action.

Above: The Super Frelon was one of several types supplied by France before she embargoed arms sales to Israel.

ISRAEL

Type	Role	Base	No	Notes
Air Force				
Mirage IIICJ	Int		30–40	Three Mirage sqns
Nesher	Int		40–50	Most probably sold to Argentina
F-15A	Int		23	More in second batch of 30 F-15A/B
F-16A	Fight		75	Includes some F-16B
Kfir	FGA	Includes Hazor	150–200	Being delivered
F-4E	FGA	Includes Hazor	204	Seven Phantom sqns
A-4E/F/H/M/N	Att		250+	Six Skyhawk sqns
RF-4E	Recce		12	
E-2C	AEW		4	
EV-1	ELINT		2	
Magister	Train/light att		70–80	
KC-130H	Tank		2	
KC-97	Tank		2	
Boeing 707	Trans/tank		5	
Boeing 707	Trans/command		5	
C-97	Trans/ECM		4	
C-130E	Trans		12	
C-130H	Trans		12	
C-47	Trans		18	
S-65C-3	Assault/trans		10	
CH-53D	Assault/trans		20	
S-61R	Assault/trans		12	
CH-47C	Assault/trans		8	
SA.321K	Assault/trans		8	
Bell 212	Assault/trans		12	
Bell 205	Assault		45	
UH-1D	Assault		25	Some carry TOW
Hughes 500M-D	Anti-tank		30	On order. Armed with TOW
AH-1	Anti-tank			
Alouette II/III	Util		30	
Bell 206	Liaison		20	
Do28D-2 Skyservant	Liaison		14	
Do27	Liaison		35	
Arava	Liaison		14	
Cessna U206C	Liaison		28	
Cessna 180	Liaison		1	
Queen Air 80	Liaison/train		16	
Super Cub	Liaison/train		20	
Islander	VIP		8–10	
Westwind	VIP		3	
F-15B	Train		2	
Mirage IIIBJ	Train		3	
TA-4E/H	Train		24	
T-41D	Train		2	
Ryan 124-1 Firebee	RPV			
Chukar	RPV			
Hawk/Improved Hawk	SAM		15 batteries	
Navy				
1124N Westwind	Mar pat		3	
Army				
Redeye	SAM		1200 rounds	
Chaparral	SAM		16 launchers	

to join the first batch from early 1981. Israel originally requested the delivery of 50 General Dynamics F-16 fighters from 1980, with a further 150 being co-produced and assembled in Israel, but the initial batch will be limited to 75 aircraft for delivery from late 1981 to the end of 1983. A letter of acceptance covering 40 of these F-16A/Bs was signed in August 1978, and negotiations for the remaining 35 covered by the first stage of this programme, code-named Peace Marble, were well advanced. No co-production will now take place, although the fighters will incorporate a substantial proportion of Israeli avionics. An additional 75 F-16s may be supplied from the mid-1980s, bringing the total to 150. IAI has been working on its own fighter-bomber for the 1980s, the Arye (Lion), although the future of this project remains in doubt. Development would cost at least $500m, and if the United States continues to meet most of Israel's defence-equipment needs then Arye is unlikely to see the light of day.

The *Heyl Ha'Avir* has achieved outstanding successes in air combat with a combination of twin DEFA 30mm cannon and Raphael Shafrir air-to-air missiles. Shafrir resembles the US Sidewinder externally but is of completely Israeli design. The missile entered service in 1969 and has a success rate of more than 50% in action.

The joint mainstays of the attack force are the McDonnell Douglas A-4 Skyhawk and the same manufacturer's F-4 Phantom. The original Skyhawk order, placed in 1967 after France's embargo on Mirage 5Js, was for 48 A-4Hs and two TA-4H trainers – basically US Navy A-4F/TA-4Fs with DEFAs replacing the Mk 12 20mm cannon, a braking parachute installed in an enlarged fin and some changes in the avionics. The first Skyhawk squadron was declared operational in 1968, and in the following year the force was expanded with the supply of 25 ex-USN A-4Es. Further deliveries have swelled the total to about 290 A-4E/H/M/N Skyhawks plus two-dozen two-seat TA-4s; this total includes 108 A-4Hs and at least ten TA-4Hs. More than 50 Skyhawks were lost during the October War, and about 250 remain in service.

The F-4 entered Israeli service in 1969, the initial contract being for 50 F-4Es and six reconnaissance RF-4Es. Deliveries by the end of 1978 had totalled 204 F-4Es and about 12 RF-4Es, although the latter figure is too small to allow each of the seven Phantom squadrons to operate its normal complement of two reconnaissance aircraft. Phantom armament includes Sparrow air-to-air missiles (although the aircraft are not generally used in the counter-air role), Shrike anti-radiation missiles, the television-guided Maverick, unguided and 'smart' bombs, and possibly the indigenously developed Luz stand-off weapon. The F-4s are expected to be retrofitted with Elta's new EL/M-2021 radar,

Below: A pair of A-4 Skyhawks take on fuel from a venerable KC-97 tanker.

Below right: Some 70 helicopters in the Bell 205/UH-1 series support ground forces.

airborne testing of which was due to begin in the autumn of 1978. The M-2021 can perform search, automatic acquisition and tracking in air combat, air-to-surface weapon delivery, ground mapping and terrain-avoidance/following.

The first two of four Grumman E-2C Hawkeye airborne early-warning aircraft, intended to replace vulnerable ground radars, were due to be delivered during 1978; the others are scheduled for handover in 1979. Electronic intelligence is performed by a pair of Grumman EV-1 Mohawks, equipped with pod-mounted sideways-looking airborne radar, which were delivered in 1976. Reconnaissance drones are also in service, including the indigenously developed Tadiran Mastiff. This high-wing RPV, powered by a 10hp piston engine, has a 4hr endurance and can carry up to 33lb of sensors.

The extensive network of ground-based radars is likely to be upgraded by the introduction of two new types developed by Elta. An example of the EL/M-2205 surveillance radar for joint military/civil use has been installed at Lod and the newer M-2215, which has the same peak power as the M-2205 but can detect a 2m² target at 60nm, was undergoing trials at an IAF base in 1978.

The large helicopter force includes French-supplied Super Frelons, which are being re-engined with General Electric T58-16 turboshafts of 1895shp in place of their original Turboméca Turmo IIICs of 1550shp, thereby improving hot-and-high performance while also achieving commonality of powerplant with the S-61/CH-53 fleet. A number of the Bell 205s are thought to be armed with the Hughes TOW anti-tank missile, and this weapon will additionally be carried by the specialist Hughes 500M-D Defender helicopter. Thirty of these are due to be delivered from mid-1979 under a deal worth $23m and the eventual force is likely to total at least 50 Defenders.

Most combat types of French origin have been retired and scrapped or sold overseas, but the licence-built Magisters

are to be updated rather than replaced so that they can remain operational until the mid-1980s. The aircraft, used for light strike tasks as well as training, are being fitted with new instruments, electrics, hydraulics and avionics.

Many transport types double up in other roles, including two Hercules, a number of Boeing 707s and some KC-97s used as tankers. Other 707s are fitted out for command-and-control duties and various transports are equipped for electronic countermeasures.

Training is carried out on Magisters after initial grading on Super Cubs, and twin conversion is carried out on Queen Airs.

The *Heyl Ha'Avir* has been involved in four major conflicts since its formation, of which the most recent at the time of going to press was the October War of 1973. The air force admits the loss of 103 aircraft in that conflict – 80% to surface-to-air missiles and guns, 10% in air combat and the rest in accidents. Only five of these were lost in the 4000 air-superiority sorties flown during that month. More than 250 Arab aircraft were claimed in air combat in this period; nearly 200 were accounted for by Shafrirs, about 60 with cannon and only seven with Sparrows. A further 90 or so were shot down by Hawks and other surface-to-air weapons. The average Israeli loss rate was 1.1%, compared with 1.4% in the 1967 Six Day War.

Navy
The Israeli Navy's first aircraft are three IAI 1124N Westwinds delivered from mid-1977 for maritime-patrol operations. These were being upgraded to 1124 Sea Scan standard during 1978 and can carry a variety of sensors in addition to the Litton LASR-2 search radar. Helicopters are also likely to be acquired.

Army
The army operates General Dynamics Redeye shoulder-launched surface-to-air missiles and the vehicle-mounted Ford Aerospace Chaparral.

LEBANON

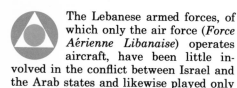

Area: 4,300 sq. miles (11,100 sq. km.).
Population: 2·65 million.
Total armed forces: 7,800.
Estimated GNP: $2·9 billion (£1·45 billion).
Defence expenditure: $167 million (£83·5 million).

The Lebanese armed forces, of which only the air force (*Force Aérienne Libanaise*) operates aircraft, have been little involved in the conflict between Israel and the Arab states and likewise played only a minor role in the civil war which racked the country in the mid-1970s. The FAéL has only some 1000 personnel and its most modern combat aircraft, Dassault Mirage IIIs, have low utilisation.

Inventory

Ten Mirage IIIEL fighters and a pair of two-seat trainers were ordered in 1965, but nearly half this force was immediately placed in storage and efforts have been made over the years to sell the surviving aircraft. Six Hawker Hunter F.6s supplied in the late 1950s have since been joined by four F.70s, three T.66 trainers and six FGA.70s. A further batch of six FGA.70s was ordered in 1975.

Additional purchases made under the five-year programme initiated in 1972 include refurbished Magisters, helicopters and Bulldog primary trainers.

LEBANON

Type	Role	No
Mirage IIIEL	Int	9*
Hunter F.70/FGA.70	FGA	17
Mirage IIIBL	Train	1
Hunter T.66C	Train	2
Magister	Train	8
Bulldog 126	Train	6
Alouette II	Train	4
Dove 6	Trans	1
Alouette III	Liaison/logistics	13
AB.212	Liaison/logistics	6

* Four in storage

Left: The nine or so surviving Mirage IIIEL interceptors operated by the *Force Aerienne Libanaise* are rarely used, and the force has tried to dispose of them. The FAL has taken little part in the conflicts which have involved Lebanon throughout the 1970s.

JORDAN

Area: 37,700 sq. miles (97,600 sq. km.).
Population: 2·7 million.
Total armed forces: 67,850.
Estimated GNP: $1·3 billion (£650 million).
Defence expenditure: $304 million (£152 million).

The Royal Jordanian Air Force, *Al Quwwat Aljawwiya Almalakiya*, originated in 1949 as the Arab Legion Air Force, the airborne logistics and communications arm of the Arab Legion. The ALAF built up its strength from its initial equipment of one de Havilland Rapide, operating almost exclusively British aircraft, and received its first combat jets, nine Vampire fighter-bombers, in 1955. The ALAF became the RJAF in the following year, and after the invasion of Suez in November the United Kingdom-Transjordan treaty was abrogated, with the result that the Royal Air Force lost its access to the air bases at Amman and Mafraq.

The RJAF continued to operate British aircraft, however, its main combat type in the 1960s being the Hawker Hunter. Jordan received two new T.66B two-seaters, together with 28 ex-RAF F.6s and a single FR.6; a total of 18 Hunters were serviceable at the outbreak of the Six Day War with Israel in June 1967, and although several flew in combat the entire force – along with three Hunters borrowed from Iraq – was subsequently destroyed. More than 40 Hunter F.73As and F.73Bs were acquired to replace the lost aircraft, and 29 of the 31 survivors were donated by King Hussein to Sultan Qaboos of Oman at the height of the air war against Dhofari guerrillas. Since the departure of the Hunters the RJAF has relied increasingly

Top: Jordan was an early customer for the CASA C.212 Aviocar light transport, which is employed on both liaison duties and to carry high-ranking personnel. The Spanish type replaced C-47s in the RJAF.

Above: Ageing F-104A Starfighters still serve as the RJAF's main interceptors, but a replacement is badly needed if the country is to be capable of defending itself from air attack.

on US equipment, both second-hand and new. Jordan's relations with her Arab neighbours are quite close, despite criticism of King Hussein's unwillingness to commit his forces unreservedly to battle in the 1973 October War.

Saudi Arabia is partially financing Jordan's arms purchases, and there is increasingly close co-operation with Syria. The possibility of a joint Jordanian-Syrian air-defence network, using Russian SA-2, SA-3 and SA-6 missiles, was discussed before the US offer of Hawks, Redeyes and Vulcans was accepted in the summer of 1976.

Inventory

The RJAF has a long-term goal of 176 fixed-wing combat aircraft, plus supporting types and helicopters. The largest part of the present front-line force comprises the Northrop F-5, used both for ground support and for interception. Thirty F-5As and four F-5B two-seaters were transferred to Jordan from Iran as the IIAF received new F-5Es to replace them. These aircraft are used mainly for visual air-to-ground operations, relying on Vortac and a fixed sight in the absence of a fire-control radar. Tip tanks are fitted as standard to the F-5As, which have four underwing pylons for rockets and bombs between 250lb and 1000lb. The F-5Bs have only two wing pylons and normally carry an SUU-20 practice pod under the fuselage. The Freedom Fighters have been supplemented by the radar-equipped F-5E Tiger II, bought new from the US. The initial order was for 30 aircraft, which have interception as their primary role. They carry AIM-9J Sidewinder air-to-air missiles and have a secondary ground-attack responsibility. The order has since been increased to 44 F-5Es, plus a pair of F-5F trainers, and then to a total of 57 and six respectively.

A single squadron of F-104 Starfighters is also allocated the interception role. The aircraft were bought second-hand in 1969 and are due for replacement; the introduction of a more modern interceptor will release the F-5Es for pure ground-attack duties, and they may be fitted with four single launchers for the Maverick air-to-surface missile. This lightweight launcher was developed specifically for the F-5E at the request of Saudi Arabia.

Modernisation is also taking place in other areas, with CASA C.212 Aviocars taking over from C-47s and additional C-130 Hercules being acquired following the withdrawal of the three C-119Ks. The fleet of Alouette III helicopters is likely to be augmented by a specialist anti-tank type and by medium-lift helicopters.

Deliveries of Raytheon Improved Hawk surface-to-air missiles began in the autumn of 1977, the weapons being part of a $540m deal which also includes 100 Vulcan 20mm anti-aircraft cannon and 300 Redeye shoulder-launched SAMs. The 14 batteries of Improved Hawks are used to defend the air bases and the capital city, Amman, where the RJAF (which is part of the Army) is headquartered. Litton and

JORDAN

Unit	Type	Role	Base	No	Notes
1 Sqn	F-5A/E	GA	King Hussein Air	30/33	Aircraft shared;
2 Sqn	F-5A/E	GA/train	Base (Mafraq)		2 Sqn acts as OCU
6 Sqn	T-37C	Train	Mafraq	12	Leased from USAF
9 Sqn	F-104A/B	Int	Prince Hassan Air Base (H5)	18/4	Due for replacement
17 Sqn	F-5E	Int/GA	Prince Hassan Air Base (H5)	24	
	F-5B/F?	Train	Mafraq/H5	4/6?	Shared by all three F-5 squadrons
	C-130B	Trans	King Abdullah Air Base (Amman)	4	More likely to be bought
	C-130H	Trans	Amman	1	
	C.212 Aviocar	Liaison/VIP	Amman	3/1	
	Boeing 727	VIP	Amman	1	For King Hussein's use
	Riley Dove	VIP	Amman	1	For King Hussein's use
	Rockwell 75A Sabreliner	VIP	Amman	2	
	Alouette III	Trans/SAR	Amman	15	Includes six SA.316Bs, four SA.316Cs. Deployed to other bases for SAR. One other lost
	Bulldog	Train	Amman	12	
	Improved Hawk	SAM	Amman & other locations	14 batteries	
	Redeye	SAM		300	

Above: Jordan's most modern combat aircraft is the Northrop F-5E Tiger II, which has bolstered the force of earlier F-5As acquired second-hand. The ground-attack aircraft double as interceptors.

Hughes submitted proposals for updating the air-defence ground environment, previously based on Marconi and Plessey surveillance radars giving virtually complete coverage of the country. The Short Tigercat SAMs have been sold to South Africa. Three new air bases and a number of satellite strips are being built well back from the border with Israel, thus overcoming a major drawback of the present bases.

The RJAF's only recent combat experience was in the 1967 Six Day War, when a number of Hunters from Mafraq flew sorties against Nantaya and engaged Israeli attackers.

Almost all pilot training is now carried out in Jordan, having formerly been conducted in Britain, the United States, Pakistan and Greece. After three months of basic officer training at the RJAF Technical School the student enters the Royal Jordanian Air Academy, based at Amman. He spends 100hr on Bulldog primary trainers, of which the academy has 12 (now transferred to Air Force control), with 70hr devoted to dual instruction and 30hr spent solo. After 55hr he is screened for suitability, and if he is successful he goes on to complete the course

after a total of some eight months. The Bulldogs fly from the air base at Mafraq as much as possible, and ground instruction alternates with flying training. A 60hr course has been devised to cut the duration to five months, with a correspondingly longer period spent on the T-37s.

After some 100hr over six to seven months on T-37s (130hr with the 60hr Bulldog course) the pilot receives his wings and converts to 2 Sqn's F-5s, spending another seven to eight months on tactics and weapon training before becoming fully operational. No 2 Sqn acts as an operational conversion unit, pilots carrying out practice interceptions under the control of ground operators at the surveillance-radar sites.

Maintenance is carried out by the RJAF itself at the air bases, and a General Electric jet intermediate maintenance (JIM) workshop has been established.

SYRIA

Area: 70,800 sq. miles (183,400 sq. km.).
Population: 8·75 million.
Total armed forces: 227,500.
Estimated GNP: $6·5 billion (£3·25 billion).
Defence expenditure: $1·12 billion (£560 million).

Traditionally one of the Soviet Union's staunchest allies in the Middle East and an opponent of Israel, Syria remains committed to collaboration with Russia despite Egypt's swing towards arms purchases from the West and a more tolerant attitude towards Israel. The Syrian Air Force, *al Quwwat al-Jawwiya al Arabia as-Suriya*, is responsible for the operation of all military aircraft; its headquarters are in Damascus, the capital, and personnel number some 25,000.

All Syrian combat aircraft have been supplied by the Soviet Union, which has replaced the heavy attrition losses virtually immediately.

Russian assistance extends to the construction of three new air bases – at Abu Dubor, Sarat, and Es Suweidiya – and provision of hardened aircraft shelters at others. The bases at Homs and in the Jezirah, between Damascus and the Iraqi border, are thought to be controlled completely by Russian 'advisers' and are out of bounds to Syrian personnel. The base at Homs, together with others at Aleppo and near the border with Lebanon, are protected by surface-to-air missiles and radar-directed cannon to provide defended staging points for Russian aircraft en route to Africa. These weapons, and others installed at Damascus airport, were supplied in an air-defence deal believed to be worth $300m. At least 3000 Russian advisers are in the country, together with Cubans, North Koreans and Vietnamese. Syria feeds back experience gained in combat with Israel's Western aircraft to other air forces in the Soviet sphere of influence.

The Syrian Air Force operates about 45 Flogger variable-geometry fighter-bombers, thought to include the ground attack variant. More of this type are likely to be supplied, especially in view of the agreement under which Iraq is building up a force of well over 80 MiG-23s. The MiG-21 remains the backbone of the interceptor element, supported by MiG-17s which also double as ground-attack aircraft alongside the Su-7s. Syria was the first overseas country to receive the Ka-25 helicopter, used for coastal anti-submarine patrols, and has turned to the West to build up the rotary-wing fleet. A requirement for between 15 and 40 Super Frelons is thought to exist, and a package deal discussed with Italy in 1977 included 24 Agusta A.109 Hirundos armed with Hughes TOW or Euromissile HOT anti-tank missiles (the latter already having been ordered to equip a large fleet of Aérospatiale SA.342 Gazelles), approximately 12 licence-built SH-3D Sea Kings and the same number of AB.212ASWs to augment the Ka-25s, a further six AB.212s for search and rescue, six AS-61A-4 transports, technical assistance and training. Additional purchases from the West include Hercules transports and MBB 223K1 primary trainers built in Spain by CASA; 32 of the latter were ordered originally and 16 more are expected to follow.

Syrian aircraft and surface-to-air missiles were engaged in sporadic fighting between the 1967 Six Day War and the 1973 October War, Syria having lost many of its aircraft in the former conflict, and formed the major part of Arab air forces operating against Israel's northeast frontier in 1973. The Syrian Air Force again

IRAQ

Area: 172,000 sq. miles (445,500 sq. km.).
Population: 12·2 million.
Total armed forces: 212,000.
Estimated GNP: $16·3 billion (£8·15 billion).
Defence expenditure: $1·66 billion (£830 million).

Iraq, the third-largest oil exporter in the Middle East, has been a socialist state since 1958 and has had close ties with the Soviet Union. Having relied on Britain for much of its early aircraft, including a large force of Hunter fighter-bombers, Iraq has since largely re-equipped with Soviet types. Orders for French helicopters indicate a relaxation of this rigid approach to arms procurement, and European combat aircraft have now been ordered, but ties with the Soviet Union were renewed early in 1977 when a further arms agreement was negotiated. The USSR is supplying Iraq with arms worth $3000–$3800m over four years in return for the number of Russian advisers in the country being increased from 5000 to 15,000, a common policy being pursued in the Mediterranean and Arabian Gulf, and MiG-25 Foxbats under Russian control being based at Shaibe. That base, the largest in Iraq, was also to come under Russian control. The Iraqi Air Force – which is responsible for all military aviation – has some 15,000 personnel.

Inventory
MiG-23 Flogger fighters and ground-attack aircraft have already been supplied by the Soviet Union, and some sources claim that up to 80 in all Floggers will be donated under the terms of the latest arms package. This deal is also thought to include 25 fire units for the SA-6 Gainful surface-to-air missile, together with a large number of transports. Deliveries of An-26s to replace Il-14s are thought to have started in early 1977, with Il-76 Candids following them from 1978. This is the first time that the Il-76 has been exported, and US sources put the eventual total of new Russian-supplied transport aircraft as high as 67, although this would seem to be excessive. In 1976 the US Department of Defense approved the sale of two Lockheed L-100-30 Hercules to Iraq, and this fleet was expected to quickly expand to ten aircraft and to reach an eventual total of 40. This deal has now fallen through, but France is supplying Iraq with military aircraft for a wide variety of roles.

An order for 36 Dassault Mirage F.1s was signed in June 1977 (the order possibly including four two-seat F.1B trainers) and up to 80 may eventually be acquired. Interest has also been expressed in about 50 British Aerospace/Dassault Jaguar attack aircraft, although the preferred version – with avionics to the standard operated by the Royal Air Force – is unlikely to be made available by the British Government. The Czech Aero L-39, apparently known locally as the Tikret, is used for both training and ground attack, although a requirement for 40 aircraft in the Alpha Jet/Hawk category may still exist. The ten SA.321H Super Frelon helicopters delivered in 1976 and 1977, at

Left: The MiG-15UTI remains in Iraqi Air Force service, though a replacement is urgently needed. This example was photographed at Prague in 1963 after a major overhaul.

Right: The An-12 is one of many transport types supplied by the Soviet Union, including – it is thought – Il-76s

Type	Role	No	Notes
MiG-23	FGA	45	One regiment of three squadrons
MiG-21PF/MF	Int	250	Four regiments of three squadrons each
MiG-17F	FGA	50	
Su-7BM	Att	60	
Su-22	Att	?	
An-12	Trans	6	
IL-14	Trans	8	
IL-18	Trans	4	
C-47	Trans	6	
PA-31 Navajo	Liaison/survey	2	
Yak-11	Train	⌉—	Being replaced by Flamingo
Yak-18	Train		
MBB 223 Flamingo	Train	48	Replacing Yak basic trainers
L-29 Delfin	Train		
Su-7U	Train		
MiG-15UTI	Train		
MiG-15UTI			
MiG-21U	Train		
Ka-25	ASW	9	Shore-based
Mi-4	Trans/assault	20	
Mi-6	Trans	10	
Mi-8	Trans/assault	Up to 50	
SA-2	SAM	⌉	
SA-3	SAM	⌉—48+ batteries	
SA-6	SAM		
SA-7	SAM	⌉—Operated by Army	
SA-9	SAM		

suffered badly, both at the hands of Israeli fighter pilots and when required to operate within its own SAM belt; a combination of over-enthusiasm by the missile operators and unreliable IFF (identification, friend or foe) equipment resulted in several Syrian aircraft being shot down. Israel claimed to have destroyed 221 Syrian aircraft during the October War; other sources put the figure at 179.

The Syrian Air Force's major tasks during the October War were support of the ground forces and base defence rather than deep strike, although some of the latter type of mission were flown. Normal practice was for the MiG-17s and Su-7s to be used in the ground-attack role at low levels with MiG-21s and Iraqi Air Force Hunters providing top cover. Some Su-7s were destroyed on the ground at Damascus, and the MiG-17s fared badly in air combat with Israeli Mirages/Neshers over the Golan Heights. Mi-8s were used for commando raids on Mount Hermon. Russian An-12s made 125 flights into Syria over a ten-day period to resupply stocks of MiG-21s, Su-7s and Mi-8s; two of the transports were destroyed on the ground on October 13. The 48 batteries of SA-2, SA-3 and SA-6 surface-to-air missiles in

the western part of the country, together with shoulder-launched SA-7s, shot down a number of aircraft and also claimed at least one Israeli Teledyne Ryan 124-1 drone. Syrian forces also saw action during the Lebanese civil war in the mid-1970s.

Syria was reported to have lost five MiG-21s on June 27, 1979, when it challenged Israeli aircraft which had attacked Palestinian targets along the Lebanese coast.

The Israeli force was being protected by F-15 Eagles. Syria claimed to have destroyed one Israeli aircraft, although Israel reported that all its aircraft returned undamaged.

Syrian Air Force training begins on the CASA/MBB Flamingo, this type replacing Yak-11s and -18s. Initial jet instruction is carried out on Aero L-29s, being joined by L-39s, before operational conversion on two-seat versions of combat aircraft.

least some of which carry the AM.39 Exocet anti-ship missile, may be followed by 30 to 40 more. Alouette IIIs and Gazelles have also been supplied.

Iraqi Air Force aircraft operated from both Egypt and Syria during the October 1973 war with Israel as well as flying from home bases. Twelve Hunters operating under Egyptian Air Force command were used from the very beginning of the conflict, and were claimed to have made three strikes without loss on the first day (October 6). Another Hunter squadron was based in Syria, the aircraft flying top-cover missions in support of ground-attack MiG-17s and Su-7s; 80 missions are reported to have been flown by IrAF aircraft based in Syria on October 7 and 8. The Egypt-based Hunters are understood

to have carried out seven raids on the second, third and fourth days of the war, in addition to supporting other missions. Iraq, although not involved in drawing up plans to attack Israel to the same extent as Syria and Egypt, committed its aircraft to action early in the war; reports of Tu-22 bombers being based in Iraq caused concern in Tel Aviv, and MiG-21s joined the Hunters. US intelligence reports put IrAF losses at 21 aircraft (Hunters and MiG-21s); some are understood to have been shot down by other Arab aircraft and/or missiles as a result of incompatibility between the British IFF (identification, friend or foe) equipment in the Hunters and the exclusively Russian-supplied radars used by Syria and Egypt.

The IrAF was also extensively involved

in action against Kurdish separatists until the end of 1975. Aircraft were lost both to ground fire in Kurdistan and to Iranian Rapier SAMs when Iraqi aircraft strayed over the border.

Basic flying training is carried out on Yak-11s, and AS.202/18A-1 Bravos have been ordered from Switzerland as replacements. Advanced instruction then takes place on Jet Provosts, L-29 Delfins and MiG-15UTIs. These types are being supplemented by the L-39 and later possibly by an aircraft in the Hawk/Alpha Jet category. Operational conversion takes place on two-seat Hunter, MiG and Sukhoi variants.

Below: The Iraqi Air Force has received a number of Czech-built L-39 Albatros light strike-trainers.

IRAQ

Type	Role	No	Notes
Mirage F.1C/B	Int/train	32/4	
MiG-23	FGA	40	Two squadrons operational. Others for delivery by 1981?
MiG-21PFM	Int	100+?	Total five squadrons; part of Air Defence Command
Hunter FGA.9/FR.10	GA	30+	Three squadrons
Su-7BM	GA	50	Three squadrons
Su-20/-22	GA	20	Two squadrons
MiG-17F	FGA	35	Three squadrons
Tu-22	Bomb	12	
IL-28	Bomb	10	
L-39	Train/GA	24	
L-29 Delfin	Train		
Jet Provost T.52	Train/GA	16	
Hunter T.66/69	Train	5	
MiG-23U	Train		
MiG-21UTI	Train		
MiG-15UTI	Train	30	
Yak-11/18	Train		
Su-7U	Train		
An-2	Trans	10	
An-12B	Trans	6–8	

Type	Role	No	Notes
An-24	Trans	10	
An-26	Trans	2	Possibly many more
IL-14	Trans	10–13	Being replaced by An-26
IL-76	Trans	?	
Heron	Trans	1	
Islander	Trans	2	
Tu-124	VIP	2	
Super Frelon	Anti-ship	10	Armed with AM.39 Exocet anti-ship missile
Alouette III		40	
Mi-1	Trans	4	
Mi-4	Trans	30	
Mi-6	Trans	15	Possibly fewer
Mi-8	Trans/assault/VIP	36	
Wessex 52	Trans	7	
SA.342 Gazelle	LOH/AT	40	
Puma	VIP	2	
Falcon 20	VIP	1	
SA-2	SAM		
SA-3	SAM		
SA-6	SAM		

IRAN

Area: 628,000 sq. miles (1·63 million sq. km.).
Population: 28·5 million.
*****Total armed forces:** 413,000.
Estimated GNP: $72·6 billion (£36·3 billion)
*****Defence expenditure:** $9·94 billion (£4·97 billion).
*Figures apply to before Khomeini's take-over.

Iran channelled much of her immense oil revenue into arms purchases in the 1970s, becoming the dominant military power in the Middle East and a major customer for the world's arms industries. However, the whole future structure of the country's armed forces has changed since the declaration in early 1979 of an Islamic

republic by the Ayatollah Khomeini, and nine-tenths of the military equipment on order has been cancelled.

At the time of writing, Iran was still in a state of disarray and accurate predictions of future military policy were impossible. Much of the existing equipment is unlikely to be made operational again, and extensive new orders are improbable. The following description applies to the Iranian armed forces as they were at the time of the revolution, although new equipment known to have been cancelled has been deleted.

The Imperial Iranian Air Force (*Nirou Havai Shahanshahiye Iran*) became an independent force under the Ministry of War in August 1955, since when the army and navy have built up their own airborne

could ultimately threaten Iran herself. A further potential adversary is Israel; Iran is not an Arab state, but after the October 1973 war the Shah said that any repetition would involve the whole Muslim nation, not just the Arabs.

Inventory

The IIAF's re-equipment programme began with 32 F-4D Phantoms, forming two squadrons, which were ordered in 1966 and a total of 104 F-5As and 22 two-seat F-5Bs, 91 of them acquired through US Military Aid Programmes, to form six fighter-bomber squadrons with a front-line strength of 16 aircraft each. A further 177 F-4E Phantoms have since been acquired, this model carrying the M61 Vulcan 20mm Gatling gun in addition to the Sparrow medium-range and Sidewinder short-range air-to-air missiles which arm both the F-4D and F-4E. The acquisition of the F-4Es has allowed the Phantom interceptor force to be increased from two

Above: The Imperial Iranian Air Force — which was expected to be renamed following the Shah's overthrow — operates Bell 214Cs on search-and-rescue duties.

and Imperial Iranian Naval Aviation. The massive increase in IIAF strength was partially sparked off by Britain's withdrawal from east of Suez in 1970. The use by the Soviet Union of Umm Qasr in Iraq as a base gave Iran cause for concern, as did the Kurdish rebellion against the Iraqi central government, a similar independence movement in Baluchistan and the Sultan of Oman's war against guerrillas infiltrating from South Yemen and supported in turn by the Soviet Union, China and various Arab states. Iranian forces took part in all three of these conflicts and Iranian military thinking in the past has been deeply affected by the domino theory – the fall of one state to external forces or even internal insurrection can result in a chain reaction which

Below: The IIAF's P-3F version of the Lockheed Orion has electronic equipment optimised for sea-surface surveillance and attack rather than anti-submarine warfare, US ASW technology being highly classified.

IRAN

Type	Role	Base	No	Notes
Imperial Iranian Air Force				
F-14	Int	Khatami AFB, Isfahan	50	Two squadrons plus reserves
F-14	Int	Shiraz	30	Two squadrons
F-4E	FGA	Mehrabad, Shiraz, Tabriz	177	Eight squadrons
F-4D	FGA	Mehrabad	32	Two squadrons
RF-4E	Recce		16	One squadron
F-5E	FGA	Bushahr, Tabriz, Chah Bahar?	141	Eight squadrons
F-5F	Train		28	
F-5A	FGA		?	Originally 104; most or all passed on to other countries
F-5B	Train		22	
T-33	Train		25+	Due for replacement
P-3F	Mar pat	Bandar Abbas	6	
Boeing 707-3J9C	Trans/tanker		13	Also freight/pax/VIP
Boeing 747/100	Trans/tanker		3/3	
Boeing 747F	Trans		4	
C-130E	Trans	Mehrabad and Shiraz	11?	Originally 15
C-130H	Trans	Mehrabad and Shiraz	42?	Originally 49
F.27 Mk 400M	Trans/aerial survey	Mehrabad	11/2	C-45 and C-47 withdrawn?
F.27 Mk 600	Trans/VIP	Mehrabad	2/3	
Caribou	Trans		1	
Falcon 20	Liaison		4	
Rockwell 690	Util		3	
Beaver	Util		7	
Bonanza F33A	Train	Faharabad	10	
Bonanza F33C	Train	Faharabad	39	Replaced T-6
Cessna 310L	Train		2	
Cessna 337IR	Train		12	
Super Frelon	Trans		16	
AB.205	Trans			
AB.206	Util		70	
Bell 214B	Util		2	
Bell 214C	SAR		39	
HH-34F	SAR		10	
Bell 212	Trans		11	
S-62	Trans		1	
CH-47	Trans		2	
AS-61A-4	VIP		2	
AB.212	VIP		5	
Rapier	SAM		?	Five squadrons plus training unit
Tigercat	SAM		25 launchers	
Imperial Iranian Army Aviation				
F.27 Mk 400M/600	Trans/target tug		1/1	
Cessna 0-2A	FAC		10	Originally 12
Cessna 180	Liaison		5	
Cessna 185	Liaison		45	
Cessna 310	Liaison		6	
CH-47C	Trans		40	
Bell 214A	Trans		287	
Bell AH-1J	AT		202	Armed with TOW missile
AB.205	Trans		70+	
AB.206	Util		115	
HH-34F	SAR		10	
Shrike Commander?	Liaison		5?	
Imperial Iranian Naval Aviation				
F.27 Mk 400M/600	Trans	Mehrabad	2/2	
Shrike Commander	Liaison		6	
SH-3D	ASW		20	Built by Agusta
RH-53D	Minesweeping		6	
HH-53	SAR		12	Order not confirmed
AB.212	Attack		6	Armed with AS.12 missile
AB.205	Trans		24?	
AB.206	Util		14	
Imperial Iranian Navy				
Seacat	SAM		5 launchers	Four frigates, one destroyer
Standard	SAM		3 launchers	Three destroyers

Note: many aircraft non-operational and may be disposed of

159

squadrons to ten. The aircraft carry APR-37 passive warning receivers to detect enemy radar emissions and missile launchings, and ALQ-72 active electronic countermeasures pods can be fitted in the rear recessed Sparrow mountings. The Phantoms have a secondary ground-attack role, weapons including Hughes Maverick air-to-surface missiles, of which 2850 have been bought.

Most of the F-5As have been passed on to other states, including Jordan, Turkey, Pakistan, Ethiopia and South Vietnam. They have been replaced on a virtually one-for-one basis by 141 F-5Es ordered in 1972 to equip eight ground-attack squadrons. The Tiger IIs have a secondary fighter role, using Sidewinder AAMs, and are supported by 28 two-seat F-5Fs. The 13 RF-5As which previously equipped one reconnaissance squadron have since been superseded by 16 RF-4Es, four of which were delivered in 1971–72 and 12 in 1976.

The ITAF received 80 F-14 Tomcat fighters armed with Phoenix long-range air-to-air missiles, but these were expected to be disposed of by the new government which took control in early 1979.

Under the Peace Crown programme a semi-automatic air-defence system has been established to cover the whole of Iran and much of the Persian Gulf. The FPS-88 surveillance radars installed under CENTO auspices as part of the Long Range Iranian Detection System (LORIDS) to give warning of air attack from the north have been supplemented by a network of new radars supplied under a US Military Assistance Programme. These include Westinghouse TPS-43 three-dimensional radars and fully mobile equipment in the Marconi S300 series. Peace Crown includes further three-dimensional surveillance radars, digital extraction of video data and computer-controlled interception facilities. All IIAF air-defence sites are linked by the Peace Scepter communication network.

The IIAF operates five squadrons of BAC Rapier low-level surface-to-air missiles for base defence. There is also a training unit at less than squadron strength, plus 25 launchers for Short Tigercat SAMs. Iran was the first export customer for the Marconi Space and Defence Systems DN181 Blindfire radar, which allows the Rapiers to be operated at night and in all weathers. Oerlikon cannon operated in conjunction with Contraves Superfledermaus fire-control radars supplement the missiles.

The range of the IIAF's tactical aircraft is extended by air-refuelling, using 13 Boeing 707-3J9C tankers. The original six, delivered in 1974, were fitted with flying-boom equipment to allow them to refuel the F-4Es. The order for F-14s, which use the US Navy's probe-and-drogue refuelling method, resulted in the addition of Beech Model 1080 hose-and-drogue pods under each wingtip of the original aircraft and the others subsequently added. The 707-3J9Cs carry more than 28,000 US Gal of fuel and are also fitted with cargo doors, tie-down points and restraint nets, allowing 13 freight pallets to be carried. The pods can be removed when not needed, and the interior can be converted to all-

Above: Imperial Iranian Army Aviation was being supplied with 202 Bell AH-1J SeaCobra attack helicopters when the Shah's overthrow halted further deliveries and cast doubt on the IIAA's future.

Below: All three branches of the Iranian forces operate Fokker-VFW F.27 transports, this example being in the colours of the IIAF. Non-combat types such as this are likely to remain in service.

passenger or VIP layouts.

The IIAF also acquired a total of 12 Boeing 747–100s from TWA and Continental Airlines. These were to be converted to all-freight layout and fitted with JT9D-7AH turbofans in place of the —3As. Some were also to be used as tankers. One crashed, however, and others have been disposed of.

The mainstay of the transport fleet is the Lockheed Hercules, of which 64 are thought to have been delivered: 49 C-130Hs and 15 C-130Es. Four of the latter were transferred to Pakistan, however, and others have been lost. The present totals may be as low as 42 and 11 respectively. Four of the Hercules have been modified by E-Systems for signal monitoring as part of the Ibex intelligence signal-gathering network developed by Rockwell Autonetics. The aircraft carry oblique long-focal-length cameras to photograph enemy radar transmitters from high altitude, plus equipment for monitoring and recording radio and radar signals emanating from neighbouring states. Ground stations are also used in Ibex; original plans called for the conversion of two 707s and a pair of

P-3Fs, but the Hercules were chosen because most operations take place from rough airfields at high altitudes.

The P-3Fs, of which the IIAF has six, are used mainly for surface surveillance in the Persian Gulf and Indian Ocean; they are also equipped to carry the McDonnell Douglas Harpoon anti-ship missile, which is in the Iranian inventory.

Imperial Iranian Army Aviation operates both helicopters and fixed-wing types, and had more than 800 aircraft on strength by 1978, with 14,000 personnel to support this force. The largest fixed-wing type is the Fokker-VFW F.27, of which two are in service: an all-passenger Mk 600 and a passenger/freight Mk 400M. These Friendships are also fitted with Marquardt target-towing equipment under the wings.

Imperial Iranian Naval Aviation likewise operates a combination of fixed-wing and rotary-wing types.

The IIAF saw action during a revolt in Fars province in 1963, and in September 1972 three helicopters were deployed to Dhofar province in Oman in support of Omani action against guerrillas supported by neighbouring South Yemen.

The Iranian presence in Oman increased to include CH-47s, AB.205 gunships and eventually F-4Es operating from Midway, later renamed Thumrayt. Iranian forces also operated a surveillance radar and provided surface-to-air artillery to protect Thumrayt, the SOAF base at Salalah and detachments near the border with South Yemen. An F-4E from Thumrayt was shot down by gunfire over South Yemen on November 25, 1976.

The IIAF's own Rapiers have shot down at least one Iraqi aircraft engaged in actions against rebels in the Kurdistan province. Iran also supplied CH-47Cs to the Pakistan Government in 1973, the helicopters being used to help quell an uprising in Kurdistan.

Basic flying and ground training is carried out at the Air Force Training Centre at Faharabad, near the IIAF's headquarters at Doshan-Tappeh. Initial flying training on Bonanzas and light twins is followed by jet instruction in Iran or overseas and then conversion on to the operational type to be flown.

Above: The IIAF was the first export customer for the Grumman F-14 fighter and its Hughes Phoenix long-range air-to-air missiles, and deliveries of the first 80 aircraft had just been completed when the Shah went into exile.

SAUDI ARABIA

Area: 927,000 sq. miles (2·4 million sq. km.).
Population: 7·2 million.
Total armed forces: 58,500.
Estimated GNP: $55·4 billion (£27·7 billion).
Defence expenditure: $9·63 billion (£4·815 billion).

Saudi Arabia, one of the oil giants of the Middle East, has built up her air force rapidly over the past decade to become a potent military and political force. In addition to financing development of her own armed services, Saudi Arabia is supplying money for neighbouring states, notably Egypt and Jordan, to bolster their own defences. The Royal Saudi Air Force, with headquarters in Riyadh, has emerged from a small air arm founded in the 1920s. By 1933 the force operated nine aircraft and British advisers were replaced by

those from other countries, including Italy. In 1950 another British military mission was sent to the country, and a number of ex-RAF Tiger Moths and Ansons were acquired. Two years later the British were again replaced, this time by the United States; a USAF base was established at Dhahran, and Buckaroo primary trainers were supplied under a Mutual Defence Assistance Programme. The US also provided C-47s and T-6s, as well as training pilots and ground engin-

eers; further instruction was given in Britain, and in Egypt under an agreement which additionally established a joint armed forces command. Combat aircraft, transports and trainers were donated by the US, Britain, Egypt and Jordan, with F-86F Sabres and T-33 trainers following from 1957. Relations with Egypt deteriorated, and in 1965 the threat to Saudi oil by Egyptian forces operating from North Yemen precipitated a decision to establish a modern air force.

Above: The Royal Saudi Air Force is obviously a satisfied Strikemaster customer, having ordered three batches totalling 46 aircraft.

Inventory

In January 1966 the Saudi Arabian Defence Consortium, comprising the British Aircraft Corporation, Airwork and AEI, was awarded a contract to supply a complete air-defence network consisting of radars, communication equipment, airfield equipment, eight launchers for 37 ex-British Army Thunderbird I medium-range surface-to-air missiles, and Lightnings, Hunters and Strikemasters for interception, ground attack and training. The RSAF received six Hunter FGA.9s and two T.66s, four ex-Royal Air Force

Below: The RSAF is the only remaining overseas operator of the BAC Lightning interceptor, Kuwait's aircraft having been replaced by Mirage F.1s. The Saudi Lightnings will give way to F-15s.

Lightning F.52s (formerly F.2s) and two ex-RAF T.54s (previously T.4s), 36 new F.53s (a ground-attack development of the F.6 armed with 1000lb bombs and rockets as well as 30mm Aden gunpacks and Red Top or Firestreak air-to-air missiles), six T.55s and 25 Strikemasters. The first Hunters and Thunderbirds entered service near the border with North Yemen in early 1967. The missiles have since been withdrawn, and the Hunters were donated to Jordan to replace those lost during the 1967 Six Day War. The Strikemasters and the main batch of Lightnings were delivered in 1968–69.

A repeat order for ten Strikemasters was placed in 1971, and a further 11 to replace attrition losses were ordered in 1976. In May 1973 the Saudi and British Governments negotiated a new support contract, worth £250m, to be managed by BAC (now British Aerospace). Extensions increased the value to £310m by 1977 and a £500m extension to the contract was signed in September 1977.

In May 1978 the United States agreed to sell 60 McDonnell Douglas F-15 fighters to replace the RSAF's Lightnings from late 1981 or early 1982, all Eagles being in service by 1984. Saudi Arabia has said that she will need more than this total to safeguard her sovereignty, and additional batches may follow. The Lightnings are expected to remain in service until 1985 in the air-defence and limited photo-reconnaissance roles, having been replaced by F-5s on ground-attack duties. Before receiving permission to buy F-15s the RSAF had also examined the Northrop F-18L (the land-based derivative of the US Navy's McDonnell Douglas/Northrop F-18 Hornet fighter) to replace F-5s from the early 1980s, although this deal is now unlikely to be concluded. Another doubtful starter is the arrangement under which Saudi Arabia was reported to be considering support for development of Dassault's private-venture Mirage 4000 twin-engined multi-role aircraft.

As part of the original contract the United States supplied Raytheon Hawk surface-to-air missiles, Cessna 172 trainers and other types. Lockheed began supplying Hercules transports, and in 1971 the RSAF placed a $130m order for 20 Northrop F-5B trainers and 30 F-5E ground-attack aircraft. The Northrop-managed part of the programme is known as Peace

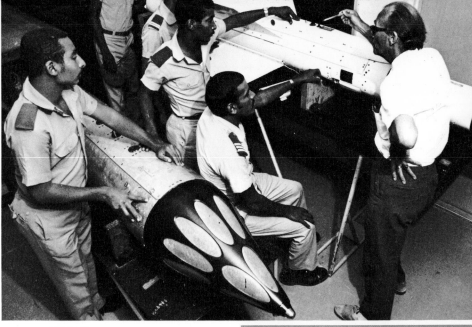

Hawk by the US, and the fifth stage of this – Peace Hawk V, negotiated in 1976 – is itself worth $1464m. This element of the programme covers training, construction, personnel support and services, plus provision for the F-5Es and two-seat F-5Fs (which double as attack aircraft) to be fitted with advanced fire-control equipment and weapons. The 20 F-5Fs and a second batch of 40 F-5Es were ordered in 1975 at a cost of $750m. The F-5E/Fs are fitted with Northrop AVQ-27 manual laser target designators, similar to the USAF's Paveway equipment, in which the pilot uses a telescopic eyepiece to align the designator with its target and then squeezes a trigger for laser illumination. This is used to deliver laser-guided bombs, and the RSAF is also receiving 650 Hughes Maverick television-guided air-to-surface missiles. Saudi Arabia sponsored development of a lightweight single-round launcher for Maverick, necessary because the original triple installation was too heavy for the F-5, produced too much drag and reduced ground clearance. The original request for 2500 Mavericks was reduced to 1000, and the US Congress granted permission for only 650 to be supplied; these are used in the anti-ship role as well as against land targets. Similarly, the planned purchase of 1000 Sidewinder air-to-air missiles was reduced to 850. The RSAF is also reported to be receiving Matra R.550 Magic AAMs to arm the F-5E/Fs.

In mid-1974 Raytheon began upgrading the Hawk surface-to-air missile system (which is operated by the Army) and two years later the company received a $1140m five-year air-defence contract. About half of the money is being spent on equipment, including Improved Hawk missiles, with the balance going on construction, maintenance and training. The Saudi Improved

SAUDI ARABIA

Unit	Type	Role	Base	No	Notes
2 Sqn	Lightning F.53'	Int	Tabuk	20	
Lightning Conversion Unit	F.52/F.53/T.54/T.55 Lightning	Train	Dhahran	3/11/ 2/5	6 Sqn has been disbanded
	F-15A	Int	Dhahran, Taif, possibly Riyadh or Khamis Mushayt	45	For delivery from 1981
	F-15B	Train		15	
7 Sqn	F-5F	Train/GA	Dhahran	24	Formerly operated F-86
15 Sqn	F-5B	Train	Dhahran	20	Formerly operated T-33
3 Sqn	F-5E	FGA	Taif	30	Two other F-5E squadrons being formed
10 Sqn	F-5E	FGA	Khamis Mushayt		(some at Tabuk?)
9 Sqn	Strikemaster	Train	Riyadh	46	Ordered in batches of 25 (1966), ten (1971),
11 Sqn	Mk 80/80A	Train	Riyadh		11 (1976). Some lost (at least five)
4 Sqn	C-130E/C-130H	Trans/tanker	Jeddah	10	
16 Sqn	KC-130H		Jeddah	25/4	
8 Sqn	Cessna 172G/H/M	Train	Riyadh	13	Basic training for KFAA
12 Sqn	AB.205/	Liaison/trans/	Taif	24	Detached to other bases
14 Sqn	AB.206	Trans/SAR	Taif	16	
	Alouette III	Trans		2	Originally six
	KV-107	SAR	various	6	
	AB.212			?	
Royal Flight (1 Sqn)	Boeing 707-320/	VIP		1/2	Boeing 747SP on order
	C-140B JetStar/AB.206/AS-61A-4			1/2	
	Improved Hawk	SAM		16 batteries	

Hawk force will total 16 batteries, to be integrated into a Litton Industries TSQ-73 air-defence system under a $1500m contract which was due to be signed at the end of 1978. The TSQ-73 will be interfaced

Left: RSAF ordnance technicians receive instruction on the Firestreak air-to-air missile which is carried by Lightnings. In the foreground is a pod for air-to-ground rockets.

Above: Saudi Arabia's great size and lack of surface transport make a long-range heavy-lift aircraft fleet a necessity. The C-130 Hercules performs this role and additionally acts as aerial tanker.

with long-range surveillance radars and troposcatter communication networks to provide automatic track initiation, target tracking, assignment of weapons to counter specific threats and exchange of information with other defence forces.

Under the terms of another deal, signed in July 1978, Thomson-CSF and the Arab Organisation for Industrialisation are setting up a jointly owned company. The concern, known as the Arab Electronics Company, has its headquarters at Al Kharj, south of Riyadh, and is to engage initially in the manufacture and maintenance of airborne radars and communications equipment. The first stage of this work is worth some $200m to Thomson-CSF and is expected ultimately to reach a value of $1000m. The French manufacturer and local production lines are both likely to contribute equipment to a new network of surveillance radars which will be established in the United Arab Emirates, Qatar and Egypt as well as in Saudi Arabia itself. Further deals involve Lockheed, which is supplying C-130 Hercules and is heavily involved in provision of joint military/civil air-traffic-control facilities. In late 1978 the Saudi Government was reported to have ordered 40 CASA C.212 Aviocar transports from P.T. Nurtanio in Indonesia, at least some of which would be operated by the RSAF.

Saudi Arabia has expressed interest in the Hughes 500D carrying the TOW anti-tank missile, and the Saudi forces will receive some of the Westland/Aérospatiale Lynx helicopters to be built under licence in Egypt. Most will be operated by the Army, although some are to be allocated to the RSAF. In June 1977 the Saudi Government ordered six Kawasaki-Vertol KV-107-1 helicopters, four for firefighting and the other two for rescue and logistics missions; these were delivered in late 1978. At that time the Saudis were negotiating for a further seven KV-107s, three fitted out as ambulances, two for firefighting and one each for rescue and command purposes. The army has ordered Thomson-CSF/Matra Shahine or Chahinn (hunting falcon) battlefield air-defence systems, using a modified version of the Crotale weapon system mounted on AMX-

30 tank chassis rather than the normal wheeled vehicles. One vehicle carries a six-round launcher and another is fitted with the surveillance radar. A broader radar aerial than that used in Crotale is employed to give a 1.4° beamwidth in azimuth compared with the normal 3.5°, in order to improve target discrimination; three or four missile vehicles will be assigned to each radar. The missile itself, designated R.460 (the standard Crotale is R.440), has a modified aerodynamic configuration, more powerful control actuators and a higher-energy propellant; these improvements increase maximum range from 8.5km to 10km and confer higher manoeuvrability. Deliveries of Shahine will begin in 1980, and the Saudis are also receiving AMX30SA radar-directed twin-barrel anti-aircraft cannon, mounted on the same type of chassis and integrated operationally with the missile fire units.

Saudi Arabia ordered 38 Mirages and the original batch of 24 Westland Commandos and six Sea Kings on behalf of Egypt (which see). The country was also a major supporter financially of the fund set up by the Arab Military Industrial Organisation to finance arms production in Egypt but withdrew in spring 1979.

Lightnings are believed to have flown in combat against Egyptian/Yemeni forces operating from North Yemen.

Pilot training is carried out at the King Faisal Air Academy (KFAA), Riyadh. After a year spent learning English and on general academic studies the student spends seven months learning to fly on the Cessna 172, then moving on to the Strikemaster for some 200hr of advanced training and weapon conversion. The complete course at the academy lasts two and a half years. Operational conversion, over about six months, takes place on the F-5Bs and two-seat Lightnings. Transport pilots are trained in the US before moving to the Hercules squadrons, and helicopter crews are instructed on AB.206s. Ground personnel are trained at the Technical Studies Institute, Dhahran, which is run – like the KFAA – by British instructors employed by BAC. Warrant-officer training takes four years, including one year learning English.

YEMEN (PDRY)

Area: 180,000 sq. miles (466,000 sq. km.).
Population: 1·8 million.
Total armed forces: 20,900.
Estimated GNP: $224 million (£112 million).
Defence expenditure: $56 million (£28 million).

The South Yemen People's Republic, or People's Democratic Republic of Yemen, was the British colony of Aden until the end of 1967, when the National Liberation Front assumed power. Plans to transfer eight ex-RAF Hunters to the new state were dropped, and British aid was curtailed, but aircraft were supplied in the form of four armed Jet Provost T.52s, the same number of Strikemaster Mk 81s and six Westland/Bell 47G-3Bs. The new air

force also acquired six Beavers from Canada and four ex-airline C-47s.

The British firm of Airwork provided support initially, but in 1969 a Russian mission arrived together with the present Soviet-supplied equipment. From 1972 the Air Force of the South Yemen People's Republic has operated the aircraft and

SOUTH YEMEN

Type	Role	Base	No
Air Force			
MiG-21F	Int		12
MiG-17F	Att		15
IL-28	Bomb		6
MiG-15UTI	Train	Khormaksar,	3
An-24	Trans	Beihan plus	3
C-47	Trans	others	4
IL-14	Trans		4
Mi-4	Util		6
Mi-8	Util		8
Army			
SA-7	SAM		

Cuban instructors have taken over from Russian personnel. The British-supplied jets are no longer in service, the Strikemasters having been passed on to Singapore, and serviceability is generally low.

South Yemen actively supported terrorists fighting for the independence of Dhofar province in Oman until the mid-1970s, and in November 1973 an AFSYPR aircraft – believed to have been an Il-28 – bombed Omani territory. The guerrillas and South Yemen regular forces were supplied with SA-7 Grail shoulder-launched surface-to-air missiles, with which several aircraft operated by the Sultan of Oman's Air Force were shot down. Surface-to-air cannon – 37mm and 57mm – together with anti-aircraft machine guns claimed further SOAF aircraft and also an Iranian Phantom operating alongside Omani forces.

YEMEN ARAB REPUBLIC

Area: 75,000 sq. miles (194,000 sq. km.).
Population: 6·5 million.
Total armed forces: 38,000.
Estimated GNP: $1·2 billion (£600 million).
Defence expenditure: $79 million (£39·95 million).

Since the 1950s the YAR has received aid successively from the United States, Czechoslovakia and the Soviet Union, Egypt, Russia again and most recently Saudi Arabia. It is unlikely that any of the combat elements remain operational, and the YARAF is now primarily a transport force with small numbers of fixed-wing aircraft and helicopters. Obsolescent ground equipment has been passed on to the YAR by Saudi Arabia and in 1977 the US State Department approved the trans-

YEMEN ARAB REPUBLIC

Type	Role	Base	No	Notes
MiG-17F	Int		12	Combat types probably not operational
IL-28	Bomb		12-16	Combat types probably not operational
MiG-15UTI	Train		4	
F-5B	Train		4	Ex-Saudi aircraft
F-5E	Fight		12	
F-5B	Train		4	
Yak-11	Train	Hodeida, Sana'a, Taiz plus others	4	Ex-Saudi aircraft
C-130E	Trans		2	
C-47	Trans		15-20	
IL-14	Trans		3	
Skyvan 3M	Trans		2	
AB.205	Util		2	
Mi-4	Util		2	

fer of four ex-RSAF Northrop F-5B trainers. Saudi Arabia has also said that it intends to buy 12 new F-5Es on behalf of the YAR, a move apparently precipitated by the Soviet Union's offer to supply MiG-21s in an attempt to re-assert its influence over Yemen. The F-5Es would not be fitted with the RSAF's specialised equipment such as the four single launcher rails for Maverick air-to-surface missiles, and by the end of 1978 the aircraft were not thought to have been delivered. Two ex-Saudi C-130E Hercules transports have also been supplied.

OMAN

Area: 120,000 sq. miles (309,600 sq. km.).
Population: 750,000.
Total armed forces: 19,200.
Estimated GNP: $2·5 billion (£1·25 billion).
Defence expenditure: $767 million (£383·5 million).

The Sultan of Oman's Air Force (SOAF), formed with British assistance in 1958, is responsible for all military air operations. The Sultan is Commander-in-Chief, and the Sultan's Armed Forces (SAF) are run from a joint headquarters outside the capital, Muscat. The Sultan is also chairman of the national defence council, established at the beginning of 1975 to lay down the basis for future action.

SOAF received its first combat aircraft in 1968 and expanded rapidly to counter insurgency by guerrilla forces in Dhofar. This province, in the south-west of the country, adjoins the People's Democratic Republic of Yemen (PDRY) – otherwise known as South Yemen – which has been supported by both the Soviet Union and China. Until 1976 guerrillas belonging to the Popular Front for the Liberation of Oman (PFLO) operated across the border from bases in PDRY. Oman is an oil-producer but is not a member of OPEC (the Organisation of Petroleum-Exporting Countries), and the war absorbed a high proportion of the country's resources. The Sultan was supported, however, by money, equipment and men from Iran, Britain,

Jordan, Saudi Arabia, the United Arab Emirates, Pakistan and India.

The major contributions came from Iran and Britain. Iranian involvement began in September 1972 with the deployment to Dhofar of three helicopters, being followed

at the end of that year by ground forces. Iran was anxious that the Straits of Hormuz, running between northern Oman and Iran, should be under friendly control. At the height of the action against the guerrillas, Iran had AB.205 gunship and CH-47

Top right: A Jaguar International lands at Thumrayt, its home base in Oman's Dhofar province.

Middle right: The BAC Strikemaster, seen here over the Omani coast near Salalah in Dhofar province, played a major part in the war against the guerrillas. Typical armament was 16 Sura rockets and two 500lb bombs, as illustrated.

Right: A SOAF Skyvan over rough Omani terrain.

transport helicopters operating in Oman and also deployed F-4 Phantoms to provide air defence. Britain contributed expertise in the form of officers and NCOs, both seconded and on contract to the SOAF.

Inventory

SOAF's first combat aircraft were 12 BAC Strikemaster Mk 82s ordered in 1967; these were followed by a batch of eight Mk 82As ordered in 1972 with an additional four being signed for in the spring of 1974. These aircraft bore the brunt of actions against the guerrillas, flying from Salalah.

The Strikemasters were supplemented from 1975 by Hawker Hunters donated to the Sultan by King Hussein of Jordan. Twenty-nine aircraft were supplied, and these are used to maintain a front-line force of about 15. Several were lost in combat, as were a number of Strikemasters and helicopters. The Hunters carry bombs and up to four 30mm Aden cannon. Five of the Strikemasters were sold to Singapore and transferred in early 1977.

The first of 12 BAC/Dassault-Breguet Jaguar International fighter-bombers were delivered in March 1977, the use of more powerful Adour Mk 804 turbofans improving the type's hot-and-high performance and allowing it to double as a fighter, Matra Magic air-to-air missiles being carried on overwing pylons for the latter role. Another three aircraft may be bought if the money is made available. The Jaguars and other aircraft are protected by 28 fire units of the BAC Rapier low-level surface-to-air missile, ordered under a £47m contract placed in 1974, with Marconi S600 radars and a tropo-scatter communications network forming the rest of the new air-defence system.

The helicopter fleet is used for transport and other non-combat duties, SOAF believing that helicopter gunships are inferior to fixed-wing types such as the Strikemaster. Similarly, neither the Defenders nor the Skyvans are armed. The older elements of the transport force, two Caribou and five Viscounts, have been withdrawn from service.

All SOAF aircraft are maintained by the British company Airwork Services. At present the great majority of pilots are British, but Omani nationals are under training to take over from them.

OMAN

Unit	Type	Role	Base	No	Notes
1 Sqn	Strikemaster Mk 82/82A	GA/train	Salalah	12	Two others in reserve
2 Sqn	Skyvan	Trans	Seeb	15	Some rotated to Salalah
3 Sqn	AB.205	Trans	Salalah	20	For transport, logistics, casevac
	AB.206	Util	Salalah	3	For transport, artillery spotting
	Bell 214	Trans	Salalah	5	
4 Sqn	BAC One-Eleven 475	Trans	Seeb	3	
5 Sqn	Defender	Trans	Seeb	8	Some rotated to Salalah
6 Sqn	Hunter	Fight/GA	Thumrayt	15	Others in store. Thumrayt formerly Midway
8 Sqn	Jaguar S	GA/fight	Thumrayt	10	Armed with R.550 Magic
	Jaguar B	Train/GA/fight	Thumrayt	2	
	Rapier	SAM	Thumrayt	28 fire units	
Police Air Wing	Turbo-Porter	Util	Seeb	2	
	Learjet 25B	Util	Seeb	1	
Royal Flight	AS.202 Bravo	VIP	Seeb	2	
	AB.212	VIP	Seeb	1	
	Bell 212	VIP	Seeb	2	
	DHC-5D Buffalo	Trans		1	

UNITED ARAB EMIRATES

Area: 32,000 sq. miles (82,900 sq. km.).
Population: 656,000.
Total armed forces: 25,900.
Estimated GNP: $7·7 billion (£3·85 billion).
Defence expenditure: $661 million (£330·5 million).

The United Emirates Air Force has been formed by amalgamating the previously allied but independent Abu Dhabi Air Force, Union Air Force and Dubai Police Air Wing. Money for the UEAF is provided by all seven emirates – Abu Dhabi, Sharjah, Ajman, Fujairah, Ras al-Khaimah, Dubai and Umm al-Qaiwain – and the force is based at Abu Dhabi, Dubai and Sharjah.

The original Abu Dhabi Air Force, which formed an integral part of the Abu Dhabi Defence Forces, contributed most to the federal UEAF in terms of equipment and personnel. A ground-attack squadron operating Hawker Hunters was formed in 1970 at Sharjah and continues to be based there, its strength at the end of 1978 being eight FGA.76s and a pair of T.77 trainers. This combat force was bolstered from 1974 by the first of 14 Dassault Mirages ordered as an initial batch. These original aircraft, a dozen Mirage 5AD strike fighters and two 5DAD combat trainers, were reinforced in 1976–77 by an additional 14 5ADs, a further two-seater and three reconnaissance 5RADs, making a total of 32 Mirages in the two-squadron interceptor-strike wing. Many of the Mirages are flown by Pakistanis, although local Arabs are planned to take over as they are trained.

In 1972 Abu Dhabi donated three AB.206s to the Air Wing of the Union

Above: A pair of MB.326K single-seat light attack aircraft operated by the United Emirates Air Force, the machines having been ordered originally by Dubai – one of seven emirates in the UAE.

Defence Force, as it then was (see below), these JetRangers being replaced by Alouettes ordered as part of the same French arms package which included the Mirages. Ten SA.319 were bought and at least three have since crashed, although these may have been replaced by new aircraft. The helicopters operate on utility duties, with a secondary responsibility for close-support and anti-tank roles using Aérospatiale AS.11 and AS.12 missiles. Five Aérospatiale SA.330 Pumas were also bought for assault and logistic support, and the UEAF was due to receive some of the Lynxes which were to be assembled under licence at Helwan in Egypt by the Arab-British Helicopter Company before financing was withdrawn. The transport fleet is likewise to be expanded.

Abu Dhabi is unusual in having bought two types of low-level surface-to-air missile, Rapier and Crotale. The Rapier order, worth £35m, includes Marconi Space and Defence Systems DN181 Blindfire radars.

The Abu Dhabi element of the UEAF has no combat experience as such, but it is likely that many of the seconded Pakistani pilots have previously flown in action against Indian aircraft and ground targets. Servicing was originally carried out by the British company Airwork, but since 1974 it has been under the supervision of Pakistani ground engineers.

The UEAF has also absorbed the jointly financed federal Union Air Force, which until October 1974 had been known as the Air Wing of the Union Defence Force. The UDF itself was founded in 1971 to replace the Trucial Oman Scouts, a paramilitary force organised and officered by the British Army with responsibility for army and interior policing duties. The Royal Air

Force and British Army Air Corps, operating from Sharjah, met the UDF's airborne requirements until their withdrawal in December 1971. The UDF Air Wing was founded in early 1972 with the three AB.206s donated by Abu Dhabi and remained an all-helicopter force until its integration into the UEAF.

In mid-1975 the UAF's AB.206s were joined by the first Bell 205A-1, and this fleet now numbers four aircraft; these, together with three AB.212s, are used for troop-transport work.

The third element which went to make up the UEAF was the Dubai Police Air Wing, which operated alongside the UAF at Dubai airport. A new helicopter hangar, workshops and offices were built at Dubai in the early 1970s and these now form the UEAF's Central Air Force Base. The DPAW's contribution to the new federal force comprised its helicopters and the light attack force of Aermacchi MB.326s. By the end of 1978 this force totalled three single-seat MB.326KDs and one two-seat MB.326LD; a further three or four may be bought. A single Aeritalia G.222 twin-turboprop transport was delivered in December 1976 and the option on a second has since been taken up, with the possibility of an additional one or two of the type following later.

The UEAF is expected to continue its gradual consolidation and expansion, and the possibility of acquiring Alpha Jet light strike-trainers from 1980–81 has been reported. The UAE is additionally planned to be included in a new network of Thomson-CSF surveillance radars being bought by and partially constructed in Saudi Arabia. Coverage will extend over Qatar and Egypt as well as Saudi and the UAE.

UNITED ARAB EMIRATES

Type	Role	Base	No	Notes
Mirage 5AD/5DAD/5RAD	GA/train/recce	Abu Dhabi	26/3/3	Two squadrons
Hunter FGA.76/T.77	GA/train	Sharjah	8/2	Two others lost
C-130H Hercules	Trans	Abu Dhabi	2	
G.222	Trans	Dubai	1	One other ordered
DHC-5D Buffalo	Trans	Abu Dhabi	4	
Caribou	Trans/casevac	Abu Dhabi	3	One other lost
Islander	Trans	Abu Dhabi	4	
MB.326KD	GA	Dubai	3	
MB.326LD	Train	Dubai	1	
SA.319 Alouette III	Util/att	Abu Dhabi	7	Originally ten. Can carry AS.11/AS.12 missiles
Puma	Assault/logistics	Abu Dhabi	10	
Bell 205A-1	Trans/liaison	Dubai	6	
AB.206	Liaison/casevac/spotting	Dubai	3	
Bell 206B	Liaison/spotting	Dubai	3	
AB.212	Trans	Dubai	3	
SF.260WD Warrior	Train	Dubai	1	
Cessna 182	Train/VIP	Dubai	1	
Lake LA-4	VIP	Abu Dhabi	1	
Pawnee Brave	Util	Abu Dhabi	1	
Rapier Crotale	⎤—SAM	various		

KUWAIT

Area: 7,500 sq. miles (19,400 sq. km.).
Population: 1·13 million.
Total armed forces: 12,000.
Estimated GNP: $12 billion (£6 billion).
Defence expenditure: $322 million (£161 million).

The Kuwaiti Air Force, which is responsible for all military aviation, is being expanded as part of a seven-year programme initiated in May 1973. The armed forces have been allocated KD422m (some $1500m at 1973 prices) for procurement of additional equipment in addition to the normal defence budget. The Air Force has approximately 1000 Kuwaiti personnel, to which must be added overseas advisers; Britain has historically had close ties with the country, and in 1960 a mission was set up to advise on the creation of the KAF from a mixed bag of British-supplied transports and light aircraft acquired by the Kuwaiti Government's Security Department during the 1950s.

The KAF's first armed aircraft – Jet Provost T.51s – were followed by Hunters flown by seconded Royal Air Force pilots and then from 1968 by Lightning interceptors, with Strikemasters following in 1969 and 1971. The transport fleet was also

KUWAIT

Type	Role	Base	No	Notes
Mirage F.1CK	Int	Kuwait City	18	Replaced Lightnings. Armed with R.550 Magic?
Mirage F.1BK	Train/int	Kuwait City	2	
A-4KU Skyhawk	GA		36	Armed with Sidewinder
TA-4KU	Train/GA		6	
Hunter T.67	Train/GA		5	
Strikemaster Mk 83	GA		9	Originally 12
L-100-20 Hercules	Trans		2	
DC-9-32CF	Trans		2	Convertible pax/cargo
SA.330H Puma	Trans/assault/logistics		10–12	
SA.342 Gazelle	AT/obs		24–30	Two squadrons, one armed with HOT anti-tank missiles
Improved Hawk	SAM		14 batteries	

QATAR

Area: 4,000 sq. miles (10,350 sq. km.).
Population: 180,000.
Total armed forces: 4,000.
Estimated GNP: $2·4 billion (£1·2 billion).
Defence expenditure: $61 million (£30·5 million).

Above: The Qatar Emiri Air Force flies a total of four Westland Commandos, the logistic and VIP derivative of the Sea King anti-submarine helicopter. All QEAF aircraft are of British origin.

The Qatar Emiri Air Force is part of the army but enjoys a measure of autonomy. Despite having substantial oil revenue, some of which is being contributed to the Arab Organisation for Industrialisation to finance the establishment of arms-manufacturing factories in Egypt (which see), Qatar has spent little on her own forces and has a combat element of only three Hunters (another was lost in 1977) compared with the once-planned force of 12 aircraft. It was originally hoped that Qatar, along with Bahrain, would form part of the United Arab Emirates; this has not materialised, however, and the QEAF acts on Qatar's own behalf as a coastal-patrol and transport force. Pilots are seconded from Britain, which is at present the sole supplier of QEAF aircraft. The air force as such has no combat experience and, being sandwiched between Saudi Arabia and Iran on the other side of the Gulf, understandably maintains a low military profile.

Inventory

The QEAF's sole base, the single-runway airport at Doha, houses the combat element of Hunters and the transport fleet comprising a variety of fixed-wing aircraft and helicopters. Dassault Mirage F.1s may be ordered to replace the Hunters, and additional Lynxes are likely to be bought directly from Westland. The five launchers for Tigercat surface-to-air missiles are also installed at Doha, although they are not permanently emplaced. The two Gazelles are flown by QEAF pilots but are used mainly on police work. Qatar is expected to benefit from a network of Thomson-CSF surveillance radars which is to be paid for by Saudi Arabia and will also cover the United Arab Emirates and Egypt. The Tigercats may be replaced by the Thomson-CSF Shahine or the Crotale system on which it is based, both weapon systems already having been ordered by Saudi Arabia.

QATAR

Type	Role	Base	No	Notes
Hunter FGA.78	FGA	Doha	2	One other lost
Hunter T.79	Train	Doha	1	
Islander	Trans/liaison	Doha	1	
Commando Mk 2A	Trans	Doha	3	
Commando Mk 2C	VIP	Doha	1	
Whirlwind III	Trans	Doha	2	
Gazelle	Police	Doha	2	Flown by QEAF pilots on police duties
Lynx	Trans	Doha	3	
Tigercat	SAM	Doha	5 launchers?	

Left: The Kuwait Air Force's Strikemasters form a light attack force to back up the main air-to-ground punch provided by the KAF's two squadrons of A-4KU Skyhawks.

expanded, and under the new equipment programme the KAF is receiving aircraft from the United States and France.

Inventory

Major purchases include Mirage F.1s, which have replaced the force of ten remaining Lightning F.53s on interception duties, and the total of 42 Skyhawks. The Lightnings – originally 12 F.53s and a pair of two-seat T.67 trainers – were armed with Firestreak and Red Top air-to-air missiles, and were being retired after only ten years' service. The Mirage order includes a pair of F.1B trainers, Kuwait being the first country to order this two-seat variant. The Mirages are thought to carry Matra R.550 Magic short-range air-to-air missiles.

The Mirages operate from the existing Lightning base, but the Skyhawks fly from one of the two new bases built with Jugoslav aid in the south and west of the country at a cost of KD70 million. The 36 A-4KUs and six two-seat TA-4KUs were ordered after the breakdown of negotia-

Above: The KAF's two McDonnell Douglas DC-9-32CFs can be converted to carry either passengers or cargo. Like the L-100-20 Hercules which they complement, the DC-9s have minimal military equipment.

tions for the Anglo-French Jaguar, Kuwait having insisted on a fixed-price contract. The Skyhawks' engines and avionics are to Kuwaiti standards, and the aircraft are armed with AIM-9H Sidewinder air-to-air missiles, 300 of which have been supplied. They may also carry laser-guided bombs and air-to-surface missiles. Teheran and Baghdad were among the targets specified before the purchase of A-4s, and Iraq is the most likely aggressor. In times of tension it is likely that the Skyhawks would be dispersed to other airfields in the Arabian Gulf, notably Bahrain.

The interceptors are integrated into an air-defence system based on two ITT Gilfillan TPS-32 three-dimensional radars plus gap-fillers, operations centres and communications links. Raytheon Improved Hawk medium-range surface-to-air missiles complement the aircraft, and a purchase of Thomson-CSF/Matra Crotale low-level SAMs for base defence has been reported.

167

Northern Africa

Both the eastern and western extremities of northern Africa have seen bitter conflict in the late 1970s, with the Somali Democratic Republic and Ethiopia clashing – during which the Soviet Union shifted its support from the former country to the latter – and Polisario guerrillas tangling repeatedly with Moroccan forces following the division of the former Spanish Sahara into areas of Moroccan and Mauretanian responsibility in 1976.

Moroccan forces took control of the ex-Spanish air bases but have been opposed by guerrillas operating with Algerian support and armed with weapons of Russian origin. These include SA-7 shoulder-launched surface-to-air missiles, to which a number of Moroccan aircraft have fallen victim. The conflict has spurred modernisation and re-equipment of the Moroccan Air Force, which has ordered Mirage F.1 fighter-bombers and an air-defence network to counter the possibility of Algeria taking a more active role in the war.

Russian support ensured that Ethiopia resisted Somali attempts to gain control of the provinces which the Somalis regard as their own, but fighting continued in Eritrea. Ethiopian aircraft have been involved extensively in attacks on forces of the Eritrean Liberation Front, which seeks independence from Ethiopia, and the sale of many Western arms to the country was embargoed. No lasting solution to the dispute seems likely as yet.

Sandwiched between Ethiopia and the Somali Republic is Djibouti, which has obtained independence from France. The French retain strong links with Djibouti, however, and continue to base combat aircraft and surface forces there. Other former French areas of influence such as Chad are now less amenable to acting as staging posts for *Armée de l'Air* transports, however.

Nigeria, with its healthy supply of natural resources and a large population, is an important influence in West Africa and has been building up its armed forces by buying from almost every conceivable arms-manufacturing country. The secession of Biafra in the 1960s showed how easily tribal and other differences can escalate into a bloody war which produces no real victors, and the possibility of such conflicts erupting remains throughout Africa. The area also has its share of despotic dictators, and the killing of schoolchildren in the Central African Empire under the orders of Emperor Bokassa in the Spring of 1979 could have far-reaching consequences.

TUNISIA

Area: 63,400 sq. miles (164,200 sq. km.).
Population: 5·7 million.
Total armed forces: 22,200.
Estimated GNP: $5 billion (£2·5 billion).
Defence expenditure: $185 million (£92·5 million).

Tunisia maintains a small defence establishment, with the air force flying mainly limited training and support duties. Manpower totals 22,000, of which some 2000 serve in the air force. Defence expenditure is a modest Dinar 77m ($185m) for 1978, of which a large slice is being devoted to new military equipment for the TAF. Headquarters and main base are at El Aouina, Tunis, and the TAF is the sole operator of military aircraft in the country.

A French protectorate since 1881, Tunisia gained its independence in 1956. The Tunisian Air Force was not officially established until 1960, however, when the first of a batch of 15 Saab Safir training aircraft arrived. Ten Swedish Air Force flying instructors and technicians were seconded to the TAF to assist in training. Two Alouette II helicopters were acquired in 1962 and the following year a French offer of assistance was accepted, resulting in the delivery of three Flamant transports and some T-6 Coin aircraft. In 1966 eight Aermacchi MB.326s arrived from Italy to form a dual strike/trainer unit. A greater combat potential accompanied the delivery in 1969 of a dozen F-86F Sabres from the United States under an MAP agree-

Top: The Tunisian Air Force operates an all-French fleet of helicopters, including this Alouette III, on various liaison duties.

Above: The TAF's Macchi MB.326B trainers are fitted with underwing hardpoints, allowing them to be used for attack as a subsidiary role.

ment for the first and, to date, only front-line combat squadron in the TAF. Italian assistance has replaced Swedish and French aid in recent years.

Tunisia is a member of the League of Arab States and has an aid agreement with Italy, while receiving limited military assistance from the US.

Inventory

Following a protracted period during which various modern combat aircraft – in particular the Northrop F-5E and McDonnell A-4 Skyhawk – were evaluated, it was reported in 1976 that the Tunisian Air Force had placed an order for 16 F-5Es. The letter of acceptance was not taken up, however, possibly due to budget problems. The air force is therefore retiring the survivors of the dozen Sabres supplied in 1969 and now employed mainly in the training role, and replacing them with a modern ground-attack type in the shape of eight Italian single-seat MB.328Ks

TUNISIA

Type	Role	No	Notes
F-5E	Int	12	Order reported placed in 1976 but later cancelled; two two-seat F-5Fs included
F-86F	FGA/train	12	Due to be replaced by single-seat MB.326Ks
MB.326B	Att/train	8	Used in dual strike/trainer role; fitted with six underwing strong points
MB.326KT	Att	8	
MB.326LT	Train	4	
Flamant	Trans	3	
S.208	Liaison	4	
Alouette II	Liaison	8	
Alouette III	Liaison	6	
Puma	VIP	1	
SF.260WT	Train	12	Secondary Coin role
SF.260CT	Train	6	Delivered in 1978
T-6	Train	12	

recently delivered. Negotiations for up to three Aeritalia G.222 STOL transports have been under way since the demonstration of a prototype to the TAF in 1974. Original plans called for the G.222s to be delivered in 1977 but the deal had yet to be finalised at the time of writing. Three ageing French Flamants are the sole fixed-wing transport type in the inventory, other liaison and communications duties being flown by some French-supplied helicopters.

Support

Initial pilot training in the Tunisian Air Force is undertaken on the SF.260s, students then progressing to the MB.326s before assignment to the single operational combat squadron. The Sabres have until recently been used for advanced training and it is expected that they will be withdrawn from use when the single-seat MB.326Ks are fully integrated into the force.

ALGERIA

Area: 855,200 sq. miles (2·21 million sq. km.).
Population: 17 million.
Total armed forces: 78,800.
Estimated GNP: $10·1 billion (£5·05 billion).
Defence expenditure: $456 million (£228 million).

The *Force Aérienne Algérienne*, with headquarters in Algiers, was established in 1962 following the country's independence from France. On July 3, 1962, France transferred sovereign power to Algeria and in November of that year the nucleus of an air force was established by the FLN, or National Liberation Army, following the gift of five MiG-15s by Egypt; a small Egyptian training mission was also sent. The following year two Beech D18S transports were delivered and 1964 saw the arrival of some Czech-built Yak-11 trainers. In 1966 the first major deliveries of Russian aircraft began, including MiG-21Fs, Il-28s and SA-2 Guideline surface-to-air missiles, together with Soviet instructors and engineers.

As well as receiving Soviet aid, Algeria is a member of the League of Arab States and has a defence agreement with Libya under which certain Libyan equipment would be put at the disposal of the Algerian armed forces. Algeria supports the Polisario Front against Morocco and Mauretania over the Spanish Sahara.

ALGERIA   Air Force

Type	Role	No	Notes
MiG-21F	Int	70	Wing of three squadrons; equipped with Atoll missiles
MiG-23	FGA		
MiG-17F	Att	60	Three squadrons (being replaced by MiG-23)
Su-7BM	Att	20	Two squadrons
Su-20	Att		
MiG-15	Train	20	
IL-28	Bomb	24	Two squadrons; low serviceability
Magister	GA/train	28	Two squadrons, delivered 1971 from Aérospatiale
Puma	Trans	5	Used for assault duties
Hughes 269	Train	6	
Mi-4	Trans	40	
Mi-6	Trans	4	
Mi-8	Trans	12	
IL-18	VIP	4	
An-12	Trans	8	
F.27-400	Trans	5	
F.27-600	Trans	1	
IL-14	Trans	6	Of about 12 originally delivered
Queen Air	Liaison	3	
Super King Air	Liaison	3	
King Air	Liaison	1	For nav-aid calibration checking
Gomhouria	Train		
Yak-11	Train		
T-34C	Train	6	
MiG-15 UTI	Train		
MiG-21U	Train		
SA-2	SAM		

Navy
| F-28 Mk 3000C | Trans | 3 | |

Above: The *Force Aérienne Algerienne* operates an extensive range of Russian-supplied types such as this An-12, one of eight in service.

Inventory

The FAA operates a wing of MiG-21 interceptors armed with AA-2 Atoll air-to-air missiles, a strike element of Su-20s, Su-7s and MiG-23s (replacing MiG-17s) plus two bomber units of doubtful serviceability flying Il-28s. In 1971 Algeria purchased some refurbished Magisters from France for light strike duties and training, and at the same time bought five SA.330 Pumas to increase the effectiveness of her helicop-

ter assault force. Elementary training is carried out on Helwan Gomhourias (Egyptian-built Bücker Bü181s), with students moving on to Yak-11s (being replaced by T-34C Mentors), and Magisters. Jet fighter conversion is flown on MiG-15UTIs and MiG-21Us. Algerian personnel are also

trained in Russia and other Arab countries. Total FAA manpower is some 4,500.

Recent combat experience involved limited participation in the 1973 Arab-Israeli war and in 1976 ground-attack sorties by MiG-21s against Moroccan targets along the border.

MOROCCO

Area: 180,000 sq. miles (466,000 sq. km.).
Population: 15·4 million.
Total armed forces: 89,000.
Estimated GNP: $9·5 billion (£4·75 billion).
Defence expenditure: $681 million (£340·5 million).

The Royal Maroc Air Force, *Al Quwwat Aljawwiya Almalakiya Marakishiya*, was formed in 1956 with French assistance as Sherifian Royal Aviation. The initial combat aircraft were four Hawker Furies donated by Iraq, and in 1961 the Soviet Union supplied the air force's first jet types in the form of MiG-17 fighters and MiG-15 trainers. Relations between Morocco and Algeria soon deteriorated, however, and the USSR, which was also supplying arms and advisers to the latter country, withdrew its support from Morocco.

The RMAF had assumed control of former French and US bases in 1963, and these countries both supplied aircraft during the 1960s. The smouldering conflict with Algeria over borders was further fuelled in the mid-1970s by Algerian support for the Polisario Front, which disputed the division of the former Spanish Sahara between Morocco and Mauretania and is supported by both Algeria and Libya. This has led to a burst of procurement for the RMAF from US and European suppliers. Personnel strength totals some 6000.

Inventory

The RMAF's major new combat type in the 1980s will be the Dassault Mirage F.1CH, of which 25 were ordered in early 1976 for delivery in 1978–79. A further 50 were taken on option at the same time, and in 1977 half this batch was converted to a firm order for service entry in 1979–80. Some of the Mirages are expected to be F.1BH two-seaters and the third batch of 25, if bought, is likely to comprise the F.1A attack variant.

The RMAF twice rejected letters of offer covering the supply of 24 Northrop F-5Es and F-5Fs to supplement the ageing force

Right: Turboprop-powered Beech T-34C-1s have taken over from the RMAF's force of T-6s and T28s used in the attack role, in addition to operating on training duties.

of F-5A/Bs, and instead ordered 24 Alpha Jet E for delivery from 1980. They will be operated in the light strike role as well as for training. Comparatively large numbers of helicopters have been acquired in the 1970s, some being operated by the army, and the transport fleet has been upgraded by the introduction of Lockheed Hercules to replace the C-119s supplied as part of the US aid programme. The US has declined, however, to provide the 24 Rockwell OV-10 Bronco counter-insurgency aircraft and Bell AH-1 gunships which Morocco requested.

Primary instruction of RMAF pilots is carried out on AS.202 Bravos, followed by basic training on Beech T-34C Turbo Mentors. The latter type also has a secondary counter-insurgency role, having superseded the mixed force of T-6Gs and

MAURETANIA

Area: 419,000 sq. miles (1·08 million sq. km.).
Population: 1·48 million.
Total armed forces: 12,500.
Estimated GNP: $210 million (£105 million).
Defence expenditure: Not known.

A limited modernisation of the Mauretanian Islamic Air Force began in 1975 to increase the effectiveness of this small force in its operations against Polisario guerrilla activity. Operated on French lines, the MIAF is closely allied with the Mauretanian army and has its headquarters in the capital, Nouakchott. It is chiefly a transport arm but in recent years has integrated a small attack and counter-insurgency element. Personnel total some 150–200.

Following independence from France in 1960, Mauretania received a Douglas C-47 and two MH-1521M Broussard light transports for supply duties, an important task in a country of its size. More French equipment arrived at irregular intervals, including some Reims Super Skymasters fitted with underwing weapon pylons for Coin, liaison and training missions. The four C-47s in service towards the end of the 1960s were replaced by two Short Skyvans.

Inventory

Economic considerations to date have prevented the acquisition of jet aircraft for the Mauretanian Islamic Air Force and the only combat machines are the Britten-Norman Defenders. These are used for attack duties, anti-smuggling patrols along the coast and liaison work complementing the Super Skymasters. The Defenders are able to carry 2300lb of weapons on four underwing pylons, including Sura air-to-ground rockets, and have provision for sideways-firing light machine guns. Further aircraft orders are likely and the procurement of Italian SF.260 trainers has been mentioned.

MAURETANIA

Type	Role	Base	No	Notes
Defender	Coin/coastal patrol	Nouakchott	7	Two each delivered in 1976, 1977 and 1978; two lost; three more ordered in 1978
Skyvan 3M	Trans	Nouakchott	2	Delivered November 1975; joint army/AF operated
DHC-5D Buffalo	Trans	Nouakchott	2	
DC-4	Trans	Nouakchott	2	
MH-1521M Broussard	Liaison/trans	Nouakchott	2	
Reims 337	Coin/FAC/train	Nouakchott	4	Possible order for four more
Caravelle	VIP	Nouakchott	1	

Below: The Britten-Norman Defenders of the Mauretanian Islamic Air Force are the Service's only combat aircraft, but they can pack a powerful punch in the form of up to ten Sura air-to-surface rockets on each pylon. The total weapon load can be as much as 2300lb, with sideways-firing machine guns providing suppressive fire. The aircraft operate on counter-insurgency and coastal-patrol missions.

Type	Role	Base	No	Notes
Air Force				
Mirage F.1CH	Int		50	
F-5A/RF-5A	FGA/recce	Kenitra	13/2	⎤ Two squadrons
F-5B	Train	Kenitra	2	⎦
MiG-17	FGA		12	⎤ In storage
MiG-15UTI	Train		2	⎦
Magister	Att		24	Two squadrons
C-130H	Trans		4	
King Air 100	Liaison/VIP		6	
MH.1512M Broussard	Liaison/VIP		12	
T-34C	Train		12	
T-6G	Train/att		up to 30	⎤ Being replaced by
T-28	Train/att		up to 20	⎦ T-34C
Alpha Jet E	Train/att		24	Delivery from 1980
SF.260MM	Train		2	
AS.202/18 Bravo	Train		14	
Do28D-2 Skyservant	VIP		1	Used by the King
AB.212	Trans		5	
AB.205	Trans		35	
CH-47C	Trans		6	
AB.206	Obs/util		5	
HH-34B	SAR		4	
SA.330	Trans		40	
Bell 47G-2	Util		4	
Crotale	SAM		—	
Army				
Alouette II	Liaison/obs/util		4	
SA.342	Liaison/obs/util		6	

T-28s. Base defence is provided by Thomson-CSF Crotale surface-to-air missiles.

The RMAF has been involved with border skirmishes with Algeria for many years and has recently been engaged in more intensive operations against Polisario guerrillas. At least three aircraft have been lost to ground fire, including one or more brought down by Russian-supplied SA-7 shoulder-launched surface-to-air missiles. The guerrillas may also have been supplied with Russian SAMs.

SENEGAL

Area: 77,800 sq. miles (201,500 sq. km).
Population: 5 million.
Total armed forces: 6,550.
Estimated GNP: $1·7 billion (£850 million).
Defence expenditure: $48 million (£24 million).

The *Armée de l'Air du Senegal*, like many of the air forces of former French colonies, is organised on French lines. Close ties continue between the two countries and an agreement exists for Senegalese air and ground crews to be trained in France. Air Force headquarters is in Dakar and the main base is at Dakar-Yoff. Some 200 personnel man the air arm, which is mainly concerned with internal security, transport and liaison duties. Total defence budget for 1978 was $48m.

Senegal became independent in November 1958, but it was 1960 before any form of national air arm was established. France supplied some C-47s, Broussard light transport/liaison aircraft and some Alouette II helicopters. *Armée de l'Air* personnel were seconded to the embryo force and Senegalese bases were used regularly during the 1960s by French air force and naval units. By 1973 all the technicians and aircrew were Senegalese except for one pilot and two groundcrew

advisers: two Bell 47Gs supplied by France some years before had been replaced by later types.

A security assistance agreement exists between the United States and Senegal, and a co-operation agreement for defence is in force with France – although training forms the main, and at present only, part of the agreement.

Inventory
Two ex-l'*Armée de l'Air* Magisters are the first jet equipment acquired by the air arm. Future procurement is likely to be slanted towards more modern transport aircraft to replace the C-47s and Broussards, and the purchase of some DHC Buffaloes is possible.

Below: The best-selling Fokker-VFW Friendship is one of only three non-French types operated by the *Armée de l'Air du Senegal*, a small air arm which relies heavily on France for training and support.

Type	Role	Base	No
Magister	Train	Dakar	2*
C-47	Trans		6
MH-1521 Broussard	Trans		4
Reims F337	Liaison		1
Alouette II	Liaison/SAR		2
Gazelle	Liaison/SAR		1
F.27 Friendship	Trans	Dakar	6
Boeing 727-200	VIP	Dakar	1

Ex-Armée de l'Air

171

GUINEA

Area: 97,000 sq. miles (251,000 sq. km.).
Population: 3·9 million.
Total armed forces: 8,850.
Estimated GNP: $740 million (£370 million).
Defence expenditure: $26·4 million (£13·2 million).

This small republic of some 4.6m people on the west coast of Africa has an air arm known as the *Force Aérienne de Guinea* which is based at the capital, Conakry. Formerly a French colony, Guinea subsequently fell within the Soviet sphere of influence both politically and militarily and strengthened the ties between the two countries some years ago by granting the Soviet Union port and airfield facilities at Conakry. In 1975–76 Russian aircraft carrying Cuban troops staged through Guinea on their way to the Angolan war. During 1978 there was a noticeable cooling in relations between Russia and Guinea. The FAG operates mainly Russian aircraft and manpower totals in excess of 600.

The country has a military assistance agreement with China, while Cuba has supplied limited military aid.

Inventory

A single fighter-bomber squadron equipped with MiG-17s forms the sole combat unit, but only about half the aircraft are thought to be operational at any one time. Future procurement is not thought to amount to any more than a few second-line types unless a change of alignment by the government occurs, as has happened in other states in Africa.

REPUBLIC OF GUINEA

Type	Role	Base	No
MiG-17	FGA		8
IL-18	Trans/VIP	Conakry	2
IL-14	Trans	Conakry	4
An-14	Trans	Conakry	4
Bell 47G	Liaison		1
MiG-15UTI	Train		2
L-29	Train		3
Yak-18	Train		7

SIERRA LEONE

Area: 27,900 sq. miles (72,200 sq. km.).
Population: 3 million.
Total armed forces: 2,200.
Estimated GNP: $720 million (£360 million).
Defence expenditure: $9·54 million (£4·277 million).

Situated on the west coast of Africa between Guinea and Liberia, Sierra Leone has only a small armed force known as the Sierra Leone Defence Forces equipped with a handful of aircraft operated mainly in the support role. Headquarters and main base are at Freetown. All the SLDF's aircraft are based at Freetown. Although independence was gained from Great Britain in 1961, the formation of an air element was not undertaken until 1973 when four Saab MFI-15s were purchased and a Swedish technical and training mission arrived to instruct Sierra Leone nationals. Three Hughes helicopters were also acquired. Current operations involve liaison and communications between outlying areas.

SIERRA LEONE

Type	Role	No	Notes
Hughes 300	Liaison	3	Supplied by Saab as
Hughes 500	Liaison	2	Hughes agent
BO105		1	

MALI

Area: 465,000 sq. miles (1·20 million sq. km.).
Population: 6·3 million.
Total armed forces: 4,200.
Estimated GNP: $615 million (£307·5 million).
Defence expenditure: $19·42 million (£9·71 million).

The *Force Aérienne du Mali* has its headquarters in the capital, Bamako, and the manpower level is not thought to be greater than 400. A token air arm comprises both Communist and Western equipment.

A French colony until independence in 1960, Mali gradually severed connections with France following the relinquishing of French bases in October 1961. An air arm was established in the mid-1960s following the donation of two Douglas C-47s by the United States. Military aid was also received from the Soviet Union and other Eastern-bloc countries. A handful of MiG-15s were supplied to the emergent air force and Mali personnel were given training in the USSR.

Mali has a military-assistance agreement with Communist China and with the Soviet Union.

Inventory

Soviet equipment predominates, although the operational status of the single combat unit flying MiG-17s is uncertain. Second-line aircraft include some Yak-12M liaison machines supplied by Poland. Future procurement is likely to involve more aircraft of Chinese or Soviet origin.

MALI REPUBLIC

Type	Role	Base	No	Notes
MiG-17	Int/GA	Bamako?	5+	Supplied by Soviet Union in mid-1960s
MiG-15UTI	Train	Bamako?	1	Supplied by Soviet Union in mid-1960s
C-47	Trans	Bamako?	2	Supplied by US
An-2	Trans		2	
An-24	Trans		2	
Mi-4	Trans		2	
Yak-12M	Liaison		1	

IVORY COAST

Area: 127,000 sq. miles (329,000 sq. km.).
Population: 5·4 million.
Total armed forces: 4,950.
Estimated GNP: $6 billion (£3 million).
Defence expenditure: $51·3 million (£25·65 million).

This West African republic has a small military air component known as the *Force Aérienne de Côte d'Ivoire* operating in support of the army, flying liaison and transport duties. Headquarters and main base are at Abidjan, the Service having a manpower of some 400 officers and men out of a total defence force of 4100. Defence expenditure for 1974 amounted to CFA Fr.6350m ($22.8m).

Formerly part of French West Africa, the Ivory Coast gained its independence in August 1960 along with a number of other French colonies and established a small army in 1961. With the help of France the country's military arm was expanded the following year with the addition of an air component for communications with

GHANA

Area: 92,100 sq. miles (238,000 sq. km.).
Population: 8·5 million.
Total armed forces: 17,700.
Estimated GNP: $4·1 billion (£2·05 billion).
Defence expenditure: $130 million (£65 million).

Essentially operating in the transport, survey and internal policing roles, the Ghana Air Force has its headquarters at Burma Camp, Accra, and comes under the control of the military government. Defence expenditure for 1977 totalled 113.5m Cedi ($130.5m) and manpower is 1400.

Ghana has no major defence alliances with other countries apart from a security assistance agreement with the US.

Following independence from Britain, the Ghana Air Force was established at Accra in 1959 with the help of Israeli and Indian Air Force personnel, operating a dozen Hindustan HT-2 basic trainers. In 1960 a British Joint Services Training Team took over the training of Ghana's armed forces, reorganising the air force along RAF lines. Chipmunk trainers replaced the HT-2s and a permanent flying training school was formed at Takoradi. Limited Soviet equipment was supplied to the GAF in the early 1960s, but Russian involvement in the country ended shortly outlying regions and transport duties as well as liaison with ground forces. A Douglas C-47 and a couple of Broussards comprised the initial equipment, these being followed by some Alouette helicopters.

Ivory Coast has a defence and military co-operation agreement with France and regional defence agreements with Niger and Benin.

Inventory

For many years a purely transport and liaison force, the *Force Aérienne de Côte d'Ivoire* has announced its intention to acquire a small strike-trainer element in the shape of a dozen Alpha Jets. These are expected to be equipped with armament for the dual role. Transport flights are conducted by C-47s and Puma helicopters, while liaison is performed by Reims Super Skymasters. Much of the equipment is French in origin, but recent acquisitions have included two Fokker-VFW F.28s and a similar number of F.27s from the Netherlands. A special government unit based at Abidjan operates six aircraft, including the F.28s and a Falcon 20 VIP jet. Future procurement could involve some modern Stol transports to replace the C-47s. Apart from Abidjan, the air force has numerous strips throughout the country, some situated around the larger townships.

Below: The President of the Ivory Coast has a wide selection of VIP aircraft to choose from, including this pair of VFW-Fokker F.28 Mk 1000 airliners.

IVORY COAST

Type	Role	No	Notes
Alpha Jet	Coin/train	12	
F.28 Mk 1000	VIP	2	For Presidential use
F.27 Mk 600	VIP	1	For Presidential use
F.27 Mk 400M	Trans	1	
C-47	Trans	3	
Falcon 20	VIP	1	For Presidential use
Gulfstream II	VIP	1	For Presidential use
Aero Commander 500B	VIP	1	For Presidential use
F337 Super Skymaster	Liaison	3	
Puma	Trans	3	One for Presidential use
Alouette II	Liaison	2	
Alouette III	Liaison	3	
Reims-Cessna 150	Train	2	*Note: All based at Abidjan*

---

afterwards and the air force placed orders for new aircraft in Canada (14 Beavers, 12 Otters, eight Caribou) and Britain (six Whirlwind, three Wessex).

Inventory

In 1973 the Ghana Air Force underwent a major re-equipment programme involving the replacement of types that had been in use for some years. The only combat aircraft in service are seven Italian MB.326 jet trainer/strike aircraft delivered in 1965 and supplemented in 1977 by six single-seat versions.

Scottish Aviation Bulldogs replaced the Chipmunks in 1973, while the helicopter squadron flies both American and French machines. Much of the GAF maintenance and support organisation is centred on Takoradi, which also has an ATC training unit. Apart from Takoradi in the west, the other main bases are Tamali in the north (completed in 1963) and Accra; numerous rough strips also exist.

GHANA

Type	Role	Base	No	Notes
MB.326F	Train/GA	Tamali	7	One squadron; delivered 1965
MB.326K	GA	Tamali ?	6	Delivered 1977
Islander	Trans/casevac	Takoradi	8	
Skyvan 3M	Trans/SAR	Takoradi	6	Two squadrons
F.27-400M	Trans	Takoradi	3	
F.27-600	Trans/SAR/VIP	Takoradi/Accra	2	
F.28-3000	VIP	Accra	1	Reportedly returned to Holland in 1977
Bulldog 120	Train	Takoradi	13	Deliveries—six in 1973, seven in 1975
Alouette III	Liaison	Accra	4	One squadron formed in 1962;
Bell 212	VIP	Accra	2	212s delivered 1975

Below: Two of the eight brightly coloured Britten-Norman Islanders flown by the Ghana Air Force on casualty-evacuation and other duties. They operate alongside Skyvans and Fokker types in the transport squadron.

UPPER VOLTA

Area: 100,000 sq. miles (259,000 sq. km.).
Population: 5·5 million.
Total armed forces: 8,000.
Estimated GNP: $800 million (£400 million).
Defence expenditure: $18·2 million (£9·1 million).

The Air Force of the Upper Volta Republic, otherwise known as the *Force Aérienne de Haute-Volta*, is administered from the capital, Ougadougou, and operates as part of the country's 3000-man Army. As a former French colony the Upper Volta received its independence in August 1960, but retains links with France by way of defence and military co-operation agreements which have provided most of the equipment used by the armed forces. Main role of the small air component is internal liaison and communications.

The Upper Volta republic did not form a military air component until 1964, when French technical assistance was provided together with a single C-47, followed by two Broussards for liaison. The insignia adopted was little removed from the French cockade and tricolour, the black, white and red representing the three arms of the Volta River.

Inventory

The population of some 5.5 million is widely scattered over the country's 100,000 square miles and relies largely on communication by air. No combat aircraft are reported in use, but the acquisition of a small number of ex-French air force Magisters is thought likely in the near future.

UPPER VOLTA

Type	Role	No
C-47	Trans	2
HS.748	Trans	1
Nord 262	Trans/VIP	2
Aero Commander 500	Trans	1
Reims 337	Liaison	1
Broussard	Liaison	3

Note: All based at Ougadougou

TOGO

Area: 21,000 sq. miles (54,400 sq. km.).
Population: 2·1 million.
Total armed forces: 3,000.
Estimated GNP: $565 million (£282·5 million).
Defence expenditure: $13·3 million (£6·65 million).

The *Force Aérienne Togolaise* has its main base and headquarters at Lomé, the country's capital. A recent equipment modernisation programme has resulted in orders for aircraft outside the traditional French sphere of influence which has existed since independence. In connection with this limited expansion, air force manpower – which was until recently only about 100 – is now likely to increase somewhat. The defence budget remains modest, however.

With the granting of independence in 1960, Togo formed a small air arm with seconded French personnel flying the standard arms package of a single Douglas

BENIN

Area: 47,000 sq. miles (121,700 sq. km.).
Population: 2·05 million.
Total armed forces: 2,250.
Estimated GNP: $650 million (£325 million).
Defence expenditure: $7·4 million (£3·7 million).

The *Force Armées Populaire du Benin* has its main base and headquarters at Cotonou, the country's largest city. Originally known as Dahomey, the country was renamed Benin in 1977. Air force manpower totals little more than 150 and little important new equipment has been obtained in recent years following independence in August 1960. In line with the practice in other ex-French dependencies, a small air arm was established with a single Douglas C-47, three MH-1521M Broussard liaison aircraft and an Alouette II helicopter. The air force currently operates mainly in the transport role, linking outlying areas in the north of the country.

BENIN

Type	Role	Base	No	Notes
F.27-600	Trans/VIP	Cotonou	1	Received mid-1978
C-47	Trans	Cotonou	3	
MH-1521M Broussard	Liaison	Cotonou	3	
Reims 337	Liaison		1	
Aero Commander 500B	Liaison		1	
Bell 47G			1	
Alouette II			1	

NIGERIA

Area: 356,000 sq. miles (922,000 sq. km.).
Population: 79·7 million.
Total armed forces: 231,500.
Estimated GNP: $34·2 billion (£17·1 billion).
Defence expenditure: $2·67 billion (£1·335 billion).

This former British colony situated on the west coast of Africa has had a relatively turbulent existence since becoming independent in 1960. The armed forces have played a major role in the country since the establishment of the Federation and a military government is at present in power. The Federal Nigerian Air Force has its headquarters in the capital, Lagos, control being exercised by the Ministry of Defence. The present armed forces manpower figure is some 230,000, but reports

NIGER

Area: 459,000 sq. miles (1·18 million sq. km.).
Population: 4 million.
Total armed forces: 2,000.
Estimated GNP: $825 million (£412·5 million).
Defence expenditure: $6·9 million (£3·45 million).

The *Force Aérienne du Niger* has its headquarters in the capital, Niamey, and operates as a transport and communications arm in this land-locked African republic. Manpower in the country's armed forces totals 2100, of which 100 constitute the air arm. Being a former French colony which gained its independence in 1960, Niger continues to maintain close links with France although military aircraft procurement is unlikely to expand over the next few years.

Following independence, France assisted in the establishment of a small Niger air force and presented a C-47 transport and three Broussard liaison aircraft. These were followed by some Noratlas transports purchased from West Germany and a single C-54 bought from France.

Niger benefits from a defence and military co-operation agreement with France under which equipment is supplied and personnel trained.

Inventory

The *Force Aérienne du Niger* is a purely transport arm with a total inventory of some nine aircraft. Future procurement is unlikely to extend to any more than a few second-line types.

NIGER REPUBLIC

Type	Role	Base	No	Notes
C-54B	Trans	Niamey	1	Ex-French Aéropostale
C-47	Trans	Niamey	2	
Noratlas	Trans	Niamey	3	
Aero Commander 500	Liaison	Niamey	1	
Cessna-Reims 337	Liaison	Niamey	2	
Broussard	Liaison	Niamey	4	Believed withdrawn from use

C-47 transport and two MH-1521M Broussards for liaison and light transports. A co-operative agreement with France was signed in 1963 but few new aircraft were delivered to Togo before the end of the decade.

Inventory

From being a purely transport and communications arm, the *Force Aérienne Togolaise* now boasts a jet strike potential following a $5.7m order for three Embraer Xavantes in late-1976. These were delivered shortly after announcement of the

Above: The *Force Aérienne Togolaise* operates a pair of DHC-5D Buffalo STOL transports originally ordered by Zaire.

order and joined five refurbished Magister jet trainers acquired in 1975 from France; a further three Xavantes were ordered in 1978. Also announced was an order for five Alpha Jets from France for the dual strike/trainer role. At the same time a requirement for a medium-sized Stol transport found Togo negotiating with France for a C.160 Transall, but talks failed and de Havilland Canada answered the need with two DHC-5D Buffaloes originally ordered by Zaire; delivery of these was made late in 1976.

TOGO

Type	Role	Base	No	Remarks
EMB.326GC Xavante	GA	Lomé ?	6	
Alpha Jet	GA/train		5	Ordered 1978
Magister	Train		5	
DHC-5D Buffalo	Trans	Lomé	2	
F.28 Mk 1000	VIP	Lomé	1	Delivered 1975
SA.330 Puma	Trans		1	Delivered 1975
Reims F.337/C.337	Liaison		2	One each, delivered 1971
C-47	Trans		2	
Alouette II	Util		1	

---

have stated that this is to be trimmed; the air force, with some 5500 officers and men, could however suffer least of all in the cutback. The air force inventory currently stands at more than 100 aircraft, significantly larger than those of other countries in the area, but serviceability is generally believed to be low.

The Federal Nigerian Air Force was formed in January 1964 following formation of the Federation in 1960. Initial assistance was provided by India and then by West Germany, the latter country sending a 42-man *Luftwaffe* training mission to Nigeria in mid-1963. The fledgling air arm equipped with ex-West German Piaggio trainers, Do27 AOP aircraft and Noratlas transport. In May 1967 Col Ojukwu, governor of the eastern region

Below: This line-up of two Nigerian L-29 Delfins in front of five MiG-21s typifies the equipment with which the Soviet bloc has supplied many African air forces.

of Nigeria, declared his intention to secede from the Federation and named the new state Biafra after the bight in the Gulf of Guinea. A bitter civil war lasted some two years, during which air power was to play a relatively small part in terms of operations. To combat offensive moves by Biafran forces, however, the FNAF accepted military aid from Russia in the form of MiG-17s and Czechoslovakia supplied a handful of L-29 Delfin armed trainers. Later six Ilyushin Il-28 bombers arrived from Egypt and Algeria and these machines formed the backbone of the FNAF for the next few years, crewed initially by Egyptian and mercenary pilots and later by Nigerian nationals.

The Biafran air force on the other hand fielded a motley, if effective, collection of types including half-a-dozen Malmo MFI-9B 'Minicons', which succeeded in causing damage to Federal forces out of all proportion to their numbers, and two B-26s which enjoyed only a short career before

being captured by the Federal Army. The war ended as a guerrilla conflict in 1970 and the FNAF gradually established itself on a more peaceful footing, placing orders for some F.27 transports and Bulldog trainers in an effort to modernise equipment and thinking.

Nigeria receives military aid from the Soviet Union but continues to place some orders with Western arms suppliers. She is also a member of the Organisation of African Unity (OAU).

Inventory

In 1975 the Federal Nigerian Air Force began a modest re-equipment programme when the first of a batch of Soviet-supplied MiG-21 interceptors arrived in the country, being the first new combat aircraft received by the arm since the war in 1967. Camouflaged for the dual strike/intercepter role and based at Kano, it is believed that they may be equipped with early versions of the Atoll air-to-air missile. The

NIGERIA

Type	Role	Base	No	Notes
MiG-21	Int/att	Kano, Kaduna	25	One sqn
MiG-17	Att	Benin	12	Total of 41 supplied by Russia and Egypt during civil war—remainder in store
IL-28	Bomb		4	Serviceability questionable
C-130H	Trans	Lagos	6	
F.27 Mk 400	Trans	Lagos	2	
F.28	VIP	Lagos	1	
Do28D-2 Skyservant	Trans	Lagos	20	Used for transport, IFR training and casevac duties
Do27	Liaison		15	
Navajo	Liaison		2	
Navajo Chieftain	Liaison		1	
Alouette II	Liaison		10	In storage
Puma	Trans		10	Used for heavy lift work
BO105C	SAR		10	
Whirlwind	Trans		3	
Bulldog 123	Train		32	Primary training
P.149D	Train		10	
L-29 Delfin	Train		16	
MiG-15UTI	Train		4	

MiG-17/Il-28 ground-attack and bomber elements are understood to have remained in service although in greatly reduced numbers. From the United States the FNAF received the first of six Lockheed Hercules in September 1975 following the announcement of an order worth $47m including spares a year earlier. Other types procured include four MBB B0105 helicopters for search and rescue duties, two Pumas for transport work and eight Dornier Skyservants for utility flights. Future procurement could see the replacement of the fleet of Do27 liaison aircraft with a more up-to-date type and the acquisition of more MiG-21s to replace the Il-28s.

Support

Training is initiated on the Scottish Aviation Bulldogs, with students then proceeding on to the Piaggios for more advanced work. Jet conversion is carried out on the L-29 with combat pilots flying the two-seat MiG-15s and MiG-21Us before the single-seat MiGs. Basic tuition on the MiG-21s was performed in the Soviet Union. The L-29s are to be replaced by Alpha Jets in 1981–82.

CHAD

Area: 488,000 sq. miles (1·26 million sq. km.).
Population: 5·8 million.
Total armed forces: 33,000.
Estimated GNP: $3·7 billion (£1·85 billion).
Defence expenditure: $98 million (£49 million).

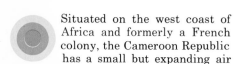

This former French colony was granted independence in 1960 and the *Escadrille Tchadienne* was established the following year. Initial equipment comprised a single C-47, three MH-1521M Broussards and an Alouette II, all supplied by France. The present Chad air force has its main base and headquarters at Fort Lamy, with a total personnel strength of some 300 officers and men. Recently it has been supporting the small Chad army fighting desultory actions against guerrillas on the border with Libya. Through a defence agreement with France, French troops and air force detachments 4re stationed in Chad on a rotational basis.

In addition to an agreement with France, Chad also has a regional defence treaty with the Central African Empire, Gabon and the Congo.

CHAD

Type	Role	Base	No	Notes
A-1D Skyraider	GA	Fort Lamy	4	Ex-Armée de l'Air
C-47/DC-4	Trans	Fort Lamy	9/3	
Caravelle	VIP	Fort Lamy	1	
Puma	Army support		• 4	Delivered 1975-76
Turbo-Porter	Liaison/trans	2	2	Delivered 1977
Reims C.337	Liaison		4	
Broussard	Liaison		2	

CAMEROON

Area: 183,000 sq. miles (474,000 sq. km.).
Population: 7 million.
Total armed forces: 6,100.
Estimated GNP: $1·7 billion (£850 million).
Defence expenditure: $51·47 million (£25·735 million).

Situated on the west coast of Africa and formerly a French colony, the Cameroon Republic has a small but expanding air force known as *l'Armée de l'Air du Cameroun*. With headquarters at Yaoundé, the AAC is mainly concerned with flying services to outlying areas, liaison with the Army and a role more recently given new emphasis, internal security and counter-insurgency. The force has a manpower strength of 300 and an inventory of about 27 aircraft.

Having been ruled by France under mandate since the First World War, when it was a German colony, the Cameroons gained its independence in January 1960 and with the assistance of France formed a small air arm for transport and internal policing duties. Broussards were delivered in 1961 and a C-47 followed in 1965. More of both types arrived later, together with Alouette helicopters and Flamant transports. In recent years the Cameroons has placed equipment orders in countries other than France, although a co-operation agreement including military clauses was signed by the two countries in 1974. China has a military assistance agreement with the republic, but the air force does not appear to have benefited from it.

Inventory

A modest combat capability exists within the *Armée de l'Air du Cameroun* following the delivery of a small batch of ex-French Magisters, the first such aircraft acquired by this previously purely transport and liaison arm. Although the force has operated French-supplied machines since its inception, recent new equipment orders have shown less reliance on this source. Two Lockheed Hercules have been received from the United States for heavy logistical work in the more inaccessible areas of the country, while two HS.748s have been purchased from Britain to further increase the capability of the air arm; one HS.748 has been fitted with an executive interior for the use of the President, although both aircraft have large side freight doors. Further orders are likely.

CAMEROON REPUBLIC

Type	Role	No
Magister	Att/train	4
C-130H	Trans	2
HS.748	VIP/trans	2
Caribou	Trans	2
C-47	Trans	4*
Do28D-2 Skyservant	Trans	2
Queen Air	Liaison	1
Broussard	Liaison	7
Alouette II/III	Liaison	2/1
Puma	VIP	1

Note: All based at Douala
* Being replaced by HS.748s

CENTRAL AFRICAN EMPIRE

Area: 234,000 sq. miles (606,000 sq. km.).
Population: 3·2 million.
Total armed forces: 1,200.
Estimated GNP: $300 million (£150 million).
Defence expenditure: $7·87 million (£3·935 million).

The *Force Aérienne Centrafricaine* flies a variety of French- and Italian-supplied aircraft from its main air base and headquarters at Bangui, capital of the republic. Approximately 100 personnel make up the air arm, which operates mainly in the transport and liaison roles.

The country's independence from France in 1960 was followed by the establishment of a small air component with a single C-47, three MH-1521M Broussards and an Alouette II as its initial equipment. A bilateral defence agreement with France was signed and the CAR currently has a regional defence treaty with the Congo, Chad and Gabon.

Above right: The *Force Aérienne Centrafricaine* flies a mixed bag of transport and liaison types, of which this Italian-supplied AL.60C-5 is a typical example.

CENTRAL AFRICAN EMPIRE

Type	Role	Base	No	Notes
AL.60C-5	Trans	Bangui	10	Bought from Italy 1971
C-47	Trans	Bangui	3	Gift from French Armée de l'Air
DC-3	Trans	Bangui	1	Gift from French Armée de l'Air
DC-4	Trans	Bengui	1	Ex-Air France. Gift from President de Gaulle 1968
Falcon 20	VIP	Bangui	1	
Caravelle	VIP	Bangui	1	Ex Sterling Airways
Broussard	Liaison		6	
Alsuette II			1	
H-34				

SUDAN

Area: 967,500 sq. miles (2·5 million sq. km.).
Population: 19·5 million.
Total armed forces: 52,100.
Estimated GNP: $4·4 billion (£2·2 billion).
Defence expenditure: $237 million (£118·5 million).

The Sudan Air Force, *Al Quyyat Al-Jawwiya As-Sudaniya*, is in the process of being re-equipped with Western aircraft in place of Russian and Chinese types. A small budget, totalling £S82.6m ($237m) in 1978, keeps procurement at a relatively low level. Headquarters of the air force, which is the sole operator of military aircraft in the Sudan, is in the Ministry of Defence in Khartoum. The 52,500 military personnel include some 2000 in the air force.

After being under Anglo-Egyptian rule for a number of years, the Sudan became independent in 1956 and the following year received four Gomhouria primary trainers from Egypt to establish an air arm. Hunting Provost T.53s were purchased from Britain for armed trainer duties, and the air force expanded during the 1960s with the acquisition of Jet Provost jet trainers, Fokker Troopship transports and Pilatus Turbo-Porter utility aircraft. A left-wing coup in 1969 saw the arrival of Soviet and Chinese technicians and equipment to reorganise the SAF, Shenyang F-4s (MiG-17Fs) and two-seat MiG-15UTIs being supplied for a combat squadron; these were later supplemented by a batch of MiG-21 interceptors from Russia to form a second unit.

The Sudan is a member of the League of Arab States and has received military assistance from the Soviet Union and China, although little new equipment has been supplied by either source for some years. The neighbouring Ethiopian-Somalia war of 1977 did not directly involve Sudanese forces, but Ethiopia consistently accused the Sudan of allowing Ethiopian Royalist factions to operate from the south-east of the country against Ethiopian Government forces. Diplomatic relations with the Soviet Union were severed in May 1977.

Below: The Sudan Air Force still operates a small number of BAC 145s in the training and light attack roles, although the earlier Jet Provosts are in storage.

177

Inventory

By the end of 1977 the Sudan Air Force still retained the two combat squadrons of MiG-21s and M8G-17s supplied in the 1960s. Soviet aid, which had been spasmodic since the supply of the MiGs, ceased altogether by the mid-1970s when President Numeiri expelled 90 Russian military advisers and half the Russian diplomatic staff as the climax of alleged Soviet backing to an attempted coup in 1971. The President then requested the supply of Northrop F-5 fighter-bombers and Lockheed Hercules transports from the United States, but the F-5s were refused because of the tension in the area with the Ethiopian-Somalia conflict. The United States did, however, agree to the sale of six Hercules at a cost of some $70–80m, and 12 F-5Es are being supplied from the US, with Saudi funding. The Sudan approached France and received tentative approval for the procurement of 15 Mirage 5s and 10 Puma assault helicopters plus attendant support and training for Sudanese personnel. The Sudan Air Force has also received four DHC Buffaloes to modernise the transport squadron and replace the An-24s.

SUDAN

Type	Role	Base	No	Notes
MiG-21PF	Int		20	Includes two MiG-21U trainers; total in-service strength not thought to be more than 12
Shenyang F-4	FGA		17	One sqn with about 12 flying
Shenyang F-2	Train			Small number supplied with the F-4s
Mirage 50	FGA		14	Order reported in 1977 to equip one sqn, replacing the MiGs; 14 on option
An-12	Trans	Khartoum	6	One sqn
An-24	Trans	Khartoum	5	
Twin Otter 300	Survey	Khartoum	1	
Buffalo	Trans	Khartoum	4	Replacing the An-24s in single transport sqn
C-130H	Trans		6	
King Air 90	Liaison		1	Flown by the Police Air Wing
Turbo-Porter	Trans		8	
Mi-4	Trans		4	
Mi-8	Trans		10	
Puma	Trans		10	
Jet Provost T.51/52	Train		8	In storage
BAC 145 Jet Provost T.55	Train/att		4	
King Air 90	Liaison		1	Operated by Police Air Wing
BO105C	Liaison		20	Operated by Police Air Wing

A single helicopter squadron has a mix of Mil Mi-4s and Mi-8s, some of the latter equipped for the assault role, and most if not all are expected to be replaced when the Pumas are delivered. The eight Jet Provosts bought in the early 1960s have long been in storage, but four of the five BAC 145s continue in use in the light strike role.

Support

Sudan Air Force personnel have undertaken training courses in the countries that have supplied equipment in the past, namely Russia and Communist China, but with the recent order for Mirage 5s and Pumas an agreement was also concluded for training courses for SAF air and ground crews in France.

ETHIOPIA

Area: 400,000 sq. miles (1·03 million sq. km.).
Population: 28 million.
Total armed forces: 93,500.
Estimated GNP: $2·9 billion (£1·45 billion).
Defence expenditure: $165 million (£82·5 million).

The Ethiopian Air Force has a current strength of 2300 personnel and a front-line combat element of some 60 aircraft. Control of the EAF comes directly from the ruling military junta with headquarters in Addis Ababa. The Army operates a small support force of helicopters and fixed-wing aircraft. Defence expenditure for 1978 amounted to $150m.

Imperial Ethiopian Aviation was formed in 1924 with Potez 25 bombers and two light aircraft. The Italian attack on Ethiopia in 1935 temporarily ended the life of the fledgling air arm, however, and it was only after the end of World War II that it was reformed. Organised with the help of Swedish Air Force personnel, the renamed Imperial Ethiopian Air Force acquired a batch of Saab Safir trainers in 1946 and two years later formed its first combat units with Saab 17 bombers and Firefly fighters. In 1960 the United States began providing military aid, including a batch of Sabres and T-33 trainers.

Ethiopia has a security assistance agreement with the US and a defence agreement with Kenya. Diplomatic relations with the US were broken off in April 1977, however.

There exists a possibility of an alliance among Ethiopia, Somalia and South Yemen.

Inventory

Before the overthrow of Emperor Haile Selassie in 1974, Ethiopia had placed military equipment orders with the US worth $100m. The orders included 16 Northrop F-5E/F interceptor/trainers, 12 Cessna A-37 strike aircraft and 15 Cessna 310 trainers. The US Government embargoed the delivery of these items, and to help maintain the EAF's combat strength against attrition a number of surplus ex-Iranian F-5As have been supplied. Bases include Bishoftu, Asmara, Behar Dar and various rough strips.

Ethiopian aircraft have been operating for some years against the Arab-backed Eritrean Liberation Front in the north of the country. F-5As and Canberras flying ground-attack sorties have accordingly suffered attrition, the ELF claiming to have destroyed at least seven aircraft. Ethiopian royalists operating from the Sudan have also prompted air action by the EAF. In the Ogaden war with Somalia in 1977–78 the Ethiopians were heavily supported by Russia, which supplied equipment and weapons to the EAF including MiG-21s and Mil helicopters. During operations against Somali forces, Cuban mercenaries flew the Soviet-supplied aircraft. A ceasefire was agreed in March 1978.

Right: The C-119 is becoming a rare sight but the Ethiopian Air Force still operates a sizeable squadron of 18 C-119Ks, which form the backbone of the transport fleet.

Right: Canberra bombers have been operated by the EAF on ground-attack sorties against the Eritrean Liberation Front, in which role a number are thought to have been lost.

Below: Saab Safirs have served the EAF for more than 30 years and are still going strong, although they must be due for replacement soon.

ETHIOPIA

Type	Role	Base	No	Notes
Canberra B.52	Bomb		2	One squadron; four delivered in 1969
MiG-21	Fight		40–50	
MiG-17	Fight		20	
F-5A/B/E	FGA	Asmara	13/2/8	One squadron
MiG-23	FGA		10	
T-28D	GA		6	One Coin squadron
C-47	Trans		10	One squadron
C-119K	Trans		18	
Dove	Trans		2	
IL-14	VIP		1	Gift from Russia
C-54	Trans		2	
Alouette III	Liaison		5	
Mi-8	Trans		2	Gift from Russia; more believed to have been received in 1977
Mi-6	Trans		2	
Puma	VIP		1	
AB.204B	Trans		6	
T-33A	Train	Bishoftu	11	
Safir 91C/D	Train	Bishoftu	13	
UH-1H	Trans		6	Army-operated
Otter	Trans		4	Army-operated
Twin Otter	Trans		3	Delivered 1977
Do28D-2 Skyservant	Trans		2	

SOMALI DEMOCRATIC REPUBLIC

Area: 246,000 sq. miles (637,000 sq. km.).
Population: 3·2 million.
Total armed forces: 51,500.
Estimated GNP: $425 million (£212·5 million).
Defence expenditure: $25 million (£12·5 million).

The Somalian Aeronautical Corps is currently Russian-equipped and trained, although moves are still being made to acquire Western material following the expulsion of Soviet and Cuban personnel in November 1977. Headquarters of the Corps is in the capital, Mogadishu, and personnel strength is approximately 2700.

Somalia became an independent repub-

Below: The Somalian Aeronautical Corps' Russian-supplied equipment includes An-24 transports, although Somalia's treaty of friendship with the USSR has ended.

lic in 1960, being formed by the unification of the former Italian-administered trust territory of Somalia and the former British Somaliland protectorate. Initial Somali Air Corps equipment comprised C-45 and C-47 transports and some F-51D Mustangs from Italian stocks. Two Gomhouria trainers were presented to the force by Egypt as part of a trade and military aid agreement, these being accompanied by a small Egyptian training mission.

From 1963 Somalia was a recipient of Soviet military equipment in return for base facilities. MiG-15s and -17s arrived to expand this previously modest force, and towards the end of the 1960s a limited strike potential was received in the form of a handful of Il-28 light bombers. A military coup overthrew the government in 1969 but Soviet aid to Somalia continued, as did the construction of a large port and missile overhaul facility at Berbera to provide a major base for the expanding

Soviet Navy. By 1974 Somalia had received a total of 54 MiG fighters from Russia, plus some An-24 transports and a few more Il-28s.

Somalia terminated the treaty of friendship with the Soviet Union in November 1977 and was expected to sign a similar agreement with Western countries in order to obtain military and economic aid. A treaty with Cuba was also cancelled.

Inventory

The front-line combat force in the Somalian Aeronautical Corps comprises some 55 Soviet-supplied MiGs and nearly a dozen Ilyushin Il-28 light bombers. Three squadrons operate the MiGs, the majority being of the older -15 and -17 variety with a single interceptor squadron of MiG-21s. According to Ehtiopian sources, the Somalis lost more than 23 MiGs in the first six months of the war between the two countries, which began in July 1977.

SOMALIA

Type	Role	Base	No	Notes
IL-28	Bomb	Mogadishu	10	One sqn
MiG-21	Int	Hargeisa	10	One sqn
MiG-17	FGA	Hargeisa	44	Two sqns
MiG-15	FGA	Hargeisa		
An-2	Trans	Mogadishu	3	
An-24	Trans	Mogadishu	3	
An-26	Trans	Mogadishu	2	One sqn
C-45	Trans	Mogadishu	1	
C-47	Trans	Mogadishu	3	
AB.204B	Trans	Mogadishu		
Mi-4	Trans	Mogadishu		One sqn
Mi-8	Trans	Mogadishu		
Piaggio P.148	Train		8	
Yak-11	Train		20	
MiG-15UTI	Train			

Somali Police Air Wing

Type	Role	Base	No	Notes
Do28	Trans	Mogadishu	2	
Cessna 150 Aerobat	Train	Mogadishu	1	

Southern Africa

Since the end of World War II virtually the whole of southern Africa has become independent of its former British, French, Portuguese and Belgian masters. In the spring of 1979 a black majority government took power for the first time in Rhodesia, renamed Zimbabwe-Rhodesia, leaving only South Africa with a white-dominated administration. The transition from colonial status to self-determination has been bloody as often as not, and conflict shows little sign of diminishing even though the desire for independence is no longer the aim.

The withdrawal of Portuguese forces in Angola precipitated a civil war, with the Marxist MPLA faction emerging as victors — thanks mainly to support from the Soviet Union and Cuba. The Cubans have become an increasingly important influence in Africa during the 1970s, having provided assistance to Mozambique and Guinea as well as Angola, and this trend is likely to continue.

Air power has played a significant role in the war between Rhodesia (and Zimbabwe-Rhodesia) and guerrillas operating from neighbouring countries, although no air combat was involved. A number of Rhodesian aircraft have been lost to surface-to-air missiles, however, including Vis-

counts operated by the country's commercial airline.

South Africa has a well established aircraft industry which has built European types under licence, although the embargo placed by France on further arms sales is beginning to bite. South Africa supplied Rhodesia with military assistance before the establishment of black majority rule in the latter country, and this involvement could well increase if South Africa feels her borders to be threatened as a result of the handover of power to a black majority government.

The spring of 1979 also saw the invasion of Uganda by Tanzanian forces, aided by Ugandan emigrés and supported by a large proportion of the population. President Amin was supplied with aid by Libya before his downfall, this assistance including the use of Libyan Tu-22s to bomb the Tanzanian forces.

Air forces outside the remaining areas of white control in southern Africa are limited in equipment and proficiency. Given the ability of shoulder-launched surface-to-air missiles to bring down even well flown modern aircraft, however, air combat is no longer necessary to inflict severe damage on a sophisticated air force.

GABON

Area: 101,400 sq. miles (262,600 sq. km.).
Population: 500,000.
Total armed forces: 1,250.
Estimated GNP: $1·8 billion (£900 million).
Defence expenditure: $20·25 million (£10·125 million).

Along with other former French possessions in Africa, Gabon continues to receive both economic and military assistance from France and the *Force Aérienne Gabonaise* is organised along *Armée de l'Air* lines. Headquarters and main base are at the capital, Libreville, and total strength is about 150 personnel. A limited expansion programme is under way.

Originally part of French Equatorial Africa, Gabon gained its independence in 1960 and established an air arm with the assistance of the French. Initial equipment comprised a C-47 transport and two MH-1521M Broussards. During the mid-1960s more ex-French aircraft were supplied and links with France were strengthened. A Bell 47G and some Alouette IIIs were also received.

As well as being a recipient of French aid, Gabon has a regional defence treaty with the Central African Empire, Chad and the Congo.

Inventory
To strengthen its position in west Africa Gabon has received five Dassault Mirage 5s from France to give the *Force Aérienne Gabonaise* its first jet combat equipment.

Negotiations were held in 1975 and delivery took place in 1978. Before this order the FAG had carried out a purely transport, liaison and training role with equipment purchased from as far afield as Japan, the US and France. Two NAMC

YS-11As are used for civil and military duties, as is the single Gulfstream II, while heavy-lift work is the job of the two 'stretched' Hercules. These are assisting in the construction of the Trans Gabon Railway and other civil engineering projects.

GABON

Type	Role	No	Notes
Mirage 5G/5DG	GA/train	3/2	
L-100-20/-30	Trans	2	-30 delivered May 1975; -20 delivered late 1976. Chiefly used for civil work
C-130H	Trans	1	Delivered in December 1977
NAMC YS-11A	Trans	2	First delivered August 1973; civil and military use
Gulfstream II	VIP	1	Delivered 1975
Falcon 20E	VIP	1	
C-47	Trans	3	Possibly withdrawn
Broussard	Liaison	4	Possibly withdrawn
Nord 262	Trans	3	
Puma	Assault/trans	4	
Alouette III	Liaison	4	
Reims C.337	Liaison	1	

Note: All based at Libreville

Above: The FAG operates a pair of NAMC YS-11s on both civil and military duties from its main base at Libreville.

180

CONGO REPUBLIC

Area: 130,000 sq. miles (336,000 sq. km.).
Population: 2·1 million.
Total armed forces: 7,000.
Estimated GNP: $610 million (£305 million).
Defence expenditure: $37·2 million (£18·6 million).

 With its title *l'Armée de l'Air du Congo* proclaiming its former French colonial connections, the Congo Air Force has its headquarters in Brazzaville and operates mainly in the liaison and transport roles. The country became independent from France in August 1960, having previously formed one of four territories of French Equatorial Africa, and for many years the new state was known as Congo-Brazzaville before the ex-Belgian Congo adopted the name Zaire. Present manpower in the AAC is more than 250, while defence expenditure for all the armed forces in 1978 amounted to £18·6 million.

With the granting of independence the Congo Republic established a small transport and communications arm with French aid. Initial equipment comprised C-47s, Broussards and Alouette helicopters, and later in the 1960s the Soviet Union supplied some Antonov and Ilyushin transports. More recently an F.28 has been purchased for the use of the President. Future procurement is likely to take the form of a small number of counter-insurgency jets of the Magister type, together with some modern liaison aircraft to replace the long-serving Broussards. Source of any new equipment could be France, with which the Congo has a military training agreement and close ties, or the Soviet Union, with which she co-operated during the 1975 Angolan civil war by acting as a staging post for Russian and Cuban supplies. At one stage about 20 MiG-21s were based at Brazzaville to support the Angolan MPLA movement.

As well as her agreement with France,

Above: The President of the Congo Republic is one of several African heads of state which use Fokker-VFW F.28s as their personal transports.

Congo has defence treaties with the Central African Empire, Chad and Gabon.

CONGO REPUBLIC

Type	Role	No
An-24RV	Trans	5
C-47	Trans	3
IL-14	Trans	5
Broussard	Liaison	3
Frégate	Trans	1
F.28 Mk 1000	Liaison	1
Alouette II/III	Liaison	3/1

Note: All based at Brazzaville

ZAIRE

Area: 905,600 sq. miles (2·35 million sq. km.).
Population: 21·6 million.
Total armed forces: 33,400.
Estimated GNP: $3·5 billion (£1·75 billion).
Defence expenditure: $164 million (£82 million).

The *Force Aérienne Zaïroise* is one of the better equipped air forces in Africa, with about 48 aircraft supported by a modern transport element procured mainly from America and Canada. FAZ headquarters is in the capital, Kinshasa (known as Leopoldville when the country was the Republic of the Congo), and the current manpower strength totals some 3000 out of total armed forces of 33,400. Defence spending for 1978 amounted to Zaïres 142m ($164m). The air force is the sole operator of aircraft and is organised into two Air Groups.

Zaïre gained its independence from Belgium in 1960, becoming the Congo Republic. Shortly after independence the state of Katanga broke away from the central government and formed its own embryo air arm, the *Force Aérienne Katangaise*, equipped with a variety of types obtained from a number of sources. To counter this, the Congo government established its own air force with the assistance of Belgian advisers and a bitter civil war ensued. Elements of the rival air arms were later combined and the *Forces Aériennes Congolaises* were reorganised. By the mid-1960s the FAC had a bomber force of

Above: The C-130 Hercules has proved just about indispensable as a medium to heavy transport, this example being one of six C-130Hs operated by the FAZ.

ZAÏRE

Group	Unit	Type	Role	Base	No	Notes
2 Groupement Aérien Tactique	21 Wing	Mirage 5M	FGA	Kamina	9	Total of 14 5M and three 5DM delivered
		Mirage 5DM	Train	Kamina	2	
		MB.326GB	Att	Kamina/ Kinshasa	12	Also used by 132 Sqn for advanced training
		MB.326K	Att	Kamina	6	Ordered in 1978
		T-6G	Att	Kamina	8	Further 10 T-6s used for basic training
		T-28D	Att	Kinshasa	10	
	22 Wing	C-46	Trans	Kinshasa	2	
		Caribou	Trans	Kinshasa	2	
		Buffalo	Trans	Kinshasa	3	
1 Groupement Aérien	19 Wing	DC-6	Trans	Kinshasa	2	
		C-54	Trans	Kinshasa	4	
		C-47	Trans	Kinshasa	10	
		C-130H	Trans	Kinshasa	6	
		MU-2J	VIP	Kinshasa	2	Believed unserviceable
		Alouette III	Liaison	Kinshasa	14	
		Puma	Assault	Kinshasa	8	
		Super Frelon	VIP	Kinshasa	1	
		Bell 47G	Train/liaison	Kinshasa	7	
	13 Wing	Cessna 310R	Train	Kinshasa	12	Total of 15 delivered for training and liaison
		Cessna 150 Aerobat	Train	Kinshasa	15	
	131 Sqn	SF.260MC	Train	Kinshasa	23	

181

Above: Zaire has been independent since 1960, but Belgian industry has contributed to FAZ re-equipment in the form of work by SABCA on its squadron of Mirage 5 fighter-bombers, including this Mirage 5DM trainer.

B-26Ks supplied by the United States, and T-28s and T-6Gs used in the counter-insurgency role were provided by the US and Italy respectively. The country received considerable assistance from outside military advisers and her ties with Italy remained strong, a training agreement being signed between the two countries and numbers of SF.260 and MB.326 trainers being bought. In 1972 the country changed its name to Zaïre; the armed forces continue to receive limited assistance from a number of countries, including Israel.

Zaïre is a member of the Organisation of African Unity, constituted in 1973, and has a security assistance agreement with the United States. She also has close ties with Italy and France, having signed a technical military assistance agreement with the latter in 1974.

Inventory

The *Force Aérienne Zaïroise* has a combat force of Mirage fighter-bombers supported by a counter-insurgency element of MB.326 armed trainers, and backed up by transport units flying Hercules, Buffaloes and Caribous. Much of the FAZ manpower is now made up of nationals, the training programme under Italian auspices having been relatively successful over the past decade. Undoubtedly the elite element in the air force is the Mirage squadron. In 1973 General Mobutu announced a requirement for up to three squadrons, but after the arrival of the first batch in 1975 the purchase of further aircraft was prevented by shortage of money. Zaïre's connections with France had already resulted in the acquisition of some Puma helicopters and a single Super Frelon for the personal use of the President. Seven Hercules had been received by the FAZ from the US by the middle of 1977. An order

for six Buffaloes was reported in 1974 but budget problems again forced this to be halved, the three aircraft arriving in 1976.

Zaïre's armed forces were mobilised in March 1977 following the declaration of a state of emergency by President Mobutu when the southern state of Shaba was invaded by mercenaries intent on re-establishing the province of Katanga. Efforts to halt the invasion by the FAZ's force of Mirages and MB.326s flying in the ground-attack role met with limited success. An appeal by the President for outside assistance was accepted by the French President, under the terms of a technical military agreement signed in 1974, and by the King of Morocco. This resulted in 'Operation Verveine', involving the use of 13 French Air Force Transall transports to airlift 1500 Moroccan troops to the Zaïre outpost at Kolwezi on the Shaba battlefront.

Limited US aid was supplied in the form of arms, equipment and a seventh Hercules to replace one that had crashed some time before. By the end of May 1977, after bitter fighting and increased strike operations by the FAZ, the invading force was officially reported as being defeated. In a second invasion in May 1978, the FAZ suffered the loss of at least six MB.326s and two Pumas before the invaders were driven back with the help of French and Belgian paratroops. A repeat order for two replacement MB.326GBs was placed by the FAZ in 1978, and to increase the effectiveness of this Coin element, six single-seat K versions are also being delivered.

Support

Initial flying training is conducted at Kinshasa on the Cessna 150s, with students moving to the Cessna 310Rs for a twin conversion course. Other aircraft in use include the SF.260s, while the MB.326s and T-6s are flown by 132 Sqn on advanced training. Under the terms of the technical assistance agreement with France, Zaïroise pilots were trained on the Mirages in France before the squadron was declared operational.

RWANDA

Area: 10,200 sq. miles (26,400 sq. km.).
Population: 4 million.
Total armed forces: 3,150.
Estimated GNP: $515 million (£257·5 million).
Defence expenditure: $6·5 million (£3·25 million).

R   This small land-locked central African republic supports a 3700-man army assisted by a modest air component used mainly in the transport and liaison roles but having a limited counter-insurgency capability. Headquarters and main base are near the capital, Kigali, and about 150 officers and men make up the air arm.

As Ruanda-Urundi the country was a United Nations trust territory administered by Belgium until independence in July 1962, when the two countries separated to become Rwanda and Burundi. An air arm for the Belgian-trained Rwanda army was established in 1972 with three Aermacchi AM.3C liaison aircraft ordered from Italy, with further plans reported for the purchase of a small number of MB.326 jet trainers. Procurement of these jets

UGANDA

Area: 91,000 sq. miles (235,500 sq. km.).
Population: 11·2 million.
Total armed forces: 21,000.
Estimated GNP: $3·2 billion (£1·6 billion).
Defence expenditure: $52 million (£26 million).

 This African republic has received military aid from a number of countries and her air arm in particular reflects this assistance. Known as the Uganda Army Air Force and controlled by the Defence Ministry of Kampala, the force was purely an army support element until the delivery of MiGs in the mid-1970s. Of the 21,000 personnel in the Ugandan armed forces, about 1000 serve in the air force. A paramilitary force, the Uganda Police Air Wing, operates alongside the UAAF on internal communications and policing duties.

Uganda gained its independence from Britain in 1962 and a small air arm was formed two years later, initially to support the Uganda Rifles. A Police Air Wing was already in existence and the Uganda Army Air Force, as the new arm was called, was intended to supplement the PAW, which was largely staffed with European personnel. Two Westland Scouts were shared between the two arms. Assistance in forming the UAAF was provided by the Israeli Government via an Israeli Air Force/Defence Force team which began training Ugandan personnel on 16 Piper Super Cubs and a couple of Magister jets at Entebbe and Jinja.

In 1966 some Soviet MiGs arrived to establish a combat unit within the UAAF, the Ugandan pilots undertaking a training failed to materialise, however, possibly due to budget problems, and instead three Fouga Magisters were acquired from France in 1975 for use as armed trainers. To increase the transport and liaison capability of the arm, two C-47s and a couple of Alouette helicopters have been bought, while to add to the Islander purchased in 1975 a further Defender was ordered by the Rwanda Defence Ministry in 1977. Future procurement is likely to be limited by the low-priority military budget allocated to the air component.

RWANDA REPUBLIC

Type	Role	Base	No
Magister	GA/train	Kigali	3
C-47	Trans	Kigali	2
Islander/Defender	Trans	Kigali	1/1
AM.3C	Liaison/obs	Kigali	3
Alouette III	Liaison	Kigali	2

Right: The small air arm of the Rwanda Republic operates the Britten-Norman Islander both in its standard transport form (seen here) and as the Defender. The only other armed type is the Fouga Magister.

course in Russia. Israeli personnel were dismissed by President Amin in 1972, followed shortly afterwards by the British pilots and technicians employed by the PAW.

Uganda is a member of the Organisation of African Unity (OAU) and had close military links with Libya as well as having received limited aid from the Soviet Union. Uganda's poor relations with neighbouring Kenya and Tanzania have resulted in minor clashes along the borders.

In the spring of 1979 the long-standing bitterness between Uganda and Tanzania escalated into open warfare, with Tanzanian forces eventually overrunning Uganda's capital, Kampala. Both sides used their air forces, and Ugandan troops received support from Libyan forces – including, it is thought, Tu-22 Blinder bombers.

Inventory

The combat strength of the Uganda Army Air Force underwent a dramatic reduction in July 1976 when almost half the MiGs in service were destroyed during an Israeli commando raid on Entebbe. Uganda's harbouring of Palestinian terrorists following the hijacking of an Air France airliner resulted in the swift Israeli action, during which seven MiG-21s and four MiG-17s were destroyed on the ground.

Libya then supplied 30 Mirages on temporary detachment to the UAAF. Subsequent Soviet arms supplies are believed to have included replacements for the destroyed MiGs. Russian SA-3 missiles are also operated. Fixed-wing transport aircraft appear limited to a couple operated by the Police Air Wing following the sale to a foreign buyer, believed to be an Israeli freight company, of the UAAF's fleet of five C-47s originally

donated by the Israeli Government. Helicopters are the main form of transport, nearly a dozen of Italian manufacture being used.

Support

Uganda ordered eight Bravos from Switzerland in early 1977 to augment and modernise the primary training element. Advanced training is performed on the Piaggios, with jet conversion flown on the Delfins and two-seat MiGs. The Magisters fly both training and light strike duties, although the lack of spares from Israel may have halved the number serviceable.

Below: This immaculate Piper Super Cub was flown in the training role by the Uganda Army Air Force under President Amin, but the future direction and equipment policy of the air arm is in doubt following his overthrow.

UGANDA

Type	Role	Base	No	Notes
MiG-21	Int	Entebbe	10	Seven lost in Entebbe raid in 1976; replacements supplied by Soviet Union
MiG-17F	Att	Entebbe	12	Four lost in Entebbe raid in 1976; replacements supplied by Soviet Union
IAI/Fouga Magister	Att/train	Jinja	8	Supplied by Israel in mid-1960s; total of 12 delivered. Serviceability unknown
Caribou	Trans	Entebbe	1	Used by the Police Air Wing
Twin Otter	Trans	Entebbe	1	Used by the Police Air Wing
Gulfstream II	VIP	Entebbe	1	Used by the President
AB.205	Liaison	Entebbe	6	Possibly more in use
AB.206	Liaison	Entebbe	4	Possibly more in use; a number of the helicopters are flown by the Police Air Wing
AS202 Bravo	Train		8	
Super Cub	Train		10	
P.149	Train		5	
L-29	Train		5	
MiG-15UTI	Train		2	
SA-3	SAM			

KENYA

Area: 225,000 sq. miles (582,000 sq. km.).
Population: 13 million.
Total armed forces: 9,100.
Estimated GNP: $3·7 billion (£1·85 billion).
Defence expenditure: $80 million (£40 million).

The Kenya Air Force, with headquarters in Nairobi, has nearly 1200 officers and men. The air arm, run on similar lines to the RAF, is chiefly involved in transport duties, border patrols and supply flights to remote areas. Defence expenditure for 1978 totalled 668m Shillings ($80m).

Established with British assistance following independence in 1963, the Kenya Air Force was officially inaugurated in June 1964 and was initially equipped with ex-RAF Chipmunks. The following year the KAF's first two operational squadrons were formed with seven DHC Beavers and four Caribou Stol transports purchased from Canada. During 1967 African pilots gradually replaced seconded RAF personnel following CFS training in the UK.

Kenya has a security assistance agreement with the United States and training, overflying and defence arrangements with Britain.

Above: The Kenya Air Force has acquired several additions to its transport force in recent years, including half a dozen Dornier Do28D Skyservants.

Above: The Beaver is one of three de Havilland of Canada types operated by the KAF. The aircraft are dispersed throughout the country.

Inventory

Kenya's air force has undergone a steady equipment modernisation since 1969 within the limits of its modest defence fund allocation. In October 1969 an order for five Bulldog basic trainers was placed with Scottish Aviation to replace the KAF Chipmunks and the following year a further order was announced, this time for six BAC Strikemasters to equip a dual-purpose attack/trainer squadron. Following its completion in 1973, Nanyuki became the KAF's main base and in July 1974 the airfield saw the arrival of the country's first combat aircraft, six refurbished Hawker Hunters. Meanwhile the transport force expanded with additional Caribou and Beavers with, more recently, the purchase of four DHC-5D Buffaloes and six Dornier Skyservants. To further update the KAF's combat potential, it was announced in 1976 that an order for 12 Northrop F-5 Tiger IIs had been placed with the US in a $75m deal, with delivery completed in 1978; 12 Hawks for the training/ground attack role were also ordered in 1978.

Training and support in the Kenya Air Force is undertaken within the force with limited assistance from a UK advisory team. There is a Ground Training School, and flying training is run along RAF lines with students progressing from basic work on the Bulldogs to advanced instruction on the Strikemasters. Twelve Hawk Mk52s have been ordered to supplement and eventually replace the Strikemasters and the new aircraft are expected to have a dual trainer-strike capability.

In recent years Kenya has enjoyed poor relations with neighbouring countries Uganda, Tanzania and Somalia, which have openly threatened action against Kenya. No air action had occurred at the time of writing, however.

KENYA

Type	Role	Base	No	Notes
F-5E	Int	Nanyuki?	10	Ordered in 1976 in $75m deal, delivery in 1978
F-5F	Train	Nanyuki?	2	
Hunter FGA.9	GA	Nanyuki	4	Delivered in 1974
Hunter T.80	GA/train	Nanyuki	1	Two delivered; one lost
Hawk Mk 52	Train/GA		12	On order
Strikemaster Mk 87	GA/train	Nanyuki	5	Six delivered in 1971
DHC-5D Buffalo	Trans	Nanyuki	4	
Do28D Skyservant	Trans	Nanyuki	6	
Caribou	Trans	Nanyuki	6	One sqn
Beaver	Trans	Eastleigh/Nanyuki	15	One sqn with detachments at various bases
Bulldog 103/127	Train	Eastleigh	14	Five delivered in 1972, nine in 1976–77
Navajo Chieftain	Liaison		1	
Rockwell 680F Commander	Liaison		1	
Alouette II			3	
Bell 47G			2	
SA.342 Gazelle			2	
SA.330 Puma				Small number in use

TANZANIA

Area: 362,800 sq. miles (939,600 sq. km.).
Population: 16 million.
Total armed forces: 26,700.
Estimated GNP: $2·9 billion (£1·45 billion).
Defence expenditure: $140 million (£70 million).

Chinese equipment dominates the Air Wing of the Tanzanian People's Defence Force, as it does in the other elements of the country's armed forces. The sole operator of aircraft in the country, the Air Wing has its headquarters in the capital, Dar-es-Salaam, and has a personnel establishment totalling some 1000 officers and men. The defence budget has been steadily increased in the mid-1970s and for 1978 stood at Shillings 1.17bn ($140m), but procurement of equipment has remained relatively low.

Tanzania was established as a republic in 1964, being formed from the British colony of Tanganyika and the British protectorate of Zanzibar. The Air Wing of the Tanzanian People's Defence Force was created as an autonomous element designed to support the army and provide air links with distant areas of the country. West Germany furnished Noratlas transports, Do28 liaison aircraft and nine Piaggio P.149 trainers, while Canada provided some seconded military personnel to establish a training base near Dar-es-Salaam.

The recognition of East Germany by Tanzania forced the cancellation of the West German aid programme although Canadian help continued until 1970, by which time five Caribou, eight Otters and six Beavers had been supplied. By 1972 Tanzania was undergoing economic expansion with Chinese assistance and at the same time her armed forces began the transition to Chinese equipment, the air force receiving a dozen Shenyang F-4s (MiG-17Fs) and a couple of F-2s (two-seat MiG-15UTIs) in 1973. Tanzanian Air Wing personnel underwent training in China on the aircraft before delivery, and a major air base was constructed at Mikumi, near Dar-es-Salaam, with Chinese technical assistance.

Tanzania is a member of the Organisation of African Unity (OAU) and retains a military-assistance agreement with China.

Inventory

The Air Wing of the Tanzanian People's Defence Force maintains a combat element of three squadrons equipped with Chinese-supplied Shenyang F-4s, F-6s and F-7s totalling some 33 machines. Most of these aircraft were delivered in 1973–74 and are maintained and flown by Tanzanian personnel with limited assistance from Chinese advisers. Apart from a single Antonov An-2 biplane transport delivered some years ago, the Air Wing relies exclusively on western transport and support aircraft, with a dozen Caribous providing essential support duties for the army. The Transport Group recently satisfied a passenger/freight aircraft requirement by ordering three HS.748s and four Buffaloes to supplement the Canadian Caribous. Other procurement includes some Cessna 310s and Cherokee trainers plus some Agusta-Bell JetRangers from Italy. Following the purchase of the HS.748s in 1977, future procurement is not thought to encompass any more large types in the short term, but an expansion of the helicopter force is likely.

Above: The Air Wing of the Tanzanian People's Defence Force operates a dozen DHC Caribou STOL transports in support of the army. They have now been joined by the turboprop Buffalo.

Support

Training to pilot status is conducted on the Cherokee 140s, followed by twin conversion on the Cessna 310s and jet instruction on the two-seat MiG-15UTIs. Some training is carried out in China.

TANZANIA

Unit	Type	Role	Base	No	Notes
Transport Group	Shenyang F-7 (MiG-21)	FGA	Mikumi	15	One sqn
	Shenyang F-6 (MiG-19)	FGA	Mikumi	8	One day fighter-bomber sqn
	Shenyang F-4 (MiG-17)	FGA	Mikumi	10	One sqn
	Shenyang F-2 (MiG-15UTI)	Train	Mikumi	2	
	An-2	Trans	Mikumi	1	
	Caribou	Trans	Mikumi	12	Eight Otters traded back to DHC as part of deal
	DHC-5D Buffalo	Trans/VIP		4	
	HS.748	Trans/VIP		4	One civilian-registered for VIP use; three ordered in 1977 for passenger/freight work
	AB.206	Liaison		2	
	Bell 47G	Util		2	
	Cessna 310	Liaison/train		6	Includes two 310Qs; used chiefly for twin training
	Cherokee Six	Liaison		1	
	Cherokee 140	Train		5	Replaced Piaggio 149Ds in primary role

MALAWI

Area: 45,700 sq. miles (118,400 sq. km.).
Population: 5·5 million.
Total armed forces: 2,400.
Estimated GNP: $925 million (£426·5 million).
Defence expenditure: $1·5 million (£750,000).

Constituted as part of the Malawi Army, the country's military air arm has its headquarters in the capital, Blantyre, and operates from a main operational base at Lilongwe. Bordered by Mozambique, Tanzania and Zambia, Malawi achieved independence in July 1964 (and became a republic in July 1966) having previously been known as the British protectorate of Nyasaland. With British help a small policing force was formed which, by the mid-1970s, stood at some 2300 personnel.

To provide communications, liaison and light transport flights an aviation section was established and, despite being in existence for more than a decade, little new equipment has been purchased following the acquisition of two Pembrokes (now withdrawn from use) and some C-47s. Future procurement is unlikely to extend to anything more than light transport types as now in use.

The country is a member of the Organisation of African Unity (OAU), which was constituted in May 1973.

MALAWI

Type	Role	Base	No
C-47	Trans	Lilongwe	4
Do28D-2 Skyservant	Trans	Lilongwe	6
Alouette III	Liaison	Lilongwe	1
SA.330L Puma	Trans	Lilongwe	1

ZAMBIA

Area: 290,600 sq. miles (752,600 sq. km.).
Population: 4·7 million.
Total armed forces: 14,300.
Estimated GNP: $2·2 billion (£1·1 billion).
Defence expenditure: $310 million (£155 million).

The Zambian Air Force, part of the Defence Forces, operates mainly in the transport and communications role. Some 1500 of the 7800 armed forces personnel make up the Air Force, whose commanding officer is of Group Captain's rank. Defence expenditure for 1978 totalled Kwacha 246m ($310m).

After the break-up of the Central African Federation in October 1964, the British colony of Northern Rhodesia became the Republic of Zambia. Under President Kenneth Kaunda the new state took over the Northern Rhodesia regular army, which included an air wing equipped with four C-47s and two Pembrokes originally transferred from the Royal Rhodesian Air Force.

British aid assisted the expansion of the force, with a seconded RAF mission helping to train Zambian nationals, but the agreement was terminated in 1970 in favour of a similar agreement with Italy and Jugoslavia. Final British involvement in the running of the ZAF ended in 1972. Zambia is a member of the Organisation of African Unity (OAU).

Inventory

In May 1977 President Kaunda declared that a state of war existed between Zambia and neighbouring Rhodesia over the latter's attitude to black-majority rule. Military operations, limited initially to desultory border incidents with virtually no Air Force involvement, erupted in 1978 when Rhodesian Air Force aircraft attacked guerrilla bases in the country. Budget limitations have kept the size of the Zambian Air Force to a relatively modest arm of some 150 aircraft, comprising a combat element flying Jastreb light strike jets and 18 MB.326s plus a combined trainer/counter-insurgency unit flying SF.260s. BAC Rapier low-level anti-aircraft missiles were bought in the early

Above: These Zambian Air Force SF.260MZs are illustrated here in their role as unarmed trainers, but they can also be fitted with underwing weapons for counter-insurgency tasks. In training, the SF.260s operate alongside Saab Safaris.

Above: Another Italian dual-role training and light attack type used by the ZAF is the Macchi MB.326, which makes up the force's main combat element.

1970s at a reported cost of some £6m ($15m), and more were supplied in 1978 following Rhodesian air attacks on the guerrilla bases.

The transport force is made up of aircraft acquired from a number of different sources. Yet another source of arms supplies was revealed in 1977 when it was confirmed that Zambia had purchased 20 Saab Safari two-seat trainer/utility aircraft in 1976.

Support

Training is conducted both within the country and overseas, with a number of Zambian personnel undergoing tuition in Jugoslavia and Italy. Basic training is flown on the SF.260s and Safaris, with advanced tuition on the Galebs and MB.326s. Twin conversion for transport pilots is performed on the Skyservants, while the Bell 47Gs are used for helicopter training.

ZAMBIA

Type	Role	Base	No	Notes
MB.326GB	Att/train		18	Total of 20 delivered; fitted with underwing weapon pylons for dual Coin role; three replacement aircraft ordered in 1978
Jastreb	Att	Mbala	4	
Galeb	Train	Mbala	2	
SF.260MZ	Train/Coin	Lusaka	8	Flown in dual role with provision for underwing weapons
Safari	Util		20	
Buffalo	Trans	Lusaka	7	
Caribou	Trans	Lusaka	5	
C-47	Trans	Lusaka	10	
Do28	Trans	Lusaka	10	Re-order probable. Also used as trainers
Beaver	Liaison	Lusaka	7	
HS.748	VIP	Lusaka	1	Used by special VIP/Presidential Flight
Yak-40	VIP	Lusaka	2	Used by special VIP/Presidential Flight
AB.205A	Liaison	Lusaka	25	
AB.212	Liaison/VIP	Lusaka	1	
Alouette III	Liaison	Lusaka	8	
Bell 47G	Liaison/train	Lusaka	17	
Mi-6	Trans		6	Unconfirmed report
Rapier	SAM		1 battery	Army-operated

ANGOLA

Area: 488,000 sq. miles (1·26 million sq. km.).
Population: 5·8 million.
Total armed forces: 33,000.
Estimated GNP: $3·7 billion (£1·85 billion).
Defence expenditure: $98 million (£49 million).

The Angolan Republic Air Force, FAPA, was formed by the Marxist MPLA Government during the civil war following the Portuguese withdrawal in November 1975. Current FAPA headquarters is in the country's capital, Luanda. Pending the appointment of Angolan nationals, major support is being provided by Cuba, the Soviet Union and to a lesser extent by East Germany. Some Portuguese mercenaries are also serving with the air force.

Portugal ceased her colonial rule in Angola in November 1975 and immediately a civil war broke out among three factions, the western-backed FNLA and UNITA, and the Marxist MPLA. After initial setbacks the MPLA gained almost complete control of the country by March 1976. At least eight MiG-21s were stationed in Congo-Brazzaville in support of the MPLA and a further 12 aircraft were assembled at Luanda airport. These latter aircraft were to form the nucleus of FAPA and began combat operations in mid-January 1976.

Inventory

Current FAPA strength includes MiG-21MF interceptor/strike aircraft flown by Cuban-trained Portuguese mercenaries and Cubans, together with MiG-17Fs and three Fiat G.91R-4s. The only other jet types in use are three MiG-15UTI two-seat trainers. Ex-Portuguese types in service include 20 Alouette IIIs and three C-47s, but it would seem likely that the air force has impressed into service for liaison duties a number of other aircraft abandoned by the Portuguese. Types left in Angola include Do27s, T-6s and OGMA-built Austers.

A Russian ground-based air defence radar is in use at Luanda. Future expansion of the Angolan air force appears centred on Cuban and Soviet help, with more deliveries of MiG-21s expected.

ANGOLA

Type	Role	Base	No	Notes
MiG-21MF	Int/GA	Luanda	12	
MiG-17F	GA	Luanda	8	
G.91R-4	GA	Luanda	3	Ex-Portuguese AF
MiG-15UTI	Train	Luanda	3	
Alouette III	Trans		20	Operated from a number of bases
An-26	Trans		5	Six received, one shot down
C-47	Trans	Luanda	3	Ex-Portuguese AF
Turbo-Porter	Trans		2	Delivered in 1976

BOTSWANA

Area: 222,000 sq. miles (575,000 sq. km.).
Population: 630,379.
Total armed forces: 2,000.
Estimated GNP: $200 million (£100 million).
Defence expenditure: $20 million £10 million).

This small independent South African state which until 1966 was a British Protectorate, established the Botswana Defence Force in 1977 as an extension of its police organisation. Its headquarters are situated in Gaberone Village and it conducts border control, casevac, communications and forward air control with two Britten-Norman Defenders. These aircraft are equipped with extra fuel tankage and have provision for Sura rockets, machine-guns and Skyshout loudspeaker equipment; a third aircraft was lost on operations early in 1978. A further two Defenders have been ordered and the force has a requirement for a basic trainer.

Below: The only aircraft operated by the Botswana Defence Force are Britten-Norman Defenders, but they can pack a potent punch. Armament can include machine guns and Sura air-to-surface rockets, with other specialised equipment comprising Skyshout airborne loudspeakers and additional fuel tanks.

RHODESIA

Area: 150,800 sq. miles (390,500 sq. km.).
Population: 6·3 million.
Total armed forces: 10,800.
Estimated GNP: $3·1 billion (£1·65 billion).
Defence expenditure: $242 million (£121 million).

At the time of writing the Rhodesian Air Force is engaged in a major counter-insurgency anti-guerrilla campaign which has been under way since 1972. A United Nations arms embargo has been in force since premier Ian Smith announced the Unilateral Declaration of Independence (UDI) in 1965, and the Rhodesian armed forces have had difficulty in acquiring aircraft and spares. The Rhodesian Chief of Air Staff has claimed, however, that by the mid-1970s the RhAF had more aircraft on strength than at the time of UDI, possibly a veiled reference to the modest number of machines which find their way into Rhodesia via South Africa. RhAF headquarters is in Salisbury and the total strength of the force is about 1300 officers and men. The war has led to increased defence spending, which in 1978–79 totalled $R149m ($242m), almost 26% of the national budget.

Southern Rhodesia, as a British colony, formed an Air Section as part of the country's Defence Force in 1935. With the advent of the Second World War the Air Section was renamed the Southern Rhodesian Air Force and formed three squadrons which were absorbed into the RAF and flew Spitfires, Typhoons and Lancasters. Flying training for the RAF took place in Rhodesia under the Empire Air Training Scheme and eventually four FTSs were operating in the country.

After the war the SRAF was re-formed with a number of second-line training and transport types plus, in 1951, 22 ex-RAF Supermarine Spitfires for two fighter squadrons. A steady expansion through the 1950s saw the introduction of Vampire jet fighters, Provost trainers and Canberra bombers. When the Federation of Rhodesia and Nyasaland was formed in 1953, the SRAF took over responsibility for defence of the territory and was granted the prefix 'Royal'. Ten years later, in 1963, the federation was disolved, with Northern Rhodesia and Nyasaland becoming Zambia and Malawi respectively with the Southern area retaining its name Rhodesia. UDI was declared in 1965 to prevent any granting of independence on terms unacceptable to white Rhodesians, and, despite efforts at mediation, UN sanctions were imposed on the country. When the Salisbury regime proclaimed a republic in 1969, the Rhodesian Air Force dropped the title 'Royal' and reverted to its present name.

Rhodesia has few announced allies but enjoys close military as well as economic ties with neighbouring South Africa. Since 1972, black guerrilla movements have been waging a steadily expanding terrorist war against the country from bases in Mozambique, Botswana and Zambia, forcing the Rhodesian armed forces on to a war footing. In Spring 1979 elections were held in Rhodesia, leading to the formation of a black-majority government, which renamed the country Zimbabwe-Rhodesia. It is anticipated that sanctions will be lifted by many countries in the near future.

Inventory
The Rhodesian Air Force maintains a high state of readiness and efficiency, and over some five years of operations is widely regarded as one of the most experienced air arms in counter-insurgency work. The majority of the front-line aircraft were obtained before UDI and, despite oil and arms embargoes, they continue in service. South Africa has played a not inconsiderable role in maintaining the RhAF at its present operational level, providing spares backing for the Canberras and Vampires together with modern armament such as Matra rocket pods for ground-attack use on the Hunters.

Attempts to acquire more combat aircraft in the early 1970s met with little success, international action preventing the sale of 31 Jordanian Hunters to the country; the purchase of 28 F-86K Sabres from Venezuela was also blocked. Notwithstanding these setbacks, the Rhodesian Air Force is believed to have received up to 34 Alouette III helicopters from South Africa. These machines have suffered more casualties on operations than any other type, being used in the gunship and assault roles as well as for liaison and casualty-evacuation duties. A batch of Agusta-built AB.205s has also been delivered (via Israel), their presence being revealed in late 1978.

A transport squadron originally operated four C-47s, but the number in service has increased and the type is used for a variety of duties including supply work and parachute missions for the army's airborne unit. The slower counter-insurgency role is undertaken by Reims Lynx push-pull twins obtained in 1976–77 and fitted with underwing weapons pylons, supported by some Italian AL-60 Trojans bought some years before for reconnaissance use. Latest aircraft acquired for road convoy escort and other support roles are SIAI-Marchetti SF.260W Warrior armed trainers.

Apart from the principal bases at New Sarum (Salisbury), Thornhill and Kariba, there are numerous advanced strips around the borders from where operations have been conducted against camps in neighbouring hostile states.

Support
Training is conducted in Rhodesia on the two-seat Provosts and Vampires, pilots then proceeding on to the Canberras. Tuition is also undertaken in South Africa on the Atlas-built MB.326 Impala. The Aircrew Selection Centre, Apprentice Training School and Photographic Establishment are at New Sarum.

RHODESIA

Unit	Type	Role	Base	No	Notes
5 Sqn	Canberra B.2	Bomb	New Sarum	8	Fitted with undernose rocket rails
1 Sqn	Canberra T.4	Train	New Sarum	3	
	Hunter FGA.9	FGA	Thornhill	9	Flown with Matra rocket pods for ground attack
2 Sqn	Vampire FB.9	Att	Thornhill	8	
2 Sqn	Vampire T.55	Att/train	Thornhill	8	
6 Sqn	Provost T.52	Recce	New Sarum	13	Fitted with underwing weapon pylons
4 Sqn	AL-60F5 Trojan	Trans	Thornhill	7	
3 Sqn	C-47	Trans	New Sarum	8	Used in paratroop role; actual inventory likely to be more than stated figure
7 Sqn	Alouette III	Liaison/gunship	New Sarum	30	
	AB.205	General purpose		13	
	Islander	Trans		3	
	F.337 Super Skymaster/Lynx	Coin		20	
	SF.260W Warrior	Coin		20 approx	
	Baron	Liaison		1	

SOUTH AFRICA

Area: 455,000 sq. miles (1·18 million sq.km.).
Population: 26 million.
Total armed forces: 65,500.
Estimated GNP: $43.8 billion (£21.9 billion).
Defence expenditure: $2·62 billion (£1·31 billion).

Backed by a home armaments industry, manned by highly trained personnel and flying a front-line force of some 250 aircraft, the South African Air Force stands as the most potent air arm on the continent. The country's apartheid policies resulted in late-1977 in a total arms embargo by United Nations countries. More than 75% self-sufficiency in arms production is claimed by South Africa, however, and this will have to be increased to nearly 100% with the implementation of the UN arms embargo. Defence spending continues at a high level to finance the arms programme.

The SAAF is the sole operator of military aircraft and is controlled from Defence HQ in Pretoria. It is organised into four operational Commands – Strike, Maritime, Light Aircraft and Air Transport –

forces in the Western Desert, East Africa and Europe. In the post-war years the SAAF was reorganised into a small, efficient arm, capable of rapid expansion should the need arise.

A small force of SAAF F-51D Mustangs was flown to Korea in 1950 to operate with UN forces against the Communists, and it was in Korea two years later that the SAAF received its first jet aircraft in the form of F-86 Sabres on loan from the USAF. De Havilland Vampires and Canadair Sabre 6s were later acquired and in 1957 the Service's long-serving Sunderland flying boats were replaced by Avro Shackletons to patrol the important Cape shipping route. South Africa withdrew from the British Commonwealth in 1961 and UN

Above: A SAAF Impala (MB.326B built under licence by Atlas Aircraft) escorts a pair of Mirage IIIs, one a two-seat trainer and the other a single-seater.

and two support Commands, Training and Maintenance. Of the 65,500 personnel in the armed forces, about 10,000 serve in the SAAF. Basic flying unit is the squadron, with an establishment of between seven and 20 aircraft each. Supporting the regular force is a strong reserve element flying locally built aircraft and known as the Active Citizen Force. Finally there are 13 Air Commando squadrons flying civilian-owned aircraft put under SAAF control in an emergency.

The South African Aviation Corps was formed in 1915, although some South African Army officers had received flying training two years previously and a number of South Africans were already serving with the RFC on the Western Front. Disbanded in 1918, the SAAC was reformed as the South African Air Force in 1920. Avro Tutors and Hawker Hinds were manufactured in the country under

licence for the SAAF before the Second World War and a major expansion programme undertaken in the late-1930s, with Britain supplying Hurricanes, Gladiators, Battles and Blenheims. During the Second World War the SAAF flew maritime patrols around the coast and conducted operations against German and Italian

Above: The Super Frelon, used as a transport and for casualty evacuation, is one of several types supplied to the SAAF before France embargoed arms sales to the country.

sanctions against the country began, although an Anglo-South African Defence Treaty (the Simonstown Agreement) enabled the SAAF to purchase 16 Buccaneer strike aircraft before the British Government finally cut off all arms dealing with the country.

With its traditional arms supplier gone, South Africa turned to other sources, particularly France and Italy, both countries establishing close ties with the Pretoria government and becoming virtually the only military-equipment suppliers through the late 1960s and 1970s. The United States supplied some Hercules and Swearingen Merlins but generally abided by the UN embargo.

South Africa retains close links with Rhodesia. SAAF forces have conducted limited operations along the country's northern borders and in Angola during the 1976 war.

Inventory

Strike Command comprises five squadrons with a mixture of Mirage fighters and

Left: An Atlas II, the South African-built version of the Macchi MB.326K single-seat light attack aircraft, flies over Johannesburg. Production in South Africa has kept the local aerospace industry in business and circumvented problems of embargoes.

Canberra and Buccaneer bombers. Command headquarters is at Waterkloof, which also forms the main base for the Mirage III/F.1 force of three units employed in the fighter-bomber and interceptor roles. Before the UN arms embargo of 1977, France and South Africa agreed that the latter would assemble F.1s under licence at the Atlas Aircraft Corp., and the announced cut-off of spares from France would seem to have little effect now that the programme is under way. The Mirage IIIs carry Matra R.530 missiles and the F.1CZs have R.550 Magics. Air-to-surface AS.30 missiles are carried by the Canberras of 12 Sqn as well as by Mirage IIIEZs. Only about eight of the 16 Buccaneer S.50s purchased in the late 1960s survive. They are fitted with auxiliary rockets for hot-and-high take-offs and with underwing fuel tanks have a typical

Above: The Canberra B(I).12s of the SA Air Force are kept in first-class condition and are among the few to remain operational with the French AS.30 missile.

strike range of some 2300 miles.

Protection for the SAAF's main bases is afforded by a version of the French Crotale surface-to-air missile known as Cactus, supplemented by Short Tigercat SAMs bought from Jordan in 1974. Strike Command also controls the Active Citizen Force, made up of five light-attack squadrons equipped with Atlas-built Impala IIs plus a further unit flying Harvards.

Cape Town is the headquarters of Maritime Command, which conducts the important task of patrolling the sea lanes around the Cape and provides a search-and-rescue element. Long overdue for replacement are the seven Shackleton MR.3s bought in 1957 and since resparred and updated to cope with the increasing submarine threat in South African waters. No replacement is likely for these aircraft in the near future, and they are expected to soldier on into the 1980s.

Italian-built Piaggio Albatross light twins equip a squadron at Ysterplaat for short-range patrol and SAR duties, and another unit operates a handful of C-47s for target towing and support work. This same base also has a flight of Westland Wasp helicopters for ASW duties from South African Navy frigates. Air Transport Command is headquartered at the large Transvaal base of Waterkloof, where the main heavy-lift element comprises 28 Sqn with Hercules and Transalls acquired in 1963 and 1969 respectively.

An additional transport force exists in the shape of the logistical 'airline' Safair, which has a fleet of 15 Lockheed L-100-30s based at Jan Smuts Airport and is known to operate in a para-military role when not flying civilian freight charters. A VIP unit flies a mixture of types from the Pretoria base of Zwartkop, the most recent acquisition being seven Swearingen Merlin transports bought in 1975. The helicopter element of the SAAF flies mostly French types. The 40 Puma tactical machines are divided into flights at Zwartkop in the

SOUTH AFRICA

Unit	Type	Role	Base	No	Notes
1 Sqn	Mirage F.1AZ	Att	Waterkloof	32	
2 Sqn	Mirage IIICZ	Int	Waterkloof	16	
2 Sqn	Mirage IIIBZ	Train	Waterkloof	3	
2 Sqn	Mirage IIIRZ/R2Z	Recce	Waterkloof	4/4	
3 Sqn	Mirage F.1CZ	Int	Waterkloof	16	Armed with R.530 and R.550 AAMs
4 Sqn	Impala II	Att	Waterkloof		
5 Sqn	Impala II	Att	Durban		Active Citizen Force
6 Sqn	Impala II	Att	Port Elizabeth	50	squadrons; licence-built
7 Sqn	Impala II	Att	Ysterplaat		MB.326Ks
8 Sqn	Impala II	Att	Bloemspruit		
11 Sqn / 43 Sqn	Cessna 185A/D	Liaison	Potchefstroom	20	
12 Sqn	Canberra B(I).12	Bomb	Waterkloof	6	Armed with AS.30 ASMs
12 Sqn	Canberra T.4	Train	Waterkloof	3	
15 Sqn	Super Frelon	Trans	Bloemspruit and Zwartkop	14	Also used for casevac
16 Sqn / 17 Sqn	Alouette III	Liaison	Bloemspruit, Durban, Port Elizabeth, Zwartkop	40+	As many as 70 reported in use. Ten operated by 87 AFS for training
19 Sqn	Puma	Trans	Zwartkop, Durban	40	
21 Sqn	Viscount 781	VIP	Zwartkop	1	
21 Sqn	HS.125 Mercurius	VIP	Zwartkop	4	
21 Sqn	Merlin IVA	VIP	Zwartkop	7	One fitted as ambulance
22 Flt	Wasp HAS.1	ASW	Ysterplaat	11	Operated from SAN frigates
24 Sqn	Buccaneer S.50	Att	Waterkloof	8	
25 Sqn	C-47	Trans	Ysterplaat	6	
27 Sqn	P.166S Albatross	SAR/Mar pat	Ysterplaat	20	Radar-equipped for coastal patrol
28 Sqn	C-130B	Trans	Waterkloof	7	
28 Sqn	C.160Z	Trans	Waterkloof	9	
35 Sqn	Shackleton MR.3	Mar pat	Cape Town	7	
41 Sqn	C4M Kudu	Liaison	Zwartkop,	40	
41 Sqn	AM.3C Bosbok	Liaison	Potchefstroom		
42 Sqn	AM.3C Bosbok	Liaison		30	
44 Sqn	DC-4	Trans	Zwartkop	5	Ex-South African Airways
44 Sqn	C-47	Trans	Zwartkop	10	
85 Advanced Flying School	Mirage IIIDZ/D2Z	Train	Pietersburg	3/10	
	Mirage III EZ	Train	Pietersburg	16	
	Sabre 6	Train	Pietersburg	12	
86 AFS	C-47	Train	Bloemspruit	6	
87 AFS	Alouette III	Train	Ysterplaat	10	Included in total for 16+ 17 Sqn
FTS	Impala I	Train	Langebaanweg	151	Licence-built MB.326B
FTS	Harvard	Train	Dunnottar	80	Being replaced by Impalas
	Cactus	SAM		18 batteries	
	Tigercat	SAM		54 launchers	

Below: One of the Mirage IIIBZ dual-control trainers used by the South African No 2 Sqn at Waterkloof, the main base of the original Mirage force. It is not known if dual Mirage F1 versions are in use.

north and Durban in the south, while the heavy-lift unit, 15 Sqn, operates Super Frelons on a similar basis with half the force based at Zwartkop and the remainder at Bloemspruit. Alouette IIIs operated in the liaison role are attached to two squadrons, with aircraft detached to bases throughout the country.

Fixed-wing liaison duties are flown by aircraft attached to Light Aircraft Command and the four squadrons under its control. Locally built AM.3CM Bosboks and the larger Kudus equip these units, flying alongside a few remaining Cessna 185s. As a back-up to the regular units and the Active Citizen Force there are 13 Air Commando squadrons equipped with civil light aircraft for use in emergencies.

Support
Training Command is steadily replacing its long-serving Harvards with Impala I/IIs, the latter equipping the Flying Training School at Langebaanweg on the west coast. There are three Advanced Flying Schools: 85 AFS at Pietersburg with 12 ex-1 Sqn Sabre 6s and 29 Mirage IIIs; 86 AFS at Bloemspruit with C-47s; and 87 AFS at Ysterplaat with Alouette IIIs. Basic flying training on the Harvards lasts for 125hr followed by the advanced stage of 120hr on the Impala I.

MOZAMBIQUE

Area: 303,074 sq. miles (786,762 sq. km.).
Population: 9,870,000.
Total armed forces: 21,000.
Estimated GNP: $2 billion (£1 billion).
Defence expenditure: $109 million (£54.5 million).

This East African country was Portugese from 1508 until 1974 when independence was achieved, along with Angola and Guinea-Bissau in the west. As with many ex-colonies, Mozambique quickly established an armed force which included a small air arm using aircraft left behind by the departing Portuguese. The country's governing body, FRELIMO, has its military headquarters at Laurenco Marques, and assistance is being provided by the Soviet Union, Cuba and some East European countries.

Mozambique has been one of the main bases for guerrillas operating against neighbouring Zimbabwe-Rhodesia and this has prompted a number of retaliatory actions by Rhodesian forces on the guerrilla bases. To deter such operations, the Soviet Union delivered eight MiG-21MF fighter-bombers to the port of Nacala in March 1977, accompanied by equipment and missiles. These aircraft are believed to have been a gift from the Soviet Government.

Inventory
The front-line combat force of the Mozambique air arm, or *Forca Popular Aerea*, is made up of 35 MiG-21MFs, delivered early in 1978, joining the eight earlier aircraft. It is thought that these machines are based at one of two new air bases built recently near Nacala and Beira. At least three Mil Mi-8 transport helicopters have also been received from the same source. Fixed-wing transport types include some ex-*Forca Aerea Portuguesa* Douglas C-47s and about four Nord Noratlases, together with a few Antonov An-24s. A training school at Maputo has some single-engined Cessnas and some North American T-6G Texans, the training programme being supervised by Cuban and East European 'advisers'. Other ex-Portugese types left behind in 1974 included some Dornier D027s and OGMA-built Auster D.5/160s and the FPA could still be flying these machines.

MADAGASCAR

Area: 228,000 sq. miles (590,000 sq. km.).
Population: 8 million.
Total armed forces: 10,500.
Estimated GNP: $1·8 billion (£900 million)
Defence expenditure: $40·2 million (£20·1 million).

The island republic of Madagascar, formerly Malagasy, has a small transport and internal policing air arm known as the *Armée de l'Air Madagascar*. Situated off the south-east coast of Africa, Madagascar was a French colony before gaining independence in 1961, and its present armed forces comprise some 4760 personnel of which only about 160 constitute the air force. Headquarters and one of the main bases are at Ivato near the capital, Tananarive.

This small air arm recently received its first combat aircraft in the form of eight MiG-17s supplied by North Korea, but the main role of the AAM is light transport and liaison, with a limited policing task facilitated by the arrival in the mid-1970s of a Britten-Norman Defender. In the late 1960s Madagascar co-operated with Britain and the United Nations against Rhodesia by providing facilities at the air base of Majunga to blockade the Portuguese port of Beira and prevent oil supplies from getting through to the illegal regime of Ian Smith. Detachments of RAF Shackletons operated from Majunga during the period of the blockade, sharing the base with a few AAM Skyraiders then in use.

Madagascar retains military co-operation with France but an agreement signed some years ago has been terminated.

Inventory
All the aircraft used by the *Armée de l'Air Madagascar* have been acquired from France with the exception of the Defender and the MiG-17s received from North Korea. Future procurement could see the updating of the force with some new transport aircraft.

Below: One of the AAM Britten-Norman Defenders, photographed in British civil registration (but Madagascar air force insignia) before delivery.

MADAGASCAR

Unit	Type	Role	No
1 *Esc* Militaire Aérienne	C-53D	Trans	1
	C-47	Trans	5
	Defender	Coin	1
	MiG-17	Int	8
	Aztec D	Liaison	1
	Reims Super Skymaster	Liaison	3
	Cessna 172M	Train	4
	Alouette II/III	Liaison	1/2
	Bell 47G	Train	1

Note: All based at Ivato

Indian Subcontinent

Religious and cultural differences have repeatedly led to conflict since British control of the sub-continent was withdrawn in the late 1940s, and what was once one country is now three independent states: India, Pakistan and Bangladesh. India has one of the world's largest air forces, operating a mixture of indigenously designed, licence-built and foreign types. Aircraft of North American, Western European, Eastern European and Soviet origin rub shoulders, and Hindustan Aeronautics has the only overseas production line for a Russian combat type (if Chinese copies are excluded).

Pakistan likewise has a mixed bag of aircraft, including the unusual combination of Chinese-built MiG-19s (F-6s) armed with US-supplied Sidewinder air-to-air missiles. India and Pakistan have fought each other in 1948, 1965 and 1971, the last of these leading to East Pakistan becoming the independent state of Bangladesh.

The unstable relationship between India and Pakistan has been further affected by the 1978 Marxist coup in Afghanistan, which removed that country from its position of buffer state between the Soviet Union and the subcontinent. The USSR has since put pressure on Pakistan to withdraw from the Central Treaty Organisation (CENTO), and continued unrest amongst Islamic revolutionaries – spurred on by the success of a similar movement in Iran – has forced Russia to become even more heavily involved in Afghanistan.

China invaded north-eastern India in 1962 but relations between the countries have since become less strained. Another incipient area of conflict is the Bangladesh/Burma border, across which Muslim refugees have been streaming from predominantly Hindu Burma.

Potential areas of unrest can also be found at India's northern and southern extremes. Some Tamils in Sri Lanka (Ceylon) want independence for their part of the country, in much the same way as Cyprus was split, and both Nepal and Bhutan fear that India has eyes on their territory.

PAKISTAN

Area: 310,400 sq. miles (803,900 sq. km.).
Population: 73·4 million.
Total armed forces: 429,000.
Estimated GNP: $17·6 billion (£8·8 billion).
Defence expenditure: $938 million (£469 million).

Controlled from the new centralised Air Headquarters at Rawalpindi, the Pakistan Air Force is an experienced military arm which has faced formidable odds in three wars against a much larger enemy. Today it operates a cosmopolitan spread of aircraft types acquired from sources both in the East and the West and is one of three aviation arms within the country's armed forces, both the army and navy having modest air elements. A front-line combat strength of more than 200 aircraft is supported by a relatively small transport force comprising a single squadron of US- and Iranian-supplied machines. Nearly half-a-million men are under arms in Pakistan – 429,000 in all, of which about 18,000 make up the air force. Defence spending in 1978 totalled $938m, the increase over previous years being due in part to a requirement to purchase a new long-range strike aircraft.

To increase the efficiency of the PAF and provide greater protection for the country, Pakistan has been divided into three defence sectors: North, Central and South. North and South have the bulk of combat squadrons, while Central has the task of early warning. Units operate on a squadron basis made up of about 16 aircraft each. Future plans call for the establishment of an indigenous industry for the production of military and civil aircraft, lessening Pakistan's reliance on foreign arms suppliers.

With the partitioning of the Indian subcontinent in August 1947, Pakistan was formed in two parts, East and West, separated by 1100 miles of Indian territory. As a dominion within the British Commonwealth the new state formed the Royal Pakistan Air Force with RAF assistance and seconded British personnel; its initial strength was two squadrons, one flying Hawker Tempest fighters and the other C-47 transports. Some Halifax bombers followed for a third squadron, and in 1950 a batch of Hawker Fury fighters entered service. The vital link between the two elements of Pakistan was an air bridge established with an eventual total of 62 Bristol 170 Freighters, a service that was to continue until the war in 1971 when East Pakistan became the independent state of Bangladesh.

Supermarine Attacker jets arrived for the RPAF in 1950, equipping 11 Sqn, and these served in the fighter-bomber role until replaced by F-86F Sabres in 1957. Martin B-57Bs joined the Service the following year as a result of a mutual Defence Aid Programme being signed by Pakistan and the US; this also yielded T-6 trainers for the school at Risalpur, HU-16 Albatross amphibians for rescue duties

Right: A pair of Pakistan Air Force Shenyang F-6 (licence-built MiG-19) fighters are scrambled on an interception mission. The aircraft are normally used for ground attack.

and some H-19 helicopters for communications work. Pakistan's second major conflict with India was fought in 1965 and lasted for 17 days. As in the first clash in 1947, the war was over the disputed Kashmir territory. Relative success accompanied PAF operations during the war with losses far fewer than those of the Indian Air Force. A ceasefire was agreed, but not before the termination of US aid which Pakistan had been receiving for almost a decade.

Pakistan therefore turned to other sources for arms and found China willing to supply aircraft in the form of Shenyang F-6s (Chinese-built MiG-19s), while France provided Mirage IIIs. Ninety ex-*Luftwaffe* Canadair Sabres 6s were received via Iran in 1966 and, to increase the efficiency of the country's early-warning system, several Plessey AR-1 low-level radars and

Marconi Condor GCI stations were purchased. In November 1971, war with India once more erupted and from the intense fighting over the Indian invasion of East Pakistan there emerged the independent state of Bangladesh.

Inventory

In 1976 the Pakistan Air Force selected 110 Vought A-7 Corsairs to meet its urgent long-range strike-aircraft requirement. US State Department approval for the deal, which was worth $700m, followed the lifting of an arms embargo the previous year but still depended on Pakistan's announced intention to buy a nuclear fuel reprocessing plant from France, a move vigorously opposed by the US Government. Consequently the deal was cancelled and the PAF was forced to examine alternative aircraft; none had been selec-

Above: The PAF's main interceptor force comprises Mirage IIIEPs, but these can be augmented by Mirage 5PAs (illustrated) which are more usually employed for ground attack.

ted at the time of writing. Late in 1978 the nuclear plant agreement was cancelled by France so the Corsair order may after all be placed. Saudi Arabia signed an aid agreement in 1976, and money for the purchase of new aircraft is likely to come in part from this source.

The present PAF bomber force is based near Karachi and comprises some 15 Martin B-57Bs, the survivors of 26 delivered in the 1950s for two squadrons. Attrition and two wars have forced the PAF to merge the surviving aircraft into one unit, and this operates mainly in the night attack role. Following the 1965 Pakistan-Indian war and the subsequent severing of US aid, the country looked for other sources of military equipment and was offered Shenyang F-6s by China. The offer was accepted and today the air force has seven squadrons flying this type. Basically the F-6 is a Chinese-built MiG-19, its main role being ground attack, but PAF technicians modified the aircraft to carry the US Sidewinder air-to-air missile in addition to its standard quartet of 30mm cannon, and in recent years Martin-Baker zero-zero ejection seats have been fitted in place of the original semi-automatic seats.

In addition to the F-6s a handful of two-seat MiG-15UTIs were supplied for pilot conversion, plus some Il-28 tactical

bombers which were never put into active service. Dassault Mirages were first procured in 1967 with 18 IIIEPs being received the following year, and subsequent batches have brought the total to 64 of several versions. France has also supplied Crotale surface-to-air missiles. Three ex-*Aéronavale* Breguet Atlantics were purchased in 1975 for long-range ASW patrols along the Arabian Sea coast, Pakistani crews undergoing training in France on the aircraft before their delivery to the squadron at Karachi. Iran has sold a number of C-130 Hercules to the PAF and the present force totals some 12 aircraft flying with No 33 Transport Wing at Chaklala. The helicopters in use include ten ex-USAF twin-rotor Kaman Huskies, more than a dozen Alouette IIIs and some Bell 47Gs.

Pakistan's air force stands today as an efficient arm struggling to rationalise its equipment. It operates from modern bases equipped with hardened shelters and installations and able, in an emergency, to call on the services and aircraft of other states with which Pakistan has good relations. These include Abu Dhabi and Libya, each of which operates Mirages and has been assisted by seconded PAF maintenance personnel. In the 1971 war Libya supplied Pakistan with a squadron of Northrop F-5s, but these were later returned following the cessation of hostilities. Both Iran and Saudi Arabia have given financial aid to the country.

Support
Flying training is conducted at the PAF Academy, Risalpur, on Supporters and T-6s, with advanced tuition provided by the T-37Cs and T-33As. Operational conversion is flown on the two-seat Mirages and MiG-15s. Personnel from other countries also undergo training with the PAF at Risalpur. The College of Engineering and the School of Aeronautics and Electronics are housed at Korangi Creek, while at Kohat is the Recruits Training School and the Ground Instructors School. Plans to set up an aircraft assembly plant at Cambellpur were announced in 1976, initial production being concentrated on Cessna T-41D trainers and Hughes 500 helicopters.

Navy
The Pakistan Navy's air arm operates helicopters, until 1975 having been equipped with a small number of Sikorsky UH-19s for search-and-rescue duties around the coast. A growing anti-submarine threat was recognised in the early 1970s and six Westland Sea Kings were ordered in 1973; AM.39 Exocet missiles are carried for anti-ship attacks.

Army
Pakistan Army Aviation operates both light fixed-wing aircraft and helicopters. Its history goes back to 1954, when US

Right: These Saab Supporters carry Pakistani Air Force markings and are used by that Service as trainers and for utility roles. A number are operated by the army, however, and have been flown in action against guerrilla forces.

194

military aid was provided before Pakistan joined the CENTO and SEATO pacts. The supply of about 60 Cessna O-1 Bird Dogs established the force as a distinct formation within the army, two squadrons being formed for AOP and liaison duties. The helicopter force was established with a batch of Bell 47Gs supplied under the Military Aid Programme in 1964 and crewed by US-trained Pakistani personnel. After the 1965 war some Alouette IIIs were purchased, followed in 1969 by the first of a dozen Mi-8s bought from the Soviet Union for supply and casevac work.

Pakistan Army Aviation remains a relatively small force totalling about 100 aircraft. Main base and home of the Army Aviation School is Dhamial, near Rawalpindi, and from here flights are detached to field formations around the country. The Cessna Bird Dog remains the largest single type numerically, and its value can be measured by the fact that the Army Aviation Workshop at Dhamial continues to manufacture examples from knocked-down components on its own production line. Alouette IIIs are also assembled and overhauled at the facility for all three Services. Budget restrictions have prevented the purchase of a Bird Dog replacement, but some of the Saab Supporters bought by the PAF have been passed to the

PAKISTAN Air Force

Unit	Type	Role	Base	No	Notes
7 Sqn	B-57B	Bomb	Masroor	15	
5 Sqn	Mirage IIIEP	Int	Sargodha	17	
? Sqn	Mirage IIIDP	Train	Sargodha	5	Total of four Mirage squadrons
? Sqn	Mirage IIIRP	Recce	Sargodha	13	
9 Sqn	Mirage 5PA	Att	Sargodha	28	
11 Sqn 14 Sqn 17 Sqn 18 Sqn 19 Sqn 23 Sqn 25 Sqn	Shenyang F-6	Att	Sargodha, Rafiqui, Masroor	140	
15 Sqn 26 Sqn	F-86F	FGA/train	Chaklala	40	
29 Sqn	Atlantic	Mar pat	Karachi	3	
6 Sqn	C-130B/E	Trans	Chaklala	7/4	
	L-100			1	
	F.27 Mk 200	VIP		1	
12 Sqn	Falcon 20	VIP		1	
	Aero Commander	Liaison		1	
2 Sqn	T-33A	Train	Mauripur	12	
	L-23	Liaison		2	
	Alouette III	Liaison		14	
	Bell 47G	Liaison		12	
	Puma	VIP		1	
	HH-43B	SAR		10	
	T-37C	Trans	Risalpur	20	
	MiG-15UTI	Train		5	
	Supporter	Train/util		45	Also anti-guerrilla and army roles
	Crotale	SAM			

Above left: This Dassault-Breguet Atlantic anti-submarine aircraft, one of three transferred from France's *Aéronavale* carries Pakistan Navy markings but is thought to be operated by the PAF.

PAKISTAN Navy

Type	Role	No
Sea King Mk 45	ASW/SAR	6
Alouette III	Liaison	4

PAKISTAN Army

Type	Role	Base	No	Notes
Alouette III	Liaison	Dhamial	30	
Puma	Trans		35	Order placed in 1977
Bell 47G	Train/liaison	Dhamial	20	
O-1	Liaison	Dhamial	50	
Mi-8	Trans	Dhamial	12	Used for casevac and supply
Supporter	Liaison	Dhamial	45	

army for liaison duties. Early in 1977 the army placed an order for 35 SA.330J Puma assault helicopters, and delivery of these were made in 1978–79.

Support

No 503 Workshop at Dhamial overhauls army machines and some of those operated by the PAF and the Navy. At the same base is the Army Aviation School, which accommodates pupils from all three Services, taking them through an initial pilot training course of some 80hr on the O-1 and Supporter followed by a similar period on the Bell 47G before conversion to the Alouette or Mi-8.

AFGHANISTAN

Area: 250,000 sq. miles (647,000 sq. km.).
Population: 16·5 million.
Total armed forces: 110,000.
Estimated GNP: $2·3 billion (£1·65 billion).
Defence expenditure: $60 million (£30 million).

The Afghan Air Force has its headquarters at the Ministry of Defence in Kabul. Since the change of power in April 1978, the country has moved firmly into the Soviet sphere of influence and a number of aid agreements have been concluded between the two countries. Personnel strength totals approximately 10,000, while defence expenditure for 1978 was $60m.

A military aid agreement with the Soviet Union signed in December 1955 continues to provide Afghanistan with modern aircraft and equipment. Civil and military airfields have been built by the Russians, and the United States has assisted the country by constructing the airport at Kandahar. Border disputes with Pakistan were rife in the 1960s but involved virtually no aerial activity.

Formed in 1924 by King Amanullah, the Afghan air force was initially a branch of the army. Its first aircraft were two Bristol Fighters supplied by Britain and flown by German crews. These were later supplemented by a squadron of R-1 reconnaissance biplanes donated by Russia. Following the civil war of 1928–29, no new equipment was received until 1937. Two squadrons of Hawker Hind bombers were supplemented by Italian-built trainers and reconnaissance aircraft; both Britain and Italy also sent training missions to the country.

In 1948, 12 Avro Anson 18s were acquired by the now officially named Royal Afghan Air Force for transport and training duties. With the withdrawal of the Hinds the Ansons became the RAfAF's sole operational type for some years.

Inventory

The Soviet Union has been the sole supplier to the RAfAF – recently renamed Afghan Air Force – since August 1957, when the fruits of a military aid agreement started reaching the country in the form of some 30 MiG-17s – later increased to about 100 – plus helicopters and transports. The current combat strength comprises some 170 aircraft, although front-line strength may be somewhat less than that. An Air Defence Division includes a brigade with three battalions of SA-2 Guideline surface-to-air missiles, totalling 48 launchers.

Training is conducted on Soviet lines, an air academy being situated at Sherpur. Initial flying training is carried out on Yak-11s and Yak-18s, with jet conversion flown on two-seat MiG-15UTIs, MiG-21UTIs and Il-28Us. A batch of Czech L-39s have been received.

Type	Role	Base	No	Notes
MiG-21	Int	Pagram	40 approx	Total of three squadrons
MiG-19	Int		12	Total of 18 supplied in 1965
MiG-17	Day int	Mazar-i-Sharif	50	One wing of four squadrons
Su-7BM	GA	Shindand	24 approx	Two squadrons
IL-28	Bomb	Shindand	45 approx	Three squadrons
An-26	Trans		10	
IL-14	Trans		25	Two squadrons
An-2	Util		10	
IL-18D	VIP		2	
Mi-4	Trans		18	One squadron; operates from several bases
Mi-8	Trans		—	
Yak-11	Train	Sherpur	—	
Yak-18	Train	Sherpur	—	
MiG-15UTI	Train		—	
MiG-21U	Train		—	
IL-28U	Train		—	
L-39	Train		—	
SA-2	SAM			48 launchers Three battalions

INDIA

Area: 1·26 million sq. miles (3·26 million sq. km.).
Population: 606 million.
Total armed forces: 1·096 million.
Estimated GNP: $101 billion (£50·5 billion).
Defence expenditure: $3·57 bill:on (£1,785 billion).

Large armed forces are maintained by this populous Asian country, and all three Services have an aviation element. During the past decades India has been involved in a number of military conflicts, with China in 1962 and with Pakistan in 1967 and 1971, and over that period Indian air power has been steadily modernised to meet the more demanding nature of a re-organised air-defence and strike capability. The Indian Air Force is the largest of the three air arms, having some 700 combat aircraft in almost 45 squadrons manned by 100,000 officers and men. A growing defence industry caters for national needs, some of the aircraft produced being of Indian design while others are built under licence. Credit is one of the main problems encountered by the Indian Government in its search for new equipment for the armed forces, hence the large number of Russian aircraft types currently in IAF service, these being procured at more favourable terms than comparable Western equipment.

Nearly 70 per cent of the equipment requirements of the Indian armed forces are now met from indigenous sources and the country is aiming at self-reliance in the manufacture of spare parts over the next few years. Defence spending for 1978 totalled Rs29.45bn ($3.57bn), and this is expected to increase in subsequent years to cater for inflation and the acquisition of more modern military equipment. The IAF is organised into three operational commands on a regional basis – Eastern Air Command with HQ at Shillong, Western

Air Command with HQ at Delhi and Central Air Command at Allahabad, the last-mentioned controlling all bomber and transport elements. In addition there are Training and Maintenance Commands.

The Indian Air Force was officially established in 1933 under RAF control, initially operating Westland Wapiti general-purpose aircraft. The gradual expansion that followed was hastened when the Second World War began and the IAF soon found itself equipped with Hawker Hurricane fighters, Vultee Vengeance bombers and Westland Lysander reconnaissance aircraft in support of Allied ground units fighting the Japanese. In recognition of the IAF's wartime work, the 'Royal' prefix was granted by King George VI in 1945. When India was divided into two countries, India and Pakistan, on being granted independence in 1947, the IAF retained seven fighter squadrons with Hurricanes, Tempests and Spitfires plus a C-47 transport unit. The piston-engined fighters gave way to Vampire jets in the years following independence and these were supplemented in 1953 by more than 100 Dassault Ouragan fighter-bombers bought from France. Mystère IVA fighters arrived three years later and a licence was obtained from Britain for the production of Folland Gnat lightweight fighters by Hindustan Aeronautics. Other types bought in substantial numbers include Hunters and Canberras for the front-line squadrons, Fairchild Packets and Ilyushin Il-14 transports, and Bell 47G helicopters and licence-built North American T-6 trainers. In the 1960s India turned to Russia for her next-generation fighter and a licence was obtained for the production of the MiG-21 interceptor, a decision heavily influenced by the favourable terms offered by the Soviet Union compared to those from Western manufacturers. With the assistance of a handful of German designers, Hindustan Aero-

nautics flew the ptototype of the HF-24 Marut fighter-bomber in June 1961. Other aircraft put into service during the decade include An-12 and Caribou transports, Mil Mi-4 helicopters and Sukhoi Su-7 strike aircraft.

India's main source of arms supply over the past 20 years has been the Soviet Union, with which she has a treaty of friendship, co-operation and mutual assistance. The country's traditional enemy is Pakistan, with which she has been at war a number of times since independence in 1947. Border clashes have also occurred with Communist Chinese forces.

Second-largest air arm in Asia, and one of the ten largest in the world, the Indian Air Force is in the throes of a re-equipment programme involving home-produced combat types as well as the purchase of Jaguar International strike aircraft to replace the Su-7. Some 200 Jaguars will eventually join the IAF, the

first 40 coming from the UK with the remainder being built under licence by HAL. Main IAF interceptor is the MiG-21, of which nearly 200 early-model MiG-21FLs built under licence by Hindustan Aeronautics were later supplemented by 150 improved MiG-21MFs and by the -21bis powered by an uprated Tumansky R25 engine and having a revised avionics fit.

The Soviet Union has delivered a number of MiG-21bis fighter-bombers to the IAF to equip two squadrons, and these are expected to be joined by the first Indian-assembled aircraft in 1979. The 15 MiG squadrons are an integral part of the nation's air-defence system, which also has a ground network of radars and communications links. Low-level defence is assigned to some 250 HAL-built Gnat F.1 fighters. This remarkable little aircraft proved particularly successful in the conflicts with Pakistan and so impressed was the Air Force that it requested the type be

Top: The IAF can boast a potent interceptor force in the shape of 15 squadrons operating MiG-21s of various marks, some received directly from the Soviet Union and others built under licence by HAL.

Below left: An Indian Air Force An-12 Cub which appears to be in the process of having its temporary camouflage paint removed by rain and/or human endeavour, revealing the normally white upper surface.

Above: Three bomber squadrons fly the ever-popular Canberra, which additionally serves in the photo-reconnaissance role. Despite its age, the aircraft continues to provide good service in air forces around the world.

Below: The indigenously developed Kiran trainer is operated by the IAF alongside Polish-built Iskras, which were ordered to overcome the problem of slow Kiran deliveries from Hindustan Aeronautics' production line.

developed further into the Gnat Mk 2 or Ajeet. This incorporated integral wing fuel tanks which allow the underwing drop tanks fitted on the older Mk 1 Gnat to be dispensed with and weapons to be carried in their place. Other modifications are also incorporated and 100 Ajeets are being delivered, beginning in 1978. The IAF has some 120 launchers for Russian-supplied SA-2 surface-to-air missiles at 20 sites, and low-level defence is provided by 40 Short Tigercat systems.

Four ground-attack squadrons have about 120 Hawker Hunters from some 230 bought from Britain, beginning in 1957–58, and three further units have HF-24 Marut Mk 1 strike aircraft. The Su-7 element comprises four squadrons and it is these units which are destined for early re-equipment when the new Jaguar Internationals begin arriving. Three strike squadrons and a single photo-reconnaissance unit operate a mix of BAC Canberras obtained in various batches since 1958.

The large IAF transport force has a heavy-lift element of two squadrons equipped with Antonov An-12s supplied by Russia for Himalayan support duties, two medium transport units with C-119G jet-boosted Packets, three units with C-47s, and a Stol squadron flying DHC Caribous bought in the mid-1960s. India has been producing British Aerospace HS.748s for a number of years. The age of some of these aircraft has caused some concern to the Air Force and replacements are being sought.

The vital communications role in some of the more remote areas of India is accomplished by helicopter; the IAF's rotary-wing force operate about 350 machines, although not all are operational. Licence-built Aérospatiale Cheetahs have replaced

indigenous Krishak fixed-wing AOP aircraft with Army co-operation units.

Future procurement for the Indian armed forces is to be diversified away from almost exclusive reliance on the Soviet Union, and French Matra Magic air-to-air missiles are being bought for the IAF's MiG-21s, which are to undergo modifications to accommodate the new armament.

Support
Pilot training in the elementary stage lasts 40hr and is flown on the HT-2s at the EFTS, Bidar. From 1981 the ageing HT-2s are due for replacement by the HPT-32, designed and developed by Hindustan Aeronautics at Bangalore. Advanced training is performed on Kirans at the Air Force Academy, Hyderabad, the course lasting a year and involving 180hr, including 20hr of weapons training. A Mk 2 version of the Kiran is being developed which will have a derated Orpheus engine and four underwing hardpoints specific-

Above: The Indian Navy's shore-based rotary-wing anti-submarine force comprises 15 Westland Sea King Mk 42s operating from Cochin. The IN's seagoing ASW helicopters are Ka-25As.

ally for weapon training. As a result of the premature retirement of the Vampire training fleet, an order for Iskra advanced trainers was placed with Poland in the mid-1970s and these are now in service. Operational conversion follows wings presentation, graduate pilots flying a course on Kirans and Iskras. There is a Tactical Development School at Ambala equipped with MiG-21s, Su-7s and some Gnats, and an Armament Training Wing at Jamnager with Hunters. A Transport Training Wing with C-47s and HS.748s is based at Yelahanka, while at Begumpet, seven specially equipped HS.748s equip the Navigation and Signals School. The Helicopter Training School is situated at Hakimpet and flies Alouette IIIs.

Navy

Indian Naval Aviation was established as part of the Indian Navy shortly after partition, personnel undergoing training in Britain, which also provided initial equipment in the shape of the former light fleet carrier HMS *Hercules* in 1957. This 16,000-ton ship was renamed INS *Vikrant* and was commissioned in March 1961 after a conversion which included the fitting of an angled deck, mirror sight and steam catapult. Hawker Sea Hawk fighter-bombers were purchased, as were some French Breguet Alizé anti-submarine bombers and Alouette III helicopters. *Vikrant* remains the Indian Navy's flagship but the force of Sea Hawks is ageing rapidly and a replacement is being sought. An order for Harriers seems likely if the ever-present credit problem can be resolved, the total requirement being expected to reach some 25–30 machines. The Alizé squadron, No 310, was expected to relinquish its aircraft for a more modern type in 1977–78 but the force is now planned to continue in use for some years, the Indians having purchased 12 refurbished ex-French naval Alizés late in 1977 to make good attrition.

Alouette III helicopters are used for both planeguard duty aboard the carrier and as torpedo-carrying strike aircraft aboard *Leander*-class frigates. Shore-based ASW duties are performed by Westland Sea Kings and two fixed-wing squadrons with Super Constellations and Soviet-supplied Ilyushin Il-38 MR aircraft. Five Britten-Norman Islander light transports were ordered in 1976 for liaison work but these machines are being fitted with nose radar and other equipment suitable for coastal patrol duty. On the strength of the Indian Navy and delivered in 1978 are two Soviet *Krivak*-class destroyers which are being equipped with Kamov Ka-25 helicopters for ASW work. Kirans, Vampires and Sea Hawks are used for training, although the Indian Air Force conducts initial pilot instruction. Short Seacat surface-to-air missiles are fitted to four frigates.

Below: The Caribou's short take-off and landing performance allows it to fly from the IAF's semi-prepared strips.

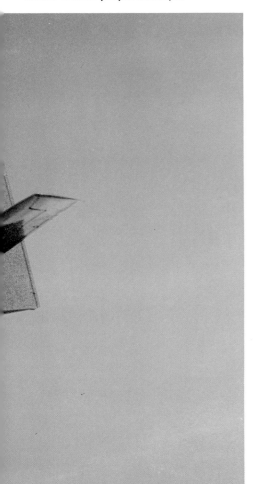

INDIA   Air Force

Unit	Type	Role	Base	No	Notes
1 Sqn, 3 Sqn, 4 Sqn, 7 Sqn, 8 Sqn, 17 Sqn, 21 Sqn, 24 Sqn, 28 Sqn, 29 Sqn, 30 Sqn, 45 Sqn, 47 Sqn, 101 Sqn, 108 Sqn	HAL/Mikoyan MiG-21FL, HAL/Mikoyan MiG-21MF, HAL/Mikoyan MiG-21bis, HAL/Mikoyan MiG-21UTI	FGA / Train / Train		150 / 160 / 200 to be acquired / 40	Two Russian-supplied squadrons from 1977, HAL deliveries from 1979 / Two aircraft with each squadron
2 Sqn, 9 Sqn, 15 Sqn, 18 Sqn, 22 Sqn, 23 Sqn	HAL Gnat F.1/ Ajeet	Int		150/100	Re-equipped with Ajeet
10 Sqn, 31 Sqn, 220 Sqn	Marut Mk 1	FGA		125	
5 Sqn, 16 Sqn, 35 Sqn	Canberra B(I).12/ B(I).58/ B.74/T.13	Bomb/train		52	
6 Sqn	Canberra	Anti-ship		6	
	Super Constellation	Trans		2	
106 Sqn	Canberra PR.57	Recce		8	
14 Sqn, 20 Sqn, 27 Sqn, 37 Sqn	Hunter F.56/T.66	Att/train		120/20	
26 Sqn, 32 Sqn, 221 Sqn, 222 Sqn	Su-7	Att		75	
25 Sqn, 44 Sqn	An-12			29	
41 Sqn, 59 Sqn	Otter			20	
33 Sqn	Caribou			19	
11 Sqn, 43 Sqn, 49 Sqn	C-47	Trans		40	
19 Sqn, 48 Sqn	C-119G			50	
	HS.748			58	
12 Sqn	Tu-124	VIP	Delhi	1	Includes three for VIP duties
104 Sqn, 112 Sqn, 114 Sqn	Alouette III/Cheetah	Liaison		180	
105 Sqn, 107 Sqn, 110 Sqn, 111 Sqn, 115 Sqn, 116 Sqn	Mi-4	Trans		80	
109 Sqn, 118 Sqn, 119 Sqn	Mi-8	Trans		35	
659 Sqn, 660 Sqn, 661 Sqn, 662 Sqn	Cheetah	Liaison		100	
	HT-2		Bidar	70	
	HPT-32	Train			To replace HT-2s from 1981
	Kiran		Hyderabad	130	
	Iskra			40	
	SA-2	SAM	20 sites	120 launchers	
	Tigercat			40 launchers	

INDIA   Navy

Unit	Type	Role	Base	No	Notes
300 Sqn	Sea Hawk	FGA	INS *Vikrant*	25	
			INS *Vikrant*	25	
310 Sqn	Alizé	ASW		12	
312 Sqn	Super Constellation	MR	Dabolim, Goa	5	Originally flown by 6 Sqn IAF until MR task taken over by IN
315 Sqn	IL-38			3	Further three aircraft on option
330 Sqn, 336 Sqn	Sea King Mk 42	ASW	Cochin	15	
321 Sqn, 331 Sqn	Alouette III	Att/SAR	INS *Vikrant*	18	Also flown by 561 Sqn training unit; eight aircraft in 331 Sqn deployed aboard frigates for special strike duties
550 Sqn	Islander	Comms/MR		5	
	Devon	Comms		2	
562 Sqn	Hughes 300		Cochin	4	
550 Sqn	Vampire T.55	Train		4	
551 Sqn	Kiran Mk.1			15	
333 Sqn	Ka-25	ASW	Krivak-class destroyers	5	
	Seacat	SAM	Frigates	4	

SRI LANKA

Area: 25,300 sq. miles (65,500 sq. km.).
Population: 14·3 million.
Total armed forces: 13,300.
Estimated GNP: $4 billion (£2 billion).
Defence expenditure: $14·3 million (£7·15 million).

The Sri Lanka Air Force operates on a low-key basis, having a small combat element for counter-insurgency duties and a transport/liaison force for a variety of military and civilian work. Responsible to the government, the Commander of the Air Force administers the arm from the headquarters in the country's capital, Colombo. Present strength comprises some 2300 personnel with an aircraft inventory totalling 60 machines. The 1978 defence budget amounted to Rs.211m ($14.3m).

The British island colony of Ceylon first established an air arm in October 1950 and shortly afterwards King George VI decreed that its title should be the Royal Ceylon Air Force. RAF instructors and technicians were seconded to the Ceylon Government to train the force; early equipment included de Havilland Chipmunks and Boulton Paul Balliols for pilot training up to wings standard, while some Airspeed Oxfords were employed on immigration patrols. In 1955 the first Dove light transport arrived, together with Scottish Aviation Pioneers and Westland Dragonfly helicopters. The air arm's first jet equipment was received in 1958 when eight Hunting Jet Provost armed trainers arrived, followed by four more. Aid from the United States began in the 1960s in the form of Bell JetRanger helicopters, but this was eclipsed in 1971 when a Soviet military mission arrived following an armed insurrection in the country. A small number of MiG-17Fs were supplied and a military aid agreement was signed between the two countries. Main bases include Katunayaka and China Bay, Trincomalee.

Inventory

In the early 1970s Ceylon changed its name to Sri Lanka and the Sri Lanka Air Force is currently the sole operator of aircraft in the country's armed forces. A single combat unit continues to operate

SRI LANKA

Type	Role	Base	No	Notes
MiG-17F	FGA	Katunayake	5	Supplied by Russia in 1971; flown irregularly
Jet Provost T.51	Armed train	Katunayake	8	Fitted with underwing armament
MiG-15UTI	Train	Katunayake	1	Supplied by Russia in 1971
Convair 440	Trans		1	Purchased in 1973 for tourist flying
DC-3	Trans		2	Ex-Air Ceylon bought in 1976 for tourist work
Riley Heron	Trans		4	
Dove	Trans		5	
Cessna Super Skymaster	Trans/train		4	
Bell 47G	Liaison		6	
Bell 206	Liaison		7	Also used for tourist flying; supplied by US
Dauphin II	Liaison		2	
Ka-26	Liaison		2	
Cessna 150	Train		6	
Chipmunk	Train		9	

the five MiG-17Fs supplied by Russia in 1971, flying them alongside the Jet Provosts. Operational emphasis in the air force is centred on maintaining internal order, communications, transport and training, the single largest element being the transport fleet. This undertakes both military and civilian flying tasks including tourist flights around the island, for which a Convair 440 and two ex-Air Ceylon DC-3s have been acquired. Helicopters and a modest training fleet complete the inventory. Future procurement is likely to be limited due to budget restrictions even though much of the equipment is ageing.

Support

Elementary pilot training begins on the Cessna 150s and Chipmunks, then divides with future jet pilots moving on to the Jet Provosts and the MiG-15UTI, while transport pilots convert to the Doves then to the larger machines.

Below: Jet Provost T.51s serve the Sri Lanka Air Force as both trainers and ground-attack aircraft, being fitted with underwing racks for the latter role.

BANGLADESH

Area: 55,100 sq. miles (142,700 sq. km.).
Population: 81 million.
Total armed forces: 73,300.
Estimated GNP: $6·9 billion (£3·45 billion).
Defence expenditure: $150 million (£75 million).

Formerly East Pakistan, Bangladesh became an independent state with Indian assistance in 1971 and currently supports a military element known as the Bangladesh Defence Force, of which the Air Wing is the sole operator of aircraft. Air Wing personnel total about 2500 from overall armed forces' strength of some 71,000; its headquarters is in the capital, Dacca. An assortment of aircraft cover the spectrum of air operations, with aid coming from India and the Soviet Union. Defence expenditure for 1976–77 amounted to Taka 746m ($51.5m).

The origins of the Bangladesh Air Wing lie in the air support accorded Mukti Bahini forces in former East Pakistan just before the outbreak of the 14-day war between India and Pakistan in December 1971. Limited operations were flown with Indian assistance. Following the cease-fire, the new state of Bangladesh officially established the Air Wing as part of the Defence Force, manned by personnel who had defected from the Pakistani forces. The wing was initially equipped with two or three ex-Pakistan Air Force Canadair Sabre 6s left behind by the PAF, plus a

single T-33 trainer. A Caribou transport and three or four Alouette III helicopters arrived from India, but the Air Wing's chief equipment supplier became the Soviet Union following a visit to Moscow in March 1972 by the Bangladesh prime minister Sheikh Mujibur Rahman. Early the following year the first of a squadron of MiG-21s arrived in the country as part of a Russian military aid agreement, and later supplies included transports and helicopters.

Bangladesh has an assistance agreement with the Soviet Union signed in 1972.

Inventory

A single MiG-21MF-equipped combat squadron provides offensive as well as defensive operations within the Bangladesh Defence Force (Air Wing). Air Wing personnel underwent training in Russia on three-month courses before converting on to the type. After delivery of the MiGs, the Sabres and T-33 originally in use were permanently grounded due to lack of spares and further standardisation resulted in the sale of two Wessex helicopters donated to the country some years before by Britain. However, since an attempted coup in 1977, the Air Wing has turned to Communist China for equipment and has received 24 MiG-19S fighters to supplement the remaining MiG-21s (at least six of which have been lost). Transport aircraft include an Antonov An-24 and some An-26s, while Mil Mi-8 helicopters joined the Alouettes for liaison and tactical duties.

BANGLADESH

Type	Role	Base	No	Remarks
MiG-21MF	FGA	Tezgaon	3	One sqn
MiG-21U	Train	Tezgaon	2	
MiG-19S (F-6)	Int	Tezgaon, Jessore	24	Two sqns
An-24	Trans	Tezgaon	1	
An-26	Trans	Tezgaon	3+	
Bell 212	Liaison		6	
Alouette III	Liaison		4	Ex-Indian AF
Mi-8	Trans		3+	
NZAI Airtourer	Train		1	Presented by New Zealand; one of two supplied
Magister	Train		8	Ten originally received

NEPAL

Area: 54,300 sq. miles (140,000 sq. km.).
Population: 13 million.
Total armed forces: 20,000.
Estimated GNP: $1·6 billion (£800 million).
Defence expenditure: $13·8 million (£6·9 million).

The Air Wing of the Royal Nepalese Army operates in the support role, using a small number of light transport aircraft. Headquarters and main base of the Wing are in the country's capital, Kathmandu, and from here the aircraft operate throughout the country in support of the 20,000-man Gurkha army which protects this Asian monarchy.

Following British and US aid in the early 1960s, a Royal Flight was established with two Scottish Aviation Twin Pioneers

and a single Ilyushin Il-14, the latter presented by Russia to the Nepalese ruler, King Mahendra. A Short Skyvan Executive followed and in 1970 the Royal Nepalese Army purchased a military Skyvan 3M for STOL operations. A DHC Twin Otter was also acquired, subsequently being supplemented by two more. In 1975 the Royal Flight took delivery of an HS.748 and two years later received one of two Pumas ordered, the other going to the Air Wing.

NEPAL

Type	Role	No	Notes
HS.748 Srs 2A	VIP	1	Fitted with side freight door. Used by Royal Flight
Skyvan 3M	Trans/VIP	2	
Twin Otter	Trans	3	
Puma	Trans/VIP	2	

Note: All based at Kathmandu

201

Central and East Asia

Asia is where the world's great powers meet head-on, with potentially catastrophic consequences. Russia, China, Korea and Japan have been locked in conflict of one sort or another for decades, with European countries and the United States — together with states in Southeast Asia and those with a European-orientated culture — becoming involved as Japan attempted to establish its own empire in the 1930s and 1940s, the north and south of Korea fought each other in the early 1950s, and China and the Soviet Union settled down to a seemingly implacable mutual enmity.

Perhaps the most important development in the area during the 1970s was the gradual rapprochement between China and the United States in particular, and the West generally. China has survived Mao's death without great changes being wrought, and is seeking to become a major industrial and economic power with the aid of Western expertise and equipment. The sacrificial lamb is Taiwan, with which the US is severing its ties in response to Chinese insistence that the island is a province of China as a whole rather than the seat of the rightful Chinese administration.

China's increasingly close relationship with Kampuchea (Cambodia) following the re-unification of Vietnam was abruptly ended in late 1978 and early 1979 by the Russian-orientated Vietnamese Government, which invaded the country with the aid of Kampuchean guerrillas and brought down the Pol Pot administration. China retaliated by attacking the north of Vietnam itself.

Japan is also in dispute with the Soviet Union over the latter's occupation of what were until 1945 the northern islands of Japan itself, and the defection of a Russian MiG-25 pilot with his aircraft in 1976 was an occasion for mild sabre-rattling by the USSR.

North and South Korea have been involved in border clashes but there is a chance of improved relations as the US reduces its active military involvement with the Seoul government.

CHINA

Area: 3·7 million sq. miles (9·58 million sq. km.).
Population: 900 million approx.
Total armed forces: 4·325 million.
Estimated GNP: $350 billion (£175 billion).
Defence expenditure: $35 billion (£17·5 billion) approx.

China's air forces are among the largest in the world, in proportion to the vast size of the country, but on the other hand are among the least modern air arms. The Maoist emphasis on "alternative technology" as the means of bringing the benefits of industrialisation to a massive peasant population meant until the late 1970s that China was extremely poor in high technology. China's increasing isolation from the rest of the Communist Bloc, as well as from the West, exacerbated this problem. In recent years China has made efforts to purchase military aircraft and equipment from the West, in some cases successfully. China's new regime, however, is making concerted attempts to import new material from the West. China is also concerned to develop and advance its own technology with outside aid.

The full title of the Chinese air arm is the Air Force of the People's Liberation Army. It is not known whether strategic missiles are under air force command, although they are thought to be operated by the army's Second Artillery. Other forces for coastal patrol and defence of port areas are under the control of the People's Navy.

China has few allies as such; with its vast population and resources it probably needs none, and it would be hard to envisage the country being anything other than an independent superpower once military maturity is attained. As yet, however, the country is kept short of this status by its limited industrialisation and lack of effective military technology. As the latter advances, though, China's importance as a supplier and patron to smaller states will increase.

The Chinese air arm was established under Soviet tutelage after the defeat of the Nationalist forces on the mainland in 1949. Very large numbers of modern aircraft were delivered during the Korean War, where Chinese forces were the means of bringing Russian technical aid and equipment into the war without the participation of Soviet combat personnel. The 1950s were a period of rapid expansion, with China receiving high priority in deliveries of modern Soviet types for tactical operations, including Mikoyan MiG-15 Fagot, MiG-17 Fresco and MiG-19 Farmer fighters and Ilyushin Il-28 Beagle light bombers. Deliveries of long-range bombers, however, were confined to Tupolev Tu-4 Bull bombers copied from the American Boeing B-29 Superfortress of the Second World War. Together with deliveries of aircraft the infrastructure of airfields and control systems was built up with the aid of technicians from the Soviet Union and European aligned nations. At this time Chinese military aviation de-

Right: The F-6 and other indigenous versions of the MiG-19 are still in production, these aircraft forming the backbone of China's air-defence force.

pended largely on the Soviet Union for training as well as for equipment. A significant development in this period was the start of licence manufacture of Soviet designs in the State Aircraft Factory at Shenyang.

A few Mikoyan MiG-21F Fishbed C supersonic day fighters and a small number of Tupolev Tu-16 Badger bombers may have been delivered to China before the break in relations with the Soviet Union, which was partially caused by the Soviet refusal to supply China with nuclear weapons and a modern delivery system. Throughout the 1960s the Chinese air forces suffered from the vacuum left by the withdrawal of Soviet back-up services, and the technical standard of equipment stagnated. Until the late 1960s the remnants of the Tu-4 force were the only long-range bombers in service, and the most modern fighter available in any numbers was the Shenyang F-6 (licence-built MiG-19).

China has supplied military and other aid to a number of countries, but has no allies among its neighbours since Kampuchea was overrun by Vietnam in late 1978. Albania and Tanzania have both taken delivery of Shenyang F-6s and F-7s (the latter the licence-built version of the MiG-21F, which does not appear to have been constructed in the same quantities as the older aircraft) and Pakistan has a number of F-6s. Bangladesh also flies F-6s. These supplies, however, have no great military significance to China itself, except in as much as the links with Albania and Tanzania serve as a minor offset to the influence of the Soviet Union in Europe and Africa.

The Chinese strategic missile forces have been the main area of technological development and expansion over the past ten years or so, following the explosion in the 1960s of the first Chinese nuclear and thermonuclear weapons. The CSS-1 medium-range ballistic missile entered service in 1966, and 50–90 missiles are reported to be in service. It is believed to be based on the Soviet SS-4 Sandal missile, and its 1800km range is sufficient to threaten Soviet border installations and China's immediate neighbours. It has been followed into service by the CSS-2 intermediate-range ballistic missile (IRBM) with a range reported as up to 5500km in one version; about 20 are operational, deliveries having started in 1971. The longer-range CSS-3 is reported to have become operational in North-West Sinkiang in 1975, and the CSS-4, with near-global range, is under development. The first of a class of ballistic-missile submarines is under construction, but the capability of the sea-launched ballistic missile (SLBM) which the new craft will carry is unknown; the US expects it to be a two-stage solid-propellant weapon comparable to the early Polaris. All the Chinese strategic missiles reported so far have been liquid-fuelled, unlike the latest US and some Soviet types, increasing reaction times and vulnerability.

The aircraft-borne component of the Chinese deterrent rests on a force of 60–65 Shenyang-built Tupolev Tu-16 Badger bombers armed with free-fall thermo-

nuclear weapons. The type is still in production. A few Tu-4s are still in service for attack duties and training.

The Chinese air-defence force is mainly dependent on a Chinese-built version of the Soviet V750VK Guideline (SA-2) surface-to-air missile (designated CSA-1 by Nato when in Chinese service) and the Shenyang F-6 (MiG-19) limited all-weather fighter. Some F-6s are equipped to carry the locally produced version of the AA-1 Alkali air-to-air missile, which arms later Soviet-built MiG-19s. F-6s in Pakistani service are day fighters armed with the US Sidewinder missile; Chinese aircraft may carry a copy of the similar Soviet AA-2-1 Atoll. About 1500 F-6s are in service with the Air Force air-defence units, a force which contrasts with the 100 F-7 Fishbed Cs armed with Atoll copies. The F-6 versions are still in production.

China's first partly indigenous supersonic fighter, the F-6bis Fantan A, was initially reported to have been a failure but is now regarded as being at least a limited success. Fantan A is a development of the F-6 with side intakes and was first reported in 1970. Development of this indicates that China has tried to produce an indigenous intercept radar.

A substantial number of obsolete Shenyang F-4 (MiG-17) fighter-bombers operate in the strike role, together with some 200 Ilyushin Il-28 Beagle light bombers, some Soviet-supplied and some Shenyang-built. The latter are designated B-5. A handful

Above: The Fantan A fighter-bomber known as F-6bis (often mis-named F-9).

Left: Fantan A appears to have an internal weapons bay, seen here with the doors open.

of Shenyang F-2s (MiG-15s) may remain in service to augment the F-4s, which could total as many as 1500 aircraft.

China has made considerable efforts since the early 1970s to acquire modern combat aircraft from the West. Repeated overtures have been made to Britain for the supply of a large force of British Aerospace Harrier V/Stol strike fighters. A 100-aircraft order is in prospect, with the possibility of further licence production. The Chinese Harrier deal may well go ahead – the aircraft are tactical types, presenting no threat to the West.

An agreement between China and Rolls-Royce in 1976 provided for the establishment of a plant to build RB.168 Spey afterburning turbofans in China. The Shenyang Aircraft Factory is working on the design of a combat aircraft with a single Spey to re-equip air-defence and fighter-bomber units of the air force. The choice of the 20,000lb-thrust Spey implies a level of capability well in advance of any aircraft now in service with the Chinese forces, and the design of such an aircraft demands far greater technical resources than are evidenced by the manufacture of the F-6bis. It is reported that the Chinese are using the Soviet technique of evaluating swept-wing and delta aircraft using common components. Known as F-12, the new aircraft will presumably be used in the tactical strike role as well as interception and close support.

A handful of Il-28s and some of the Tupolev Tu-16s fly in the reconnaissance role. Transport aircraft include a fairly large number of Shenyang-built Antonov An-2s (about 250), some piston-engined Ilyushin Il-14s and the remains of some 100 Lisunov Li-2s (licence-built, Soviet-powered C-47s) delivered by the Soviet Union. Five Ilyushin Il-18s are also operated. The Chinese military presumably has access to the ten Boeing 707–320s (including six freighter-convertible aircraft) operated by the Civil Aviation Administration of China. Thirteen Aérospatiale Super Frelon medium-lift helicopters have been delivered. A fairly large number (about 200) of Mil Mi-1 and Mi-4 helicopters are still in service; the latter is built in China as the H-5. Canadian PT6 turboprops have been acquired by China, possibly for a new small transport aircraft.

Training is carried out on the standard Eastern Bloc range of trainers: the Yakovlev Yak-18 (Chinese-built as the BT-6) primary trainer, the more powerful piston-engined Yak-11 and the veteran MiG-15UTI. Also in service is a Chinese-developed trainer version of the F-4 (MiG-17). A few Il-28Us are also used, and the last of the Tu-4s still serve as trainers for the Tu-16 force.

No full information is available on the deployment of Chinese air units, although the country is thought to be divided into six Air Regions for command purposes. The detail structure is division/regiment/squadron, as in the Soviet Union, with each regiment comprising three or four 12-aircraft squadrons. The Chinese Air Force still uses the bases in Western and Central China built by the US Army Air Force for the bombing of Japan in the 1939–45 war.

The Aviation of the People's Navy operates an assortment of aircraft for maritime reconnaissance and coastal defence, including a mixed force of Shenyang F-2s, F-4s and F-6s and a number of torpedo-carrying Il-28s. A few Beriev Be-6 flying-boats remain in service for the anti-submarine role. A training force operates the same mixture of aircraft as the Army Air Force units, and a few Il-14s and An-2s provide transport. About 50 Mil Mi-4s are operated from shore bases.

The People's Navy includes a number of missile-armed warships, mostly carrying Chinese copies of the SS-2 Styx surface-to-surface anti-shipping missiles delivered to China by the Soviet Union before the breakdown in relations between the two states. A large number of Hoku (Komar copy) and Hola (Osa copy) boats are operated, as well as a number of Chinese-built destroyers. Missile boats of the new *Hai Dua* class carry six of a new, Chinese-developed SSM. The new *Kiangtung*-class frigates carry a SAM system which may use a missile based on the Soviet SA-N-1 Goa.

Chinese military personnel have been involved in recent conflicts in South-East Asia as advisers, and close relations were established with the Pol Pot regime in Kampuchea (Cambodia). The Pol Pot government was overthrown by Vietnamese forces – aided by a Kampuchean guerrilla organisation – in early 1979, however, and China responded in February by sending troops into Vietnam. Chinese forces were beaten back by the Vietnamese, and by the end of March 1979 all the invading troops had returned to China. Further conflict along the Sino-Vietnamese border can be expected.

MONGOLIA

Area: 600,000 sq. miles (1·55 million sq. km.).
Population: 1·5 million.
Total armed forces: 30,000.
Estimated GNP: $2·8 billion (£1·4 billion).
Defence expenditure: $120 million (£60 million).

 One of the smallest in the Eastern Bloc, the Air Force of the Mongolian People's Republic is entirely run and commanded by a cadre of Soviet personnel as a small extension of the Soviet forces in the adjacent Soviet Union military districts.

Combat equipment comprises a single battery of V750VK (SA-2 Guideline) surface-to-air missiles and a unit of ten Mikoyan MiG-15 fighter-bombers. Some 30 Antonov An-2s provide transport support for the army, together with a handful of Mil Mi-1 and Mi-4 helicopters. Air Mongol operates Ilyushin Il-14s and Antonov An-4s, which are presumably available to the Air Force. The SAMs and the MiG-15s are probably deployed at Ulan Bator.

Training to basic standard is carried out locally on a few Yak-11s and Yak-18s, but advanced training is performed by the Soviet Union.

KOREA (NORTH)

Area: 47,200 sq. miles (122,400 sq. km.).
Population: 14·5 million.
Total armed forces: 512,000.
Estimated GNP: $9·8 billion (£4·9 billion).
Defence expenditure: $1 billion (£500 million).

 The People's Republic of North Korea has been engaged in an arms race with the South since the disarmament provisions in the post-war armistice of 1953 broke down in the late 1950s. The Korean People's Army Air Force is accordingly not only numerically strong but fairly modern in its equipment.

The Air Force is responsible for air defence and support of ground forces under direct army command. There is no naval force. Korea tended to follow its ally of the early 1950s, China, in its drift away from the Soviet Union in the early 1960s, but more recently has returned to the Soviet camp. The main potential enemy is South Korea, and relations between the two countries are generally regarded as unstable.

The Mikoyan MiG-21 Fishbed is probably the most important weapon in the Air Force inventory. Together with a reported 20 battalions of V750VK (SA-2 Guideline) surface-to-air missiles, first-generation and second-generation variants of the standard Soviet fighter, armed with AA-2 Atoll air-to-air missiles, form the backbone of the air-defence force. Deliveries of third-generation versions, with multi-role capability, have probably increased the total force past the 200 mark. So far there have been no reports of MiG-23/27 deliveries to Korea, but this move is likely if South Korea receives General Dynamics F-16As. Some Mikoyan MiG-19 Farmers continue in service in the air-defence role, but these will probably be retired over the next few years and replaced by MiG-21s.

Fighter-bombers comprise a fairly large force of obsolescent MiG-17s, which are presumably due for replacement by third-generation MiG-21s, and one regiment of Sukhoi Su-7 Fitter A strike fighters. There are over 300 aircraft in all. Some Ilyushin Il-28s remain in service in the light-bomber role.

Conversion trainers are attached to Il-28, MiG-21 and Su-7 units, and advanced training is carried out on the standard MiG-15UTI. Basic training is performed on Yakovlev Yak-11s and Yak-18s.

A mongrel transport force is shared with the Civil Aviation Administration of Korea (CAAK). It includes three Tupolev Tu-154Bs used for government transport duties, and a small fleet of twin-turboprop Antonov An-24s recently augmented by purchases from East Germany. A pair of Ilyushin Il-18s are in service, together with a number of Ilyushin Il-14s and some Chinese-built Antonov An-2s. Mi-4s and Mi-8 helicopters are used for tactical transport.

In preparation for the Korean War the Air Force adopted an organisational pattern based on that of the Chinese air arm. The division is the basic command formation, comprising two regiments which are larger than the 36–40 aircraft usually included in the Soviet regiment. Details of deployment are not available, although principal bases are identified.

Right: The KPAAF's interceptor force is based on large numbers of MiG-21s backed up by SA-2 Guideline surface-to-air missiles – a familiar combination.

KOREA (SOUTH)

Area: 38,500 sq. miles (99,700 sq. km.).
Population: 35·8 million.
Total armed forces: 642,000.
Estimated GNP: $31·5 billion (£15·75 billion).
Defence expenditure: $2·6 billion (£1·3 billion).

Korea remains a divided country following the ceasefire between the North and South in 1953. The South supports an air arm, the Republic of Korea Air Force, equipped largely with American machines and trained along similar lines to those of the US Air Force. It is the only major military aviation arm although the other Services – the Army, Navy and Marines – each have small support elements. President Carter's withdrawal of some 40,000 American troops from South Korea is being tempered by the creation of a major industrial defence complex in the country, which is expected to make it self-reliant in the production of all equipment except fighter aircraft and nuclear weapons by the early 1980s. Already munitions and military vehicles are being built in the South, and light helicopters are beginning to come off an assembly line at a new plant being constructed near Seoul.

More than half a million men are in the armed forces and defence spending for 1978 was Won 1.26bn ($2.60bn), although future defence budgets are likely to be increased to meet the country's military needs. Headquarters of the ROKAF is at Yong Dong Po City and manpower totals some 30,000 officers and men. Combat Air Command, with HQ at Osan AB, controls the Air Force's tactical elements as well as a sophisticated air-defence system with integrated radars, troposcatter and microwave links. Basic operational unit is the wing, within which are up to three squadrons, each having an establishment of some 25 aircraft.

The Republic of Korea Air Force was formed in May 1949, its first aircraft being some Piper L-4 Cubs supplied by the United States. With the advent of the Korean War the ROKAF expanded rapidly with US assistance and was soon flying Mustangs against North Korean targets. In 1956 the first of 112 F-86F Sabres arrived to form the country's first jet fighter squadrons, followed by some T-33 trainers. Under a US aid programme Korea received many second-line types as well as some F-86D Sabre all-weather fighters.

The Republic of Korea benefits from a military and economic alliance with the United States under which the US has provided more than $12,000m in aid since the end of the Korean War. A military facility agreement is also in force under which the US bases air and ground forces in the country. Although the US is reducing its presence in Korea, as noted above, some 8300 Americans serving with USAF units in the country are expected to remain. The republic's major enemy is the Communist-ruled North Korea and minor incidents along the border continue to cause unrest between the two countries.

Inventory

Under a major re-equipment programme aimed at strengthening its defensive and striking power, the Republic of Korea Air Force is receiving quantities of F-5Es and F-4 Phantoms from the United States and has requested General Dynamics F-16s from the same source. Combat Air Command controlling the main combat wings is steadily replacing older types such as F-86F and F-86D Sabres with F-4E Phantoms and F-5E Tiger IIs. The US Government has approved the sale of a further batch of 18 F-4Es to add to the 19 Es and 18 Ds already in service, and this can be interpreted as a move to placate the South Korean Government following the announcement by President Carter in 1977 that the US was withdrawing most of its military personnel and equipment from the country.

The new aircraft were due for delivery in the first half of 1979. Four squadrons of F-5E fighter-bombers are replacing, on a one-for-one basis, the older force of F-5A and B aircraft delivered in the 1960s. Nearly 50 Sabres remain in use but the last of these are expected to be withdrawn from front-line service by 1980. To provide the ROKAF with a counter-insurgency/light strike element, an order for 24 OV-10 Broncos was announced in 1977 in an arms package worth $58.2m and including 733 Sidewinder missiles. The OV-10s will supplement the jet types and replace the F-86Fs. In a further move aimed at increasing the country's self-sufficiency, a subsidiary of Korean Airlines known as the Hanjin Corporation has established a helicopter production plant near Seoul and is building Hughes 500M-D Defender multi-role helicopters for the armed forces. There have been major military overhaul facilities in South Korea for many years, but this is seen as a considerable technical advance and could lead to other types being built under licence. A total of 100 Defenders are being contracted for and many will be fitted with Tow missiles for anti-tank use, while others are being assigned to the Navy for ASW duties. Nike Hercules and Hawk surface-to-air missiles are operated; the latter being upgraded to Improved Hawk standard.

To modernise the transport force, which at present comprises about 40 obsolete aircraft, the US is delivering six C-130 Hercules which were ordered in 1977. As well as the Hughes 500s from Hanjin, the air force has a selection of US-supplied machines flown mainly in the SAR role and supported by a fixed-wing liaison and communications force of Cessna O-1 Bird Dogs, U-17s and Beavers.

The navy's maritime-patrol and ASW force with about 20 Trackers is flown by the Air Force, while both the Army and Marine Corps fly an assortment of helicopters and light aircraft including Cessna O-1s and a few Beavers for AOP and support missions.

Flying Training School is based on US thinking and procedures. Primary grading is performed on the T-41Ds, with basic flying conducted on the fleet of T-28Ds. Jet conversion is flown on the T-33As.

KOREA (SOUTH)

Unit	Type	Role	No	Notes
	F-16A/B	Int	80	Requested from USA in 1977
1st FW	F-4D/E	FGA	18/37	Four sqns
10th FW	F-5E/F	FGA	126/9	Replaced 87 F-5As and 35 F-5Bs in four sqns
	RF-5A	Recce	12	
11th FW	F-86F	FGA	50	
12th FW	F-86D	Int	18	
	S-2A/F	ASW	20	South Korean Navy
	OV-10G	Coin	24	
	Aero Commander 680F	Liaison		
	C-46D	Trans	20	
	C-54	Trans	4+	
	C-123K	Trans		
	HS.748	VIP	2	Presidential Flight
	C-130H	Trans	6	
	O-1	Obs	30+	Some flown by the Army and Marine Corps
	O-2A	Obs	12	
	U-17A	Liaison		
	Beaver	Liaison		Flown by all three Services
	UH-1D	Liaison	5	
	AH-1S	Att		
	Bell 212	Trans	2	
	UH-IN	VIP	1	Some used by the Army
	Hughes 500M-D Defender	Att	100	Assembled by Hanjin Corp, Seoul, from 1977
	OH-23	Liaison		Used by the Army
	UH-19	Trans	6	
	T-41D	Train	20	
	T-33A	Train	30	
	T-28D	Train	24	
	Nike Hercules	SAM		1 battalion
	Improved Hawk	SAM		3 battalions

Below: The ROKAF is one of many air forces being supplied with F-5E Tiger II fighters by the United States to replace its earlier F-5S Freedom Fighters. The aircraft is armed with Sidewinders.

JAPAN

Area: 143,000 sq. miles (370,000 sq. km.).
Population: 112 million.
Total armed forces: 240,000.
Estimated GNP: $677 billion (£338·5 billion).
Defence expenditure: $8·57 billion (£4·285 billion).

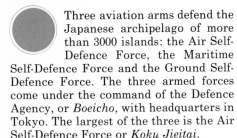

Three aviation arms defend the Japanese archipelago of more than 3000 islands: the Air Self-Defence Force, the Maritime Self-Defence Force and the Ground Self-Defence Force. The three armed forces come under the command of the Defence Agency, or *Boeicho*, with headquarters in Tokyo. The largest of the three is the Air Self-Defence Force or *Koku Jieitai*.

Administrative control of the JASDF comes under the Air Command (*Koku Sohtai*), which has its headquarters alongside that of the USAF's 5th Air Force at Fuchu. Subordinate to the AC are three regional air commands comprising the Northern AC (*Hokubu Koku Homen-tai*) covering Hokkaido and the northern part of the Japanese 'mainland' island of Honshu, with HQ at Misawa; the Central AC (*Chubu Koku Homen-tai*) covering central Honshu with HQ at Iruma; and the Western AC (*Seibu Homen-tai*) covering south-west Honshu and a large part of the islands of Shikoku and Kyushu. Integrated into the Western Air Command is the South-West Mixed Air Wing (*Nansei Koku Kansei-dan*) based at Naha on Okinawa. The regional Air Commands also have direct control over the surface-to-air missile units of both the JASDF and JGSDF which are linked into the Hughes Badge (Base Air Defence Ground Environment) computer-controlled defence system.

In the post-war years Japan's recovery was aided by the United States and a National Police Reserve was established in 1950 as a para-military ground force. Two years later, in August 1952, the Security Agency was created to control a National Security Force and a Maritime Safety Force equipped with a variety of light aircraft and helicopters. The Korean War was a lesson to Japan that the anti-militarism in the country could hardly guarantee the country's safety in the future, and the 1954 Self-Defence Law led to the establishment of a Defence Agency to administer three simultaneously created Self-Defence Forces.

The Japanese Air Self-Defence Force was initially equipped with Beech T-34A Mentors for training, the force being increased by the addition of Curtiss C-46 Commandos, North American T-6s and some Lockheed T-33As transferred from USAF stocks. In December 1955 the Service received North American F-86F Sabres for training purposes, this type also equipping the JASDF's first tactical fighter unit formed in October 1956 at Chitose. The following year Japan initiated its first five-year Defence Build-up Programme, which called for 33 squadrons and a total inventory of 1300 aircraft but in the event saw only 17 squadrons formed with 1127 aircraft by 1962. During

the 1960s Mitsubishi undertook licence production of the Lockheed F-104J Starfighter as a Sabre replacement, and 210 were delivered over the second five-year DBP to be integrated into the Badge air-defence system installed by Hughes. The third five-year DBP initiated the development of a new fighter known as the F-X, a transport known as C-X and the T-X trainer. The McDonnell Douglas F-4E Phantom, for which a licence-manufacturing agreement was concluded between the parent company and Mitsubishi, was selected as the F-X. The transport requirement was met by NAMC/Kawasaki with the C-1A, while the trainer requirement resulted in the Mitsubishi T-2 supersonic twin-jet aircraft.

Although a dominant economic force in the Pacific, Japan is not a member of a military alliance or economic grouping. Instead, she relies for her defence on the Japan–United States Security Treaty signed in 1951, under which Japan defends her territory against external attack until US forces come to her assistance. She is also a signatory of the nuclear non-proliferation treaty and has renounced the manufacture or possession of nuclear weapons.

Inventory

Amid increasingly strident calls for a more independent defence policy, the Japanese Air Self-Defence Force continues through its current Defence Build-up Programme, the fifth since 1958, to modernise its equipment. Economic problems and inflation have steadily reduced the approvals, which in 1978 amounted to only 56 aircraft for the JASDF out of 96 requested by the Service. Actual totals were 23 McDonnell Douglas F-15C/Ds, 15 Mitsubishi F-1s, three Mitsubishi T-2As, 14 Fuji T-3s and a Kawasaki-built Boeing KV-107-IIA. Under the 1979 defence budget the JASDF will receive eight F-1 strike

fighters, six Grumman E-2C Hawkeye early-warning aircraft, two Kawasaki C-1A transports, 22 Fuji T-3 trainers, two Mitsubishi MU-2 SAR twins and three KV-107-IIA helicopters. The present fighter-interceptor squadrons total 14 flying in eight wings – 2, 3, 5, 6, 7, 8, Southwest and General HQ – equipped with 169

Below: Mitsubishi also builds the Phantom under licence as the F-4EJ and is now moving on to assembly of F-15s.

Bottom: Yet another Mitsubishi responsibility is the MU-2, again developed in Japan and used for various utility roles.

Above: The indigenously developed Mitsubishi T-2 trainer was the first Japanese-designed aircraft to exceed the speed of sound. It has since been developed into the single-seat F-1 attack fighter.

Mitsubishi-built Lockheed F-104J Starfighters armed with Sidewinder missiles, together with more than 160 North American F-86F Sabres. The Sabres operate mainly in the advanced trainer role and are due to be retired by about 1980. The McDonnell Douglas F-4EJ Phantom equips the remaining units, with 140 aircraft purchased for five air wings each having one squadron. There is also a reconnaissance unit flying 14 RF-4EJs with a further 14 aircraft required.

In 1977 the JASDF selected the McDonnell Douglas F-15C/D Eagle as a replacement for the Starfighters and 100 have been ordered from Mitsubishi, which will build the type under licence between 1980 and 1988. A possible follow-on order for a further 23 may be placed to meet the JASDF's requirement for five squadrons. A replacement for the Sabre force is the home-designed Mitsubishi F-1 close-support fighter, orders for 67 of which have been placed for three squadrons. Deliveries of the F-1 are under way, with the first squadron having been formed in March 1978. To meet the Air Force's long-standing requirement for an early-warning aircraft, six Grumman Hawkeyes were ordered in the Fiscal 79 budget and a further batch are expected to be ordered in 1980.

The Transport Wing of the Air Force is made up of three squqdrons equipped with Kawasaki C-1A twin-jet freighters and NAMC YS-11As. When Fiscal 79 deliveries are complete a total of 30 C-1As will be in service with the JASDF, the type having replaced the long-serving Curtiss

Commando fleet. A flight-check unit operates a variety of aircraft including YS-11s and MU-2Js, while an Air Rescue Wing – based primarily at Iruma but having detachments at seven other bases – operates nearly 60 aircraft, of which more than half are helicopters. There is also an Air Proving Wing at Gifu charged with testing and pre-service evaluation of all Japanese military aircraft.

Air Training Command (*Hikoh Kyoiku Shuhdan*) of the JASDF has its headquarters at Hamamatsu and is organised along conventional Western lines. Ab initio flying begins on the Fuji-built KM-2B, which is replacing the long-serving and reliable Beech T-34A Mentor.

Students complete some 30hr before being streamed into transport, fighter and helicopter categories. Future transport pilots undergo a further 40hr tuition on the KM-2B/T-34, followed by 150hr on the Beech King Air and culminating with 70hr on the YS-11/C-1A at squadron level. The fighter pilot progresses from the T-34 to the Fuji T-1A/B, Japan's first post-war jet, for 70hr followed by the advanced stage flying T-33As for a further 100hr. The new T-2 comes next, with 80hr advanced flying plus 60hr combat training before assignment to a squadron. The Service is planning to replace the T-33As with an Alpha Jet/Hawk-type aircraft in the 1980s. Helicopter students proceed from the T-34 to

JAPAN Air Force

Unit	Type	Role	Base	No	Notes
—	F-15C/D	Int	—	100	Ordered in 1978
301 Sqn 302 Sqn 303 Sqn 304 Sqn 305 Sqn	F-4EJ	Int	Hyakuri, Chitose, Tsuiki	140	Licence-built by Mitsubishi
501 Sqn	RF-4EJ	Recce	Hyakuri	14	Bought from the USA; further 14 required
202 Sqn 203 Sqn 204 Sqn 205 Sqn 207 Sqn	F-104J	Int	Nyutabaru, Chitose, Chitose, Komatsu	169	Licence-built by Mitsubishi; 207 Sqn based at Naha on Okinawa as part of South-West Mixed Air Wing
6, 8 and HQ Sqn	F-86F	Train	Tsuiki, Komaki, Iruma, Hamamatsu	160	Used as advanced trainers; being replaced by F-1s
3 Sqn	F-1	Att	Misawa	67	Ordered for three sqns to replace Sabres
—	E-2C	AEW	—	6	Ordered in FY79; six more planned
401– 503 Sqn	C-1A	Trans	Miho, Iruma	30	Total includes four pre-series aircraft
	YS-11A	Trans	Miho, Iruma	13	Total includes one YS-11E ECM trainer and two flight-check machines
11th & 12th Train. Wings	T-34A	Train	Shizuhama, Bofu	80	Built by Fuji; being replaced by KM-2B
	Fuji KM-2B (T-3)	Train	Shizuhama, Bofu	54	Ordered to replace T-34s; total of 60 required
13th FT Wing	T-1A/B	Train	Ashiya, Gifu	50	Flown by the Air Proving Wing, Gifu
33 Sqn, 1 Wing; 7 & 35 Sqn, 4 Wing	T-33A	Train	Hamamatsu, Matsushima	186	Used for advanced training; also flown on liaison and target-towing duties; total of 210 licence-built by Kawasaki
21 Sqn, 4 Wing, 22 Sqn	T-2A	Train	Matsushima	66	Total of 49 ordered up to 1978; entered service in 1976; replacing F-86s
	MU-2S	Rescue	Iruma	23	Three MU-2Js flown by flight-check unit at Iruma; seven rescue detachments at other bases with MU-2S
	KV-107-II	Rescue	Iruma	33	Licence-built Boeing Vertol 107
	Mitsubishi S-62A	Rescue	Iruma	7	Licence-built from Sikorsky

207

60hr on the King Air before proceeding on to Hughes TH-55Js of the JGSDF and finally to the S-62.

Navy

The Japanese Maritime Self-Defence Force (JMSDF) is the country's anti-submarine arm, comprising five shore-based air groups; an independent unit assigned to the nation's Defence Fleet; and three independent smaller units controlled by each district command. Headquarters of the JMSDF's Air Command (*Koku-Shuhdan*) is at Atsugi, which in turn is subordinate to the Self-Defence Fleet Headquarters at Yokosuka. Formed in 1956 alongside the JASDF, the Maritime SDF was helped during its early years by the United States, which supplied a number of Sikorsky HSS-1s and Bell 47s for helicopter training plus some Beech T-34As for fixed-wing work. The first major combat type received by the fledg-ling Service was the Lockheed P2V-7 Neptune, followed by a batch of Grumman S-2A Trackers in 1958. Three years later the first Grumman Albatross amphibians arrived for the important air-sea rescue task, and Japan's faith in the flying-boat prompted the development of a replace-ment for these aircraft in the shape of the indigenous Shinmeiwa US-1. By the begin-ning of the third Defence Build-up Pro-gramme in 1967 the JMSDF possessed a total of 228 aircraft, rising at the end of the DBP to 280 of which 64% were anti-submarine warfare aircraft. In the early 1970s, the United States returned the administration of the Okinawa islands to Japan and the JMSDF established the *Okinawa Koku-tai* at Naha Air Base on the main island.

Inventory

The JMSDF is numerically the smallest of the three Services, with 297 aircraft in the inventory and more on order in the fifth five-year DBP (1977–82). Tasked with maritime surveillance and anti-submarine warfare, the main front-line aircraft cur-rently in JMSDF service is the Kawasaki P-2J Neptune. A conversion of the basic P-2H, this aircraft is powered by two GE/Ishikawajima-Harima T64 turbo-props, with two underwing-mounted IHI J3 turbojets providing auxiliary power. To replace these aircraft, the Government approved the purchase of 45 Lockheed P-3C Orions in April 1978, Kawasaki being named prime licence contractor between 1981 and 1988. The JMSDF will form four squadrons, beginning in 1982. Japan re-mains almost the only country in the world to have flying-boats in the military inventory in the form of the PS-1/US-1. Designed and built by Shinmeiwa, succes-sor to the wartime Kawanishi concern which produced the fine H8K Emily flying-boat, a total of 20 PS-1 patrol air-craft will be in use by March 1980 plus six US-1 SAR machines. To expand the Mari-time force the Fiscal 1979 budget calls for the delivery of one US-1 amphibian, two Beech King Air trainers, three Fuji KM-2 trainers, 12 Mitsubishi-Sikorsky SH-3 Sea Kings and two S-61As.

The current equipment organisation comprises three groups (Nos 1, 2 and 4) and

the independent unit on Okinawa flying 83 Kawasaki P-2Js and seven Grumman S-2As. The Lockheed P-2H fleet was due for retirement in 1978 but the S-2As are expected to soldier on until 1982, there being a total of 28 in the JMSDF inventory. The fourth group (21st) and the other two independent units at Ohmura and Ohmi-nato air bases have 83 Sikorsky SH-3As, while the fifth group (31st) operates the PS-1 fleet. Plans for the procurement of some Sikorsky RH-53D minesweeping heli-copters have been delayed and the Service continues to operate nearly a dozen Kawasaki KV-107-11As in this role. There is a single transport unit equipped with YS-11s. With the selection of the Orion for the ASW role, the JMSDF has only one other urgent requirement and that centres on a new helicopter to equip its frigates; the type is expected to be the

Above: The Kawasaki C-1 medium transport has replaced the JASDF's C-46s.

Sikorsky UH-60 following its selection for the US Navy's Lamps programme.

Training

The JMSDF's Air Training Command was formed at Shimofusa in September 1961 and currently comprises three Air Train-ing Groups. For basic pilot training, the Ozuki ATG controls one flying squadron (201st) with KM-2s and a ground school designated 221 Sqn. The Tokushima Air Training Group is responsible for twin and instrument training using Beech Queen and King Airs, while 203 Sqn conducts P-2J conversion flying; helicopter tuition comes under the wing of the Kanoya ATG, using Hughes OH-6s and Bell 47s. ASW training is flown on specially adapted versions of the YS-11 transport.

Army

An integral part of the Japanese Army, the Ground Self-Defence Force (*Rikujoh Jiei-tah*) was established in 1954 with the help of the United States. Early equipment included Beech T-34 Mentor trainers and Sikorsky S-55 helicopters, to be supple-mented and eventually replaced by Fuji-built Bell UH-1Bs and Kawasaki-built Boeing Vertol KV-107-IIAs – types which

remain in service today. The current JGSDF organisation is based on the allo-cation of units to regional commands. The Northern District Air Command (*Hokubu Homen Koku-tai*) has its headquarters at Okadama, together with its Air Squadron. Each command has a District Command Helicopter Squadron (*Hokubu Homen Herikoputai-tai*) and a minimum of two numbered air squadrons, those in the Northern Command comprising 2, 5, 7 and 11 Sqns (*Hiko-tais*) at Asahigawa, Obihiro and Okadama.

The North-East (*Tohoku*) District Air Command has an Air Squadron and its headquarters at Kasuminome. There is also the District Command Helicopter Sqn at the same base, while 6 and 9 Sqn are based at Jinmachi and Hachinoe respec-tively. Eastern (*Tobu*) District Air Com-mand has its Air Squadron, Helicopter

Sqn, 1 and 12 Sqn all based at Tachikawa; Central (*Chubu*) DAC has its Air Sqn, Helicopter Sqn and No 3 Sqn based at Yao, with 10 and 13 Sqn at Akeno and Hofu respectively; Western (*Seibu*) DAC has its headquarters at Kumamoto, the District Command Helicopter Sqn at Metabaru with 4 Sqn, while 8 and 101 Sqn operate from Kumamoto and Naha on Okinawa respectively.

The equipment in these units comprises in the main helicopters – Kawasaki KV-107-IIAs, Bell UH-1Hs and Hughes OH-6As – while a diminishing number of fixed-wing types continue to be used such as the Cessna O-1 Bird Dog, Fuji LM-1 and Mitsubishi LR-1. Each District Command Helicopter Squadron usually has ten KV-107s and UH-1Hs for assault purposes, while the Headquarters element uses around ten Bird Dogs (being replaced by helicopters), Fuji LM-1s and Bell UH-1s. The numbered squadrons have OH-6Js and UH-1s and there are plans to purchase a total of 80 Bell AH-1S Cobra gunships to increase the combat potential in the Service. In the Fiscal 79 defence budget the JGSDF is receiving five more MU-2 liaison aircraft, 15 Kawasaki-Hughes OH-6D light observation helicopters, six Fuji-Bell UH-1H transport helicopters and two more KV-107-IIA assault machines. Future procurement centres on a replacement for the OH-6 and a more modern transport helicopter in place of the KV-107.

Training

Training in the JGSDF is undertaken on TH-55Js at the primary stage at Kasumigaura, alongside some OH-6Js. Two other establishments are situated at Akeno and Iwanuma, the JGSDF also being responsible for the primary training of Air Force helicopter crews.

JAPAN Navy

Unit	Type	Role	Base	No	Notes
1 Sqn, 1 Grp; 2 Sqn, 2 Grp; 3 Sqn, 4 Grp	Kawasaki P-2J	Mar pat	Atsugi, Kanoya,	83	Also used by units on Okinawa and 203 Training Sqn
	P-3C	Mar pat	—	45	Ordered in 1978 for four sqns from 1982
11 Sqn, 1 Grp; 14 Sqn, 4 Grp	S-2A	ASW	Atsugi, Kanoya	28	Not all in service, due for retirement in 1982. Flown by 61 Sqn for transport duties
31 Sqn, 32 & 51 Sqn	PS-1	Mar pat	Atsugi, Iwakuni	20	Three lost in accidents
71 Sqn	US-1	SAR	Tateyama	6	Named Ohtori
101 Sqn, 21 Grp; 121 Sqn	Mitsubishi SH-3A	ASW	Tateyama	95	Built under licence from Sikorsky; some flown from frigates
71 Sqn	S-61A	SAR	Tateyama	5	Total includes two aircraft ordered in 1979; type used by South Pole Observatory Group
71 Sqn	S-62A	SAR	Tateyama	8	Used for base rescue duties
61 Sqn	YS-11	Trans/train	Atsugi	10	Total includes six YS-11T ASW trainers with the 205th Sqn at Atsugi
202 Sqn	Queen Air	Train	Tokushima	28	Used for instrument training
	King Air	Train	Tokushima	8	Includes two ordered in FY79
201 Sqn	KM-2B	Train	Shimofusa	34	Replaced T-34s with Ozuki Air Training Group
211 Sqn	Kawasaki OH-6J	Train	Kanoya	3	Unit forms part of the Kanoya Air Training Group; type built under licence from Hughes
	Bell 47G	Train	Kanoya	8	Unit forms part of the Kanoya Air Training Group

JAPAN Army

Unit	Type	Role	Base	No	Notes
2, 3, 4, 5, 9, 10, 11, 12 Sqn	Kawasaki OH-6J	LOH	Kofu, Metabaru, Yao, Obihiro, Asahikawa, Okadama, Hachinoe, Tachikawa, Iwanuma,	117	Used for training and LOH duties
	Kawasaki OH-6D	LOH	Kasumigaura	25	Total includes aircraft ordered in FY79
1 Heli. Wg; 2 Sqn	KV-107-IIA	Assault	Tachikawa, Yao, Kisarazu, Kasuminome, Okadama	56	Licence-built from Boeing Vertol; flown on Okinawa by 101 Sqn
2–5, 8, 9, 11, 12, 13, Sqn	Fuji UH-1B/H	Trans	Asahikawa, Hachinoe, Hofu, Kasuminome, Kumamoto, Metabaru, Obihiro, Okadama, Tachikawa, Yao	82/60	Licence-built from Bell
	TH-55J	Train	Kasumigaura, Akeno, Iwanuma	38	Entered primary training use in 1972
	AH-1S	Gunship		2	Purchased in 1977 and 78 for evaluation; up to 80 required
	O-1A/E	Obs		27	Being replaced by helicopters
	LM-1/2	Liaison	Tachikawa, Okadama, Kasuminome, Yao, Metabaru, Kumamoto	13	Being replaced by helicopters
	LR-1 (MU-2C/K)	Recce		15	
	T-34A	Train		2	Withdrawal imminent

Above: A Shin Meiwa US-1 search-and-rescue amphibian poses in front of the PS-1 anti-submarine flying boat from which it was developed. The PS-1 is unusual in being a fixed-wing aircraft equipped with a dipping sonar, which allows it to sit on the sea surface while searching for submarines.

Left: The Kawasaki P-2J, the final development of Lockheed's Neptune anti-submarine aircraft, ended production only in 1979.

TAIWAN

Area: 13,800 sq. miles (35,750 sq. km.).
Population: 16·7 million.
Total armed forces: 475,000.
Estimated GNP: $20 billion (£10 billion).
Defence expenditure: $1·67 billion (£835 million).

The Chinese Nationalist Air Force, one of two Taiwanese military air arms, has been established with American help and operates almost exclusively aircraft of US origin. Its headquarters are in the capital, Taipei, and the combat strength totals some 400 aircraft. Personnel strength is about 70,000, with the total armed forces amounting to 474,000. Military spending remains high, the 1977 figure being $NT63.47bn ($1.67bn). The CNAF is organised into wings of three squadrons, each with an establishment ranging from eight to 30 aircraft per squadron.

Chinese aviation dates back to 1914, when an Army Air Arm was established at Nan Yuan in the form of a flying school under the tutelage of an American pilot. By the mid-1920s the force was equipped with Avro 504K trainers and Handley Page O/400, Breguet 14 and Ansaldo A.300 bombers, with much practical assistance being provided by British and US personnel. A Central Government Air Force was formed in 1934 following Japan's invasion of Manchuria, the United States supplying Boeing 218 and Curtiss Hawk fighters together with a number of other miscellaneous types.

A year later an Italian Air Mission brought with it some of the country's latest aviation products, including Fiat fighters and Breda bombers. The six squadrons of the CGAF, with supplies of Soviet aircraft swelling their ranks, acquitted themselves well against what was generally accepted as being a superior enemy but failed to prevent the Japanese from occupying all of central China, Manchuria and some smaller regions. In 1941 an American Volunteer Corps was formed to fly Curtiss

P-40s within the CGAF. With the total involvement of the USA in World War II this force received the blessing of the US Government and there followed a steady stream of aircraft and equipment to China until 1945.

The CGAF was reorganised and renamed the Chinese Air Force after the end of the war, and received more US aid to supplement the surviving material. In 1949 the central government was forced by the communist forces to withdraw to the island of Formosa.

Taiwan enjoyed close ties with the United States from 1951 to the end of 1978, receiving large amounts of military and economic aid. The increasingly warm relations between the US and mainland China led to diplomatic relations between the United States and Taiwan being broken off in December 1978, however, and the last combat aircraft which Taiwan can expect from its erstwhile benefactor are 48 additional F-5Es. Even before the severing of relations, the US soft-pedalled in supplying advanced weaponry to Taiwan. Requests for McDonnell Douglas F-4 Phantoms or the more advanced General Dynamics F-16 and Northrop F-18 were refused, as was an application for the proposed Northrop F-5G armed with Sparrow air-to-air missiles.

Taiwan's intention of ordering Israel Aircraft Industries Kfir fighter-bombers had already been approved by the US (such permission is necessary because the Israeli aircraft is powered by General Electric J79 engines), but Taiwan is unlikely to proceed with the purchase. The Kfir's range apparently proved too short for Taiwan's needs and the country was faced with an oil embargo by Arab

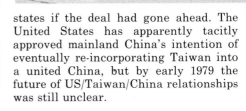

states if the deal had gone ahead. The United States has apparently tacitly approved mainland China's intention of eventually re-incorporating Taiwan into a united China, but by early 1979 the future of US/Taiwan/China relationships was still unclear.

Inventory
The Chinese Nationalist Air Force has been on a war footing since the communists under Mao Tse-tung took over mainland China. In terms of numbers the CNAF is no match for the Air Force of the People's Liberation Army, but technically it stands way above its opponent – thanks largely to the United States, which has supplied the vast majority of aircraft since its formation. In order to gain some self sufficiency the Taiwan Government set up a manufacturing facility known as the Aero Industry Development Centre at Taichung in the centre of the country, and 120 Northrop F-5Es are being assembled under licence. Six squadrons of ageing F-86 Sabres are due to be re-equipped with a total of 180 of the new aircraft, which will supplement 70 older F-5As currently flying in three squadrons.

Super Sabres and Starfighters complete the front-line combat force, although the CNAF has stated that it is seeking a replacement for the Starfighters. The single operational reconnaissance squadron conducts regular flights over or near the mainland and adjoining areas, using RF-104Gs. The CNAF previously operated several Lockheed U-2 high-altitude aircraft but, following some years of successful use, a number were shot down over the mainland and the type was withdrawn. The CNAF transport fleet of Providers and Packets, C-46s and C-47s is ageing rapidly, but no replacement has been announced for these machines and it is expected that they will continue in service for some years to come.

The AIDC at Taichung also produces Bell UH-1Hs under licence for both the air force and the army, and the same plant additionally builds Taiwanese-designed TCH-1B basic trainers to replace

TAIWAN Army

Type	Role	No
UH-1H	Trans	50
Kawasaki KH-4	Liaison	2
CH-34	Trans	7
Nike Hercules	SAM	72 launchers
Improved Hawk	SAM	24 launchers

TAIWAN Air Force

Unit	Type	Role	No	Notes
1st Fighter Wing	F-5A	FGA*	70	Three sqns; includes some two-seat F-5Bs
2nd Fighter Wing	F-5E	Int/FGA ⎤	180	120 being assembled by the Aero Industry Development Centre, plus 60 from US; to replace six sqns of Sabres
3rd Fighter Wing	F-5E	Int/FGA ⎦		
4th Fighter Wing	F-100A/D	FGA	90	Three sqns supplied in 1960 and 1970 from USAF stocks; includes some two-seat F-100Fs
5th Fighter Wing	F-104G	Int/FGA	63	Three sqns
	RF-104G	Recce	8	One sqn
	S-2A	Mar pat	9	Attached to navy
	HU-16B	SAR	10	One rescue unit
	Boeing 720B	VIP	1	
	C-46	Trans	30	
	C-47	Trans	50	
	C-119	Trans	40	
	C-123	Trans	10	
	UH-1H	Trans	60	AIDC-assembled
	Bell 47G	Liaison	10	
	Hughes 500	Liaison	6	
	UH-19	Trans	7	
	AIDC PL-1B Chienshou	Train	50	
	AIDC TCH-1B	Train	30	
	T-6	Train		
	T-28D	Train		
	T-33A	Train		
	T-38A	Train	30	

*Based at Kung Kuan

Above: An F-104 undergoes maintenance.

Left: Having depended on the United States for its combat aircraft – such as this F-5 – for so long, Taiwan is likely to have difficulty in adjusting to the absence of US aid.

the force of T-6s and T-28s. Attached to the Navy is a reconnaissance and rescue unit flying S-2 Trackers and a handful of Albatross amphibians. These latter aircraft are used to supply the forward Nationalist island bases of Quemoy in the south and Matsu in the north.

Support

Primary training in the CNAF is under-taken on the locally built AIDC Chienshou (an improved Pazmany PL-1 light two-seater), students then proceeding on to the T-6 and T-28. These ageing aircraft are being replaced with the AIDC-developed TCH-1B. Advanced training is performed on the Lockheed T-33As and T38A Talons, after which conversion is carried out onto two-seat versions of the main combat types.

Army

The Chinese Nationalist Army has a large helicopter force engaged in support and liaison work with the ground forces. Main type in use is the Bell UH-1H, a batch of 50 being assembled by the Aero Industry Development Centre at Taichung. The army also has 72 launchers for Nike Hercules SAMs and 24 launchers for the medium-range Improved Hawk.

HONG KONG

Area: 403·8 sq. miles (1045 sq. km.).
Population: 4·56 million.
Total armed forces: 800
Estimated GNP: $4·023 billion (£2·012 billion).
Defence expenditure: Not known.

The Royal Hong Kong Auxiliary Air Force (RHKAAF), which is a department of the Hong Kong Government, has its headquarters and main base near RAF Kai Tak alongside the International airport. Full-time pilots and servicing personnel are supported by a volunteer Auxiliary Establishment of some 110 members including pilots, crewmen and air-traffic controllers. The unit's main duty is to provide helicopter or fixed-wing aircraft to conduct tasks for other government sections such as the Police and the Medical and Health Departments.

The unit began life in 1949 as the air arm of the Hong Kong Defence Force, equipped with four Austers, four Harvards and four Spitfire fighters. Granted the title 'Royal' in 1951, the force acquired two Westland Widgeon helicopters in 1958 and phased out of service the Harvards and Spitfires. French Aérospatiale Alouette III helicopters in turn replaced the Wid-geons and in 1970 the force was given its present title.

Helicopters form the vital element in the Royal Hong Kong Auxiliary Air Force; the three Alouette IIIs perform important casualty-evacuation tasks among the colony's 236 islands and in the mountains as well as numerous other roles. Largest type in use is a Britten-Norman Islander used for aerial survey work, advanced pilot training and SAR duties. For survey flights a Wild RC10 aerial camera is fitted into the floor of the aircraft and operated by remote control. The Auster light aircraft were retired in 1971 and replaced in the pilot training role by two Beech Musketeers, observation being a secondary role. In 1977 two Scottish Aviation Bulldogs were ordered and the remaining Musketeer is due for retirement in 1980. Apart from Kai Tak, the only other base used for fixed-wing operations in the colony is Sek Kong airstrip in the rural New Territories.

Training performed on the Musketeer and the two new Bulldogs at Kai Tak and Sek Kong is to RAF standards. These aircraft are also used by the Dept of Civil Aviation to train air-traffic controllers to private-pilot standard.

HONG KONG

Type	Role	Base	No	Notes
Islander	Survey/SAR	Kai Tak	1	Delivered March 1972; also used for storm recce and anti-smuggling patrols
Alouette III	Casevac/SAR	Kai Tak	3	Two delivered 1965, one in 1970
Bulldog 128	Train	Kai Tak	2	Ordered in 1977 to replace Musketeers
Musketeer	Train	Kai Tak	1	Two delivered, one in use; due for withdrawal in 1980

Right: Half the RHKAAF on the apron at once – a pair of Alouette III helicopters used for casualty evacuation and SAR work, the single Islander and one of the two Bulldog trainers.

Southeast Asia

With such a history of racial, religious, cultural and other deep-seated differences, coupled with a long line of dictators of one political colour or another, it is hardly surprising that Southeast Asia is a region of virtually ceaseless turmoil. The Vietnamese have been continuously at war with the Chinese, Japanese, French, US or neighbouring countries in Indochina for nearly half a century, and they at last seem to have achieved many of their aims: a re-united country, varying degrees of control over some immediate neighbours and the proven ability to give the Chinese more than a good run for their money in any border conflict short of outright war. The Vietnamese depend heavily on Soviet support, however, and may have to be more flexible in their future relations with the USSR.

Kampuchea (Cambodia) has been under the control of the Vietnamese since they invaded the country in late 1978 and early 1979, with the support of Kampuchean guerrillas. Vietnam also has considerable influence over Laos and supports guerrillas operating in Thailand. The Thai borders with Burma in the north and Malaysia in the south are both areas of conflict, with dissidents and smugglers operating in the former region and

communist Malays in the latter, but these are minor problems compared with the more pressing disputes which abound in the area.

Malaysia and Singapore are probably closer now than at any time since their ways parted in 1965, and both enjoy reasonably good relations with neighbouring Indonesia and the Philippines. Indonesia has been re-equipping its armed forces on a modest scale, with the possibility of much larger orders being placed in the future. Oil-rich Brunei achieves full independence in 1983, and many of its neighbours are waiting to see how this will affect relations.

The United States retains its bases in the Philippines, and the present regime of President Marcos is likely to continue – with his wife, the so-called Iron Butterfly, a favourite to succeed him. With the loss of bases and facilities in Vietnam, coupled with a running down of involvement in Taiwan and Korea, the US presence in the Philippines is increasingly important.

US policy towards Southeast Asia as a whole remains unclear, following the undignified withdrawal from Indochina. The region is an important and expanding market, however, so this state of affairs is unlikely to be allowed to persist.

BURMA

Area: 262,000 sq. miles (678,500 sq. km.).
Population: 31·5 million.
Total armed forces: 170,000.
Estimated GNP: $4·2 billion (£2·1 billion).
Defence expenditure: $164 million (£82 million).

The Union of Burma Air Force is primarily concerned with internal policing and the support of Burmese Army units in operations against Communist guerrillas in the north and east of the country. Continued limited budget allocations have restricted the modernisation of the UBAF, and comparatively obsolete equipment forms the bulk of the force. Total defence expenditure for 1976–77 was 787m Kyat ($113m).

Founded in 1955 with British assistance, the UBAF acquired a number of ex-RAF aircraft – including some Spitfires, Airspeed Oxfords and Tiger Moth trainers.

Further small numbers of second-line aircraft were bought during the late 1950s, the first jet aircraft comprising a few de Havilland Vampire T.55s which arrived in 1957–58. In 1960 six Kawasaki-Bell 47G helicopters were supplied by Japan as war reparations. American aid was initiated in 1967 as a result of increased Communist guerrilla activity.

No major defence agreements involve Burma, although US aid continues on a limited basis. Potential enemies are China and Laos. Small private armies have been established in the eastern Shan state to protect large opium-growing areas stretching into northern Thailand and Laos; the UBAF ostensibly operates anti-narcotics patrols to curb the drug traffic.

Inventory

Total UBAF manpower is about 7000 and front-line strength is approximately 40 aircraft, although not all are operational. Burma's sole interceptor squadron comprises six Lockheed AT-33As supplied by the USA in the 1960s while ten Siai Marchetti SF.260s are in service for strike-trainer duties, delivery having been made curing 1976. A year earlier 18 Bell UH-1s were supplied by the US.

Right: A Pilatus PC-7 Turbo-Trainer of the Union of Burma Air Force formates on the manufacturer's demonstrator. The UBAF PC-7s are replacing Chipmunks.

BURMA

Type	Role	No	Notes
AT-33A	Recce/train	6	A number of T-33 trainers are also operated; supplied by US
SF.260MB	Coin	10	Used on anti-narcotics work
FH-227	Trans	4	
Porter/Turbo-Porter	Trans	3/4	
UH-1	Trans	18	
PC-6/B2 Turbo-Porter	Trans	7	
Alouette III	Trans/liaison	13	
Kawasaki-Bell 47G	Train	13	
HH-43B	SAR	12	
Cessna 180	Liaison	10	
T-37C	Train	12	
Chipmunk	Train	10	
PC-7 Turbo-Trainer	Train	18	Delivered 1978–79

THAILAND

Area: 198,200 sq. miles (513,300 sq. km.).
Population: 42 million.
Total armed forces: 212,000.
Estimated GNP: $18·1 billion (£9·05 billion).
Defence expenditure: $746 million (£373 million).

The Royal Thai Air Force, Navy, Army and the para-military Thai Border Police are all equipped with a variety of mainly US-supplied aircraft. The RTAF is organised into four basic groups; Combat Group, providing the operational elements; Logistic Support Group, encompassing the transport units; Training Group; and Special Services Group. Personnel strength totals some 42,000.

The origins of the Royal Thai Air Force can be traced back to the days when the country was known as Siam and three officers of the Royal Siamese Engineers were sent to France to study flying. That was in 1911; three years later the Royal Siamese Flying Corps was officially established with a strength of eight French aircraft. In the 1930s the Curtiss Hawk III was adopted for the single-seat fighter units and in 1939 the country was renamed Thailand, with the subsequent adoption of the armed forces' present titles.

Thailand was invaded by Japan in December 1941 and the RTAF subsequently operated a variety of Japanese aircraft in a restricted non-combatant role. After the war the RTAF was given assistance by the Royal Air Force and a number of Miles Magister and Texan trainers were supplied, together with C-47 transports. In 1950 the rising tide of communism in South-East Asia prompted the United States to send an advisory group to Thailand to help modernise and expand the RTAF. The US supplied 200 Grumman Bearcat fighters and 138 Texan armed trainers and undertook the training of Thai personnel. Thirty F-84G Thunderjets were delivered in February 1957 to form the country's first jet fighter squadron, and T-33As arrived for pilot training. These were followed by F-86 Sabres and F-5 fighter-bombers in the 1960s, while

THAILAND Air Force

Wing	Unit	Type	Role	Base	No	Notes
1st Wing	11 Sqn	T-33A	Train	Don Muang	20	
		RT-33A	Recce	Don Muang	4	
		F-5E	Int/FGA	Don Muang	17	
		F-5F	Train	Don Muang	3	
	13 Sqn	F-5A	FGA	Don Muang	24	
		F-5B	Train	Don Muang	2	
		RF-5A	Recce	Don Muang	4	
2nd Wing	21 Sqn	T-28D	Coin	Chieng Mai, Ubol, Udon	45
	22 Sqn	T-28D	Coin		
	23 Sqn	T-28D	Coin		
	53 Sqn	T-6G	Coin	Prachuab, Satahip	30
	62 Sqn	T-6G	Coin		
	73 Sqn	T-6G	Coin		
	12 Sqn	A-37B	Coin		16	
		AU-23A	Coin		31	
		OV-10C	Coin		38	Two sqns
6th Wing	61 Sqn, 62 Sqn	C-47	Trans	Don Muang	20	
		C-123B/K	Trans	Don Muang	40+	Includes ex-USAF and South Vietnamese aircraft
		C-45	Trans	Don Muang	5	
		HS.748	VIP	Bangkok	2	
		Merlin IVA	VIP	Bangkok	4	Five received, one lost in November 1978
	63 Sqn	UH-1H	Trans		63	
		HH-43B	SAR		3	
		H-34			18	Converted to S-58T standard
		CH-34C	Trans		40	
		UH-19	Trans		13	
		U-8F	Liaison			
		Islander	Trans		1	
		Cessna 310F	Liaison			
		Beaver	Liaison		4	
		U-10A	Liaison		5	
		Turbo-Porter	Liaison		10	
		Airtrainer	Train	Korat	24	
		T-41D	Train	Korat	4	Ex-South Vietnamese
		Chipmunk	Train	Korat	10	Re-engined with R-R/Continental IO-360 engines
		SF.260MT	Train	Korat	12	
		T-37B	Train	Korat	14	

the counter-insurgency role flown by the Bearcats was taken over by substantial numbers of T-28D light attack aircraft, the Bearcats having been withdrawn from use in 1962.

Thailand was one of the signatories of the South-East Asia Collective Defence Treaty which came into force in 1955 and brought SEATO into being. By late 1977, however, SEATO had virtually ceased to exist. The country's close association with the United States, highlighted during the Vietnam war when US military operations were conducted from Thai bases, has undergone a reassessment and there are virtually no US military personnel now based there, although procurement is still made from US sources. A joint border agreement was signed with Malaysia in March 1977 to help reduce the communist

guerrilla threat and operations have been mounted to this effect.

Inventory
The RTAF's basic operating formation is the wing, which may contain as many as six squadrons. Equipment is generally of US origin, and apart from a small force of pure interceptors, is made up of counter-insurgency types and tactical transports. Recent procurement includes a squadron of 20 F-5E/Fs ordered in 1976 and operated alongside the force of older F-5As first delivered in 1966. The squadron of F-86L

Below: The Royal Thai Navy uses a pair of Canadair CL-215 amphibians on search-and-rescue tasks, operating alongside the two remaining HU-16 Albatrosses.

Sabres was retired shortly before the arrival of the F-5Es. Other recent purchases include two Swearingen Merlin IVAs bought for the use of the Royal family, three Merlin IVAs equipped for reconnaissance and six more Rockwell OV-10C Broncos to join 32 bought in two batches from 1971 onwards. These aircraft have been responsible for much of the anti-guerrilla flying in recent years, and the re-order placed late in 1977 is to partially make up for attrition. At least six Coin units are equipped with North American T-6Gs, T-28Ds and Cessna A-37Bs, and 31 Fairchild AU-23A Peacemakers have been procured in two batches.

The RTAF has more than 60 transports equipping two squadrons. A large helicopter force is concentrated in a single squadron; a further 13 UH-1Hs were delivered by the US Government in 1977–78. Liaison and communications duties are performed by a mixed force of Beech U-8s, Cessna 310s, Helio U-10s and some DHC Beavers.

Support

Training, both primary and advanced, is conducted from Korat – a former USAF strike base returned to Thailand in the mid-1970s. Pupils receive primary grading for some 20hr on the CT/4s and SF.260s, followed by jet training for 85hr, on the T-37s. They then continue to 11 Sqn for 60–100hr on the T-33As. Transport pilots proceed from primary to advanced training, which is flown on the C-47s of 61 Sqn for some 120hr. Helicopter pilots complete their course on Bell 47Gs.

Navy

The Royal Thai Navy has a small air element with headquarters in the capital, Bangkok. Since the early 1950s, when a handful of Fairey Firefly fighters were purchased from Britain, the Thai Navy has maintained an ASW and SAR force for patrol duties in the Gulf of Siam and to the west of the Isthmus of Kra, adjacent to Malaysia. A single ASW squadron is equipped with Grumman Trackers acquired from US Navy stocks, while rescue flights are flown by a few Albatross amphibians and two Canadair CL-215 amphibians. With the increase in Communist infiltration along the coastline, further procurement by this arm is likely.

Army

The Royal Thai Army has a substantial air element made up of mostly US-supplied types. Helicopters predominate, although there is a fixed wing liaison and communications force of some 90 Cessna 0-1 Bird Dogs.

Police

Operated in close association with the Army is the para-military Thai Border Police, which carries out anti-drug smuggling patrols along the extensive borders with Burma, Laos and Cambodia. The present strength is three Caribous, three Skyvans, five Peacemakers and four Porters, three Skyservants and one CT/4 Airtrainer. The helicopter force has ten Bell 204Bs, 11 Bell 205s, two Bell 205As and four Bell 206Bs.

THAILAND   Navy

Type	Role	No
S-2F	Mar pat/ASW	10
HU-16B	SAR	2
CL-215	SAR	2

THAILAND   Army

Type	Role	No
Beech 99	Trans	1
O-1	Liaison/AOP	90
Bell 206	Liaison	3
UH-1B/D	Trans	90
CH-47A	Trans	4
FH-1100	Liaison	16
OH-23F	Liaison	6
Bell 214B	Util	2

LAOS

Area: 90,000 sq. miles (233,000 sq. km.).
Population: 3 million.
Total armed forces: 48,500.
Estimated GNP: $256 million (£128 million).
Defence expenditure: $42 million (£21 million).

The Air Force of the People's Liberation Army is, like the other branches of the armed forces, controlled by the country's pro-communist military government. From its Vientiane headquarters, the AFPLA took over air operations from the Western-orientated Royal Lao Air Force in the spring of 1975 when the communist Pathet Lao overran the country and ousted the government of Prince Souvanna Phouma. The present air force is believed to have a strength of some 2000 officers and men and 90 combat aircraft, mostly ex-US types inherited from the RLAF and with a questionable degree of serviceability. At least some of the remaining T-28s are abandoned and rotting.

Formerly part of French Indo-China and bounded by China, Vietnam and Kampuchea (Cambodia), Laos established an aviation offshoot of the army in 1955 with the help of loans from the French treasury and the United States.

The Laotian Army Aviation Service became the Royal Laotian Air Force in August 1960 and was equipped almost entirely with US-supplied aircraft, more arriving in January 1961 in the form of ten T-6 Texans to counter supply drops by Soviet Ilyushin transports to northern rebel forces. French aid and training declined as US aid increased. Limited US air operations against the Ho Chi Minh trail were flown towards the end of the 1960s, while an increasing number of North American T-28s arrived for the RLAF, gradually replacing T-6s in the ground-attack role and again the result of a US military grant arrangement. The collapse of the "neutralist" government in 1975 halted further Western aid.

Laos maintains links with Vietnam and Kampuchea, the former country assisting the Pathet Lao during the guerrilla war and currently deploying regular forces in Laos. Recent equipment has come from the Soviet Union.
Late in 1977 the Air Force of the Liberation Army received its first supersonic combat aircraft when ten MiG-21s arrived at Vientiane's Wattay airport from the Soviet Union. From the same source came some Mi-8 helicopters. An-2 liaison aircraft and An-26 transports.

At the time of the collapse, the Royal Lao Air Force had some 227 aircraft, the majority being second-line types used in the ground-support role. A total of 97 aircraft were flown out of Laos to neighbouring Thailand by RLAF crews at the time of the collapse but many of these aircraft were subsequently returned.

With so much US equipment in use, the mode of RLAF operations was closely based on USAF thinking, the various elements in use flying similar missions to those in the US Air Force. Chief among the combat aircraft was the T-28D, of which some 80 were in use in the fighter-bomber role. These were supported by a squadron of ten AC-47 gunships flown mainly at night. Transports included 18 C-47s and nearly a dozen C-123 Providers plus 42 UH-34s supplied from US and South Vietnamese sources and a small number of UH-1s, both helicopter types flown in the casevac and Army support role as well as for transport duties. Forward air control missions were carried out by Cessna O-1s and U-17s. What percentage of the aircraft listed below are serviceable with the Air Force of the People's Liberation Army is unknown at the time of writing, but reports have stated that some Soviet Yak-40 tri-jet transports have been supplied to the country.

LAOS

Type	Role	Base	No	Notes	Remarks
T-28D	Att	Vientiane	80		Fifty flown to Thailand in 1975 but many returned to Laos
MiG-21	Int	Vientiane	10		
AC-47D	Night gunship	Vientiane	10	One sqn	
C-47	Trans	Vientiane	18		Includes three supplied by Australia in 1971
C-123	Trans	Vientiane	10		
Yak-40	VIP	Vientiane	1+		
Aero Commander 500	Liaison	Vientiane	1		
Beaver	Liaison	Vientiane	1		
U-17A	FAC	Vientiane	4		
O-1	FAC	Vientiane	24		
AU-24A Stallion	Liaison	Vientiane	14		
Alouette III	Liaison	Vientiane	4		
UH-1	Trans	Vientiane	13		
UH-34D	Trans	Vientiane	42		Twenty-four transferred from S. Vietnamese AF
T-41D	Train	Savannakhat	6	Pilot training school	
An-2	Liaison				
An-26	Trans		3		
Mi-8	Trans				

KAMPUCHEA

Area: 70,000 sq. miles (181,300 sq. km.).
Population: 7·735 million.
*Total armed forces: 70,000.
Estimated GNP: Not known.
Defence expenditure: Not known.
*Before Vietnamese take-over.

Vietnamese forces and Vietnamese-backed guerrillas (similar to the Viet Cong) overthrew the Pol Pot regime in Kampuchea in late 1978 and early 1979, and the country was expected to be absorbed into Vietnam. The overthrown regime had been in power for less than three years, having assumed control in the spring of 1975. After that take-over some 90-odd of the Khmer Air Force's aircraft – mostly US-supplied – were flown to neighbouring Thailand by defectors and refugees. Most of these are thought to have been returned, however. The Pol Pot regime increasingly allied itself with China, and Vietnamese-led forces – with backing by the Soviet Union – launched an all-out attack on Christmas Day 1978. The capital, Phnom Penh, fell in the first week of January 1979 and the Kampuchean Air Force's main base at Kompong Chang was captured. The base was the home of the single Shenyang F-6 (MiG-19) squadron supplied by China, and the Kampuchean Air Force has therefore virtually ceased to exist. Any undamaged material was expected to be adopted by the Vietnamese Air Force, which is likely to have established a semi-permanent presence in the country.

Cambodia was part of French Indo-China and became an independent kingdom within the French Union in 1949, gaining independence in September 1955. An air arm was established with the title *Aviation Nationale Khmere* and was equipped with seven Fletcher Defender light attack aircraft, followed later by 14 ex-French Morane-Saulnier Alcyon trainers. Some second-line types were received from the United States, including Cessna Bird Dogs and C-47s, but most aid continued to come from France – including 30 Skyraiders and some Magisters.

During the 1960s the air force operated about 20 MiG-15/17 fighter-bombers alongside a large number of US-supplied machines. The left-wing government was overthrown in 1970 and US aid increased to help combat the increasing threat from Communist Khmer Rouge guerrillas. By the beginning of 1975, however, it became apparent to the United States that further military and economic aid to the country was unlikely to change the course of events, and in April the besieged capital of Phnom Penh fell to the Khmer Rouge.

Virtually no combat operations were flown by the Khmer Air Force, the only active part played by aircraft being numerous liaison and reconnaissance missions flown by 30 Bell UH-1s in support of the Army. A total of 97 KAF aircraft were flown to Thailand just prior to the collapse. Shortly after the take-over, Khmer Rouge forces captured the American cargo ship *Mayaguez*, prompting US retaliatory action which resulted in the successful re-possession of the ship and its crew. In attacks by US Navy carrier-borne aircraft during the operation it was reported that as many as 17 T-28s had been destroyed on airfields around the coast.

VIETNAM

Area: 129,000 sq. miles (334,000 sq. km.).
Population: 55 million.
Total armed forces: 615,000.
Estimated GNP: $7·1 billion (£3·55 billion).
Defence expenditure: $584 million (£292 million).

The Vietnamese People's Air Force is still in process of adjustment following the unification of the Northern and Southern zones of Vietnam in the early part of 1975, but is likely to settle down as a force concentrating on air defence, transport support and internal policing.

The Air Force is under direct Government control; unlike many Eastern Bloc forces it does not appear to have a close-support role under Army command. Vietnam has been extensively supplied by the Soviet Union, but has also taken delivery of aircraft from China. Vietnam retains close ties with Laos, which was instrumental in keeping open a supply route for Viet Cong forces in the southern zone.

The Vietnamese Army had the strongest surface-to-air missile force in the region at the time of the US Linebacker II offensive, although some sites may have been demobilised. There are some 300 SA-2 Guideline launchers, and these are backed up by the shorter-range SA-3 Goa. Vietnam

Below: Vietnamese MiG-21 pilots at a briefing session in front of their MiG-21 interceptors. The VPAF has about 60 of the type.

also operates the advanced rocket/ramjet-powered SA-6 Gainful with the associated Straight Flush and Long Track radars.

The Vietnamese interceptor force comprises some 60 MiG-21s, mostly of the second-generation MiG-21PF Fishbed D type. All the MiG-21s and at least some of the 75 MiG-17 Frescos (employed effectively in the air-superiority role, despite their age) carry AA-2 Atoll air-to-air missiles. A few Chinese Shenyang F-6s (MiG-19s) and Sukhoi Su-7 Fitter are in service, about 30 examples of each having been delivered. The Su-7s and some ten Ilyushin Il-28 light bombers were not frequently encountered in combat.

A small transport force includes a number of Antonov An-2s, Ilyushin Il-14s and some Lisunov Li-2s. More modern aircraft include four Antonov An-24s and a single Ilyushin Il-18. The Air Force and the People's Navy each operate about a dozen Mil Mi-4 helicopters, for transport and search-and-rescue duties respectively. Five Mil Mi-6 heavy-lift helicopters are also on strength.

In addition to the above-mentioned aircraft, an extremely large quantity of material was captured by the North Vietnamese forces in their advance on Saigon (now Ho Chi Minh City) in 1975. South Vietnamese and US aircraft included Northrop F-5E Tiger fighters, Cessna A-37B Dragonfly light strike aircraft, AC-47 and AC-119 gunships and some 500 helicopters, including Huey Cobra gunships and Chinook medium-lift helicopters. Some of the F-5Es are thought to have been incorporated into at least one composite fighter squadron with MiG-21F Fishbed Cs flying alongside the US-built aircraft; nine Tigers and six MiG-21s were operating from Hanoi in early 1979. The North Vietnamese also captured a number of C-130 Hercules transports (which may have included some AC-130 Spectre gunships), and these were used both to bomb targets in Kampuchea during the Vietnamese invasion in late 1978 and to carry reinforcements to Vietnam's northern borders during the resulting skirmishes with Chinese forces

in early 1979. Up to 25 AH-1 HueyCobra attack helicopters were also used by the Vietnamese against Chinese tanks during this conflict.

The North Vietnamese Air Force took little part in the surface war against South Vietnamese and US forces in the 1960s and early 1970s, concentrating on resisting US Air Force and US Navy attempts to establish air superiority over the North and in intercepting US bombing raids. The

Above: MiG-17s photographed at Phuc Yen by a US reconnaissance aircraft in 1966.

Right: This F-5 was one of many ex-South Vietnamese aircraft captured in 1975 and since absorbed by the Vietnamese Air Force.

Below: C-7A Caribou and C-47 Skytrains at Tan Son Nhut as it was overrun.

Below right: SA-3 surface-to-air missiles complement Vietnam's interceptors.

North Vietnamese pilots' efforts were instrumental in forcing the US to re-introduce air-combat training in its syllabus and in leading the USAF to specify air superiority as the primary role for its next fighters.

The North Vietnamese Air Force acquired the world's only recent experience in intercepting heavy bombers during the Vietnam war, and in late 1978 and early 1979 both Kampuchea and China

were reminded of the expertise which Vietnamese pilots and missile crews had acquired. Vietnamese-led forces launched an all-out offensive against Kampuchea on Christmas Day, 1978, after several months of preparation. Large numbers of captured ex-South Vietnamese F-5Es were reported to have been used against Kampuchean targets, and the Kampuchean Air Force was effectively destroyed in January 1979 when Vietnamese and rebel forces overran the Shenyang F-6 (MiG-19) base at Kompong Chang. Vietnamese MiG-19s and MiG-21s based inside Kampuchea then carried out mopping-up operations against any remaining resistance.

On February 17, Chinese forces invaded northern Vietnam and captured a number of towns. Vietnamese troops withdrawn from Kampuchea soon reinforced the border region, and six Aeroflot/Soviet Air Force An-22 transports began an airlift via Dacca in Bangladesh to resupply Vietnam with weapons including missile reloads. Only limited air action is thought to have taken place over the battlefields, and Chinese forces were withdrawn in mid-March. The efficiency of Vietnamese anti-aircraft defences is thought to have deterred China from escalating the air war, and this limitation is likely to apply in any following engagements between the two states.

MALAYSIA

Area: 130,000 sq. miles (336,000 sq. km.).
Population: 10·4 million.
Total armed forces: 64,500.
Estimated GNP: $12·3 billion (£6·15 billion).
Defence expenditure: $700 million (£350 million).

The Royal Malaysian Air Force has recently undergone a modernisation programme to help maintain its varied commitments. Split as it is between West and East Malaysia, the air force operates a strong air transport arm which regularly links bases in both areas across hundreds of miles of mountainous and jungle terrain. To combat the increasing guerrilla menace along the Thai border, the Air Force has also updated its fighter-bomber elements and at the same time expanded its air-defence potential. RMAF headquarters are in the Ministry of Defence building in Kuala Lumpur; the Air Force is one of three Services in a unified defence organisation and is manned by some 6000 Malay, Chinese and Indian personnel. Defence budget for the armed forces in 1978 totalled $M1.65bn ($700).

The origins of the RMAF stretch back to 1936 and the Straits Settlements Volunteer Air Force, which in turn gave way to the Malayan Auxiliary Air Force established in 1950 to help combat Communist guerrilla activity. On June 1, 1958, the Royal Malayan Air Force came into being with RAF assistance; it was equipped with a handful of Scottish Aviation Pioneer transports and de Havilland Chipmunk trainers. With the creation of Malaysia on September 16, 1963, the air arm adopted the title it uses today. A variety of transport, liaison and training aircraft were obtained in the early 1960s, including Twin Pioneers, Handley Page Heralds, DH Caribous, Provosts and some Alouette III helicopters. Seconded RAF personnel established the RMAF on British lines and this experience helped the force when, following the formation of the Federation, the confrontation with Indonesia began. RAF and Fleet Air Arm units were sup-ported by Malaysian elements during that period. Jet equipment arrived in 1967 in the shape of some Canadian CL-41 Tebuan (Wasp) armed trainers which currently double in the ground-attack role.

Malaysia and Thailand signed a border accord during March 1977 in an attempt to reduce Communist guerrilla operations along the border, which in recent years have increased considerably, and some success has been forthcoming during these joint operations. A five-power defence agreement relating to the defence of Malaysia and Singapore came into effect on November 1, 1971, and also involves Australia, New Zealand and the UK. Notwithstanding Great Britain's withdrawal from Singapore early in 1976, the agreement remains in force, with Australia still contributing to air defence with two squadrons of Mirage fighters in Western Malaysia. The country receives limited military aid from the United States. The 14-pointed yellow star national insignia indicates the 13 Malay states, the 14th being Singapore which is no longer a member of the Federation.

Inventory

The RMAF has some 12 squadrons currently operational, eight of them equipped with transport aircraft, both fixed- and rotary-wing. From Kuala Lumpur, the main transport base, supply flights are regularly made to bases in West and East Malaysia using the six Lockheed C-130H Hercules ordered in a $47.9m deal announced in 1974. These aircraft replaced the long-serving Heralds. Equally important are the two Sikorsky S-61A units (these machines are called Nuris, a Malay word for a small bird) which form the RMAF's heavy-lift helicopter element and are used to support counter-insurgency operations in the north of the country, where their ability to lift up to 28 troops or 5,000lb of freight on short-range flights makes them an ideal vehicle. Both Nuri squadrons operate from coastal bases and have frequently been called upon to fly air-sea rescue missions. The other helicopters in use are Alouette IIIs which have been used for a variety of tasks, not least as forward-air-control aircraft directing air and ground operations along the Malaysia-Thailand border. About 20 Aérospatiale Gazelles, together with some Bell 206B JetRangers, have entered service to supplement and possibly eventually replace these machines.

The combat units in the RMAF comprise the Butterworth-based F-5E squadrons,

Below: C-130H Hercules of 14 Sqn.

Bottom: CL-41G Tebuan, used by 6 Sqn and 9 Sqn from Kuantan.

Right: The Caribou STOL transport is flown by two RMAF squadrons, Nos 1 and 8, based at Labuan.

which are fully integrated into the defence system. Butterworth is also headquarters of the tri-national Integrated Air Defence System (IADS) which operates in conjunction with Singapore Air Force Bloodhound 2 missiles and radars; it also houses the two Australian Mirage squadrons, Nos 3 and 75. As part of the five-power defence agreement a detachment of eight Mirages is permanently based at Tengah in Singapore. At Penang a Marconi S600 radar forms a major part of the IADS.

Training

The main RMAF training centre is Alor Star in West Malaysia, where student pilots undergo basic training on Bulldogs. They may also undertake courses overseas in the UK, Canada or India. Jet training is carried out on the Tebuans at Kuantan, and when the pilots are qualified they proceed on to the F-5Es. Multi-engine training is performed on Cessna 402s, with pilots then going to Caribou squadrons. A helicopter FTS operates at Labuan in East Malaysia.

MALAYSIA

Unit	Type	Role	Base	No	Notes
1 Sqn	Caribou	Trans	Labuan	17	
8 Sqn	Caribou	Trans	Labuan		
2 Sqn	Dove, Heron, HS.125, F.28 Mk 1000	VIP	Kuala Lumpur	2/3/2/2	Operates regular services; three other Doves in store
3 Sqn	Alouette II	Liaison/FAC	Kuala Lumpur	25	
5 Sqn	Alouette II	Liaison/FAC	Labuan		
6 Sqn	CL-41G Tebuan	GA	Kuantan	16	20 originally delivered
9 Sqn	CL-41G Tebuan	Train	Kuantan		
7 Sqn	S-61A-4	Trans	Kuching	36	
10 Sqn	S-61A-4	Trans	Kuantan		
11 Sqn	F-5E/B	Int/train	Butterworth	12/2	Replaced Sabres; 14 F-5Es originally delivered
12 Sqn	F-5E/B	Int/train	Butterworth		
14 Sqn	C-130H	Trans	Kuala Lumpur	6	
FTS	Bulldog 102	Train	Alor Star	15	Replaced Provosts
	Cessna 402B	Train	Alor Star	12	Also liaison and survey
Heli FTS	Bell 47G	Train	Labuan	9	
	AB.212	Trans		3	
	Bell 206B	Train/liaison		5	
	SA.341K Gazelle	Liaison		20?	

BRUNEI

Area: 2,220 sq. miles (5800 sq. km.).
Population: 150,000.
Total armed force: 2,750.
Estimated GNP: $380 million (£190 million).
Defence expenditure: $128 million (£64 million).

The Royal Brunei Malay Regiment has a small Air Wing, mainly helicopter-equipped, for support work. Situated on the north-western coast of Borneo in two enclaves in the state of East Malaysia, the Sultanate of Brunei opted out of joining the Federation of Malaysia in 1963, preferring to remain a separate state under the protection of Britain. The Air Wing is commanded by a seconded British officer; the headquarters, and one of the few airfields of any size in the state, are situated at Berakas Camp, Brunei Town. Royal Air Force pilots have been posted to the Wing and more than a dozen Brunei nationals have gained their pilot's wings, ten being qualified on the helicopters.

The Air Wing was formed in 1965 as a support force for the Royal Brunei Malay Regiment, initial equipment comprising three Westland Whirlwind Mk 10 helicopters flown by seconded RAF pilots and maintained by World Wide Helicopters of Nassau. Two Wessex replaced the Whirl-

winds in 1967 and were supplemented by a single Bell 206A JetRanger for communications duties. The Wessex were subsequently sold and the force standardised on Bell types.

Under a military agreement, Britain is responsible for the defence of Brunei and the running of the airport at Muara Port.

Inventory

A fleet of ten Bell helicopters currently forms the bulk of the Sultan's Air Wing. Largest type in use is the twin-turbine Bell 212 used for heavy-lift duties and for over-water and IFR operations; one is fitted with a VIP interior for the use of the Sultan and government officials. The three Bell 206A JetRangers have been retro-

spectively fitted with up-rated Allison 250-C20 engines, taking them to Model 206B standard, to improve their performance in the tropics. In 1971 the Air Wing took delivery of its first fixed-wing aircraft, a Hawker Siddeley 748 for use by the Sultan in addition to passenger and freight work. It also has a limited SAR capability in the shape of a special aft chute through which a 16-man dinghy can be dropped.

BRUNEI SULTANATE

Type	Role	Base	No
HS.748	VIP/trans	Muara Port	1
Bell 205A	Trans	Berakas Camp	3
Bell 206B JetRanger	Comm	Berakas Camp	3
Bell 212	Trans/VIP	Berakas Camp	4

Right: One of the three Bell 206 JetRangers operated by the Royal Brunei Malay Regiment's Air Wing. Brunei is due to become independent in 1983 and could well expand its air arm.

219

SINGAPORE

Area: 225·6 sq. miles (584 sq. km.).
Population: 2·3 million.
Total armed forces: 36,000.
Estimated GNP: $6·5 billion (£3·25 billion).
Defence expenditure: $410 million (£205 million).

 The Republic of Singapore Air Force, part of this small island's tri-Service organisation, has some 3000 personnel. Its headquarters are in the Ministry of Defence building at Tanglin, and the force operates from two main air bases on the island, Tengah and Seletar. Defence expenditure for the Singapore Armed Forces in 1977–78 increased by more than $S200m from 1975–76 to $S820 ($410). The total tri-Service organisation has some 36,000 personnel supplemented by reserve forces.

Singapore seceded by consent from the Federation of Malaysia in 1965 and from a small internal security force established the Singapore Air Defence Command in 1968. As part of the unified Singapore Armed Forces, the SADC joined Maritime Command which controlled naval units. To aid recruitment and select prospective aircrew for the embryo air force, Cessna 172s borrowed from the Singapore Flying Club at Paya Lebar were used with civilian instructors from Britain and seconded RAF instructors. Successful students were then sent to the UK for basic flying training on RAF Jet Provosts in preparation for the take-over of the SADC's first major equipment: comprising 16 BAC Strikemasters and 20 HS Hunters. Eight Cessna 172Ks were acquired as interim training equipment pending the arrival of a batch of Italian SF.260s. Some Alouette IIIs were the Force's first helicopters, the aircrew to fly them being trained in France. In 1971 more Hunters were purchased for a second squadron and a year later a contract for Short Skyvans was announced. On April 1, 1975, SADC was restyled Republic of Singapore Air Force and the then familiar red-white-red roundel gave way to the present stylised 'S' insignia. RAF assistance accompanied the steady expansion of the Air Force until 1976, when only a handful of advisers remained.

Singapore forms an important element in the five-power defence agreement which

Above: One of six RSAF Skyvans, this and two others being used for SAR work. The aircraft is fitted with a nose-mounted radar and flare dispensers on the fuselage sides.

came into effect on November 1, 1971, and involves Malaysia, Singapore, Britain, Australia and New Zealand. Although the RAF withdrew from the island in 1976, close ties continue between the forces involved.

Inventory

With the final British withdrawal from Singapore in March 1976, the RSAF undertook responsibility for defence of the island. The BAC Bloodhound 2 SAM unit at Seletar was handed over to the RSAF, together with the associated early-warning radar and reporting centre at Bukit Gombak. These elements form Singapore's prime contribution to the tri-national Integrated Air Defence System headquartered at RMAF station Butterworth in west Malaysia and including the two Australian Mirage interceptor squadrons based there. Together with the Malaysian Marconi S600 radar situated on Penang island, IADS covers most of the Malayan peninsula. To increase the effectiveness of IADS, a detachment of eight RAAF Mirage IIIs is permanently based at Tengah for interception duties. The RSAF's two Hunter ground-attack squadrons are also integrated into IADS, and 18 Northrop F-5Es were ordered in 1976 following a protracted evaluation involving such types as the F-4 Phantom, Saab Draken, Dassault Mirage and Israeli Kfir. A major reason for the selection of the F-5E was the fact that the type was already operated by Malaysia. The $109.7m deal announced in September 1976 for these aircraft also covered three two-seat F-5Fs, 200 AIM-9J Sidewinder air-to-air missiles plus spares and other equipment.

Lockheed Aircraft Service Singapore has in recent years provided maintenance support for the RSAF as well as acting as prime contractor for the conversion of ex-US Navy A-4B Skyhawks to A-4S standard. More than 100 modifications are incorporated, including the provision of outboard weapon pylons, a drag-chute housing, split flaps and the substitution of an 8400lb J65-W-20 engine for the original 7700lb J65-W-16A. Two 30mm Aden cannon replace the original 20mm mounting and a comprehensive navy-attack system is incorporated. A unique two-seat T1-4S conversion is also flown, these aircraft being converted by the Lockheed parent company in the United States.

Servicing is performed by LASS mainly acting in a depot role but responsible for Skyhawk conversion programme. The increasing self-sufficiency of the RSAF means that air force personnel are now carrying out the majority of first- to fourth-line servicing, in particular for the SF.260s and Skyvans. LASS is likely to maintain the Changi servicing facility while running down Seletar activities.

Training begins on the SF.260s, involving approximately 110hr, followed by 85hr jet training on the Strikemasters to Wings standard. Jet conversion is performed on two-seat A-4s and Hunters in the respective squadrons.

SINGAPORE

Unit	Type	Role	Base	No	Notes
140 (Osprey) Sqn	Hunter FR.74A	Recce	Tengah	4	
	Hunter Mk 74/T.75	GA/train	Tengah	31/7	
141 (Merlin) Sqn	Hunter Mk 74/T.75	GA/train	Tengah		Includes some Mk 74B
142 Sqn	A-4S/TA-4S	GA/train	Tengah	40/6	
143 Sqn	A-4S/TA-4S	GA/train	Tengah		
	F-5E/F	Int/train	Tengah?	18/3	Armed with AIM-9J Sidewinder missiles; ordered in 1976
130 (Eagle) Sqn	Strikemaster Mk 84/81/82 and Jet Provost T.52	Train/GA	Seletar	25	
150 (Falcon) Sqn	SF.260MS	Train	Seletar	14	Two lost en route; replaced Cessna 172Ks
120 (Condor) Sqn	UH-1H	Liaison	Tengah	17	US types replacing the Alouettes
	Bell 212	Liaison	Tengah	3	
121 Sqn	Skyvan 3M	SAR/trans		6	Also used for anti-smuggling patrols
	C-130B Hercules	Trans		4	Two ex-Jordanian, two ex-USAF
	Bloodhound 2	SAM		28 launchers	

INDONESIA

Area: 735,000 sq. miles (1·90 million sq. km.).
Population: 129 million.
Total armed forces: 247,000.
Estimated GNP: $43·1 billion (£21·55 billion).
Defence expenditure: $1·69 billion (£845 million).

Above: One of eight Fokker-VFW F.27 Mk 400M transports operated by the AURI from its Halim base.

This Asian republic comprising some 2000 islands receives military and economic aid from Western countries, notably Australia and the United States, and supports a modest air force known as *Tentara Nasional Indonesia – Angkatan Udara* or Indonesian Armed Forces – Air Force. Manned by some 28,000 personnel, the TNIAU has its headquarters in Jakarta and, until the recent re-organisation which trimmed the force down to a handful of units, was organised into Operations, Air Defence, Logistic, Ground Defence and Training Commands. There were also six regional commands covering the multitude of islands which make up Indonesia, from Sumatra in the north to Timor in the south.

All three Services have aviation elements. Defence spending reflects the country's gradual expansion and for 1978–79 stood at Rupiah 701.8bn ($1.69bn) although this is expected to increase over the next few years as the armed forces recover from the period in the 1960s when Soviet equipment predominated. The Naval Air Arm and Army Aviation are known respectively as the *TNI-Angkatan Laut* and *TNI-Angkatan Dorat* and are described separately below.

Formerly a Dutch colony, Indonesia was granted independence in December 1949 after a stormy period following the end of the Second World War when the islands were occupied by the Japanese. A small air

arm was established in 1945 with a few surviving Japanese aircraft and a handful of Indonesian nationals, but these were reduced to an ineffective force by Dutch air attack and it was only after independence that the *Angkatan Udara Republik Indonesia* was officially recognised as the country's air force.

Ex-Royal Netherlands Air Force aircraft supplied to the AURI included F-51 Mustangs and Mitchell bombers, while India provided both assistance and some HT-2 trainers. With the help of seconded Indian Air Force personnel, the AURI expanded during the mid-1950s and in 1958 reached an agreement with the Czech Government for the supply of 60 MiG-17 fighters and two-seat MiG-15UTI trainers, 40 Il-28 bombers and 28 Ilyushin Il-14 transports. Some surplus B-26s were also bought from the United States, together with some new C-130B Hercules. Operations Command received Tu-16 strategic bombers from Russia in 1961, these form-

ing the equipment of 41 and 42 Sqn, and as well as having a long-range strike role these machines were also employed in the anti-shipping task, for which they were equipped with Soviet-supplied Kennel air-to-surface missiles.

The creation of the Federation of Malaysia in 1962 brought Indonesia into confrontation with Great Britain, Australia, New Zealand and Malaysia and cut the AURI off from Western arms supplies. Russia continued to support Indonesia during this period and supplied MiG-19 and MiG-21 fighters plus some SA-2 Guideline anti-aircraft missiles, An-12 transports and a variety of Mil helicopters.

The conflict ended in 1966 and was followed by the toppling from power of President Sukarno, the pro-Communist head of state. By 1970 most of the Soviet equipment had been grounded from lack of spares and the Government sought assistance from Western countries, in particular Australia, with which she

Below: As part of its gradual expansion programme the AURI has acquired 16 Rockwell OV-10Fs to equip a counter-insurgency squadron.

signed a military aid agreement, and the United States, which promised to help with both military and economic aid. The former supplied a squadron of ex-RAAF Sabres in 1972 and later the same year the US sent some Sikorsky H-34s, refurbished F-51 Mustangs and T-33 trainers.

Indonesia is currently a recipient of US military aid and also receives defence equipment, assistance and training facilities from Australia.

Inventory

Considered the most powerful air arm in South-East Asia during the early 1960s, the Indonesian Air Force is currently re-establishing itself with Western help. The Soviet equipment received during the Sukarno regime lies grounded or scrapped for the lack of spares, and the front-line combat force now numbers 46 machines; this is compared with a strength of more than three times that number in 1965, including two squadrons of Tu-16 strategic bombers equipped with air-to-surface missiles.

Main combat aircraft in present use is the Commonwealth Sabre, a squadron of 16 being supplied by Australia in 1972 and currently based in East Java. To supplement rather than replace a ground-attack unit of F-51 Mustangs, the air force received a squadron of Rockwell OV-10F Bronco counter-insurgency aircraft in 1977 through the US military sales programme. Also on order from the US is a squadron of Northrop F-5E/Fs which are expected to replace the Sabre unit from

Above: The TNIAU received 16 Beech T-34C-1 Turbo-Mentors in 1978 to replace its Fuji-Beech T-34As, the aircraft being ordered via Hawker de Havilland.

Below: Locally assembled B0105 utility helicopters are expected to serve with all three branches of the Indonesian armed forces, this one being operated by the naval air arm.

Bottom: The TNIAU flies about eight HU-16B Albatrosses in the search-and-rescue role.

1980. A reduction in the number purchased was agreed within the air force in order that funds could be allocated to the acquisition of more Hercules transports and Bell 205 helicopters. Two transport squadrons comprise a heavy-lift element (31 Sqn) equipped with eight Lockheed C-130B Hercules and a second unit (2 Sqn) flying Fokker F.27s, Nomads, C-47s and a few locally built Aviocars.

A JetStar is used for VIP transport and a variety of smaller types are flown on liaison and communications work, including some Otters and about a dozen Cessnas. Locally assembled BO105 helicopters are joining the TNIAU to strengthen that element which for some years has been relatively small, standing at 11 French and US aircraft. Future procurement could involve a replacement for the Sabres and Mustangs and a type such as the A-7 Corsair has been suggested as a likely candidate.

Support

Training Command at Halim houses the Flying Training School and conducts tuition from primary to advanced grade. The British Aerospace Hawk is on order to replace the T-33s following its evaluation in competition with the Italian MB.326 and MB.339 during 1976–77. In the meantime, the ageing Fuji-Beech T-34s are being replaced by 16 Turbo-Mentors delivered in 1978 following an order placed

INDONESIA Navy

Type	Role	No
Nomad Search Master B	Util/MR	12
HU-16B	SAR	5
C-47	Trans	6
Alouette II/III	Liaison	3/3
Bell 47G	Train	4

Note: All based at Surabaya

INDONESIA Army

Type	Role	No
C-47	Trans	2
Beech 18	Trans	1
Aero Commander 560	Liaison	2
Beaver	Liaison	1
Cessna 185	Liaison	—
Cessna 310P	Liaison	2
Alouette III	Liaison	7
O-1 Bird Dog	Obs	2+
Wilga 32	Obs	20+
Bell 205	Util	16

INDONESIA Air Force

Unit	Type	Role	Base	No	Notes
	CA-27 Sabre	Int	Iswahjudi, East Java	16	Armed with Sidewinder AAMs
	OV-10F	Coin		16	
	F-51D	Att		14	
5 Sqn	HU-16B	SAR		8	
31 Sqn	C-130B	Trans	Halim	8	
	C.212 Aviocar	Trans	Halim	25	Some allocated to other Services
2 Sqn	C-47	Trans	Halim	12	Two used for nav training
	F.27 Mk 400M	Trans	Halim	8	
	Skyvan 3M	Trans	Halim	3	One believed lost
	Nomad Mission Master	Util		6	
	JetStar	VIP		1	Flown by Presidential Flight
	Cessna 401/402A	VIP		5/2	
	Cessna T207	Liaison		5	
	Otter	Liaison		7	
	Alouette III	Liaison		4	
	Bell 204B	Liaison		2	
	Bell 47G	Train?		—	Ex-Australian
	H-34D	Trans		4	
	Puma	Trans		6	
	S-61A	VIP		1	
	T-34C	Train	Halim	16	
	Hawk Mk 53	Train	Halim	8	
	Fuji-Beech T-34A	Train	Halim	15	
	T-6	Train	Halim	—	
	T-33A	Train	Halim	10	
	NZAI Airtourer	Train	Halim	—	

Soviet aircraft in storage include 22 Tu-16s, ten IL-28s, 35 MiG-19s, 15 MiG-21s, 40 MiG-15/17s, ten IL-14s, ten An-12s, 20 Mi-4s and nine Mi-6s; the Guideline missiles are no longer in use.

via the Australian Beech agent Hawker-de Havilland. Navigation training is undertaken on two specially equipped ex-RAAF C-47s.

Navy

The Indonesian Naval Air Arm, *Tentara Nasional Indonesia – Angkatan Laut*, was established in 1958 with 16 Fairey Gannet ASW aircraft plus two Gannet T.5 trainers. In the mid-1960s the force was expanded to include two fighter squadrons flying MiG-19s and MiG-21Fs. A lack of spares gradually forced the withdrawal from use of the Gannets during the confrontation with Malaysia, and with the ceasefire and the ousting of the pro-Communist Government the Soviet equipment too was eventually grounded. The present force totals some 1000 personnel flying a modest ASW/MR/SAR fleet of aircraft originating chiefly from Australia. Main type in use is the Nomad Search Master B utility aircraft which has been acquired by the TNIAL under the aid agreement with Australia. The 12 aircraft are flown on coastal-patrol and transport duties. Older machines in service include a handful of Albatross amphibians bought from the United States and six C-47 transports.

Army

A year after the naval air arm was formed the army followed suit and established a small supporting aviation element. Known as *Tentara Nasional Indonesia – Angkatan Dorat*, the force operates a mixture of fixed- and rotary-wing types, the largest being a couple of Douglas C-47 transports. A recent delivery has been 16 Bell 205s for utility duties supplied under the aid agreement with the United States.

PAPUA NEW GUINEA

Area: 178,200 sq. miles (461,500 sq. km.).
Population: 2·8 million.
Total armed forces: 3,500.
Estimated GNP: $1·22 billion (£610 million).
Defence expenditure: $22·3 million (£11·15 million).

The Papua New Guinea Department of Defence was officially established on October 1, 1974, to prepare for the transfer of defence responsibility from Australia to its own forces the following year. As part of a $A16.4m Australian defence aid programme, four ex-RAAF Douglas C-47s were delivered early in 1975 to form the nucleus of the air component of the Papua New Guinea Defence Force. The present personnel strength of the country's military forces totals some 3,500 men. A Joint Services College at Lae operates with the assistance of instructors from the RAAF and the Royal New Zealand Air Force. Air training is conducted by the RAAF but future plans call for the formation of a Flying Training School within the PNGDF, while equipment expansion has involved the purchase of three Australian-built Nomad utility aircraft.

Right, above: The PNGDF's air arm was formed with ex-RAAF C-47s.

Right: Also from Australia, the Nomad.

223

PHILIPPINES

Area: 114,800 sq. miles (297,300 sq. km.).
Population: 42·7 million.
Total armed forces: 99,000.
Estimated GNP: $20 billion (£10 billion).
Defence expenditure: $793 million (£396·5 million).

The Philippines Air Force has its HQ at Nichols Air Base, Pasay City. A small but efficiently run force, the PAF has received regular military aid from the United States and currently operates US-supplied aircraft. The main flying elements comprise squadrons of some 15 aircraft each formed into wings. The PAF is responsible for the majority of military flying in the country, although the Navy has a small air component for search-and-rescue and liaison duties. A greater autonomy in the acquisition and operation of aircraft is planned with the establishment of production facilities in the country. The Air Force has some 16,000 personnel and about 150 aircraft, and the increasing defence budget now stands at Pesos 5.85bn ($793m).

Military aviation began in the Philippines in 1920 when the Curtiss Aeroplane Company established a flying school at Paranaque, Rizal, equipped with JN-4s and HS-1L seaplanes. Known as the Philippine Air Service, the unit was short-lived and through lack of government support it disbanded in 1923. Thirteen years later, in January 1936, the Philippine Army Air Corps was finally established as a major military arm. Initial equipment comprised Stearman trainers and Stinson Reliant liaison aircraft and, by the outbreak of war in 1941, some Boeing P-26 fighters and

Martin B-10 bombers. The Philippines were invaded by Japan in December 1941.

After the Second World War, US aid assisted the formation of the new Philippine Air Force, which was officially reactivated on July 3, 1947, and was equipped with P-51Ds and C-47s. A year later the PAF began what was to be a four-year battle against communist guerillas in Central Luzon. In 1957 the Air Force received its first jet equipment with the arrival of F-86F Sabres to replace the Mustangs and in 1959 Japanese-built Beech Mentor trainers were acquired.

The Philippines has a defence treaty with the United States dating from 1951, allowing the US base rights in the country, plus a Military Assistance Pact signed in 1947. The aid agreement is being renegotiated.

Inventory

The Republic of the Philippines consists of an archipelago of about 7100 islands, of which only 11 have an area of more than 1000 square miles. The PAF's prime roles have been and still are transport/communications and counter-insurgency. For these tasks there are two airlift wings flying a variety of standard transport aircraft including C-130 Hercules, C-123 Providers, F.27 Friendships and C-47s. The COIN role is currently performed by a number of piston-engined types including converted T-34 Mentor trainers, T-28Ds and SIAI-Marchetti SF.260 Warriors. Although a long-time recipient of US aircraft, the Philippines has purchased Nomad light transports from Australia, SF.260 trainers from Italy and a number of locally assembled MBB B0105 helicopters. The PAF is planning a greater autonomy

in the future with the establishment of aircraft production facilities. As part of the Philippine Aerospace Development Corp., the Self-Reliance Development Wing of the Air Force is preparing for production of the American Jet Industries Super Pinto jet trainer/COIN aircraft, for which all the jigs and production rights plus the single prototype have been obtained. In addition it is planned to produce an indigenous piston-engined trainer design known as the XT-100.

The PAF's front-line fighter force consists of one squadron with Northrop F-5As backed up by a second unit equipped with the survivors of the F-86F Sabres supplied by the United States in 1957. To modernise this element of the force, a batch of 35 LTV F-8H Crusaders has been received. Radar protection of Philippine air space is organised under the 580th Aircraft Control and Warning Wing composed of the 581st, 582nd and 583rd ACW Sqns. The 581st Sqn mans the site at Parades Air Station in Northern Luzon, which guards the northern part of the country; the 582nd operates the site at Gozar Air Station; while the 583rd is based at Paranal guarding the central Philippines. As well as early warning, the ACWW operates tropo-scatter communications systems, air surveillance and air-defence monitoring.

A small but active Moslem guerrilla movement in the south of the country has found the PAF operating ground-attack sorties in support of ground troops. The guerrillas have been supplied with SA-7 infantry SAMs by Libya and in 1972 the PAF lost an AC-47 to one of these missiles. To combat these dissidents, the 15th Strike Wing was formed and, despite recent reduction in guerrilla activity, the

224

Above: The Philippine Air Force is one of several air arms in South-East Asia which operate the Australian-developed N22B Nomad transport.

Right: The Fokker-VFW F.27 used as a personal transport by the President of the Philippines has the distinction of being the 500th Friendship.

Below, left: The Philippine Navy was formed with ten Britten-Norman Defenders used for search-and-rescue and coastal-patrol duties.

wing continues to conduct limited missions.

Support

Training is performed under the auspices of the 100th Wing at Fernando AB, Lipa City. Initial instruction on T-41Ds and SF.260MPs is followed by jet conversion on T-33As at Basa AB. Future requirements are expected to be met by licence-produced Super Pintos and piston-engined XT-100s. Other PAF units with a support role include the 560th Composite Tactical Wing at Mactan AB, Manila, and the 533rd Air Base Sqn at Edwin Andrews AB, Zamboanga City. Reserve Airlift and Tactical Support Service operates second-line types on SAR, reconnaissance, air transport and liaison duties. The Philippines also houses support units of the 13th US Air Force plus the 3rd Tactical Fighter Wing flying F-4 Phantoms from Clark Field AB. US Naval elements operate from Cubi Point, Bataan.

Navy

Philippine Naval aviation was formed in 1975 with ten Britten-Norman Defender light transports for SAR and coastal patrol duties; by 1977 ten were in use, together with at least three locally assembled MBB B0105 helicopters for liaison purposes.

PHILIPPINES

Unit	Type	Role	Base	No	Notes
5th Fighter Wing					
6th Tac Ftr Sqn	F-5A/B	Int	Basa AB	19/3	
7th Tac Ftr Sqn	F-86F	Int	Basa AB	20	
	F-8H	Int		35	Ex-USN
9th Tac Ftr Sqn	T-34A		Basa AB		
105th Combat Crew Trg Sqn	T-33A/RT-33A	Train	Basa AB	10/3+	
15th Strike Wing					
16th Attack Sqn	T-28D	Coin	Sangley Point	12	
17th Attack Sqn	SF.260WP	Coin	Sangley Point	16	
25th Attack Sqn	T-28D	Coin	Sangley Point	12	
37th Search, Rescue	HU-16B	SAR	Sangley Point	4	
and Recce Sqn	HU-16E	SAR	Sangley Point	6	
100th Training Wing					
101st Pilot Trg Sqn	T-41D	Train	Fernando AB	12	
102nd Pilot Trg Sqn	SF.260MP	Train	Fernando AB	32	
205th Airlift Wing					
204th Tact Airlift Sqn	F.27 Mk 100	Trans	Nichols AB	9	
206th Air Trans Sqn	C-47	Trans	Nichols AB	15	
207th Air Trans Sqn	C-47	Trans	Nichols AB	15	
505th Air Rescue Sqn	UH-1H	SAR/trans	Nichols AB	15+	
505th Air Rescue Sqn	BO105	SAR/trans	Nichols AB		
220th Heavy Airlift Wing					
221st Heavy Airlift Sqn	C-123B	Trans	Mactan AB	15	Ex-USAF and South Vietnamese
222nd Heavy Airlift Sqn	C-130H	Trans	Mactan AB	6	
222nd Heavy Airlift Sqn	L-100-20	Trans	Mactan AB	3	
223rd Tact Airlift Sqn	Nomad	Trans	Mactan AB	12	Delivered 1976–77
240th Composite Wing					
291st Special Air Mission Sqn	Beaver	Liaison	Nichols AB/ Sangley Point		
303rd Air Rescue Sqn	AC-47	Att	Nichols AB/ Sangley Point	11	Ex-USAF from South Vietnam
601st Liaison Sqn	U-17A/B	Liaison	Nichols AB/ Sangley Point	6	
901st Weather Sqn	Cessna 210 Centurion/C-47	Liaison/ recce	Nichols AB/ Sangley Point		
700th Special Mission Wing					
702nd Presidential Airlift Sqn	UH-1H/UH-1N/S-62/ YS-11/One-Eleven/ Boeing 707	VIP	Nichols AB	-/-/2/ 4/1/1	

Australasia

Australia and New Zealand, as members of the British Empire and more recently the Commonwealth, have fought in two world wars despite their geographical isolation. In the case of World War II, the aggressor – in the form of Japan – was close enough to home for Australia at least to be concerned about the possibility of the northern part of the country being invaded. More recently, Australian troops fought in Vietnam in an effort to stem the tide of communism sweeping through Southeast Asia, which again posed a threat.

Australia in particular has over the past two decades moved away from ties with Britain and closer to the United States for the purchase of military equipment. Australia still has the ability to develop technically advanced weapons, with the Barra sonobuoy – which will be used to feed information to both Royal Air Force Nimrods and RAAF P-3C Orions – as the latest example of co-operation with British industry. The Jindivik drone and Ikara anti-submarine missiles are two well established examples of Australian-developed equipment which has been adopted by Britain and other countries.

In other fields, however, the country is feeling the pinch as much as any other. The RAN is being re-equipped with the US-designed and -built missile-armed FFG7-type frigate in preference to the Australian alternative, which proved too costly. Some Australian planners may now be regretting this decision, though, since the US frigate has increased in price alarmingly, just as the F-111C bomber did more than a decade earlier. New transport and anti-submarine aircraft have been bought to replace older types, but at the time of writing the big and expensive decisions – what should replace the large force of Mirage III fighters, and what future the RAN's seagoing air arm has when the carrier *Melbourne* runs out of life – had yet to be taken.

Australia is assisting Papua New Guinea in setting up and equipping its air arm, and the indigenously designed Nomad transport and maritime-patrol aircraft had additionally been supplied to neighbouring countries such as Indonesia and the Philippines. Australia's and New Zealand's cultural and ethnic ties with Asia are tenuous at best, however. Neither country can hope to achieve much more than a token presence in the vast areas of ocean which surround them, but the possibility of Russian task forces operating regularly in the Indian Ocean must cause some concern.

AUSTRALIA

Area: 2·97 million sq. miles (7·69 million sq. km.).
Population: 14·1 million.
Total armed forces: 70,000.
Estimated GNP: $92 billion (£46 billion).
Defence expenditure: $2·68 billion (£1·34 billion).

Australia has three aviation arms: the Royal Australian Air Force, Royal Australian Navy and the Army Aviation Corps. Together with an army of some 31,500 men and a navy with 50 ships, including an aircraft carrier, the air elements help in protecting this land mass of three million square miles. Still a member of the British Commonwealth, Australia provides an important military presence in the South-West Pacific area as well as deploying forces in Malaysia as part of a five-power defence pact.

The country has a unified defence structure involving all three military arms, the overall headquarters being situated at the Ministry of Defence in Canberra. Command of the armed forces is exercised by the Chief of Defence Force Staff under the direction of the Minister of Defence. The CDFS, in conjunction with the Secretary of the Department of Defence, is responsible for the administration of the armed forces and for delegating operational command in the army, navy and air force through their respective Chiefs of Staff. A seven-man Council of Defence – which includes the CDFS, the Chiefs of Staff and the Minister of Defence

– meets on a monthly basis and decides on policy.

The RAAF is the single largest aviation arm and has been in existence for more than half a century. Current RAAF organisation involves two main commands. The first, Operational Command at RAAF Penrith, New South Wales, controls all operational units except those overseas, which remain under the control of the Department of Defence. The second, Support Command at RAAF Victoria Barracks, Melbourne, controls all supply, training, maintenance and administrative units. The stated roles of the air force are:

1 To defend Australia, its territories and Australian forces against attack;

2 To maintain an air striking capability against enemy forces and installations;
3 To provide strategic and tactical air reconnaissance, including maritime air rescue for the Australian forces;
4 To provide strategic military air transport support;
5 To provide close offensive and tactical transport support for the army;
6 To provide anti-submarine support in conjunction with the RAN.

Below: The RAAF's F-111C variable-geometry attack aircraft have an important anti-ship role, for which they may be equipped with stand-off weapons to replace the present iron bombs.

Above: The mixed bag of transports operated by the RAAF includes two squadrons of DHC Caribous based at Richmond. One unit supported Australia's involvement in Vietnam during the 1960s.

Left: Two squadrons of C-130 Hercules provide the RAAF's heavy-lift element. The 12 C-130Es have been joined by the same number of C-130Hs, replacing C-130As.

Below: The venerable C-47 soldiers on both in Australia and with a support flight based at Butterworth in Malaysia. Dakotas have served with the RAAF since the 1940s.

Although the present Royal Australian Air Force was officially formed on March 31, 1921, with the prefix "Royal" granted later that year, the origins of military aviation in Australia go back some ten years earlier, to 1911, when the formation of an Aviation Corps was sanctioned. A Central Flying School was established at Point Cook, Victoria, in 1913. Australia's involvement in the First World War extended to all three Services, with the Air Corps supplying a detachment to the Royal Flying Corps in Mesopotamia to fight the Turkish forces in 1915 and four squadrons flying S.E.5s and F.E.2bs in England and Egypt. Although fighting with distinction, the small Australian Flying Corps was disbanded after the war, only to be reformed two years later as the RAAF.

The Service underwent a steady expansion over the next 15 years, flying a mixture of first- and second-line types acquired from Britain including Bulldog and Demon fighters, Southampton flying-boats and Anson reconnaissance aircraft. In 1936 the Commonwealth Aircraft Corporation was formed by Australian industrialists, the first aircraft produced being the North American NA-33 trainer, known as the Wirraway. With war clouds gathering, the RAAF began a three-year expansion programme in 1937 but at the outbreak of war in 1939 only 12 squadrons of a planned 32 were operational. Australia contributed to the Allied war effort by organising and operating the Empire Air Training Scheme, under which the RAAF undertook to train some 780 aircrew a month, plus basic training of a further number who completed their training in Canada. As well as serving with units based in England and the Middle East, Australians also fought the Japanese. In 1943, Darwin in Australia's Northern Territory was bombed by Japanese aircraft and this prompted the rapid development of the indigenous Commonwealth Boomerang fighter, which also became an excellent ground-attack aircraft. Other types built in the country during the war included Mosquitoes, Beaufighters and Tiger Moths.

After the war the RAAF underwent a re-equipment programme involving the acquisition of P-51D Mustangs, Avro Lincolns and de Havilland Vampires, the three types being licence-built in the country and supplementing other aircraft in service such as the Mosquito, Dakota, Catalina and Wirraway. To provide permanent officers the RAAF College (later RAAF Academy) was established at Point Cook in 1948, and to ensure the flow of skilled personnel in the various trades the RAAF School of Technical Training was established at Wagga, New South Wales. Although reduced in size, the air force contributed men and machines to various emergencies that arose around the world, beginning in 1948–49 with ten transport crews despatched to assist with the Berlin airlift and continuing in 1950 with the transfer of a Dakota transport squadron and a Lincoln bomber squadron to Malaya for operations against Communist terrorists. The Korean War also found the RAAF involved, this time flying Meteor F.8s in the ground-sttack role.

Re-equipment occurred in the 1950s with F-86 Sabres joining the fighter squadrons, Canberras arriving for the bomber units and Vampires re-equipping training elements. All three types were built in Australia, by Commonwealth Aircraft, Government Aircraft Factory, and de Havilland Aircraft respectively. The next decade saw the arrival of supersonic Mirage IIIs, Lockheed C-130 Hercules and DH Caribous, plus a substantial quantity of Italian Aermacchi MB.326 jet trainers. From 1964 Australia commited elements of her armed forces to the war in Vietnam, the RAAF involvement amounting to a Canberra bomber squadron, a Caribou transport squadron and a helicopter unit equipped with UH-1 Iroquois.

Australia currently has a number of defence treaties in operation. She is a member of the tripartite arrangement known as ANZUS involving Australia, New Zealand and the United States which was signed in 1951. Under this arrangement each agrees to 'act to meet the common danger' in the event of armed attack on any one of them. A similar agreement exists within the five-power defence treaty relating to the defence of Malaysia and Singapore and involving those two countries together with Australia, New Zealand and the United Kingdom. Under this treaty, which came into effect on November 1, 1971, Australia bases two Mirage fighter squadrons at the Commonwealth Strategic Reserve base at Butterworth in west Malaysia, and a detachment of eight aircraft at Tengah, Singapore, Australia also provides defence assistance to Indonesia.

Australia does not anticipate any immediate external threat to its sovereignty, although Soviet and Chinese interests could influence future defence spending.

Below: Australia's well established aircraft industry provides equipment for all three branches of the armed forces, such as this Government Aircraft Factories Nomad Mission Master used by the Army.

AUSTRALIA   Air Force

Unit	Type	Role	Base	No	Notes
1 Sqn	F-111C	Att/recce	Amberley	17/4	
6 Sqn	F-111C	Att/recce	Amberley		
2 Sqn	Canberra B.20	Att/recce/target tug	Amberley	8	Total of 48 built in Australia 1953–58
2 Sqn	Canberra T.21	Train	Amberley	4	
3 Sqn	Mirage IIIO/DO	FGA	Butterworth (Malaysia)		100 single-, 16 two-seat IIIDOs built in Australia.
75 Sqn	Mirage IIIO/DO	FGA	Butterworth		No 2 OCU has approx
77 Sqn/S OCU	Mirage IIIO/DO	FGA	Williamtown		12 aircraft
10 Sqn	P-3C	Mar pat	Edinburgh	10	Replaced Neptunes
11 Sqn	P-3B	Mar pat	Edinburgh	10	
34 Sqn	One-Eleven/Mystère 20/HS.748 Srs 2/ Boeing 707	VIP	Fairbairn	2/3/ 2/2	
35 Sqn	Caribou	Trans	Richmond		23 aircraft divided between 35 and 38 Sqns
38 Sqn	Caribou	Trans	Richmond		
36 Sqn	C-130H	Trans	Richmond	12	Replaced C-130As
37 Sqn	C-130E	Trans	Richmond	12	
38 Sqn	C-47	SAR/util	various	17	Some with Support Flt at Butterworth
5 Sqn	UH-1B	Train/army supp	Fairbairn	31	
9 Sqn	UH-1D/H	Army supp	Townsville	16	
12 Sqn	CH-47C	Army supp	Amberley	12	
1 FTS	CT/4A	Train	Point Cook	31	Six used by CFS at East Sale for instructor training
2 FTS	MB.326H	Train	Pearce	50	Further 15 used by CFS East Sale and some by 2 OCU
No 4 Flt	Winjeel	Train	Williamtown		Small number used for FAC training
No 4 Flt	HS.748 T.2	Train	East Sale	8	Flown by School of Air Navigation

Inventory

Today the Royal Australian Air Force maintains a modern force of some 150 front-line aircraft in 16 operational squadrons. Plans are in hand for the replacement over the next few years of a number of the types currently in service, foremost among these being the Mirage III. The RAAF assessed a number of proposals submitted by European and American aircraft companies during 1978 and a decision was likely during 1979. The type is expected to be more of a multi-role aircraft than a pure interceptor as at first envisaged, and aircraft under consideration have included the McDonnell Douglas F-15 Eagle, General Dynamics F-16 and Dassault Mirage 2000.

Home base of the Mirage Wing, which forms the main fighter defence of Australia, is on the east coast at RAAF Williamtown near Sydney; the base also houses the Mirage Conversion Unit, with fighter

training conducted at the northern base of Tindal near Darwin. The Mirages operate in conjunction with surveillance radars and Plessey air-transportable early-warning radar and control systems.

Backbone of the strike force are the two squadrons of General Dynamics F-111Cs based at Amberley and assigned both overland and maritime long-range attack roles. The aircraft were delivered in 1973, ten years after the order was placed. Much controversy accompanied the order as a result of serious fatigue problems and the steadily rising cost of the 24 aircraft, which increased from $146m to nearly $345m. To increase the effectiveness and widen the tasks undertaken by the F-111s, four aircraft are being fitted with infra-red cameras, a sensor pod and other equipment at a cost of $A19m for the long-range reconnaissance role. A third strike squadron still operates the venerable Canberra, which in 1978 saw its 25th year of service with the RAAF.

Being the smallest of the world's continents and surrounded by sea, Australia

has an obvious need for a strong maritime reconnaissance element and there are two squadrons that operate in the role, one currently being updated for the job. The ten long-serving Neptunes are being replaced by the Lockheed P-3C Orion with Update II electronics. The ten aircraft are joining the present force of ten P-3Bs at the main maritime base at Edinburgh, with detachments to other bases covering the extensive coastline.

The transport arm totals five fixed-wing squadrons and three helicopter units. The former have received 12 new C-130H Hercules to increase the RAAF's airlift capability and to replace an identical number of older C-130A models delivered in 1958. The Hercules – like its short-range companion, the Caribou – did sterling work during the Vietnam war, and its long range provided the lifeline for Australian servicemen. These aircraft now assist in civil emergencies such as the relief of Darwin in December 1974, fly regular services throughout RAAF bases in Australia, and support overseas bases

and operations including those in Malaysia and Singapore. For Government and VIP use there is a special unit based at Fairbairn using a number of British and French aircraft; two Boeing 707s are being added in 1979.

The helicopter squadrons are mainly used for army support work, operating large Chinook twin-rotor machines for medium-lift operations and more than 40 Bell UH-1 Iroquois for troop transport and gunship duties. Rotary-wing training is undertaken by the RAAF for both itself and the Navy.

RAAF training is controlled by Support Command and comprises No 1 Flying Training School, No 2 FTS, the Central Flying School, the School of Air Navigation and No 2 Operational Conversion Unit.

No 1 FTS at Point Cook is charged with training all RAAF, navy and army pilots in preliminary ground and air work, as well as pre-basic flying training of the second-year cadets of the RAAF Academy. Student pilots of the RAAF and navy spend a 24-week training period where they log 60hr flying CT/4A Airtrainers. They then go to Pearce for an advanced course on MB.326Hs. The army pilots' course is over a 22-week period designed to produce a pilot proficient in basic flying; 96hr are logged. Point Cook has always been the principal flying training base for the air force, and in latter years has trained pilots from other countries including Malaysia and Papua New Guinea.

Pearce is the home base of No 2 FTS, which is the RAAF's jet training school. Initially flying Vampires in 1958 and known as No 1 Applied Flying Training School, the unit introduced the Aermacchi MB.326H into service in 1968 and was renamed No 2 FTS. The school trains students in advanced levels of flying skills, such as instrument flying, formation flying and aerobatics. Upon graduation pilots are assigned to operational squadrons. Apart from air force pilots, 2 FTS trains all navy pilots and those from other countries.

Left: The GAF Jindivik target drone, seen here in RAN colours, has proved a big dollar-earner for Australia.

Below: The RAAF has been studying a replacement for its licence-built Mirage IIIs.

Located at East Sale is the Central Flying School, which uses MB.326s for training flying instructors and providing refresher flying for pilots who have been in staff appointments. Student instructors are selected from operational units, and after completion of the instructors' course are posted to 2 FTS at Pearce. East Sale is also the base for the RAAF MB.326 aerobatic team, the Roulettes.

The School of Air Navigation was formed at East Sale in February 1946. Today the school operates eight HS.748 T.2s fitted out as flying classrooms for navigation and air electronics officer training for the RAAF and observer training for the navy. Up to 200hr of practical flying is undertaken by the students before graduating. The length of courses for RAAF navigators and AEOs is 47 weeks, while that for navy observers is 38 weeks.

Williamtown houses two smaller training elements, Bo 2 OCU and No 4 Flt. The former trains all RAAF fighter pilots, using MB.326s for ground-attack and air combat training prior to flying the Mirage IIIs. No 4 Flt was established by the air force following experience with forward air controllers in Vietnam, and this small unit now trains FACs.

Navy

The Royal Australian Navy formally commissioned its first Fleet Air Arm unit in 1948, just over a year after approval was granted for the new arm by the then Commonwealth Government. Two aircraft carriers were purchased, HMAS *Sydney* (formerly HMS *Terrible*) in 1948 and HMAS *Melbourne* (formerly HMS *Majestic*) in 1956. Initial equipment comprised two squadrons each of Hawker Sea Fury fighter-bombers and Fairey Firefly anti-submarine aircraft, later replaced by Sea Venoms and Gannets respectively. HMAS *Sydney* was decommissioned in 1974 leaving *Melbourne*, extensively modernised, as the RAN's sole carrier and flagship of the Service.

Above: The Macchi MB.326H is one of many overseas designs which have been built under licence in Australia. The type is used for a wide variety of tasks, including advanced instruction, refresher flying and training in combat techniques.

Below: The Lockheed P-3C Orion Update II greatly enhances the RAAF's ability to locate and destroy submarines.

Bottom: The RAN also has modern ASW aircraft – Sea Kings from *Melbourne*.

for the CAG when not embarked. A hangar fire at Nowra in December 1976 all but destroyed the fleet of 14 S-2E Trackers, leaving three aircraft operational. Urgent negotiations with the United States Navy resulted in the acquisition of 16 replacement Trackers which were delivered to Australia early in 1977; this almost doubled the navy's fixed-wing ASW capability. The helicopter anti-submarine force received a much-needed shot in the arm in 1975 when the first of ten Westland Sea King HAS.50s arrived to replace the Wessexes on board *Melbourne*. The Sea Kings, costing some $A16m, joined HS-817 (No 817 Sqn); the unit's Wessexes transferred to a second-line training, fleet requirements and SAR unit, HS-723 at Nowra. Consideration is being given to updating the Wessex fleet, which among other things would include replacement of the Gazelle engines with a new twin-engine installation. In addition the RAN has a requirement for more Sea Kings, but a firm order has still to be placed. A light shipboard helicopter is also under consideration for use aboard two new missile ships being built in the US, while of greater portent is the possible procurement of up to three 'Harrier carriers' designed by Vosper Thornycroft to replace *Melbourne* in the late 1980s.

Basic Fleet Air Arm pilot training is undertaken at No 1 Flying Training School, RAAF Point Cook, on CT/4s; students log 60hr in 24 weeks. They then go to Pearce for an advanced flying course on MB.326s. Observers are trained on RAAF HS.748 T.2s at the School of Air Navigation, East Sale, the course taking some 38 weeks. Students then transfer to Nowra for conversion on to the Trackers.

Above: The ageing S-2E Tracker still performs valuable service as the RAN's fixed-wing anti-submarine patroller from the aircraft carrier *Melbourne*.

Inventory

The RAN's 20,000-ton light aircraft carrier *Melbourne* is scheduled to continue in service until the 1980s and with her complement of Skyhawks, Trackers and Sea King helicopters she maintains a formidable presence in the area with the ability to conduct maritime operations or ground support for the army. *Melbourne* underwent a modernisation refit in the late 1960s to handle the Skyhawks and Trackers ordered by the RAN in 1965 and delivered two years later. In 1969 the ship embarked her aircraft and the 347 personnel of the Carrier Air Group, and in 1976 took aboard new Sea King anti-submarine helicopters to replace the Wessexes which had been in use for 13 years. The mix of aircraft in the carrier group is varied according to the task, either emphasising the strike role by carrying more Skyhawks or the anti-submarine role by carrying more Trackers and Sea Kings. The number of aircraft accommodated is normally about 16.

Main shore base is Nowra Naval Air Station, New South Wales, known as HMAS *Albatross* and acting as home base

Above: Bell OH-58A light observation helicopters assembled by the Commonwealth Aircraft Corporation are flown by the Australian Army Aviation Corps in a variety of utility roles.

Army

In 1968 the Australian Army Aviation Corps was established to organise and train the aviation units in the army. An Aviation Centre was formed at Oakey in 1971 and currently accommodates the School of Army Aviation, the 1st Aviation Regiment and No 5 Base Workshop Battalion, a major maintenance and repair facility. Helicopter units are located with major field formations in Townsville and Sydney. In addition to providing support for the army in Australia, aircraft are being used extensively in survey operations in Indonesia and Papua New Guinea.

Fixed-wing types in use with the AAC are Pilatus Turbo-Porter single-engined Stol aircraft and twin-engined Nomad transports designed and built by the Government Aircraft Factory at Fisherman's Bend. Helicopters consist of Commonwealth-assembled Bell JetRangers, the last of a batch of 53 having been delivered early in 1977. For the defence of Army units, British Aerospace Rapier SAMs are being delivered.

AUSTRALIA   Navy

Unit	Type	Role	Base	No	Notes
VF-805	A-4G	FGA	*Melbourne*/Nowra	14	
VF-805/VC-724	TA-4G	Train	*Melbourne*/Nowra	2	
VS-816	S-2E/G	ASW	*Melbourne*/Nowra	19	
HS-817	Sea King Mk 50	ASW	*Melbourne*/Nowra	8	
HS-723	UH-1B	Fleet supp/comm/SAR	Nowra	4	
	Bell 206B-1	Fleet supp/comm/SAR	Nowra	2	
	Wessex HAS.31B	Fleet supp/comm/SAR	Nowra	20	
VC-724	MB.326H	Train/trials	Nowra	8	Unit also has A-4G/TA-4Gs
VC-851	HS.748	ASW/ECM train	Nowra	2	S-2E/Gs also used by unit
	Tartar	SAM	Destroyers	5	
	Standard	SAM	and frigates	vessels	

AUSTRALIA   Army

Unit	Type	Role	Base	No	Notes
171 Air Cavalry Flight	Turbo-Porter	Liaison	Holsworthy		Total of 17 Turbo-Porters in use of
	Turbo-Porter	Liaison	Oakey		19 delivered in late 1960s
173 Support Sqn	Nomad	Liaison/trans	Oakey	11	
	Commonwealth- Bell 206B-1	LOH		53	Operated with Army units at various locations. Delivery complete in 1977
	Redeye	SAM	various		
	Rapier	SAM	various		

NEW ZEALAND

Area: 103,700 sq. miles (268,600 sq. km.).
Population: 3·14 million.
Total armed forces: 12,600.
Estimated GNP: $13·6 billion (£6·8 billion).
Defence expenditure: $242 million (£121 million).

The Royal New Zealand Air Force forms an integral part of the country's tri-Service defence organisation and the Air Staff, alongside those of the Army and Navy, has its headquarters in the Ministry of Defence in Wellington. About 4300 personnel make up the Air Force, which has an establishment of eight operational squadrons in the strike, transport and maritime roles, and training elements necessary to support these roles. Spread over seven bases, the units come under the command of either Operations Group at Auckland (strike, transport and maritime) or Support Group at Christchurch (Flying Training Wing and Ground Training Wing). Despite a limited defence budget, which for all three Services in 1978 totalled $NZ254m ($242m), the RNZAF continues to maintain units overseas in support of its defence commitments in the southwest Pacific area and South East Asia while at the same time providing a small home defence force. A nominal naval air component of two helicopters is maintained for use aboard two frigates.

Left: A Bell UH-1H Iroquois of the RNZAF's 3 Sqn, which has a secondary search-and rescue task in addition to its main role of supporting the army.

Below left: An A-4K Skyhawk of No 75 Sqn formates on a TA-4K trainer of the same unit during exercises off Hawaii.

Although the Royal New Zealand Air Force was constituted on April 1, 1937, its antecedents can be traced back to 1909 when the Hon Henry Wigram urged the formation of an air arm for the country's defence. No arm was formed, however, and New Zealand personnel served instead in the RFC and RNAS during the First World War. The New Zealand Permanent Air Force, established in the 1920s, was equipped with ex-RAF aircraft including Avro 504K trainers, DH Puss Moths and Gloster Grebes. With its separate status established in 1937, as opposed to its previous identity as a branch of the Army, the RNZAF expanded with RAF assistance just before the outbreak of the Second World War. Thirty Vickers Wellington bombers ordered by New Zealand became the nucleus of the country's war effort in Europe, while at home some 10,000 aircrew were trained via Canada under the Empire Air Training Scheme. More than 20 operational squadrons fought in the Pacific against the Japanese, and after the war the RNZAF formed part of the occupation force in Japan. The postwar Air Force was reduced to five regular and four reserve squadrons flying Mosquitoes, Vampire 9s and F-51D Mustangs. Short

Sunderland flying boats gave valuable SAR/MR service, as did a squadron of Canberra B (I).12 jet bombers; the reserve units were disbanded in 1957.

Above: The RNZAF's 40 Sqn provides a long-range transport arm in the form of five C-130H Hercules.

Left: No 14 Sqn contributes 16 British Aerospace (BAC) Strikemaster Mk 88s to supplement Skyhawks in the attack role. The aircraft have a secondary training task.

Below: CT/4 Airtrainers built by New Zealand Aircraft Industries have replaced Harvards in the primary-training role. The type has also been exported.

New Zealand is a member of a five-nation defence agreement with Australia, Malaysia, Singapore and the UK, and retains a support unit in Singapore as part of her commitment. There also exists a tripartite mutual-defence treaty, known as ANZUS, signed by Australia, New Zealand and the United States in 1951.

Inventory

Economic restrictions have limited the expansion of the RNZAF over the past two decades, with the result that the front-line combat force consists of only two squadrons flying A-4K Skyhawks and BAC Strikemasters. Five Lockheed Orions replaced the Sunderlands in 1966 to provide the country with an updated maritime reconnaissance element. To replace the veteran Bristol Freighter transports and the small force of Douglas C-47s the RNZAF purchased ten HS Andovers for £8m in 1976; these ex-RAF aircraft were flown to New Zealand the following year and currently undertake a variety of duties including trooping, continuation training and VIP flights. Long-range transport and supply missions are performed by a squadron of Lockheed Hercules purchased in 1966. A modest helicopter force of Iroquois lend support to the Army and have a secondary SAR role, while the same unit also maintains the Navy's two Wasp shipboard helicopters when shore-based. The Navy has a single four-round launcher for Seacat SAMs on each of four frigates.

The RNZAF Support Group includes all units not concerned with front-line operational flying. On South Island is situated Wigram, the force's main training base and home of the Flying Training Wing and Central Flying School equipped with New Zealand Aircraft Industries CT/4 primary trainers. Near Wigram is Weedons, the RNZAF southern stores depot; the North Island stores depot is Te Rapu. The Air Force's main repair depot is at Woodbourne near Blenheim, where major overhauls are carried out on such types as the A-4Ks and where No 4 Technical Training School is situated. The Flying Training Wing at Wigram takes pupils up to Wings standard in the Pilot Training Sqn and at the same base is the Navigators and Air Electronics Training Sqn. Helicopter training is performed on four Bell 47Gs.

NEW ZEALAND

Unit	Type	Role	Base	No	Notes
1 Sqn	Andover C.1	Trans	Whenuapai	6	Replaced Bristol Freighters and C-47s from 1977; initial crew training in UK at Brize Norton
3 Sqn	UH-1D/UH-1H/Bell 47G	Army support/SAR/trans	Hobsonville	5/5/6	
5 Sqn		MR	Whenuapai	5	
14 Sqn	Strikemaster Mk 88	Strike/train	Ohakea	16	
40 Sqn	C-130H Hercules	Trans	Whenuapai	5	One Devon attached to sqn
42 Sqn	Andover C.1/Devon	Trans/VIP	Ohakea	4/2	
75 Sqn	A-4K/TA-4K	Strike/train	Ohakea	9/4	Replaced Vampires
	UH-1H	Army support	Tengah	4	Designated RNZAF Support Unit
FTW	CT/4 Airtrainer	Train	Wigram	13	Delivered 1976; replaced Harvards
	Wasp HAS.1	Liaison	Hobsonville	2	For use aboard HMNZS *Waikato* and *Canterbury*; maintained by 3 Sqn
	Devon	Trans/train		9	30 delivered; now used by 40 and 42 Sqns and Wigram FFW
Royal New Zealand Navy					
	Seacat	SAM		4	Four frigates launchers

Inventory of Aircraft

This inventory gives basic details of the major aircraft types operated by the world's air forces. Further information about the role, deployment and armament of each type can be obtained by referring to the appropriate air force in the main section of the book, using the lists of operators as a guide.

AERITALIA AM.3C

Utility aircraft.
Dimensions: Span 41ft 5½in (12·64m); length 29ft 3¾in (8·93m); height 8ft 11in (2·72m).
Maximum speed: 173mph (278km/h).
Range: 615 miles (990km).
Users: Italy, Rwanda, South Africa.

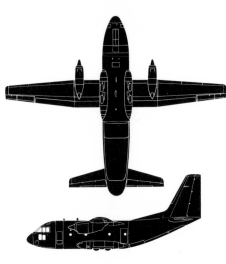

AERITALIA G222

Tactical transport.
Dimensions: Span 94ft 6in (28·8m); length 74ft 5½in (22·7m); height 32ft 1¾in (9·8m).
Maximum speed: About 370mph (595km/h).
Range: With 11,023lb (5000kg) load, 1,833 miles (2950km).
Users: Argentina, Dubai, Italy, Libya.

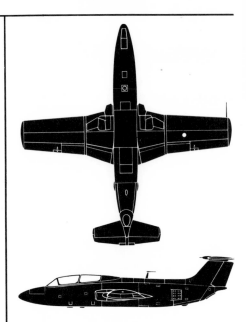

AERO L-29 DELFIN

Trainer.
Dimensions: Span 33ft 9in (10·29m); length 35ft 5½in (10·81m); height 10ft 3in (3·13m).
Maximum speed: 407mph (655km/h).
Range: Max internal fuel, 397 miles (640km).
Users: Bulgaria, Czechoslovakia, Egypt, East Germany, Guinea, Hungary, Indonesia, Iraq, Nigeria, Romania, Soviet Union, Syria, Uganda.

AERITALIA G91Y

Ground-attack fighter.
Dimensions: Span 29ft 6½in (9·01m); length 38ft 3½in (11·67m); height 14ft 6in (4·43m).
Maximum speed: Clean, most heights, 690 mph (1110km/h).
Combat radius: Low-level, clean, 372 miles (600km).
Users: Angola, West Germany, Italy, Portugal.

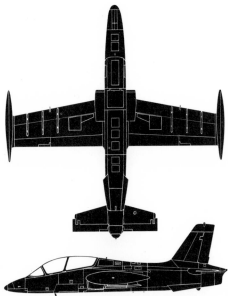

AERMACCHI M.B.339

Trainer and light attack aircraft.
Dimensions: Span (over tanks) 35ft 7½in (10·86m); length 36ft (10·97m); height 13ft 1¼in (3·99m).
Maximum speed: 558mph (898km/h).
Range: Max internal fuel, 1,093 miles (1760km).
Users: With 326, Argentina, Australia, Bolivia, Brazil, Dubai, Ghana, Italy, South Africa, Togo, Tunisia, Zaïre, Zambia.

AERO L-39 ALBATROS

Trainer
Dimensions: Span 31ft 0½in (9·46m); length 40ft 5in (12·32m); height 15ft 5½in (4·72m).
Maximum speed: 485mph (780km/h).
Range: Max internal fuel, 5 per cent reserves, 528 miles (850km).
Users: Afghanistan, Czechoslovakia, Iraq, Soviet Union; deliveries in 1979 to other WP air forces with exception of Poland.

AEROSPACE AIRTRAINER CT4

Trainer.
Dimensions: Span 26ft (7·92m); length 23ft 2in (7·06m); height 8ft 6in (2·59m).
Maximum speed: 178mph (286km/h).
Range: Max fuel (no tip tanks) plus unstated reserves, 815 miles (1311km).
Users: Australia, Hong Kong, Indonesia, New Zealand, Singapore, Thailand.

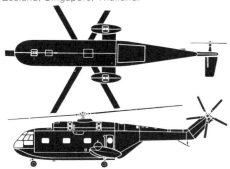

AEROSPATIALE 321G SUPER FRELON

Offshore patrol and ASW helicopter.
Dimensions: Main-rotor diameter 62ft (18·9m); length (rotors turning) 75ft 7in (23m); height 21ft 10in (6·66m).
Maximum speed: 171mph (275km/h).
Range: 509 miles (820km).
Users: China, France, Iran, Israel, Libya, South Africa, Syria.

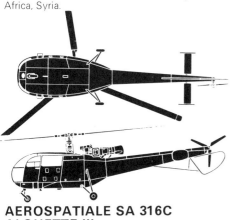

AEROSPATIALE SA 316C ALOUETTE III

Utility helicopter.
Dimensions: Main-rotor diameter 36ft 1¾in (11·02m); length (rotors turning) 42ft 1½in (12·84m); height 9ft 10in (3m).
Maximum speed: 130mph (210km/h).
Users: Abu Dhabi, Austria, Bangladesh, Burma, Cameroun, Congo, Denmark, Dominica, Ecuador, Ethiopia, Gabon, Ghana, Hong Kong, India, Indonesia, Iraq, Ireland, Israel, Ivory Coast, Jordan, Laos, Lebanon, Liberia, Malagasy, Mexico, Netherlands, Pakistan, Peru, Portugal, Rhodesia-Zimbabwe, Romania, Rwanda, Singapore, South Africa, Spain, Switzerland, Venezuela, Yugoslavia, Zaïre, Zambia.

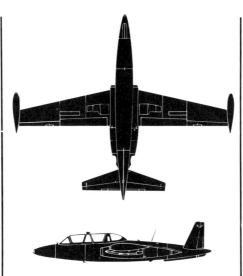

AEROSPATIALE MAGISTER

Basic trainer and (some air forces) light attack.
Dimensions: Span (no tip tanks) 37ft 5in (11·4m); length 33ft (10·06m); height 9ft 2in (2·8m).
Maximum speed: 403mph (650km/h).
Range: 735 miles (1250km).
Users: Algeria, Belgium, Brazil, Cameroun, Finland, France, Ireland, Israel, Kampuchea, Lebanon, Morocco, Rwanda, Salvador. Togo, Uganda.

AEROSPATIALE/WESTLAND SA 330 PUMA

Tactical helicopter transport.
Dimensions: Main-rotor diameter 49ft 2½in (15m); length (rotors turning) 59ft 6½in (18·15m); height 16ft 10½in (5·14m).
Maximum speed: 168mph (271km/h).
Range: 360 miles (580km).
Users: Abu Dhabi, Algeria, Belgium, Cameroun, Chad, Chile, Ecuador, France, Gabon, Ivory Coast, Kuwait, Morocco, Nepal, Nigeria, Pakistan, Portugal, South Africa, Togo, Tunisia, UK (RAF), Zaïre.

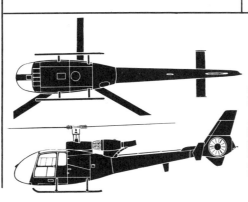

AGUSTA A 109

Utility helicopter.
Dimensions: Main-rotor diameter 36ft 1in (11m); length (rotors turning) 42ft 10in (13·05m); height 10ft 10in (3·3m).
Maximum speed: 193mph (311km/h).
Range: Normal fuel, no reserve, 351 miles (565km).
Users: Argentina, Italy and nine unannounced military and naval customers.

ANTONOV (WSK-MIELEC) AN-2 COLT

Multi-role utility aircraft.
Dimensions: Span 59ft 8½in (18·18m); length 41ft 9½in (12·74m); height 13ft 1½in (4m).
Maximum speed: 160mph (258km/h).
Range: With 1,102lb (500kg) payload, 560 miles (900km).
Users: 44 countries including air forces of at least the following: Afghanistan, Algeria, Bulgaria, China, Cuba, Egypt, Ethiopia, East Germany, Hungary, Iraq, North Korea, Mali, Mongolia, Poland, Romania, Somalia, Soviet Union, Sudan, Syria, Tanzania, Tunisia, Vietnam, Yemen, Yugoslavia.

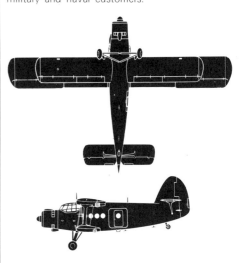

AEROSPATIALE/WESTLAND GAZELLE

Multi-role utility helicopter.
Dimensions: Main-rotor diameter 34ft 5½in (10·5in); length (rotors turning) 39ft 3¾in (11·97m); height 10ft 2½in (3·15m).
Maximum speed: Not stated but roughly the same as max cruise, 164mph (264km/h).
Range: With 1,102lb (500kg) payload, 223 miles (360km).
Users: Egypt, France, Kuwait, Qatar, Senegal, UK (Army, RAF, RN), Yugoslavia.

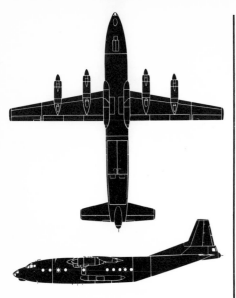

ANTONOV AN-12 CUB

Tactical transport.
Dimensions: Span 124ft 8in (38m); length 121ft 4½in (37m); height 32ft 3in (9·83m).
Maximum speed: 482mph (777km/h).
Range: With full payload 2,236 mile (3600km).
Users: Algeria, Bangladesh, Egypt, India, Indonesia (stored), Iraq, Poland, Soviet Union, Sudan, Syria, Yugoslavia.

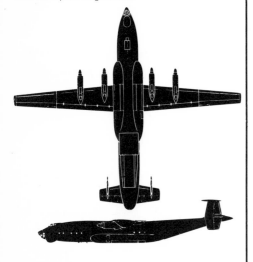

ANTONOV AN-22 ANTEI COCK

Heavy freight transport.
Dimensions: Span 211ft 4in (64·4m); length 189ft 7in (57·8m); height 41ft 1½in (12·53m).
Maximum speed: 460mph (740km/h).
Range: Max fuel and 99,200lb (45 000kg) payload, 6,800 miles (10 950km).
User: Soviet Union.

ANTONOV AN-26 CURL

Tactical transport.
Dimensions: Span 95ft 9½in (29·2m); length 78ft 1in (23·8m); height 28ft 1½in (8·58m).
Maximum speed: Not stated; typical cruising speed 267mph (430km/h).
Range: With 9,920lb (4500kg) payload, 559 miles (900km).
Users: Bangladesh, Congo, Czechoslovakia, Egypt, East Germany, Hungary, Iraq, Laos, Mongolia, Poland, Romania, Somalia, Soviet Union, Sudan, Vietnam, South Yemen, Yugoslavia.

BAe 125-400/DOMINIE

Trainer, communications and utility transport.
Dimensions: Span 47ft (14·33m); length 47ft 5in (14·45m); height 16ft 6in (5·03m).
Maximum speed: 570mph (917km/h).
Range: Typical 1,796 miles (2891km).
Users: Argentina, Australia, Brazil, Ghana, Malaysia, South Africa, UK (RAF).

BAe (AVRO/HSA) 748

Multi-role transport.
Dimensions: Span 98ft 6in (30·02m); length 67ft (20·42m); height 24ft 10in (7·57m).
Maximum speed: (Also max cruise) 281mph (452km/h).
Range: With 14,027lb (6363kg) payload and full reserves, 1,474 miles (2372km).
Users: Argentina, Australia, Belgium, Brazil, Brunei, Cameroun, Colombia, Ecuador, India, South Korea, Malaysia, Nepal, New Zealand, Tanzania, Thailand, UK (RAF), Upper Volta, Venezuela, Zambia.

BAe (BLACKBURN/HSA) BUCCANEER S.2B

All-weather attack and reconnaissance.
Dimensions: Span 44ft (13·41m); length 63ft 5in (19·33m); height 16ft 3in (4·95m).
Maximum speed: Low, internal bomb load, 645mph (1038km/h).
Range: Hi-lo-hi with internal bomb load, 2,300 miles (3700km).
Users: South Africa, UK (RAF).

BAe (HAWKER) HUNTER FGA.9

Single-seat attack fighter.
Dimensions: Span 33ft 8in (10·26m); length 45ft 10½in (13·98m); height 13ft 2in (4·26m).
Maximum speed: Clean, sea level, 710mph (1144km/h).
Range: Internal fuel 490 miles (689km), max fuel 1,840 miles (2965km).
Users: Abu Dhabi, Chile, India, Iraq, Kenya, Kuwait, Lebanon, Oman, Peru, Qatar, Rhodesia, Singapore, Switzerland, UK (RAF, RN).

BAe (ARMSTRONG WHITWORTH/HAWKER) SEA HAWK 6

Carrier-based fighter/bomber.
Dimensions: Span 39ft (11·89m); length 39ft 8in (12·08m); height 9ft 9½in (3m).
Maximum speed: Low, clean, 599mph (958km/h).
Range: High, no drop tanks, 740 miles (1191km).
Users: India (Navy).

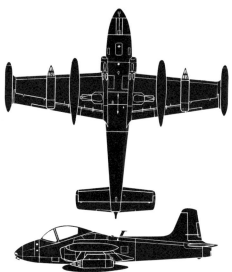

BAe (BAC) STRIKEMASTER Mk 80

Light ground attack and trainer aircraft.
Dimensions: Span 36ft 10in (11·23m); length 33ft 8½in (10·27m); height 10ft 11½in (3·34m).
Maximum speed: Clean, medium altitude, 481mph (774km/h).
Combat radius: With maximum weapon load, low-profile, 145 miles (233km).
Users: Ecuador, Kenya, Kuwait, New Zealand, Oman, Saudi Arabia, Singapore, South Yemen (status uncertain), Sudan.

BAe (HSA) HARRIER GR.3

Tactical attack and reconnaissance.
Dimensions: Span 25ft 3in (7·7m) (ferry tips 29ft 8in/9·04m); length (laser) 45ft 8in (13·91m); height 11ft 4in (3·45m).
Maximum speed: Low, clean, over 737mph (1186km/h).
Combat radius: Low, bombs but no drop tanks, 260 miles (418km).
Users: Spain, UK (RAF, RN), US (Marines).

BAe (EEC/BAC) LIGHTNING F.6

Single-seat all-weather interceptor.
Dimensions: Span 34ft 10in (10·6m); length 53ft 3in (16·25m); height 19ft 7in (5·95m).
Maximum speed: Clean, high altitude 1,500 mph (2415km/h, Mach 2·27).
Range: Without overwing tanks, 800 miles (1290km).
Users: Saudi Arabia, UK (RAF).

BAe (HANDLEY PAGE) VICTOR K.2

Air refuelling tanker.
Dimensions: Span 117ft (35·66m); length 114ft 11in (35·05m); height 30ft 1½in (9·2m).
Maximum speed: High, about 640mph (1030 km/h).
Range: 4,600 miles (7400km).
User: UK (RAF).

BAe (AVRO/HSA) SHACKLETON AEW.2

Airborne early warning.
Dimensions: Span 119ft 10in (36·53m); length 93ft 1in (28·38m); height 23ft 4in (7·11m).
Maximum speed: 300mph (482km/h).
Range: 4,200 miles (6760km).
User: UK.

BAe NIMROD MR.2

Maritime reconnaissance and ASW aircraft.
Dimensions: Span 114ft 10in (35m); length 126ft 9in (38·63m); height 29ft 8½in (9·08m).
Maximum speed: 575mph (926km/h).
Range: Ferry, unrefuelled, 5,755 miles (9265 km).
User: UK (RAF).

BAe (EEC/BAC) CANBERRA B.6

Multi-role attack and reconnaissance.
Dimensions: Span 63ft 11in (19·5m); length 65ft 6in (19·95m); height 15ft 7in (4·72m).
Maximum speed: High, clean, 580mph (933 km/h).
Combat radius: Low interdiction mission, 805 miles (1295km).
Users: Argentina, Ecuador, Ethiopia, France (trials), West Germany, India, New Zealand, Peru, Rhodesia-Zimbabwe, South Africa, Sweden (trials), Venezuela.

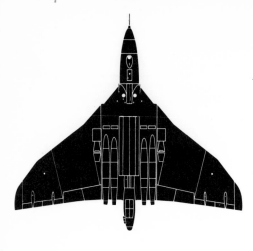

BAe (AVRO/HSA) VULCAN B.2

Bomber and reconnaissance aircraft.
Dimensions: Span 111ft (33·83m); length (including FR probe), 105ft 6in (32·15m); height 27ft 2in (8·26m).
Maximum speed: High, 640mph (1030km/h).
Range: With full bomb load, about 4,600 miles (7400km).
User: UK (RAF).

BELL 206 JETRANGER/OH-58 KIOWA

Observation and utility helicopter.
Dimensions: Main-rotor diameter 35ft 4in (10·77m); length (rotors turning) 40ft 11¾in (12·49m); height 9ft 6½in (2·91m).
Maximum speed: 138mph (222km/h).
Range: Sea level, 10 per cent reserves, 299 miles (481km).
Users: Argentina, Australia, Austria, Brazil, Brunei, Canada, Chile, Colombia, Dubai, Iran, Israel, Italy, Jamaica, Japan, Liberia, Malaysia, Malta, Mexico, Morocco, Oman, Saudi Arabia, Spain, Sri Lanka, Sweden, Tanzania, Thailand, Turkey, Uganda, United Arab Emirates, US (Army), Venezuela.

BAe HAWK T.1

Trainer and multi-role tactical aircraft.
Dimensions: Span 30ft 10in (9·4m); length (excluding probe) 36ft 7¾in (11·17m); height 13ft 1¼in (3·99m).
Maximum speed: 630mph (1014km/h).
Range: Without drop tanks 1,510 miles (2433 km).
Users: Finland, Indonesia, Kenya, UK (RAF).

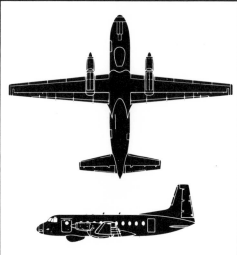

BAe COASTGUARDER

Maritime patrol and multi-role transport.
Dimensions: Span 98ft 6in (30·02m); length 67ft (20·42m); height 24ft 10in (7·57m).
Typical patrol speed: 161mph (259km/h); cruising speed 268mph (431km/h).
Range: Max, high-altitude cruise, 2,648 miles (4262km).

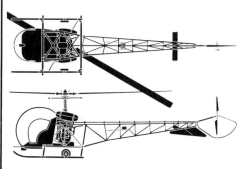

BELL 47G

Utility and training helicopter.
Dimensions: Main-rotor diameter 37ft 1½in (11·32m); length (rotors turning) 43ft 4¾in (13·2m); height 9ft 3½in (2·83m).
Maximum speed: 105mph (169km/h).
Range: 210 miles (338km).
Users: Argentina, Australia, Austria, Brazil, Burma, Canada, Chile, Dahomey, Ecuador, Greece, Guinea, Honduras, India, Indonesia, Iran, Italy, Jamaica, Japan, Kampuchea, Kenya, Libya, Malagasy, Malaysia, Malta, Mexico, Morocco, New Zealand, Pakistan, Paraguay, Peru, Spain, Sri Lanka, Taiwan, Tanzania, Turkey, UK (Army).

BAe BULLDOG

Trainer.
Dimensions: Span 33ft (10·06m); length 23ft 3in (7·09m); height 7ft 5¾in (2·28m).
Maximum speed: 150mph (241km/h).
Range: 620 miles (1000km).
Users: Ghana, Hong Kong, Jordan, Lebanon, Kenya, Malaysia, Nigeria, Sweden, UK (RAF).

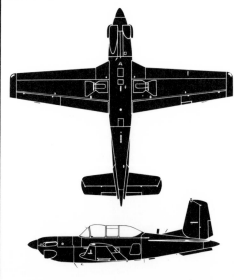

BEECH T-34C-1

Trainer and light attack aircraft.
Dimensions: Span 33ft 4in (10·16m); length 28ft 8½in (8·75m); height 9 ft 11in (3·02m).
Maximum speed: 257mph (414km/h).
Range: 749 miles (1205km).
Users: Algeria, Argentina, Ecuador, Indonesia, Morocco, Peru, USA (Navy).

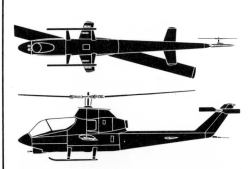

BELL AH1G HUEYCOBRA

Multi-role armed helicopter.
Dimensions: Main-rotor diameter 44ft (13·41 m); length (rotors turning) 52ft 11½in (16·14m); height 13ft 6¼in (4·12m).
Maximum speed: 172mph (277km/h).
Range: 357 miles (574km).
Users: Iran, Saudi Arabia, Spain (Navy), US (Army, Marines).

BELL 212 UH-1N

Multi-role transport helicopter.
Dimensions: Main-rotor diameter 48ft 2¼in (14·69m); length (rotors turning) 57ft 3¼in (17·46m); height 14ft 10½in (4·53m).
Maximum speed: 115mph (185km/h).
Range: Standard fuel, no reserve, 248 miles (400km).
Users: Argentina, Australia, Austria, Brazil, Brunei, Burma, Canada, Chile, Colombia, Dubai, Ethiopia, West Germany, Ghana, Greece, Guatemala, Indonesia, Iran, Israel, Italy, Jamaica, Japan, Kampuchea (status uncertain), South Korea, Kuwait, Lebanon, Malaysia, Mexico, Morocco, Netherlands, New Zealand, Norway, Oman, Panama, Peru, Philippines, Saudi Arabia, Somalia, Spain, Sweden, Taiwan, Thailand, Turkey, Uganda, United Arab Emirates, Uruguay, US (AF, Navy, Marines, Army, Coast Guard), Venezuela, Yemen, Yugoslavia, Zambia.

BOEING B-52H STRATOFORTRESS

Strategic bomber and missile launcher.
Dimensions: Span 185ft (56·4m); length 157ft 7in (48m); height 40ft 8in (12·4m).
Maximum speed: High altitude 630mph (1014km/h).
Range: High altitude 12,500 miles (20 150km).
User: US Air Force.

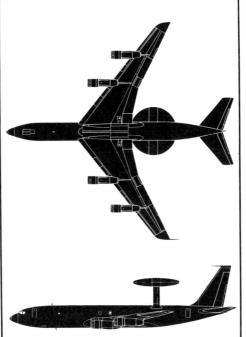

BOEING E-3A SENTRY

AWACS.
Dimensions: Span 145ft 9in (44·42m); length 152ft 11in (46·61m); height 42ft 5in (12·93m).
Maximum speed: Varies from 300mph (482 km/h) at low altitudes to almost twice this at high altitude, but normal on-station speed is about 400mph (645km/h).
Range: Not stated, but on-station endurance about 12 hours without refuelling.
Users: US (AF) and consortium of 11 NATO nations.

BERIEV BE-12 MAIL

Patrol and ASW amphibian.
Dimensions: Span 97ft 6in (29·7m); length 107ft 11¼in (32·9m); height on land 22ft 11½in (7m).
Maximum speed: 379mph (610km/h).
Range: High, with military load, 2,485 miles (4000km).
User: Soviet Union (possibly Egypt or Syria).

BOEING KC-135B STRATOTANKER

Air-refuelling tanker and strategic transport.
Dimensions: Span 130ft 10in (39·7m); length 136ft 3in (41m); height 38ft 4in (11·6m).
Maximum speed: 600mph (966km/h).
Range: Typically 4,000 miles (6437km).
Users: France, US Air Force.

BOEING T-43A

Navigation trainer.
Dimensions: Span 93ft (38·35m); length 100ft (30·48m); height 37ft (11·28m).
Maximum speed: 586mph (943km/h).
Range: With military reserves, 2,995 miles (4820km).
Users: Including 737, Brazil, US (AF), Venezuela.

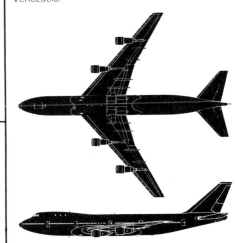

BOEING E-4B

Airborne national command post.
Dimensions: Span 195ft 8in (59·64m); length 231ft 4in (70·5m); height 63ft 5in (19·33m).
Maximum speed: 608mph (978km/h).
Range: Normally 6,500 miles (10 460km).
User: US (AF).

BOEING VERTOL CH-46D SEA KNIGHT

Multi-role utility helicopter.
Dimensions: Main-rotor diameter 50ft (15·24 m); length (rotors turning) 84ft 4in (25·7m); height 16ft 8½in (5·09m).
Maximum speed: 166mph (267km/h).
Range: With 2400lb (1088kg) payload, 633 miles (1020km)
Users: Canada, Japan, Sweden, US Marine Corps, US Navy.

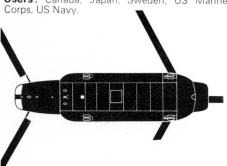

BOEING VERTOL CH-47C CHINOOK

Transport helicopter.
Dimensions: Main-rotor diameter 60ft (18·29 m); length (rotors turning) 99ft (30·2m); height 18ft 8in (5·7m).
Maximum speed: 189mph (304km/h).
Users: Argentina, Australia, Austria, Canada, Iran, Italy, Libya, Morocco, Spain, Thailand, US Army (1980, UK RAF).

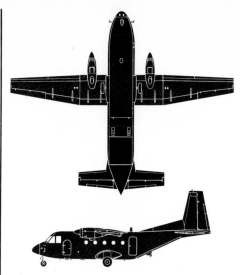

CASA C-212 AVIOCAR 200

Tactical transport.
Dimensions: Span 62ft 4in (19m); length 49ft 10½in (15·2m); height 20ft 8in (6·3m).
Maximum speed: 242mph (389km/h).
Range: With max fuel and reduced payload, 1,093 miles (1760km).
Users: Chile (?), Indonesia, Jordan, Nicaragua, Portugal, Saudi Arabia, Spain, Thailand, Turkey.

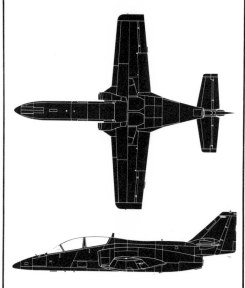

CASA C-101 AVIOJET

Trainer and light attack aircraft.
Dimensions: Span 34ft 9¾in (10·6m); length 40ft 2¼in (12·25m); height 13ft 11¼in (4·25m).
Maximum speed: 478mph (769km/h).
Range: Ferry, max fuel, 2,485 miles (4000km).
User: Spain (on order).

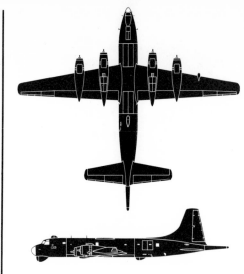

CANADAIR CP-107 ARGUS 2

Maritime patrol and ASW.
Dimensions: Span 142ft 3½in (43·37m); length 128ft 3in (39·1m); height 36ft 9in (11·2m).
Maximum speed: Sea level, 288mph (463 km/h).
Range: 5,900 miles (9495km).
User: Canada.

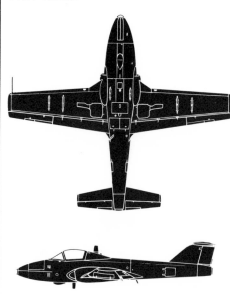

CANADAIR CL-41 TUTOR

Trainer and light attack aircraft.
Dimensions: Span 36ft 6in (11·13m); length 32ft (9·75m); height 9ft 9¾in (2·84m).
Maximum speed: 498mph (801km/h).
Range: 944 miles (1520km).
Users: Canada, Malaysia, Netherlands.

BRITTEN-NORMAN MARITIME DEFENDER

Coastal multi-role aircraft.
Dimensions: Span (extended tips) 53ft (16·15 m); length 36ft 3¾in (11·07m); height 13ft 8¾in (4·18m).
Maximum speed: Sea level, clean, 176mph (283km/h).
Range: Standard internal fuel, clean, 1,260 miles (2027km).
Users: Abu Dhabi, Belgium, Botswana, Brazil, Egypt, Ghana, Guyana, Hong Kong, India, Iraq, Israel, Jamaica, Lesotho, Liberia, Malagasy, Malawi, Mauritania, Mauritius, Mexico, Nigeria, Oman, Panama, Philippines, Qatar, Rhodesia, Rwanda, Thailand, Turkey, UK (Army, bought by club), Venezuela, Zaïre, Zambia.

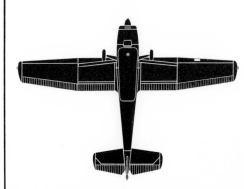

CESSNA U-3 MODEL 310

Light transport.
Dimensions: Span (over tanks) 36ft 11in (11·25m); length (not original U-3A) 31ft 11½in (9·73m); height 10ft 8in (3·25m).
Maximum speed: 238mph (383km/h).
Range: Typical, max fuel, 1,740 miles (2800km).
Users: Tanzania, Thailand, US (AF), Zaïre.

CESSNA O-2/REIMS 337G

Observation and utility aircraft.
Dimensions: Span 38ft 2in (11·63m); length 29ft 9in (9·07m); height 9ft 2in (2·79m).
Maximum speed: 206mph (332km/h).
Range: Maximum 1,325 miles (2132km).
Users: Benin, Ecuador, Gabon, Upper Volta, Iran, Ivory Coast, Malagasy, Mauretania, Niger, Senegal, US Air Force, Venezuela.

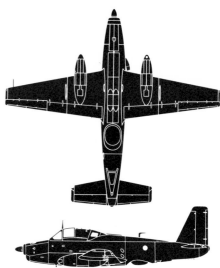

DASSAULT BREGUET
Br.1050 ALIZE

Three-seat carrier-based ASW.
Dimensions: Span 51ft 2in (15·6m); length 45ft 6in (13·86m); height 15ft 7in (4·75m).
Maximum speed: 285mph (460km/h).
Range: With auxiliary fuel, 1,785 miles (2850 km).
Users: France, India.

CESSNA A-37B DRAGONFLY

Light ground attack.
Dimensions: Span (over tanks) 35ft 10½in (10·93m); length (excluding FR probe) 29ft 3in (8·92m); height 8ft 10½in (2·7m).
Maximum speed: Medium altitude, clean, 507mph (816km/h).
Range: Max weapons, 460 miles (740km).
Users: Brazil, Chile, Ethiopia, Guatemala, Honduras, Peru, US (ANG), Vietnam.

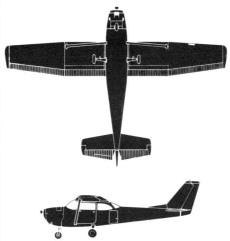

CESSNA 172/T-41 MESCALERO

Trainer and utility aircraft.
Dimensions: Span 35ft 10in (10·92m); length 26ft 11in (8·2m); height 8ft 9½in (2·68m).
Maximum speed: 144mph (232km/h).
Range: 737 miles (1186km).
Users: Argentina, Colombia, Ecuador, Honduras, Pakistan, Peru, Philippines, Saudi Arabia, Singapore, Thailand, Turkey, US (AF/A).

DASSAULT BREGUET MIRAGE 5

Ground-attack fighter.
Dimensions: Span 27ft (8·22m); length 51ft (15·55m); height 13ft 11½in (4·25m).
Maximum speed: High, clean, Mach 2·2, equivalent to 1,450mph (2335km/h).
Radius of action: Hi-lo-hi with 2,000lb (907 kg) load, 805 miles (1300km).
Users: Abu Dhabi, Belgium, Colombia, Egypt, France, Gabon, Libya, Pakistan, Peru, Sudan, Venezuela, Zaïre.

CESSNA 185/U-17A

Utility transport.
Dimensions: Span 35ft 10in (10·92m); length 25ft 7½in (7·81m); height 7ft 9in (2·36m).
Maximum speed: 178mph (286km/h).
Range: (185) without reserve, 660 miles (1062km).
Users: Bolivia, Costa Rica, Greece, Honduras, Indonesia, South Korea, Iran, Jamaica, Laos, Panama, Paraguay, Peru, South Africa, Turkey, Vietnam.

DASSAULT BREGUET/DORNIER
ALPHA JET

Trainer and light attack/reconnaissance aircraft.
Dimensions: Span 29ft 11in (9·12m); length 40ft 3¾in (12·29m); height 13ft 9in (4·2m).
Maximum speed: 576mph (927km/h).
Range: Ferry, with two drop tanks, 1,725 miles (2780km).
Users: Belgium, France, West Germany, Morocco, Nigeria.

DASSAULT BREGUET Br.1150 ATLANTIC

Maritime patrol and ASW.
Dimensions: Span 119ft 1in (36·3m); length 104ft 2in (31·75m); height 37ft 2in (11·33m).
Maximum speed: 409mph (658km/h).
Range: 5,592 miles (9000km).
Users: France, West Germany, Italy, Netherlands, Pakistan.

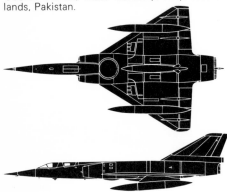

DASSAULT BREGUET MIRAGE IVA

Limited-range strategic bomber.
Dimensions: Span 38ft 10½in (11·85m); length (including FR probe) 77ft 1in (23·5m); height 17ft 8½in (5·4m).
Maximum speed: Dash, clean, at 40,000ft (13 125m), 1,450mph (2335km/h).
Combat radius: Unrefuelled, one bomb, dash to target, high-subsonic return, 770 miles (1240km).
User: France.

DASSAULT BREGUET MIRAGE F1

Interceptor and fighter/bomber.
Dimensions: Span 27ft 6¾in (8·4m); length 49ft 2½in (15m); height 14ft 9in (4·5m).
Maximum speed: High, clean, Mach 2·2, equivalent to 1,450mph (2335km/h).
Radius of action: With 3,520lb load, low profile, 400 miles (644km).
Users: Ecuador, Egypt, France, Greece, Iraq, Kuwait, Libya, Morocco, South Africa, Spain.

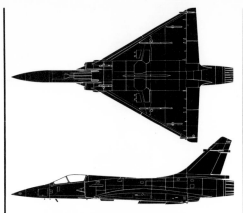

DASSAULT BREGUET MIRAGE 2000

Single-seat interceptor and air-superiority fighter.
Dimensions: Span 29ft 6in (9m); length 50ft 3½in (15·32m); height about 18ft (5·5m).
Maximum speed: High altitude, Mach 2·2, equivalent to 1,450mph (2335km/h).
Radius of action: Conditions not specified, 434 miles (700km).
User: Intended for France (Armée de l'Air).

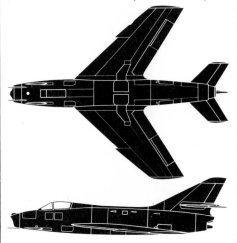

DASSAULT SUPER MYSTERE B2

Single-seat fighter-bomber.
Dimensions: Span 34ft 5¾in (10·5m); length 46ft 1¼in (14m); height 14ft 10¾in (4·53m).
Maximum speed: Clean, sea level, 686mph (1104km/h); high altitude 745mph (1200km/h).
Range: Clean, high altitude, 540 miles (870km).
Users: France, Honduras, Israel.

DASSAULT BREGUET SUPER ETENDARD

Single-seat carrier-based attack.
Dimensions: Span 31ft 5¾in (9·6m); length 46ft 11½in (14·31m); height 12ft 8in (3·85m).
Maximum speed: Clean, sea level, 745mph (1200km/h).
Range: Clean, high altitude, 1,243 miles (2000km).
User: France.

DHC-5D BUFFALO

Tactical transport.
Dimensions: Span 96ft (29·26m); length 79ft (24·08m); height 28ft 8in (8·73m).
Maximum speed: (Also max cruise) 290mph (467km/h).
Range: With max payload of 12,000lb (5443kg) from rough strip, 259 miles (416km).
Users: Abu Dhabi, Brazil, Canada, Ecuador, Kenya, Mauretania, Oman, Peru, Sudan, Tanzania, Togo, United Arab Emirates, Zaïre, Zambia.

DHC-4 CARIBOU

Tactical transport.
Dimensions: Span 95ft 7½in (29·1m); length 72ft 7in (22·1m); height 31ft 9in (9·7m).
Maximum speed: 216mph (347km/h).
Range: With max payload of 8,740lb (3965kg), 242 miles (390km).
Users: Abu Dhabi, Australia, Cameroun, Ghana, India, Kenya, Kuwait, Malaysia, Oman (?), Spain, Tanzania, Uganda, Zaïre, Zambia.

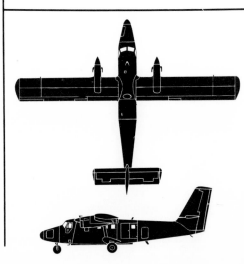

DOUGLAS DC-3/C-47

Utility transport.
Dimensions: Span 95ft (28·96m); length 64ft 5½in (19·64m); height 16ft 11in (5·16m).
Maximum speed: 230mph (370km/h).
Range: Max 2,125 miles (3420km).
Users: Argentina, Australia, Bolivia, Brazil, Burma, Canada, Central African Emp., Chad, Chile, Colombia, Congo, Benin, Denmark, Dominica, Ecuador, Ethiopia, Finland, France, Gabon, East Germany, West Germany, Greece, Guatemala, Haiti, Upper Volta, Honduras, India, Indonesia, Israel, Italy, Ivory Coast, Kampuchea, Laos, Libya, Malagasy, Malawi, Mali, Mauretania, Mexico, Morocco, New Zealand, Nicaragua, Niger, Nigeria, Norway, Oman, Pakistan, Panama, Papua/NG, Paraguay, Peru, Philippines, Portugal, Rhodesia-Zimbabwe, Rwanda, Salvador, Senegal, Somalia, South Africa, Soviet Union (Li-2), Spain, Sri Lanka, Sweden, Syria, Taiwan, Thailand, Togo, Turkey, Uganda, Uruguay, Venezuela, Vietnam, Yemen Arab, Yugoslavia, Zaire, Zambia.

DORNIER DO 28D SKYSERVANT

Utility transport.
Dimensions: Span 51ft 0¼in (15·55m); length 37ft 5¼in (11·41m); height 12ft 9½in (3·9m).
Maximum speed: 202mph (325km/h).
Range: Max fuel, 1,786 miles (2874km).
Users: Cameroun, Ethiopia, West Germany, Israel, Kenya, Malawi, Morocco, Nigeria, Somalia, Thailand, Turkey, Zambia.

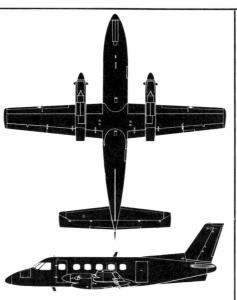

DOUGLAS A-1J SKYRAIDER

Ground attack and utility.
Dimensions: Span 50ft 9in (15·47m); length 38ft 10in (11·84m); height 15ft 8¼in (4·77m).
Maximum speed: Clean, 318mph (512km/h).
Range: Max fuel, 3,000 miles (4828km).
Users: Chad, Central African Emp. (non-operational), Kampuchea (status uncertain), Vietnam.

EMBRAER EMB-110P-2 BANDEIRANTE

Transport.
Dimensions: Span 50ft 2½in (15·3m); length 49ft 6½in (15·1m); height 15ft 6¼in (4·73m).
Maximum speed: 286mph (460km/h).
Range: Max fuel, 45 min reserves, 1,180 miles (1900km).
Users: Brazil, Chile, Uruguay.

FAIRCHILD AU-23A (PILATUS PORTER similar)

Utility.
Dimensions: Span 49ft 8in (15·14m); length 36ft 10in (11·23m); height 12ft 3in (3·73m).
Maximum speed: 174mph (280km/h).
Range: 558 miles (898km).
Users: Angola, Australia, Austria, Bolivia, Chad, Colombia, Ecuador, Israel, Oman, Peru, Sudan, Switzerland, Thailand.

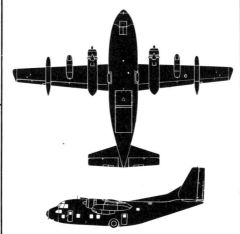

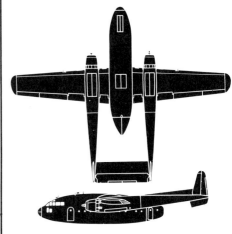

DH CANADA TWIN OTTER

Utility transport.
Dimensions: Span 65ft (19·8m); length (long-nose Series 300) 51ft 9in (15·77m); height 19ft 6in (5·94m).
Maximum cruising speed: 210mph (338 km/h).
Range: Max cruising speed with 2,550lb (1156kg) payload, 892 miles (1435km).
Users: Argentina, Canada, Chile, Ecuador, Ethiopia, Jamaica, Norway, Panama, Paraguay, Peru, USA (Army, AF).

FAIRCHILD C-123K PROVIDER

Tactical transport.
Dimensions: Span 110ft (33·5m); length 76ft 3in (23·9m); height 34ft 1in (10·6m).
Maximum speed: With jets operating 228mph (367km/h).
Range: With max payload 1,035 miles (1666km).
Users: Kampuchea, Saudi Arabia, South Korea, Taiwan, Thailand, Venezuela, Vietnam.

FAIRCHILD C-119G FLYING BOXCAR

Tactical transport.
Dimensions: Span 109ft 3in (33·3m); length 86ft 6in (26·36m); height 26ft 6in (8·07m).
Maximum speed: 243mph (391km/h).
Range: With max payload 990 miles (1595km).
Users: Ethiopia, India, Italy, Kampuchea, Taiwan, Vietnam.

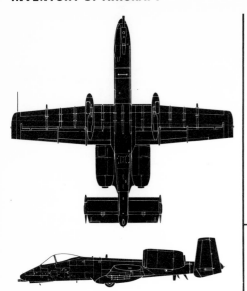

FOKKER-VFW F27 MARITIME

Maritime patrol and multi-role transport.
Dimensions: Span 95ft 2in (29m); length 77ft 3½in (25·56m); height 28ft 7¼in (8·71m).
Typical patrol speed: 168mph (270km/h); cruising speed 265mph (427km/h).
Range: Max, high-altitude cruise, 2,763 miles (4447km).
User: Peru.

FAIRCHILD REPUBLIC A-10A THUNDERBOLT II

Single-seat close-support.
Dimensions: Span 57ft 6in (17·53m); length 53ft 4in (16·26m); height 14ft 8in (4·47m).
Maximum speed: Clean, sea level, 449mph (722km/h).
Range: 2,647 miles (4200km); combat radius (2 hour loiter, all low level) 288 miles (463km).
User: United States.

GD CONVAIR F-102A DELTA DAGGER

All-weather interceptor.
Dimensions: Span 38ft 1½in (11·6m); length 68ft 5in (20·83m); height 21ft 2½in (6·45m).
Maximum speed: 825mph (1328km/h).
Range: 1,350 miles (2172km).
Users: Greece, Turkey.

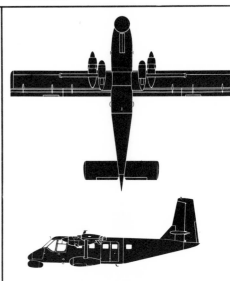

GAF SEARCH MASTER L

Maritime patrol and utility aircraft.
Dimensions: Span 54ft (16·46m); length 41ft 2½in (12·56m); height 18ft 1½in (5·52m).
Cruising speed: At 193mph (311km/h).
Range: 161mph (259km/h) with reserves, 1,300 miles (2092km).
Users: Australia, Indonesia, Papua/New Guinea, Philippines.

FAIRCHILD REPUBLIC F-105D THUNDERCHIEF

All-weather fighter-bomber.
Dimensions: Span 34ft 11¼in (10·65m); length 64ft 3in (19·58m); height 19ft 8in (5·99m).
Maximum speed: Clean, high altitude, 1,480 mph (2382km/h).
Range: Max fuel, high altitude, 2,390 miles (3846km).
User: US (ANG).

GD CONVAIR F-106A DELTA DART

All-weather interceptor.
Dimensions: Span 38ft 3½in (11·67m); length 70ft 8¾in (21·55m); height 20ft 3¼in (6·15m).
Maximum speed: 1,525mph (2455km/h).
Range: 1,700 miles (2735km).
User: US (AF, ANG).

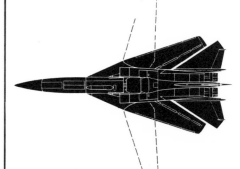

FMA PUCARA

Ground attack and Co-In aircraft.
Dimensions: Span 47ft 6¾in· (14·5m); length 46ft 9in (14·25m); height 17ft 7in (5·36m).
Maximum speed: Clean, 310mph (500km/h).
Range: Max fuel, 1,890 miles (3042km).
User: Argentina (unconfirmed, Bolivia).

GENERAL DYNAMICS F-111E

All-weather interdiction.
Dimensions: Span (16° sweep) 63ft (19·2m). (72·5°) 31ft 11½in (9·74m); length 73ft 6in (22·4m); height 17ft 1½in (5·22m).
Maximum speed: High, clean, 1,450mph (2335km/h).
Range: Max internal fuel, no ordnance, 2,925 miles (4707km).
Users: Australia, US (AF).

GENERAL DYNAMICS F-16A

Multi-role fighter.
Dimensions: Span (with AAMs) 32ft 10in (10·01m); length 47ft 7¾in (14·52m); height 16ft 5¼in (5·01m).
Maximum speed: High, with Sidewinders only, 1,300mph (2090km/h, Mach 1·95).
Range: Max ferry 2,418 miles (3892km); combat radius (CAP mission, AAMs only) 575 miles (927km).
Users: Belgium, Denmark, Israel, Netherlands, Norway, US (AF).

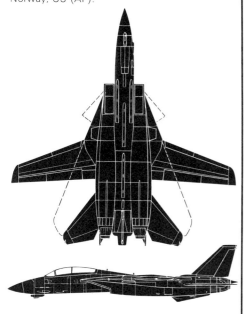

GRUMMAN F-14A TOMCAT

Multi-role carrier-based fighter.
Dimensions: Span (68° sweep) 38ft 2in (11·63m), (20°) 64ft 1½in (19·54m); length 62ft (18·89m); height 16ft (4·88m).
Maximum speed: 1,564mph (2517km/h).
Range: Not disclosed.
Users: Iran (not operational), US Navy, Marines.

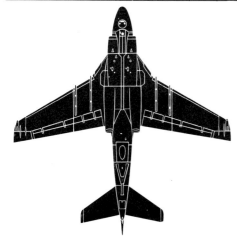

GRUMMAN A-6E INTRUDER

Carrier-based all-weather attack.
Dimensions: Span 53ft (16·15m); length 54ft 9in (16·69m); height 16ft 2in (4·93m).
Maximum speed: Low, clean, 648mph (1043 km/h).
Range: Max internal and external fuel, 2,723 miles (4382km).
User: US (Navy, Marines).

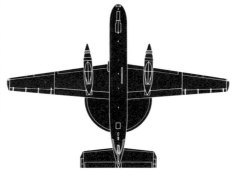

GRUMMAN E-2C HAWKEYE

Carrier-based AEW platform.
Dimensions: Span 80ft 7in (24·56m); length 57ft 7in (17·55m); height 18ft 4in (5·59m).
Maximum speed: 374mph (602km/h).
Range: About 1,700 miles (2736km).
Users: Israel, Japan, US (Navy).

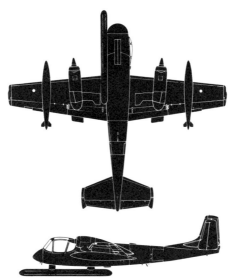

GRUMMAN OV-1D MOHAWK

Multi-sensor battlefield reconnaissance.
Dimensions: Span 48ft (14·63m); length (with SLAR) 44ft 11in (13·69m); height 12ft 8in (3·86m).
Maximum speed: 298mph (480km/h).
Range: 1,011 miles (1627km).
Users: Israel, US Army.

GRUMMAN S-2D TRACKER

ASW patrol aircraft.
Dimensions: Span 72ft 7in (22·13m); length 43ft 6in (13·26m); height 16ft 7in (5·06m).
Maximum speed: 267mph (430km/h).
Range: Typical mission with weapons, 1,300 miles (2095km).
Users: Argentina, Australia, Brazil, Canada, Italy, Japan, Peru, Taiwan, Thailand, Turkey, Uruguay, Venezuela.

HAL AJEET

Single-seat fighter and ground attack aircraft.
Dimensions: Span 22ft 1in (6·73m); length 29ft 8in (9·04m); height 8ft 10in (2·69m).
Maximum speed: 714mph (1150km/h) at altitude, about the same (clean) at sea level.
Range: With maximum fuel 1,000 miles (1610 km); tactical radius at low level with two 500lb (227kg) bombs, 127 miles (204km).
User: India.

HAL HJT-16 KIRAN

Training and light attack aircraft.
Dimensions: Span 35ft 1¼in (10·7m); length 34ft 9in (10·6m); height 11ft 11in (3·64m).
Maximum speed: 432mph (695km/h).
Range: 464 miles (747km).
User: India.

HINDUSTAN HF-24 MARUT

Ground-attack fighter.
Dimensions: Span 29ft 6¼in (9m); length 52ft 0¾in (15·87m); height 11ft 9¾in (3·6m).
Maximum speed: High, clean, 675mph (1086 km/h).
Range: 898 miles (1446km).
User: India.

HUGHES OH-6A CAYUSE

Observation and multi-role helicopter.
Dimensions: Main-rotor diameter 26ft 4in (8·03m); length (rotors turning) 30ft 3¾in (9·24m); height 8ft 1½in (2·48m).
Maximum speed: 152mph (244km/h).
Range: 380 miles (611km).
Users: Argentina, Bolivia, Colombia, Congo, Denmark, Dominica, Italy, Japan, Pakistan, Philippines, Sierra Leone, Spain, Taiwan, United States (Army).

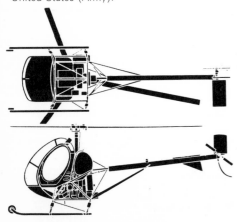

HUGHES TH-55A OSAGE

Training helicopter.
Dimensions: Main-rotor diameter 25ft 3½in (7·7m); length (rotors turning) 28ft 10¾in (8·8m); height 8ft 2¾in (2·5m).
Maximum cruising speed: 75mph (121km/h).
Range: (Without reserve) 204 miles (328km).
Users: Brazil, Guyana, Italy, Japan, Nicaragua, Sierra Leone, US (Army).

HUGHES AH-64A

Attack helicopter.
Dimensions: Main-rotor diameter 48ft (14·63 m); length (rotors turning) 57ft 9in (17·6m); height 12ft 7in (3·83m).
Maximum speed: 192mph (309km/h).
Range: Max, internal fuel, 380 miles (611km).
User: USA (Army).

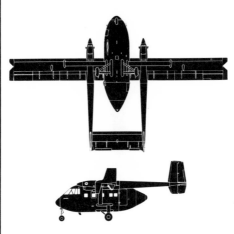

IAI-201 ARAVA

Utility transport.
Dimensions: Span 68ft 9in (20·96m); length 42ft 9in (13·03m); height 17ft 1in (5·21m).
Maximum speed: 203mph (326km/h).
Range: With max fuel, 812 miles (1306km).
Users: Bolivia, Ecuador, Guatemala, Honduras, Israel, Mexico, Nicaragua, Salvador.

IAI KFIR C2

Ground-attack fighter.
Dimensions: Span 26ft 11½in (8·2m); length 53ft 7¾in (16·35m); height 13ft 11¼in (4·25m).
Maximum speed: Over 1,450mph (2335km/h).
Combat radius: Ground attack, hi-lo-hi, 807 miles (1300km).
Users: Israel, possibly others.

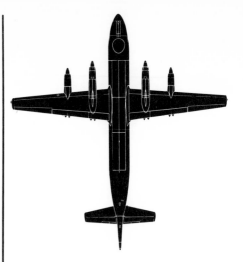

ILYUSHIN IL-28 BEAGLE

Bomber, trainer and utility aircraft.
Dimensions: Span 70ft 4¾in (21·45m); length 57ft 10¾in (17·65m); height 22ft (6·7m).
Maximum speed: 559mph (900km/h).
Range: With bomb load, 684 miles (1100km).
Users: Afghanistan, Algeria, Bulgaria, China, Cuba (not operational), Czechoslovakia, Egypt, Finland, East Germany, Indonesia, Iraq, North Korea, Nigeria, Poland, Romania, Somalia, Soviet Union, South Yemen, Syria, Vietnam, Yemen.

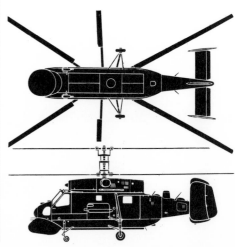

ILYUSHIN IL-38 MAY

Maritime reconnaissance and ASW aircraft.
Dimensions: Span 122ft 8½in (37·4m); length 129ft 10in (39·6m); height 33ft 4in (10.17m).
Maximum speed: About 450mph (724km/h).
Range: Estimated with full mission payload, 4,500 miles (7240km).
Users: Egypt (?), India, Soviet Union.

KAMAN SH-2F SEASPRITE

Shipboard multi-purpose helicopter.
Dimensions: Main-rotor diameter 44ft (13·41 m); length (rotors turning) 52ft 7in (16·03m); height 15ft 6in (4·72m).
Maximum speed: 165mph (265km/h).
Range: Max fuel, 422 miles (679km).
User: US (Navy).

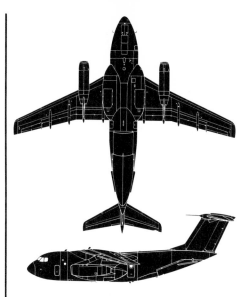

KAWASAKI C-1

Tactical transport.
Dimensions: Span 100ft 4¾in (30·6m); length 95ft 1¾in (29m); height 32ft 9¼in (9·99m).
Maximum speed: 501mph (806km/h).
Range: With maximum (17,416lb, 7900kg) payload, 807 miles (1300km).
User: Japan (ASDF).

ILYUSHIN IL-76T CANDID

Heavy tactical freighter.
Dimensions: Span 165ft 8in (50·5m); length 152ft 10½in (46·59m); height 48ft 5in (14·76m).
Maximum speed: About 560mph (900km/h).
Range: With max (88,185lb, 40 000kg) payload, 3,100 miles (5000km).
Users: Iraq, Soviet Union.

KAWASAKI P-2J

Maritime reconnaissance and ASW aircraft.
Dimensions: Span 101ft 3½in (30·87m), length 95ft 10¾in (29.32m); height 29ft 3½in (8·93m).
Maximum speed: 403mph (649km/h).
Range: Max fuel, 2,765 miles (4450km).
User: Japan (MSDF).

LOCKHEED C-5A GALAXY

Heavy strategic freighter.
Dimensions: Span 222ft 8½in (67·88m); length 247ft 10in (75·54m); height 65ft 1½in (19·85m).
Maximum speed: 571mph (919km/h).
Range: With maximum (220,967lb, 100 228kg) payload, 3,749 miles (6033km).
User: US (AF).

KAMOV KA-25 HORMONE

Shipboard helicopter for either ASW or search/rescue.
Dimensions: Main-rotor diameter 51ft 8in (15·74m); length (rotors turning) same as diameter above, fuselage about 32ft (9·75m); height 17ft 8in (5·4m).
Maximum speed: 137mph (220km/h).
Range: Max fuel, with reserves, 405 miles (650km).
Users: Soviet Union (AV-MF), Syria.

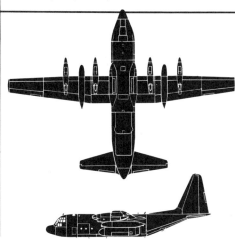

LOCKHEED C-130K HERCULES

Tactical transport.
Dimensions: Span 132ft 7in (40·41m); length 97ft 9in (29·78m) (RAF stretched 112ft 9in/34·37m); height 38ft 3in (11·66m).
Maximum speed: 384mph (618km/h).
Range: With maximum (43,811lb/19 872kg) payload, 2,487 miles (4002km).
Users: Abu Dhabi, Angola, Argentina, Australia, Belgium, Bolivia, Brazil, Cameroun, Canada, Chile, Colombia, Denmark, Ecuador, Egypt, Gabon, Greece, Indonesia, Iran, Israel, Italy, Jordan, Kuwait, Libya, Malaysia, Morocco, New Zealand, Nigeria, Norway, Pakistan, Peru, Philippines, Portugal, Saudi Arabia, Singapore, South Africa, Spain, Sudan, Sweden, Turkey, Uganda, UK (RAF), US (AF, Navy, Marines), Venezuela, Vietnam, Zaïre.

LOCKHEED SP-2H NEPTUNE

Maritime patrol and ASW.
Dimensions: Span 103ft 10in (31·65m); length 91ft 8in (27·94m); height 29ft 4in (8·94m).
Maximum speed: With jets operating 403mph (648km/h).
Range: 2,500 miles (4000km).
Users: Argentina, Australia, Brazil, Chile, France, Japan, Netherlands, Portugal, US (Naval Reserve).

LOCKHEED C-141B STARLIFTER

Heavy freight transport.
Dimensions: Span 159ft 11in (48·74m); length 168ft 4in (51·3m); height 39ft 3in (11·96m).
Maximum speed: 571mph (919km/h).
Range: With maximum (70,847lb/32 136kg) payload, 4,080 miles (6565km).
User: US (AF).

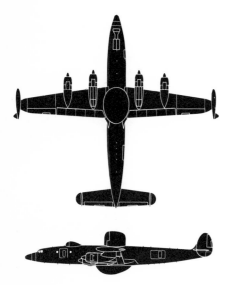

LOCKHEED S-3A VIKING

Carrier-based ASW aircraft.
Dimensions: Span 68ft 8in (20·93m); length 53ft 4in (16·26m); height 22ft 9in (6·93m).
Maximum speed: 506mph (814km/h).
Combat radius: Two drop tanks, no time on station, 1,540 miles (2480km).
User: US (Navy).

LOCKHEED EC-121H WARNING STAR

Electronic reconnaissance, early warning and communications.
Dimensions: Span 126ft 2in (38·45m); length 116ft 2in (35·41m); height 24ft 8in (7·52m).
Maximum speed: 321mph (517km/h).
Range: 4,600 miles (7400km).
User US (Air Force) (Navy and Indian Navy basically similar).

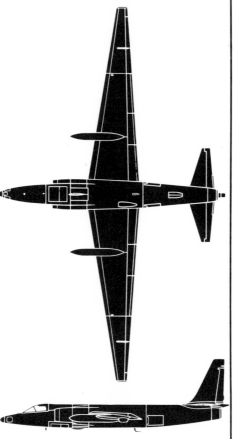

LOCKHEED U-2D

EW and special reconnaissance aircraft.
Dimensions: Span 80ft (24·38m); length 49ft 7in (15·1m); height 13ft (3·96m).
Maximum speed: High, clean, 528mph (850km/h).
Range: High profile, typical, 4,000 miles (6437km).
User: US (AF, NASA).

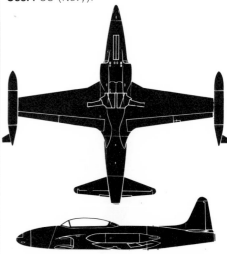

LOCKHEED T-33A

Advanced trainer.
Dimensions: Span (no tip tanks) 38ft 10½in (11·85m); length 37ft 9in (11·48m); height 11ft 8in (3·55m).
Maximum speed: 590mph (950km/h).
Range: 1,345 miles (2165km).
Users: Belgium, Bolivia, Brazil, Burma, Canada, Chile, Colombia, Denmark, Ecuador, Ethiopia, France, Greece, Guatemala, Honduras, Indonesia, Iran, Japan, South Korea, Libya, Mexico, Netherlands, Nicaragua, Pakistan, Peru, Philippines, Portugal, Spain, Taiwan, Thailand, Turkey, Uruguay, Yugoslavia.

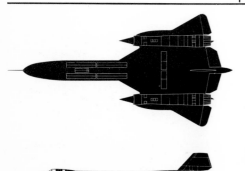

LOCKHEED SR-71A

Strategic reconnaissance aircraft.
Dimensions: Span 55ft 7in (16·95m); length 107ft 5in (32·74m); height 18ft 6in (5·64m).
Maximum speed: High, 2,200mph (3540 km/h).
Range: At Mach 3 at high cruise, 2,982 miles (4800km).
User: US (AF).

LOCKHEED (AERITALIA) F-104S

All-weather interceptor
Dimensions: Span (no tip tanks or missiles) 21ft 11in (6·68m); length 54ft 9in (16·69m); height 13ft 6in (4·11m).
Maximum speed: Clean, high altitude, 1,450 mph (2335km/h).
Range: With max fuel, high altitude, 1,815 miles (2920km); radius with max weapons, 300 miles (483km).
Users: (All versions) Belgium, Canada, Denmark, West Germany, Greece, Italy, Japan, Jordan, Netherlands, Norway, Pakistan, Spain, Taiwan, Turkey, US (ANG).

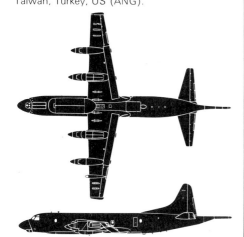

LOCKHEED P-3C ORION

Maritime reconnaissance and ASW aircraft.
Dimensions: Span 99ft 8in (30·4m); length 116ft 10in (35·6m); height 33ft 8½in (10·29m).
Maximum speed: 473mph (761km/h).
Range: Max fuel, high, 4,800 miles (7725km).
Users: Australia, Canada, Iran, Japan, Netherlands, New Zealand, Norway, Spain, US (Navy).

McDONNELL DOUGLAS C-9A NIGHTINGALE

Aeromedical airlift transport.
Dimensions: Span 93ft 5in (28·47m); length 119ft 3½in (36·37m); height 27ft 6in (8·38m).
Maximum cruising speed: 564mph (907 km/h).
Range: 1,923 miles (3095km).
User: US Air Force (Marines and Navy versions similar size).

McDONNELL DOUGLAS A-4M SKYHAWK

Carrier-equipped attack bomber.
Dimensions: Span 27ft 6in (8·38m); length (excluding FR probe) 40ft 4in (12·29m); height 15ft (4·57m).
Maximum speed: Low, with 4,000lb (1814kg) bomb load, 646mph (1040km/h).
Combat radius: With max external fuel, high, 691 miles (1112km).
Users: Argentina, Australia, Israel, Kuwait, New Zealand, Singapore, US (Navy, Marines, formerly AF).

McDONNELL DOUGLAS F-15A EAGLE

All-weather interceptor and air-superiority fighter.
Dimensions: Span 42ft 9¾in (13·05m); length 63ft 9in (19·43m); height 18ft 5½in (5·63m).
Maximum speed: High, clean, over 1,650mph (2655km/h).
Ferry range: With FAST pack, over 3,450 miles (5560km).
Users: Israel, Japan, Saudi Arabia, US (AF).

McDONNELL DOUGLAS F-18A HORNET

Multi-role carrier-based fighter.
Dimensions: Span (with AAMs) 40ft 8in (12·41m); length 56ft (17·07m); height 15ft 3½in (4·66m).
Maximum speed: High, clean, 1,190mph (1912km/h).
Combat radius: Internal fuel over 460 miles (740km).
User: US (Navy, Marines).

McDONNELL DOUGLAS F-4E PHANTOM

Multi-role fighter.
Dimensions: Span 38ft 7½in (11·77m); length 63ft (19·2m); height 16ft 5½in (5·02m).
Maximum speed: With AAMs only, high, 1,450mph (2335km/h).
Combat radius: Interdiction mission, hi-lo-hi, 712 miles (1145km).
Users: West Germany, Greece, Iran (status uncertain), Israel, Japan, South Korea, Spain, Turkey, UK (RAF), US (AF, Navy, Marines).

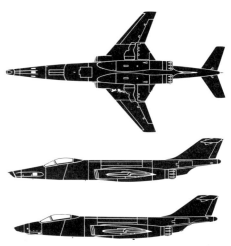

McDONNELL DOUGLAS RF-101G VOODOO (plan, RF-101C)

Single-seat reconnaissance aircraft.
Dimensions: Span 39ft 8in (12·09m); length 67ft 6in (20·57m); height 18ft (5·49m).
Maximum speed: At high altitude, 1,200mph (1930km/h, Mach 1·84); at sea level, subsonic.
Range: At altitude with two drop tanks, 1,700 miles (2736km).
User: United States (ANG).

McDONNELL DOUGLAS/BAe AV-8B

V/STOL ground-attack fighter.
Dimensions: Span 30ft 3½in (9·23m); length 45ft 6in (13·87m); height 11ft 3½in (3·44m).
Maximum speed: Clean, most heights, Mach 0·95, equivalent to 722mph at sea level (1162 km/h).
Combat radius: Low, seven Mk 82 bombs and drop tanks, 748 miles (1204km).
User: US (Marines, possibly Navy, but not funded 1979).

McDONNELL DOUGLAS EKA-3B SKYWARRIOR

Air refuelling tanker and ECM platform.
Dimensions: Span 72ft 6in (22·1m); length 76ft 4in (23·3m); height 23ft 6in (7·16m).
Maximum speed: 610mph (982km/h).
Range: Typical 2,000 miles (3220km).
User: US Naval Reserve.

MIG-15UTI MONGOL

Trainer.
Dimensions: Span 33ft 0¾in (10·08m); length 31ft 11¼in (10·04m); height 12ft 1½in (3·7m).
Maximum speed: 630mph (1015km/h).
Range: High, with slipper tanks, 885 miles (1424km).
Users: Afghanistan, Albania, Algeria, Angola, Bulgaria, China, Cuba, Czechoslovakia, Egypt, Finland, East Germany, Guinea, Hungary, Iraq, North Korea, Mali, Mongolia, Morocco, Nigeria, Poland, Romania, Somalia, Soviet Union, South Yemen, Sri Lanka, Syria, Tanzania, Uganda, Vietnam.

MIG-19SF (SHENYANG F-6)

Day fighter.
Dimensions: Span 29ft 6½in (9m); length (excluding pitot boom) 41ft 4in (12·6m); height 13ft 2¼in (4·02m).
Maximum speed: High, clean, 902mph (1452 km/h).
Combat radius: With two drop tanks, 426 miles (685km).
Users: Afghanistan, Albania, Bulgaria, China, Cuba, Czechoslovakia, East Germany (not in front-line service), Hungary, Iraq, North Korea, Pakistan, Poland, Romania, Soviet Union, Tanzania, Uganda (status uncertain), Vietnam.

MBB BO 105

Multi-role helicopter.
Dimensions: Main-rotor diameter 32ft 2½in (9·82m); length (rotors turning) 38ft 11in (11·86m); height 9ft 9½in (2·98m).
Maximum speed: 167mph (270km/h).
Range: Normal fuel, no reserve, 408 miles (656km).
Users: West Germany, Indonesia, Netherlands, Nigeria, Philippines.

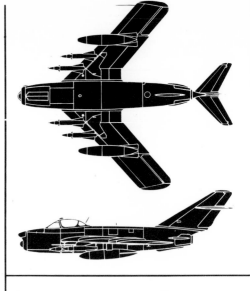

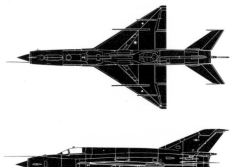

MIG-21bis FISHBED-L, -N

Ground-attack fighter.
Dimensions: Span 23ft 5½in (7·15m); length (excluding pitot boom) 49ft (14·9m); height 14ft 9in (4·5m).
Maximum speed: High, clean, 1,385mph (2230km/h).
Range: Internal fuel, high, about 700 miles (1127km).
Users: Afghanistan, Albania, Algeria, Angola, Bangladesh, Bulgaria, China, Cuba, Czechoslovakia, East Germany, Egypt, Ethiopia, Finland, Hungary, India, Iraq, Laos, Mozambique, Nigeria, North Korea, Poland, Romania, Somalia (status uncertain), Soviet Union, Sudan, Syria, Tanzania, Uganda (status uncertain), Vietnam, Yemen, Yugoslavia.

MIG-25 FOXBAT-A

All-weather interceptor.
Dimensions: Span 45ft 9in (13·95m); length 73ft 2in (22·3m); height 18ft 6in (5·63m).
Maximum speed: Never-exceed limit with AAMs, Mach 2·8 (1,848mph/2975km/h).
Combat radius: Internal fuel, 700 miles (1130km).
Users: Algeria (?), Libya, Soviet Union, plus others unknown in mid-1979.

MIG-17F FRESCO (SHENYANG F-4)

Single-seat all-weather fighter.
Dimensions: Span 31ft (9·45m); length 36ft 3in (11·05m); height 11ft (3·35m).
Maximum speed: Clean, medium altitude, 711mph (1145km/h).
Range: High altitude, two drop tanks, 913 miles (1470km).
Users: (all MiG-17 versions) Afghanistan, Albania, Algeria, Angola, Bulgaria, China, Cuba, Czechoslovakia, Egypt, E. Germany, Guinea, Hungary, Indonesia (stored), Iraq, Kampuchea, N. Korea, Mali, Morocco (stored), Nigeria, Poland, Romania, Somalia, S. Yemen, Soviet Union, Sri Lanka, Sudan, Syria, Tanzania, Uganda, Vietnam, Yemen Arab.

MIG-23S FLOGGER-B

Fighter.
Dimensions: Span (16° sweep) 46ft 9in (14·25m), (72°) 26ft 9½in (8·17m); length 55ft 1½in (16·8m); height 14ft 4in (4·37m).
Maximum speed: High, AAMs only, Mach 2·3 (1,520mph/2443km/h).
Combat radius: Intercept mission 575 miles (925km).
Users: Algeria (?), Bulgaria, Cuba, Czechoslovakia, Egypt (Flogger D/E/F), Ethiopia, Iraq, Libya, Syria, Soviet Union.

MIG-27 FLOGGER-D

Close-support, attack and reconnaissance.
Dimensions: Span (max) 46ft 9in (14·25m), (min) 26ft 9½in (8·17m); length 55ft 1½in (16·8m); height about 15ft (4·6m).
Maximum speed: Sea level 722mph (1162 km/h); high altitude 1,155mph (1860km/h).
Range: Max, three drop tanks, 1,550 miles (2500km).
Users: Soviet Union, Libya, Egypt.

MIL MI-4 HOUND

Transport or ASW helicopter.
Dimensions: Main-rotor diameter 68ft 11in (21·0m); length (fuselage only) 55ft 1in (16·8m); height 17ft (5·18m).
Maximum speed: 130mph (210km/h).
Range: Full payload, 155 miles (250km).
Users: Afghanistan, Albania, Algeria, Bulgaria, China, Cuba, Czechoslovakia, Egypt, Finland, East Germany, Hungary, India, Indonesia, Iraq, Kampuchea (status uncertain), North Korea, Mali, Mongolia, Poland, Romania, Somalia, Soviet Union, Syria, Vietnam, Yemen, Yugoslavia.

MIL MI-6 HOOK

Heavy transport helicopter.
Dimensions: Main-rotor diameter 114ft 10in (35m); length (rotors turning) 136ft 11½in (41·74m); height 32ft 4in (9·86m).
Maximum speed: 186mph (300km/h).
Range: With 13,228lb (6000kg) payload, 404 miles (650km).
Users: Bulgaria, Egypt, Iraq, Soviet Union, Syria, Vietnam, Zambia.

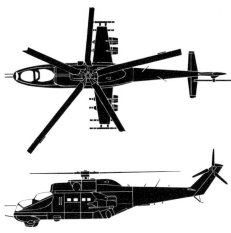

MIL MI-28 HIND-D

Gunship helicopter.
Dimensions: Main-rotor diameter about 55ft 9in (17m); length (ignoring rotors) 55ft 9in (17m); height 14ft (4·25m).
Maximum speed: About 170mph (275km/h).
Range: Probably about 300 miles (483km).
Users: Soviet Union, probably soon other WP and foreign countries.

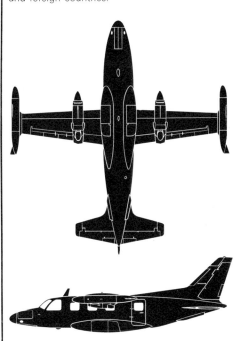

MITSUBISHI MU-2J

Multi-role transport.
Dimensions: Span 39ft 2in (11·94m); length 39ft 5in (12·01m); height 13ft 8in (4·17m).
Maximum cruising speed: 340mph (550 km/h).
Range: Max fuel, reserves, 1,450 miles (2330 km).
User: Japan.

MIL MI-8 HIP

Transport or ASW helicopter.
Dimensions: Main-rotor diameter 69ft 10½in (21·29m); length (rotors turning) 82ft 9¾in (25·24m); height 18ft 6½in (5·65m).
Maximum speed: 161mph (260km/h).
Range: Cargo, max weight, 285 miles (460km).
Users: Afghanistan, Bangladesh, Czechoslovakia, Egypt, Ethiopia, Finland, East Germany, Hungary, India, Iraq, North Korea, Libya, Pakistan, Peru, Poland, Romania, Somalia, Soviet Union, South Yemen, Sudan, Syria, Vietnam, Yugoslavia.

MITSUBISHI F-1

Close-support fighter.
Dimensions: Span 25ft 10½in (7·88m); length (including pitot probe) 58ft 6¼in (17·84m); height 14ft 4¼in (4·38m).
Maximum speed: High, clean, Mach 1·6, equivalent to 1,056mph (1700km/h).
Range: With eight 500lb (227kg) bombs, 700 miles (1126km).
User: Japan (ASDF).

NORTHROP T-38A TALON

All-through trainer.
Dimensions: Span 25ft 3in (7·7m); length 46ft 4½in (14·13m); height 12ft 10½in (3·92m).
Maximum speed: 812mph (1305km/h).
Range: 1,093 miles (1760km).
Users: West Germany (in USA), US (AF/Navy/NASA).

PANAVIA TORNADO IDS

All-weather multi-role aircraft
Dimensions: Span (25° sweep) 45ft 7¼in (13·9m), (68°) 28ft 2½in (8·6m); length 54ft 9½in (16·7m); height 18ft 8½in (5·7m).
Maximum speed: High, clean, considerably greater than Mach 2 (1,320mph/2125km/h).
Combat radius: Not disclosed, but considerably greater than other aircraft in this category.
Users: West Germany (Luftwaffe. Marineflieger), Italy, UK (RAF).

MYASISHCHEV M-4 BISON-C

Bomber, reconnaissance and EW aircraft.
Dimensions: Span 170ft (51·82m); length (including FR probe) 162ft (49·38m); height 45ft 3in (13·8m).
Maximum speed: Estimated at optimum medium height, 659mph (1060km/h).
Range (estimated): With 9,920lb (4500kg) of weapons or electronics, 6,835 miles (11 000km).
User: Soviet Union.

NORTHROP F-5E TIGER II

Light ground-attack fighter.
Dimensions: Span (with AAMs) 28ft (8·53m); length 48ft 2in (14·68m); height 13ft 4in (14·06m).
Maximum speed: High, clean, Mach 1·64 (1,082mph), 1742km/h).
Combat radius: With 5,200lb (2358kg) weapons, max fuel, lo-lo-lo mission, 138 miles (222km).
Users: Brazil, Canada, Chile, Egypt, Ethiopia, Greece, Indonesia, Iran, Jordan, Kenya, South Korea, Libya, Malaysia, Morocco, Netherlands, Norway, Parkistan, Philippines, Saudi Arabia, Singapore, South Korea, Spain, Sudan, Switzerland, Taiwan, Thailand, Turkey, US (AF, Navy, NASA), Vietnam.

PZL-SWIDNIK (MIL) MI-2 HOPLITE

Utility helicopter.
Dimensions: Main-rotor diameter 47ft 6¾in (14·5m); length (rotors turning) 57ft 2in (17·42m); height 12ft 3½in (3·75m).
Maximum speed: 130mph (210km/h).
Range: Max payload plus reserves 105 miles (170km).
Users: Include Bulgaria, Czechoslovakia, Hungary, Poland, Romania, Soviet Union.

NORTH AMERICAN F-86F SABRE

Single-seat fighter bomber.
Dimensions: Span (F-86F-40, as shown) 39ft 1½in (11·9m); length 37ft 6in (11·43m); height 14ft 8¾in (4·47m).
Maximum speed: Clean, 678mph (1091km/h).
Range: High altitude, two tanks, 850 miles (1368km).
Users: Argentina, Bangladesh, Bolivia, Burma, Ethiopia, Indonesia, Japan, Malaysia, Pakistan, Peru, Philippines, Portugal, Saudi Arabia, S. Africa, S. Korea, Taiwan, Thailand, Tunisia, Venezuela, Yugoslavia.

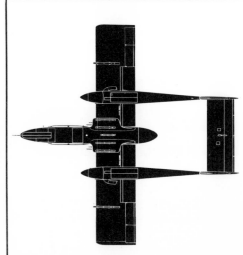

ROCKWELL INTERNATIONAL T-2C BUCKEYE

Trainer.
Dimensions: Span (over tanks) 38ft 1½in (11·62m); length 38ft 3½in (11·67m); height 14ft 9½in (4·51m).
Maximum speed: 522mph (840km/h).
Range: 1,047 miles (1685km).
Users: Greece, Morocco, US (Navy), Venezuela.

ROCKWELL INTERNATIONAL RA-5C VIGILANTE

Carrier-based reconnaissance and EW aircraft.
Dimensions: Span 53ft (16·15m); length 75ft 10in (23·11m); height 19ft 5in (5·92m).
Maximum speed: High, clean, 1,385mph (2230km/h).
Range: With four drop tanks, 3,200 miles (5150km).
User: US (Navy).

ROCKWELL INTERNATIONAL T-28D/FENNEC

Trainer and light attack/utility.
Dimensions: Span 40ft (12·19m); length 33ft 8in (10·26m); height 12ft 8in (3·86m).
Maximum speed: 380mph (611km/h).
 sange: 1,000 miles (1610km).
Users: Argentina, Bolivia, Brazil, Dominica, Echador, Ethiopia, Haiti, South Korea, Laos, Mexico, Morocco, Nicaragua, Philippines, Taiwan, Thailand, Vietnam, Zaire.

ROCKWELL INTERNATIONAL T-39 SABRELINER

Pilot-proficiency and combat-support aircraft.
Dimensions: Span 44ft 5in (13·53m); length 43ft 9in (13·33m); height 16ft (4·88m).
Maximum speed: 563mph (906km/h).
Range: 2,000 miles (3219km).
User: US Air Force (Argentina and US Navy similar).

SAAB JA37 VIGGEN

All-weather interceptor.
Dimensions: Span 34ft 9½in (10·6m); length (including probe, pitot boom) 52ft 9¾in (16·4m); height 19ft 4¼in (5·9m).
Maximum speed: High, clean, over Mach 2 (1,320mph/2125km/h).
Combat radius: High, external weapons, over 620 miles (1000km).
User: Sweden.

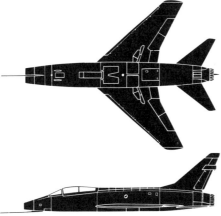

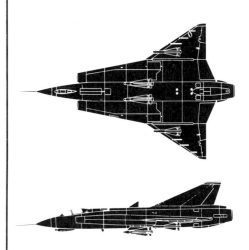

ROCKWELL INTERNATIONAL OV-10A BRONCO

Multi-role Co-In aircraft.
Dimensions: Span 40ft (12·19m); length 41ft 7in (12·67m); height 15ft 2in (4·62m).
Maximum speed: 281mph (452km/h).
Range: With external weapons, 600 miles (966km).
Users: West Germany, Indonesia, South Korea, Thailand, US (AF, Marines), Venezuela.

ROCKWELL INTERNATIONAL F-100F SUPER SABRE

Multi-role fighter/attack/trainer.
Dimensions: Span 38ft 9½in (11·81m); length (excluding boom) 52ft 6in (16m); height 16ft 2¾in (4·96m).
Maximum speed: Clean, high altitude, 864 mph (1390km/h).
Range: High altitude, two drop tanks, 1,500 miles (2414km).
Users: Taiwan, Turkey.

SAAB J35F DRAKEN

All-weather interceptor.
Dimensions: Span 30ft 10in (9·4m); length 50ft 4in (15·4m); height 12ft 9in (3·9m).
Maximum speed: High, clean, 1,320mph (2125km/h).
Range: High, max fuel, 2,020 miles (3250km).
Users: Denmark, Finland, Sweden.

SAAB 105 SK 60A

Trainer and light tactical aircraft.
Dimensions: Span 31ft 2in (9·5m); length 34ft 5in (10·5m); height 8ft 10in (2·7m).
Maximum speed: 447mph (720km/h).
Range: 1,106 miles (1780km).
Users: Austria, Sweden.

SAAB SUPPORTER

Trainer and light tactical aircraft.
Dimensions: Span 29ft 0½in (8·85m); length 22ft 11½in (7m); height 8ft 6½in (2·6m).
Maximum speed: 146mph (236km/h).
Range: 672 miles (1080km).
Users: Denmark, Pakistan.

SEPECAT JAGUAR GR.1

Close-support and reconnaissance aircraft.
Dimensions: Span 28ft 6in (8·69m); length (including pitot probe) 55ft 2½in (16·83m); height 16ft 0½in (4·89m).
Maximum speed: Original low-thrust engines, high, 990mph (1593km/h).
Combat radius: With weapons and external fuel, hi-lo-hi, 818 miles (1315km).
Users: Ecuador, France, India, Oman, UK (RAF).

SHENYANG F-6 BIS FANTAN

Ground attack fighter.
Dimensions: Span 33ft 5in (10·2m); length 50ft (15·24m); height 13ft 2in (4m).
Maximum speed: about 1000mph (1610km/h).
Range: High altitude on internal fuel, about 800 miles (1290km).
User: People's Republic of China.

SHORTS SKYVAN 3M

Utility transport.
Dimensions: Span 64ft 11in (19·79m); length (with radar) 41ft 4in (12·6m); height 15ft 1in (4·6m).
Maximum speed: 250mph (402km/h)
Range: 670 miles (1075km).
Users: Argentina, Austria, Botswana, Ecuador, Ghana, Indonesia, Lesotho, Mauretania, Nepal, Oman, Panama, Singapore, Thailand, Yemen Arab.

SHIN MEIWA US-1

Air-sea rescue amphibian.
Dimensions: Span 108ft 8¾in (33·14m); length 109ft 11in (33·5m); height 32ft 2¾in (9·82m).
Maximum speed: 340mph (547km/h).
Range: With military load, 1,347 miles (2168 km).
User: Japan (MSDF).

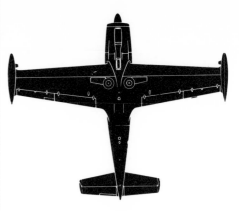

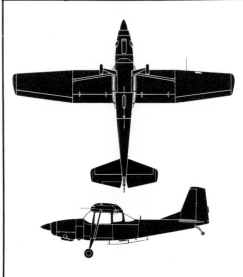

SIAI-MARCHETTI SM.1019E

Utility and light attack aircraft.
Dimensions: Span 36ft (10·97m); length 27ft 11½in (8·52m); height 9ft 4½in (2·86m).
Maximum cruising speed: 186mph (300 km/h).
Range: With two external tanks, 840 miles (1352km).
User: Italy.

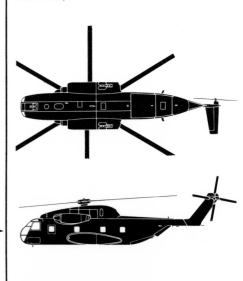

SIKORSKY CH-53A/CH-53D SEA STALLION

Transport helicopter.
Dimensions: Main-rotor diameter 72ft 3in (22·02m); length (rotors turning) 88ft 3in (26·9m); height 24ft 11in (7·6m).
Maximum speed: 196mph (315km/h).
Range: Two auxiliary tanks 540 miles (869km).
Users: Austria, West Germany, Iran, Israel, US Army/Marine Corps/Navy.

SIAI-MARCHETTI SF.260M

Trainer.
Dimensions: Span 27ft 4¾in (8·35m); length 23ft 3½in (7·1m); height 7ft 11in (2·41m).
Maximum speed: 211mph (340km/h).
Range: Max fuel 1,025 miles (1650km).
Users: Belgium, Bolivia, Burma, Ecuador, Italy, Libya, Morocco, Philippines, Singapore, Thailand, Tunisia, Zaïre, Zambia.

SIKORSKY CH-54 TARHE

Heavy-lift crane helicopter.
Dimensions: Main-rotor diameter 72ft (21·95 m); length (rotors turning) 88ft 6in (26·97m); height 18ft 7in (5·67m).
Maximum speed: 105mph (169km/h).
Range: 230 miles (370km).
User: US Army.

SIKORSKY S-58/H-34

Multi-role utility transport helicopter.
Dimensions: Main-rotor diameter 56ft (17·07 m); length (rotors turning) 65ft 9in (20·03m); height 14ft 3½in (4·36m).
Maximum speed: 123mph (198km/h).
Range: 280 miles (450km).
Users: Argentina, Belgium, Brazil, Canada, Central African Empire, Chile, France, West Germany, Haiti, Indonesia, Israel, Italy, Japan, Laos, Netherlands, Nicaragua, Philippines, Thailand.

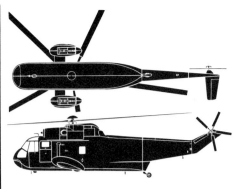

SIKORSKY S-61B/SH-3H

ASW helicopter.
Dimensions: Main-rotor diameter 62ft (18·9m); length (rotors turning) 72ft 8in (22·15m); height 16ft 10in (5·13m).
Maximum speed: 166mph (267km/h).
Range: Max fuel, 10 per cent reserves, 625 miles (1005km).
Users: Argentina, Brazil, Canada, Denmark, Indonesia, Iran, Israel, Italy, Japan, Malaysia, Spain, US (AF, Navy, Marine Corps).

SIKORSKY UH-60A BLACK HAWK

Assault transport helicopter.
Dimensions: Main-rotor diameter 53ft 8in (16·36m); length (rotors turning) 64ft 10in (19·76m); height 16ft 10in (5·13m).
Maximum speed: 184mph (296km/h).
Range: 30-min reserve, 373 miles (600km).
User: US (Army, Navy).

SUKHOI SU-7B FITTER-A

Ground-attack fighter.
Dimensions: Span 29ft 3½in (8·93m); length (including pitot boom) 57ft (17·37m); height 15ft (4·57m).
Maximum speed: High, clean, 1,055mph (1700km/h).
Combat radius: Details not specified, 200/300 miles (320/480km).
Users: Afghanistan, Algeria, Czechoslovakia, Egypt, Hungary, India, Iraq, North Korea, Poland, Romania, Soviet Union, Syria, Vietnam.

SOKO/CIAR IAR-93 ORAO

Fighter and close-support attack.
Dimensions: Span 24ft 9¾in (7·56m); length (excluding probe) 40ft 10¼in (12·45m); height 12ft 4¾in (3·78m).
Maximum speed: (production version, clean, high altitude) 1,089mph (1752km/h); sea level 762mph (1226km/h).
Combat radius: With 4,410lb (2000kg) external stores hi-lo-hi, 404 miles (650km).
Users: Romania, Yugoslavia.

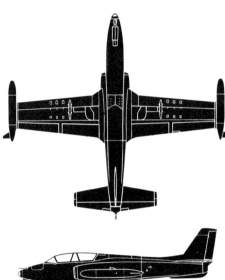

SOKO G2-A GALEB

Armed trainer.
Dimensions: Span (no tip tanks) 34ft 4½in (10·47m); length 33ft 11in (10·34m); height 10ft 9in (3·28m).
Maximum speed: 505mph (812km/h).
Range: 770 miles (1250km).
Users: Libya, Yugoslavia, Zambia.

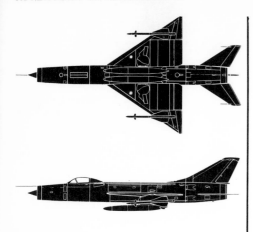

SUKHOI SU-11 FISHPOT-C

All-weather interceptor.
Dimensions: Span 27ft 8in (8·43m); length (including pitot boom) 56ft (17m); height 16ft (4·9m).
Maximum speed: High, clean, 1,190mph (1915km/h).
Range: About 700 miles (1125km).
User: Soviet Union.

SUKHOI SU-20 FITTER-C

Ground-attack fighter.
Dimensions: Span (28° sweep) 45ft 11¼in (14m), (62°) 34ft 9½in (10·6m); length (including pitot boom) 61ft 6¼in (18·75m); height 15ft 7in (4·75).
Maximum speed: High, clean, Mach 2·17, equivalent to 1,432mph (2305km/h).
Combat radius: With 4,409lb (2000kg) bomb load, hi-lo-hi, 391 miles (630km).
Users: Egypt, Peru, Poland, Soviet Union.

TUPOLEV TU-16 BADGER-B

Strategic bomber.
Dimensions: Span 110ft (33·5m); length 120ft (36·5m); height 35ft 6in (10·8m).
Maximum speed: High, clean, 587mph (945 km/h).
Range: High, 6,615lb (3000kg) of weapons or electronics, 3,975 miles (6400km).
Users: China, Egypt, Indonesia (not active), Iraq, Soviet Union.

SUKHOI SU-15 FLAGON-F

All-weather interceptor.
Dimensions (estimated): Span 34ft 3in (10·44m); length 70ft (21·34m); height 16ft 6in (5m).
Maximum speed: High, with AAMs, 1,520 mph (2445km/h).
Combat radius: Intercept mission, 450 miles (725km).
User: Soviet Union.

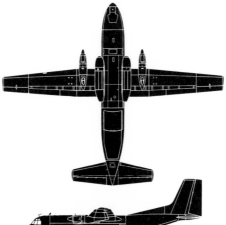

TRANSALL C.160

Tactical transport.
Dimensions: Span 131ft 3in (40m); length 106ft 3½in (32·4m); height 40ft 6¾in (12·36m).
Maximum speed: 368mph (592km/h).
Range: With maximum (35,274lb/16 000kg) payload, and full reserves, 1,056 miles (1700 km).
Users: France, West Germany, South Africa, Turkey.

TUPOLEV TU-22 BLINDER-C

Maritime reconnaissance and EW aircraft.
Dimensions: Span 90ft 10½in (27·7m); length 132ft 11½in (40·53m); height 35ft (10·67m).
Maximum speed: Estimated, high, clean, 920mph (1480km/h).
Range: Estimated, high, internal fuel only, 1,400 miles (2250km).
Users: Libya, Soviet Union.

SUKHOI SU-19 (provisional)

Two-seat close-support, interdiction and reconnaissance.
Dimensions: Span (max) 56ft 10in (17·3m), (min) 32ft 8in (10m); length 72ft 10in (22·2m); height 18ft (5·48m).
Maximum speed: Clean, sea level, 913mph (1470km/h), (high altitude) 1,452mph (2340 km/h).
Range: Max ferry 3,680 miles (5925km), combat radius with full bomb load, hi-lo-hi, 1,050 miles (1690km).
User: Soviet Union.

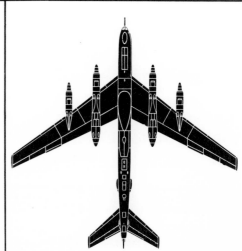

TUPOLEV TU-26(?) BACKFIRE-B

Bomber and reconnaissance aircraft.
Dimensions (estimated): Span (min sweep) 113ft (34·45m), (max sweep) 86ft (26m); length 132ft (40m); height 33ft (10m).
Maximum speed: High, clean, over Mach 2 (1,320mph/2125km/h).
Combar radius (estimated): Unrefuelled, hi-lo-hi, max weapons, 1,600 miles (2600km).
User: Soviet Union.

TUPOLEV TU-126 MOSS

Strategic AEW platform.
Dimensions: Span 168ft (51·2m); length 181ft 1in (55·2m); height 52ft 8in (16·05m).
Maximum speed: Probably not over 500mph (805km/h).
Range: At least 6,000 miles (9650km).
User: Soviet Union.

WESTLAND SEA KING HAS.1

ASW helicopter.
Dimensions: Main-rotor diameter 62ft (18·9m); length (rotors turning) 72ft 8in (22·15m); height 16ft 10in (5·13m).
Maximum speed: 143mph (230km/h).
Range: Max fuel, 937 miles (1507km).
Users: With Commando, Australia, Belgium, Egypt, West Germany, India, Norway, Pakistan, Qatar, UK (RAF, Royal Navy).

TUPOLEV TU-28P FIDDLER

All-weather interceptor.
Dimensions (estimated): Span 65ft (20m); length 85ft (26m); height 23ft (7m).
Maximum speed: High, clean, 1,150mph (1850km/h).
Range: Max fuel, high patrol, 3,100 miles (4898km).
User: Soviet Union.

VOUGHT F-8J CRUSADER

Single-seat fighter bomber.
Dimensions: Span 35ft 8in (10·87m); length 54ft 6in (16·6m); height 15ft 9in (4·8m).
Maximum speed: Clean, at altitude, 1,105mph (1780km/h).
Range: High-altitude, internal fuel only, 1350 miles (2173km).
Users: United States (Naval Reserve), Philippines.

WESTLAND LYNX AH.1

Multi-role tactical helicopter.
Dimensions: Main-rotor diameter 42ft (12·8m); length (rotors turning) 49ft 9in (15·16m); height 12ft (3·66m).
Maximum speed: Cruising speed 175mph (282km/h).
Range: Troop-carrying, with reserves, 336 miles (540km).
Users: Argentina, Belgium, Brazil, Denmark, France, West Germany, Netherlands, Norway, Qatar, UK (RAF, Royal Navy, Army).

TUPOLEV TU-95 BEAR-D

Maritime reconnaissance and missile guidance aircraft.
Dimensions: Span 159ft (48·5m); length 155ft 10in (47·5m); height 39ft 9in (12·12m).
Maximum speed: High, clean, 540mph (870 km/h).
Range: With 25,000lb (11 340kg) weapon or electronics load, 7,800 miles (12 550km).
User: Soviet Union.

VOUGHT A-7E CORSAIR II

Carrier-based attack aircraft.
Dimensions: Span 38ft 9in (11·8m); length 46ft 1½in (14·06m); height 16ft 0¾in (4·9m).
Maximum speed: Low, 12 Mk 82 bombs, 646mph (1040km/h).
Combat radius: With 8,000lb (3630kg) bombs, hi-lo-hi, 675 miles (1085km).
Users: Greece, Pakistan (?), US (AF, Navy, Marines).

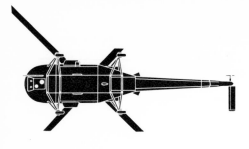

WESTLAND WASP

Shipboard ASW helicopter.
Dimensions: Main-rotor diameter 32ft 3in (9·83m); length (rotors turning) 40ft 4in (12·29 m); height 11ft 8in (3·56m).
Maximum speed: 120mph (193km/h).
Range: With weapon load and reserves, 270 miles (435km).
Users: With Scout: Australia, Bahrain, Brazil, Jordan, New Zealand, Netherlands, South Africa, Uganda, UK (Royal Navy, Army).

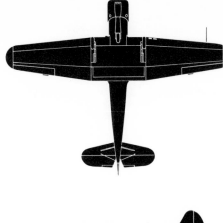

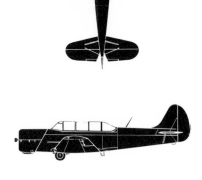

YAKOVLEV YAK-18A

Trainer.
Dimensions: Span 34ft 9$\frac{1}{4}$in (10·6m); length 27ft 4$\frac{3}{4}$in (8·35m); height 11ft (3·35m).
Maximum speed: 186mph (300km/h).
Range: 435 miles (700km).
Users: China, East Germany, North Korea, Mongolia, Romania, South Yemen, Soviet Union, Syria.

WESTLAND WESSEX HC.2

Transport helicopter.
Dimensions: Main-rotor diameter 56ft (17·07 m); length (rotors turning) 65ft 9in (20·03m); height 16ft 2in (4·93m).
Maximum speed: 133mph (214km/h).
Range: 390 miles (630km).
Users: Australia, Brunei, Ghana, Iraq, UK, (RAF, Royal Navy, Marines).

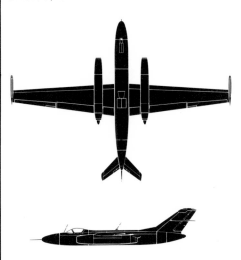

YAKOVLEV YAK-36 FORGER-A

Day attack VTOL aircraft.
Dimensions (estimated): Span 23ft (7m); length 49ft 3in (15m); height 13ft 3in (4m).
Maximum speed (estimated): At optimum medium-altitude height, clean, 860mph (1380 km/h).
Combat radius (estimated): Internal fuel, hi-lo-hi, not greater than 200 miles (320km).
User: Soviet Union.

YAKOVLEV YAK-25R
MANDRAKE

Reconnaissance and EW aircraft, and RPV.
Dimensions: Span (estimate) 71ft (22m); length 51ft (15·5m); height 13ft (4m).
Maximum speed (estimate): 470mph (755 km/h).
Range (estimate): 2,500 miles (4000km).
User: Soviet Union.

WSK-MIELEC TS-11 ISKRA

Trainer and light attack aircraft.
Dimensions: Span 33ft (10·06m); length 36ft 7in (11·15m); height 11ft 5$\frac{1}{2}$in (3·5m).
Maximum speed: 447mph (720km/h).
Range: 776 miles (1250km).
Users: India, Poland.

YAKOVLEV YAK-28P
FIREBAR

All-weather interceptor.
Dimensions: Span 42ft 6in (12·95m); length (long radome) 74ft (22·56m); height 12ft 11$\frac{1}{2}$in (3·95m).
Maximum speed: High, with AAMs, 735mph (1183km/h).
Combat radius: High, max fuel, 575 miles (925km).
User: Soviet Union.

INDEX

The page numbers of the aircraft listed in the captions are given in *italics*, as are the page numbers of the other items illustrated, such as missiles and aircraft carriers; and the page numbers of the aircraft given in the Inventory appear in **bold** type. Aircraft are mentioned both by maker and designation, viz 'Boeing KC-10A' and 'KC-10A, Boeing', except for the Russian makers Antonov, Ilyushin, Tupolev and Yakovlev, where the designations are listed under the makers' names.

PICTURE CREDITS

The publishers wish to thank the following organisations and individuals who have supplied photographs for this book. Photographs have been credited by page number.

Jacket: Saab-Scania. **½ title:** P. Steinemann. **Title page:** US Navy. **Credits:** Sepecat.

Foreword (p6–7): British Royal Navy. **21:** Swiss Air Force.

22-24: Canadian Dept of Defence **25:** Top, USAF; bottom left, Boeing; bottom right, Lockheed **26-27:** General Dynamics **28-31:** USAF **32:** Top, US Navy; centre, Grumman; bottom, LVT **33:** Top, US Navy; bottom, McDonnell Douglas **34:** Top, Lockheed; bottom, USMC **35:** Top, Vought; bottom, BAC **38:** US Army **40:** Top and bottom, USAF **41:** Top right, US Army **46:** Cessna **47:** Top, Simon R P Thompson; bottom, Peter J Bish **48:** Denis Hughes **49:** M Hooks **50:** Dassault **51:** Simon Thompson **52:** Cessna **53:** Rockwell **54:** Cessna **55:** Top, and centre, BAC; bottom, VFW-Fokker **56:** Top, M Hooks; bottom, Forca Aerea Brasileira **57:** Top, M Hooks; bottom, Westland **58:** Forca Aerea Brasileira **59:** Gates Learjet **60-61:** Top left, Cessna; top right, Bell Helicopter Co; bottom, Fuerza Aerea Argentina **62:** Top, Beech; bottom, Bell Helicopter Co **63:** Fokker-VFW **64:** Top, de Havilland Canada; bottom, Cessna **65:** Cessna **66:** P Steinemann **67:** Top, Rolls Royce; bottom, British MOD **68-69:** BAC **70-71:** M Hooks **72:** Westland **73-76:** British MOD **77-78:** Royal Norwegian Air Force **79:** Swedish Air Force **80-81:** Saab-Scania **82:** BAC; bottom, Saab-Scania **83:** C Ballantyne **84-85:** Top, Saab-Scania; bottom, German MOD **87:** Top, MBB; bottom, German MOD **88-89:** Contraves **90:** Top, Dutch MOD; bottom, Rolls Royce **91:** Top, Dutch MOD **92:** Top, General Dynamics; bottom, Belgian Air Force **93:** French Air Force **94:** Dassault-Breguet **95-97:** French Air Force **98:** Dassault-Breguet **99:** SNIAF **100:** French Air Force **101:** SNIAF **102:** Dassault-Breguet **103:** Swiss Air Force **104:** Swiss Air Force **105-106:** Austrian MOD **107:** Italian MOD **108:** Aerofan **109:** Left, Aerofan **109:** Italian

MOD **110:** Top, Italian MOD; bottom, Spanish MOD **111:** Top, M Hooks; centre and bottom, Spanish MOD **112-113:** Top, Canadair **113:** Bottom, Portugese Air Force **115:** Top, Greek Air Force; bottom, Vought **116:** Canadair **120:** Top, Institut Lotnictwa; bottom, Pezetel **121:** Top, Institut Lotnictwa; bottom, Pezetel **137:** P Steinemann **146:** Top, Flight International; bottom, US Navy **148:** Top, Interinfo; bottom, Westland **149:** Top, Interinfo **150:** Top, Pilot Press; bottom, Rex Features **151:** Top, Pilot Press; bottom, David Eshel **152-153:** Israeli Embassy, London **154:** Top, Dassault-Breguet; centre and bottom, Pilot Press **155-157:** Pilot Press **160:** Top, Bell Helicopter Co; bottom, Fokker-VFW **161:** Top, Grumman; centre and bottom, BAC **162-163:** BAC **164:** Top and centre, BAC; bottom, Pilot Press **165:** Pilot Press **166:** Top, Pilot Press; bottom, BAC **167:** Top, Westland; bottom, R Kunert **170:** Simon R P Thompson **171:** Bottom, M Hooks **173:** Bottom, Simon R P Thompson **175:** Top, de Havilland (Canada); bottom, C Ballantyne **176:** Top, BAC; bottom, J Guthrie **177:** Top, Stephen P Peltz; bottom, Pilot Press **178:** Stephen P Peltz **179:** Top, BAC; centre, Saab-Scania; bottom, C Ballantyne **181:** Top, Fokker-VFW; bottom, C Ballantyne **183:** Simon R P Thompson **184:** Top, Barry C Wheeler; bottom, M Hooks **185:** M Hooks **187:** Simon R P Thompson **189:** Top, South African Air Force **190:** Top, M Hooks; bottom, Dassault-Breguet **191:** Top, M Hooks; bottom, Simon R P Thompson **192-193:** Pakistan Air Force **196:** C Ballantyne **197:** Centre, BAC; bottom, US Navy **198:** Top, Westland; bottom, de Havilland (Canada) **200:** M Hooks **202-203:** Hsinhua **204:** US Govt **206-207:** Japanese Self Defence Force **208-209:** Mitsubishi **210:** Taiwan Govt **211:** Top, Taiwan Govt; bottom, Hong Kong Govt **213:** Canadair **216:** Top, USAF; bottom, DRVN **217:** DRVN **218:** Top, Lockheed; bottom, M Hooks **219:** Top, de Havilland (Canada); bottom, Bell Helicopter Co **222:** Top, Beech; centre, MBB; bottom, Grumman **223:** Top, Papua New Guinea DOD; bottom, GAF (Australia) **224:** Simon R P Thompson **225:** Top, GAF (Australia); bottom, VFW-Fokker **226-227:** Australian Ministry of Defence **228-229:** GAF (Australia) **229:** Bottom, Rolls Royce **230:** Australian Ministry of Defence **231:** Top, Grumman; bottom, Bell Helicopter Co **232:** Top, BAC; centre and bottom, New Zealand Ministry of Defence **233:** New Zealand Ministry of Defence